Pocket Guide to

GENE LEVEL DIAGNOSTICS

in Clinical Practice

Victor A. Bernstam
BioSol Ltd.
Ann Arbor, Michigan

CRC Press
Taylor & Francis Group
Boca Raton London New York

CRC Press is an imprint of the
Taylor & Francis Group, an **informa** business

CRC Press
Taylor & Francis Group
6000 Broken Sound Parkway NW, Suite 300
Boca Raton, FL 33487-2742

First issued in hardback 2019

ISBN-13: 978-0-8493-4485-5 (pbk)
ISBN-13: 978-1-138-43012-9 (hbk)

Visit the Taylor & Francis Web site at
http://www.taylorandfrancis.com

and the CRC Press Web site at
http://www.crcpress.com

Library of Congress Card Number 92-23745

Library of Congress Cataloging-in-Publication Data

Bernstam, Victor A.
Pocket guide to gene level diagnostics in clinical practice / Victor A. Bernstam.
p. cm.
Companion v. to: Handbook of gene level diagnostics in clinical practice / Victor A. Bernstam. 1992.
Includes index.
ISBN 0-8493-4485-9
1. Medical genetics—Handbooks, manuals, etc. I. Bernstam, Victor A. Handbook of gene level diagnostics in clinical practice. II. Title.
[DNLM: 1. Genetic Techniques—handbooks. 2. Genetics, Medical—methods—handbooks. 3. Hereditary Diseases—diagnosis—handbooks. 4. Hereditary Diseases—genetics—handbooks. 5. Prenatal Diagnosis—methods—handbooks. QZ 39 B531p]
RB 155.B483 1992
616′.042—dc20
DNLM/DLC
for Library of Congress 92-23745
CIP

PREFACE

Developments in the Human Genome Project, the avalanche of changes in and progressive sophistication of nucleic acid technology occur at lightning speed. This cannot justify waiting for a better time to take stock of this everchanging field — if anything, just the opposite is true. The formerly undreamed of possibilities open up in molecular diagnostics. An overview of such a vast amount of data **can only be selective** at any given time. It may, however, provide a basis for better understanding of the new dimension that contemporary medicine is assuming.

Both the Handbook and Pocket Guide are intended for students and residents, physicians of various disciplines, forensic and laboratory scientists who are interested in applying molecular biological technology to patient care. The objective of the Handbook is to provide an overview of options and spectacular advances in disease evaluation at the gene level. The condensed Pocket Guide incorporates additional data on

- representative manufacturers of instruments and reagents useful in molecular diagnostics, including descriptions of selected products (Supplement 1)
- representative small and larger commercial laboratories offering services in molecular genetic testing (Supplement 2)
- a listing of the U.S. laboratories already licensed to perform polymerase chain reaction for human diagnosis at the time of this writing (Supplement 3). Listing of such laboratories elsewhere in the world was not available at the time of this writing
- description of the current efforts of three professional groups in the U.S. practicing gene level evaluation — human geneticists, pathologists and forensic scientists — that includes an Executive Summary of the recent recommendations of the Committee on DNA Technology in Forensic Science of the National Research Council and Legislative Guidelines for DNA Databases (Supplements 4 and 5). The latter define not only the procedural, scientific and legal issues involved, but also address ethical aspects of DNA testing in forensic practice

Both publications are being continually updated to reflect changes in this exciting field. I welcome any contributory information, constructive criticism, comments, and suggestions helpful in making these vast databases as useful as possible for patient care.

The production of the Pocket Guide has only become possible due to the invaluable support and patient cooperation of the CRC staff, particularly Jeff Holtmeier and Suzanne Lassandro.

I am grateful to all those who appreciated my efforts and found time and energy in their busy schedule to contribute material for the Supplements. My special thanks go to Carl W. Bell, George Bers, Joseph A. Chimera, Jacqueline Cossmon, Ellen Daniell, Paul B. Ferrara, Robert C. Giles, Linda H. Masterson, Robert M. Nakamura, Michael Mechanic, Nancy Mulzoff,

Patricia Murphy, Joshimi Munch, Deborah J. Oronzio, Roberta A. Pagon, John P. Richard, Barbara E. Thalenfeld, Stephen N. Thibodeau, Annette Short, Steve S. Sommer, John M. Sorvillo, Michael S. Watson, and Frank Witney.

I cannot express enough gratitude to Elmer V. Bernstam for his immense and creative involvement also with this book, accepting with humor and stoicism my reliance on his talents at all stages of this project as can only be expected from a true co-author.

Victor A. Bernstam
Ann Arbor, Michigan
October 1992

The Author

Victor A. Bernstam, President of BioSol, Ltd. (Ann Arbor, MI) received his M.D. from the Leningrad (now St. Petersburg) Medical Institute, USSR in 1966, and his Ph.D. in cellular molecular biology from the USSR Academy of Sciences Komarov Botanical Institute in Leningrad in 1973. Since 1960 he has conducted research on brain biochemistry, molecular mechanisms of hormonal induction of gene expression, heat shock effects on protein biosynthesis, membrane phenomena in experimental and clinical atherosclerosis, and laboratory diagnostics. Dr. Bernstam has held positions as a Research Scientist at the Komarov Botanical Institute and the Institute of Cytology of the USSR Academy of Sciences in Leningrad, as well as in the Cellular Chemistry Laboratory at the University of Michigan School of Public Health in Ann Arbor. His medical experience includes work as an Emergency Physician in Leningrad, and later following American Pathology Board Certification as a Pathologist concentrating on clinical special chemistry and molecular pathology. Dr. Bernstam served as a Medical Director of Chelsea Medical Laboratory, MDS Laboratories, and later was one of the organizers and, subsequently, Medical Director of Warde Medical Laboratory (Ann Arbor, MI) where he established and headed a Molecular Diagnostics Group. He has published and lectured on GABA metabolism, heat shock research and apolipoprotein testing for coronary risk assessment. He has developed and patented a novel ligand-removing, clarifying reagent, Liposol®. He is a member of the College of American Pathologists, American Association for Clinical Chemistry, New York Academy of Sciences, and the honorary society Sigma Xi. He continues research on laboratory diagnostics and consults in molecular diagnostics.

To
Elmer, Luda, and Inna, Luba and Shura

CONTENTS

Part 1
General Aspects of Gene Level Evaluation: Survey of Techniques and Applications

Chapter 1.1
Introduction

New diagnostic possibilities at the gene level present a challenge in selecting reliable molecular tools that can benefit patient management. Clearly, a "wholesale" transfer of molecular biological tools to clinical laboratories may not offer better diagnostic modalities than the existing methods. However, in some areas limitations of the "conventional" tools are apparent. Thus, although **serum tumor markers** are useful for management of cancer patients, their use in cancer **detection** is limited. The majority of antigenic determinants used as tumor markers are not organ specific and cannot be used for definitive diagnosis, but only in **monitoring** therapy of cancer patients. In a number of conditions gene level diagnostics allow for a substantially higher level of patient care.

Diagnostic modalities of exquisite sensitivity and specificity, such as the polymerase chain reaction (**PCR**), introduced **new definitions** of

- **infection**
- **etiology**
- **the role of viral factors**
- **the efficacy of treatments**

The PCR has established **new criteria of diagnostic sensitivity** and **performance standards** for other diagnostic methods.

In **genetic disorders**, new molecular probes offer not only diagnostic, but also **predictive**, possibilities. Practical **guidelines** for the acceptability of a given molecular biological test as a clinical diagnostic tool and appropriate **standards** are being developed.

The *Handbook of Gene Level Diagnostics in Clinical Practice* provides an overview of gene level evaluation of a disease with emphasis placed on the **demonstration of various methodological choices and possibilities.** No attempt is made to offer comprehensive coverage of the molecular pathology of all the conditions discussed. This **pocketbook version** represents a convenient reference source for quick review of those subjects of interest covered in greater detail in the handbook and in the references provided therein.

1.1.1 HUMAN GENOME PROJECT (HGP) (p. 4)

HGP represents an international, concerted effort toward deciphering the nucleotide sequence of the entire human genome with the following major objectives:

1. The development of **high-resolution genetic maps** of the human genome, eventually at the resolution of 1 centimorgan (cM)
2. The establishment of overlapping physical (**contig**) maps of all human chromosomes at the resolution of reference DNA landmarks spaced at 100-kb intervals
3. The determination of the complete sequence of human DNA
4. The development of appropriate adaptation of achievements of **informatics** in collecting, storing, distributing, as well as analyzing and retrieving the data generated by physically discrete research teams
5. The creation and utilization of **appropriate technologies** to achieve these goals

By now almost 2000 genes of the estimated 50,000 to 100,000 genes in the human have been mapped to **specific chromosomal locations**. Deciphering of nucleotide sequences encoding proteins and controversial efforts to patent those sequences proceed at breakneck speed, despite the fact that the significance of this information will be evaluated only later. The most productive approach to human **gene mapping** has been **somatic cell hybridization** followed by ***in situ*** **hybridization (ISH)**. The **location of the gene** is further defined by **restriction fragment length polymorphism (RFLP)** with respect to other established DNA markers.

The methodologies helpful in gene mapping include:

- **somatic cell hybrids**, also combined with PCR analysis
- **assignment** of specific genes to chromosomes **by PCR**
- **radiation-induced gene segregation (radiation mapping)**
- **yeast artificial chromosomes (YACs)**
- **microcell-mediated gene transfer**
- **chromosome-** and **DNA-mediated gene transfers**

These will not be employed to any great extent in **clinical gene level diagnosis** in the near future. On the other hand, such strategies as **physical mapping by repetitive sequence fingerprinting** yield approaches applicable to identity studies.

Other techniques may be and some have already been used as clinical diagnostics:

- **RFLP**
- **sequence-tagged sites (STSs) and expressed sequence tags (ESTs)**
- **allele-specific oligonucleotide (ASO)** screening procedures, especially using **PCR**
- **allele recognition by restriction enzyme analysis**
- **cleavage mismatch detection**
- **deletion detection by multiplex PCR**
- **competitive oligonucleotide priming (COP)**
- analysis of the **genomic distribution of α satellite DNA sequences** and many others (see below)

Nonradioactive PCR-based detection systems applicable to clinical diagnostics are being introduced. For example, **sulfonated cytosines** and

electrochemiluminescence (ECL)-based detection systems allow enzyme immunoassay formats suitable for **automation**. Nucleic acid **handling procedures** (e.g., RNA-DNA hybridization using **magnetic beads** coupled with streptavidin) are being simplified for use in clinical and forensic diagnostic laboratories.

Quantitation of DNA hybridization in routine clinical laboratories and even in field studies uses, for example, the **slide immunoenzymatic assay (SIA-DNA)**. In this method, a cloned probe p17h8 (D17Z1) for the highly repetitive, primate-specific **α satellite DNA** quantitates minute amounts of human genomic DNA in forensic specimen extracts within less than 4 h with subnanogram sensitivity.

1.1.2 SEQUENCE-TAGGED SITES (STSs) (p. 7)

The **STS proposal** allows merging of the genetic and physical maps of the human genome generated in different laboratories into a **single consensus physical map**. The STS concept **uses short tracts of single-copy DNA** in a given region of DNA that can be easily recognized and recovered at any time by the PCR. The **physical location** of an STS on the genome map can be identified by the PCR using a 200- to 500-bp sequence unique to a given DNA fragment.

The advantages of STSs are as follow:

- The **identification code** of a particular segment of DNA can be stored in a **database**
- Anyone interested in the particular stretch of DNA can use a **set of PCR primers specifically defined by the appropriate STS**

STSs can also serve as **reference points** for testing mapped disease genes. Application of automated sequencing technology yielded sequences of coding regions of the genome — **expressed sequence tags (ESTs)** — of thousands of as yet undefined genes in a very short time. Application of this approach and substantial increase in the speed of sequencing make complete genome sequencing a realistic goal achievable sooner than originally planned.

1.1.3 REPETITIVE DNA (p. 8)

Repetitive DNA is composed of **tandemly repeated sequences**, which show an **extraordinarily high degree of diversity** (see also Section 1.2.5). They apparently play a role in regulating the expression of genes as suggested by the **variable degree of their methylation**. The so-called "middle-repetitive" DNA comprises up to 40% of the DNA in eukaryotic genomes and consists of numerous families of lower-copy-number repetitive sequences.

Repetitive DNA is distributed in the human genome essentially in two ways:

- **Type I** repeated sequences are **large tandem block structures** (up to several thousand nucleotides) composed of head-to-tail arrangements of various sequence units with a highly variable copy number.

- **Type II** repeated sequences are composed of **small variable tandem sequence repeats** where the same basic sequence, or **core unit**, can be either clustered or interspersed in the genome.

Three main classes of repetitive DNA recognized by the reassociation kinetics are as follow:

- the **highly repetitive DNA** with more than 10^5 copies per haploid genome
- the **middle repetitive** sequences — 10^2 to 10^5 copies per haploid genome
- the **low repetitive** sequences — all sequence families are represented in 2 to 100 copies per haploid genome

Analysis of the genome by **restriction mapping** and **sequence determinations** yields a substantially different picture compared to that gained by reassociation kinetics only. In CsCl density gradients repetitive DNA usually forms four additional satellite peaks separate from the rest of the DNA.

ISH indicates that repetitive DNA is primarily located in the **constitutive heterochromatin** of the centromeric regions of chromosomes. Some **chromosomes (2, 3, 4, 6, 8, 11,** and **18)** show **no evidence of satellite DNA**, and **chromosomes 10, 12, 16, 19,** and **X** contain only small amounts of satellite DNA. Certain satellite sequences are **characteristic of a given chromosome** (such as those defined for chromosome 16, and the Y chromosome). An extrachromosomal **alphoid** DNA ***Sau*3A subfamily** is important for its high degree of RFLP, and has been used in molecular genetic identifications.

Type II repetitive sequence families are composed of **microsatellites**, or **simple sequence repeats (SSR)** such as **$(T)_n/(A)_n$, $(G)_n/(C)_n$, $(TA)_n$, $(CA)_n$, $(TG)_n$, $(GC)_n$, $(TC)_n$, $(GGA)_n$, $(GCA)_n$,** and **minisatellite sequence families** such as **$(TTTTA)_n$ and $(TTAGGG)_n$**. Certain similarities have been detected in comparing the structure of repetitive sequences such as ***Alu*** and **L1**, belonging to the **short interspersed repeat (SINE) family** in human DNA, with those of **retroviral elements** and **transposons**. The highly conserved nature of tandemly repeated DNA sequences is apparently associated with specific, apparently regulatory, and so far not fully disclosed, functions in the genome. Definite **tissue-specific changes** in repetitive DNA have been observed during the developmental process.

Repetitive DNA is the **area of more frequent mutational events**, such as

- **insertions**
- **deletions**
- **amplifications**
- **oncogene translocations**

Furthermore, repetitive DNA is thought to play a role in **influencing the conformation of chromosome loops**. Repetitive DNA sequences in the vicinity of deletion breakpoints show **nonrandom sequence elements** both at the nucleotide and dinucleotide sequence levels. This observation underscores the influence exerted by local primary and secondary DNA structures on the mutational process.

1.1.4 RECOMBINATIONAL HOT SPOTS (p. 9)

Rearrangements commonly follow integration of DNA sequences. This leads to **instability of the genome** associated with loss or amplification of the integrated sequences and neighboring DNA. **Integration frequently occurs in regions containing repetitive cellular DNA, although random integration events also happen**. Amplification and the formation of episomal intermediates often accompany the destabilization of the area of integration, leading to transcription deregulation of the adjacent cellular gene. An area of instability can be detected by an RFLP near the integration site, with the majority of examples so far demonstrated in hematological malignancies.

In **lymphoid malignancies**, some genes activated by chromosomal translocation encode **helix-loop-helix (HLH) proteins**. These bind to DNA and form complexes with heterologous members of this family of proteins possessing enhancer regulatory activities. The prototypical HLH protein is the ***myc*** protooncogene. Examples of other genes encoding oncogenic regulatory proteins include the ***Hox 11*** homeobox gene, ***Ttg/Rhom*-1**, and ***Rhom*-2**.

Translocations can also occur in introns and **chimeric proteins** are formed with unique new functions not present in wild-type cells, as seen, for example, in **acute promyelocytic leukemia (APL)**. In APL **translocation t(15;17)** leads to the formation of a truncated product of **the retinoic acid receptor α (RAR-α)**. Chimeric RAR-α exhibits altered trans-activation properties compared with the wild type. Chimeric proteins may also act as **competitive inhibitors** of their normal cellular counterparts. These leukemias proved to respond dramatically to retinoid treatments. The identification of the site of fusion using molecular or immunological techniques has important implications now for the management of patients with APL.

Altered transcription factors can also change gene expression and lead to oncogenic conversion. Frequently affected are c-*jun*, c-*fos*, c-*rel*, and c-*myc* — the so-called "immediate early" genes activated early in mitogenic stimulation of cells. Other protooncogenes, such as c-*erb*A, may be involved in promoting differentiation — an opposing influence. Yet other genes, such as those encoding **AP-1** and **NF-κB**, appear to have a wide range of targets, acting as general second messengers with a pleiotropic mode of action.

1.1.5 TUMOR SUPPRESSOR GENES (p. 10)

A number of genes display **recessive characteristics** in tumorigenesis and are referred to as **recessive oncogenes, antioncogenes**, or **tumor**

suppressor genes (TSGs). Retinoblastoma, Wilms' tumor, osteosarcoma, and **colon, renal, lung,** and **breast cancers** are some of the best known examples (see appropriate sections). In fusion experiments, the phenomenon of **suppression of malignancy** by a wild-type TSG has been discovered, indicating that loss of genetic material may lead to malignancy. It appears that imposition of **terminal differentiation** leads to suppression of malignancy.

A number of common characteristics of TSGs are recognized. TSGs appear to encode proteins controlling fundamental cellular functions:

- **signal reception**
- **transduction**
- **propagation**
- **progression through the cell cycle**
- **initiation of DNA replication** and **transcription**

Diverse oncoproteins share their

- **similarity in their amino acid sequences**
- interaction with specific **target genes**
- **ability to interact with other oncoproteins** in bringing about **immortalization** of cells

A **cooperation** between dominantly acting oncogenes and inactivating TSGs occurs in the **multistage process of carcinogenesis** demonstrable even in a single tumor.

- **Loss of function of TSGs** may be brought about by
  - **focal deletions**
  - **larger "interstitial" deletions**
  - the **loss of an entire chromosome**
  - **chromosomal translocations**

  Mutations of TSGs are more likely to occur on the paternal rather than the maternal chromosome.
- Some of the **TSG products** are localized to the **nucleus (*p53*, *RB*-1, *WT*-1)**, whereas others resemble **cell adhesion molecules** such as **DCC**. Loss or mutations of *p53* predispose cells to malignancy although the role of *p53* in embryonal development and differentiation of the organism still remains unclear.
- **Ectopic TSGs** introduced into transformed or cancerous cells, with their counterparts either defective or absent, **lead to apparent restoration of normalcy**, at least in part.
- TSGs apparently may trigger a **cascade-type action**, interfering with the process of transformation eventually targeting similar effector mechanisms. Examples of these convergent general pathways are the *ras*-dependent transformation interrupted by **Krev-1, RAP-2**, and the **18-kb DNA fragment** isolated from preneoplastic rat 208F cells transformed with the human EJ-Ha-*ras* gene and pSV2*neo*.

DNA methylation and **chromatin topology** play a role in the expression of transformation-related genes. Those regulatory mechanisms affected in the process of transformation and carcinogenesis are in essence similar to the regulatory activities operating in normal development and differentiation.

Experimental and clinical evidence supports the notion of **multistage carcinogenesis**. The **"two-hit" Knudson model** of the development of retinoblastoma envisioned the alteration of both alleles of the *RB* gene as necessary for the disease to occur. Although this concept was fundamental in establishing the existence of putative recessive genes, or TSGs, a **multihit** (three-, four-, or five-hit) **process** leading to TSG alteration has also been proposed.

Identification and **isolation of TSGs** can be achieved by **differential** or **subtractive hybridization**, which is based on the comparison of gene expression in two related cell populations. Examples of the use of subtractive hybridization are the isolation of TSGs from Syrian hamster embryo (SHE) cells undergoing carcinogen-induced neoplastic transformation, and from human breast tissue where the genes uniquely or preferentially expressed in one of a pair of closely related cell populations were selected for and recovered from breast carcinoma cells.

Among the TSGs isolated from breast carcinoma were the following:

- those encoding the **gap-junction protein connexin 26**
- two different **keratins**
- **glutathione *S*-transferase**

Also isolated was an **unknown gene** expressed in normal mammary epithelial cell strains, but absent from tumor-derived cell lines. This appeared to be a Ca^{2+}-binding protein of the S-100 gene family, tentatively **related to calcyclin**.

While traditionally the loss of TSG expression has been related only to alterations of the gene itself, the **regulatory events** serving to maintain that gene integrity can also determine its expression. In this regard **two classes of TSGs** are distinguished:

- **Class I** includes genes that lost their function by mutation.
- **Class II** includes genes whose expression is lost due to a mutation occurring elsewhere.

According to this model, it is expected that the **regulatory gene must be a positively activating regulator**, which is operational in normal cells and is nonfunctional in tumor cells. In cell hybrids, this regulatory gene is active and allows for the expression of class II genes that maintain tumor suppression. For example, *RB*, *WT*-1, and *p53* belong to class I TSGs, which are oncogenic when their function is lost, most likely affecting a whole subset of downstream class II genes. In fact, *p53* has been shown to control the cell cycle progression through a selective down regulation of the expression of the **proliferating cell nuclear antigen** mRNA and protein.

With respect to **clinical utility** class II genes appear to be of particular interest, because they are not lost and their function may be restored or up regulated. **Differential expression of some genes** in normal cells and in tumors where they are absent **may serve as a useful diagnostic marker**.

1.1.6 GENOMIC IMPRINTING (p. 14)

Genomic imprinting (GI) encompasses the disease phenomenology common in humans stemming from chromosomal abnormalities such as translocations, inversions, and duplications **only if these occur in chromosomes transmitted from either parent** (the mother or the father) and are consistently manifested for a given genetic disorder. GI leads to modification of the phenotype determined by a particular allele depending on the origin of the gamete. Some cases formerly interpreted as expression of autosomal recessive inheritance proved to be traceable to specific molecular events such as **submicroscopic deletions**. A striking example is the male-to-male transmission of hemophilia A, where both the X and the Y chromosome came from the father.

In **maternal imprinting**, the affected gene is not phenotypically expressed regardless of the sex of the mother's offspring. The male line of the family may carry the gene without expression and the offspring of the male line who inherit the gene will express it and manifest the phenotype. On the other hand, the children of the mother's nonmanifesting daughter, like the mother, will not express the trait phenotypically. Thus, it is the **transmitting ("skipped"), nonmanifesting individuals** who are the clue to the type of imprinting — maternal or paternal:

- in **maternal imprinting**, a nonmanifesting male transmits the genes to manifesting offspring; and vice versa
- in **paternal imprinting**, females transmit the trait, being nonmanifesting carriers themselves

Imprinting occurs during gametogenesis and, in contrast to mutation, this modification **is reversible** as demonstrated by nuclear transfer experiments in differentiated somatic cells. Imprinting amounts to **functionally turned-off segments of chromosomes** and appears to be a dominant function. Examples of GI include

- **Beckwith-Wiedemann syndrome**
- **narcolepsy**
- **embryonal rhabdomyosarcoma**
- **Wilms' tumor**
- **Prader-Willi and Angelman syndromes**

Evidence points to the **critical role of modifier genes,** alleles at tumor suppressor loci that **inactivate the allele generating the imprint**. The imprinting of the paternal tumor suppressor allele has been demonstrated in a number of tumors — Wilms' tumor, osteosarcoma, bilateral retinoblastoma, and embryonal rhabdomyosarcoma.

Variability of the expressivity of disease phenotypes such as observed in Huntington's disease, myotonic dystrophy, neurofibromatosis, and possibly in a number of other conditions displaying variable severity of disease manifestations can be explained by GI. Prenatal evaluations **in some conditions** may be erroneous if based solely on chromosomal counts without the recognition of the phenomenon of GI.

1.1.7 DNA METHYLATION (p. 15)

DNA in terminally differentiated somatic cells has a fixed methylation pattern reproducible after replication. On the other hand, the methylation pattern of genes expressing tissue-specific and housekeeping activities varies during development. The process of specific gene **demethylation** is rapid, cell-type specific, and involves a variety of **CpG island** sequences.

The **extent of methylation** and **transcriptional inactivation** do **correlate**, but not in absolute terms. **Satellite DNA** that is transcriptionally inactive contains several times more **methyl-5-cytosine (m5C),** predominantly in CpG islands, than does bulk DNA; it is also enriched in methylated **CpT, TpC,** and **CpC dinucleotides.** CpG islands are known to occur in all cell types and mark the 5′ end of genes, so that methylation-free CpG islands occur in transcriptionally active DNA.

DNA methylation may exert its transcriptional inactivation effect indirectly through **methyl-CpG-binding proteins**. Even partial methylation affecting either the coding or noncoding DNA strand is sufficient to block expression of the hemimethylated chromatin. The methylation/demethylation switch may also function at the protein level (e.g., in repeatedly activatable cellular systems such as the **G proteins**).

The **dynamic nature of methylation** has been implicated in the **inheritance of epigenetic defects,** some of which may be **associated with carcinogenesis and aging**. Sperm-specific methylation patterns have been demonstrated in the c-Ha-*ras*-1, insulin, and *RB* genes representing imprinting of the parental chromosomes.

A **variable state of methylation** observed in **human testicular cancer** is detectable in hypoxanthine phosphoribosyltransferase and the phosphoglycerate kinase gene loci by *Hpa*II/*Msp*I analysis. **Teratocarcinomas** show generalized hypermethylation and areas of complete demethylation in testicular tumors of germ cell origin, and in female somatic and embryonal carcinomas.

The human **DNA methyltransferase gene** codes for the enzyme catalyzing DNA methylation. The expression of this gene is low in normal human cells, increased up to 50-fold in virally transformed cells, and markedly (several hundredfold) elevated in human cancer cells.

Hypomethylation also occurs in some forms of autoimmune diseases (systemic lupus erythematosis and rheumatoid arthritis). **Demethylation** of regulatory genes is considered to constitute a possible mechanism leading to cellular oncogene activation in autoimmune diseases. Determination of the state of methylation of specific genes may be important in clinical disease evaluation.

1.1.8 MITOCHONDRIAL DNA (p. 17)

The mitochondrial genome is located in the cytoplasm and is **almost exclusively maternally transmitted,** with the occasional paternal transmission of phenotypic abnormalities. This could be explained on the assumption that certain enzyme subunits may be encoded by nuclear DNA. The **circular mitochondrial DNA (mtDNA) molecule** is transcribed and the posttranscriptional processing is performed with the help of proteins and RNA molecules encoded by the nuclear DNA. Analysis of mtDNA provides insights into the molecular pathology of

- **neuromuscular disorders**
- **degenerative diseases**
- **aging**
- **liver diseases**
- **ophthalmoplegia**
- **encephalopathies**
- **cardiomyopathies**
- various aspects of **endurance training**

A whole range of diseases traceable to alterations in mtDNA has been uncovered and evaluation of mtDNA appears to be acquiring a progressively larger role in human disease. The best known examples are:

- a base change in mtDNA associated with **Leber's hereditary optic neuropathy**
- large deletions in various **encephalomyelopathies** (e.g., **Kearns-Sayre syndrome**)

Mitochondrial tRNA genes appear to be the hot spots for point mutations causing neuromuscular diseases.

mtDNA defects can **result from pharmacological therapies**; examples include **zidovudine-induced myopathy** in acquired immunodeficiency syndrome (AIDS) patients and the mitochondrially inherited **susceptibility to aminoglycoside antibiotic-induced ototoxicity**.

mtDNA-nuclear DNA regulatory interactions are exemplified by the selective inhibition of the expression of mitochondrial genes by interferon. This inhibition appears to be mediated by an interferon-responsive nuclear gene that encodes a product that, in turn, regulates mitochondrial gene expression.

Analysis of mtDNA for clinical diagnostic purposes can be performed following its isolation by simplified procedures, subjecting it to restriction digestion, Southern analysis, or evaluation by PCR-based assays.

Chapter 1.2

A Survey of Newer Gene Probing Techniques

1.2.1 INTRODUCTION (p. 19)

Molecular genetic techniques are being used more and more in:

- **identifying pathogens**
- attempting to define **distinction** between norm and pathology
- in tracing the **genealogy** of individuals
- establishing **identity characteristics**
- definitively identifying the **source of the biological material in forensic practice**

Because most **genomic DNA sequences** are shared among humans, informative genetic markers allow investigators to meet these challenges. Until relatively recently almost all **polymorphic markers** were limited to **gene products** identified by electrophoretic or serological techniques. Molecular biological technology now allows the polymorphisms to be defined at the level of the genome and used for diagnostic purposes.

1.2.2 DETECTION OF KNOWN MUTATIONS (p. 19)

Existing molecular biological techniques allow us to establish gene level diagnosis, at least tentatively, for **over 500 genetic disorders** for which chromosomal assignments have been determined. The database of established DNA alterations associated with human disease is being constantly updated. Furthermore, because **close to 10,000 genes have been defined** by now at least in some detail, direct evaluation of some of these for clinical diagnosis is also possible.

The search for **disease susceptibility genes** proceeds along different routes, including the epidemiological approach. The pace of introduction of gene level diagnosis in real medical practice, however, lags behind for a number of reasons. The enormous speed of accumulation of practical, usable information on human genes associated with disease makes the widespread introduction of gene level evaluation in routine clinical practice technologically a realistic possibility in the near future.

For simplicity of presentation, all of the methods to be briefly described here are organized in the following groups:

- those primarily utilized for the **detection of known alterations of DNA**
- **methods for scanning DNA** when the target alteration or mutation has not been defined

- techniques employed in **linkage analysis**
- approaches currently used predominantly in **gene mapping** studies, but which may eventually find application in clinical practice

New developments in the methodology of gene analysis have been increasing almost exponentially in number and versatility of analytical potential. This discussion gives only a brief survey of representative technologies. [*See also Supplements 1 and 2.*]

The **detection of variations in DNA structure** is undergoing a continuing evolution toward a progressively higher degree of precision, simplification, automation, and informative content. The polymerase chain reaction (**PCR**) and other gene amplification methods have revolutionized the very concept of the **molecular diagnostic field**.

1.2.2.1 SOUTHERN AND NORTHERN BLOTTING (p. 20)

DNA sequence alterations associated with disease can be produced by

- focal **single-base modifications**
- **deletions**
- **duplications**
- **insertions**
- **translocation**
- **methylation**, etc.

Southern blotting is capable of identifying such DNA alterations, revealing relatively major sequence differences detectable by **restriction endonuclease digestion**. The classic example of this approach is the detection of the sickle cell β-globin allele responsible for sickle cell anemia. This approach is the basis for **restriction fragment length polymorphism (RFLP) analysis** as well.

If transcription is not affected, an altered gene can be detected by using **northern analysis** to identify and study the **mRNA** produced.

Numerous modifications of **nucleic acid extractions** that are adaptable to clinical diagnostic laboratories, including **automated methods for DNA extraction** from patient specimens, have been appearing almost daily. Efforts to bring recombinant DNA technology closer to the clinical diagnostic laboratory are continuously being made. One example is the development, for specific gene detection, of an automated chemical system similar to **Southern analysis based on solution hybridization** and **solid-phase capture chemistry**. An even further procedural simplification allows the detection of specific mutations by **PCR directly from dried blood spots** on Guthrie cards, eliminating the requirement for extraction of genomic DNA.

1.2.2.2 ALLELE-SPECIFIC OLIGONUCLEOTIDE (ASO) PROBES (p. 20)

Changes in DNA sequences at the **single-base** level can be detected by **ASO probes** that form a duplex with the nucleic acid to be tested; the duplex is extremely unstable even if only a **single nucleotide mismatch** occurs.

The absence of a duplex in the expected position signals the presence of a mutation. The efficiency of this now widely used approach was initially demonstrated by the detection of sickle cell β-globin allele. In general, the ASO technique can be used to detect **point mutations** that do not create a recognition site for restriction enzyme digestion, provided an appropriate oligonucleotide probe is constructed.

The **reverse ASO** technique offers the benefit of easier standardization through the simultaneous testing of many alleles by manufacturing uniform lots of **immobilized (diagnostic) oligonucleotides** to which the unknown material is added. Enhancement of the selectivity of the ASO probe, and thus specificity of the assay, can be achieved by **oligonucleotide competition** and an **increase in the stringency** of the hybridization and washing conditions.

1.2.2.3 GENOMIC SUBTRACTION (p. 21)

This technique isolates the **DNA that is absent from deletion mutants**, and subtracts the DNA present in both the wild-type and the deletion mutant genomes from a mixture of the wild-type and mutant DNA. The mutant DNA is biotinylated, allowed to reassociate with the wild-type DNA, and the hybrids are then "**subtracted**" from the mixture by capturing them on avidin via the biotin attached to the mutant DNA. The **DNA not present in the deletion mutant is thereby left behind** in the solution and, after several cycles of removal of the DNA corresponding to the mutant DNA, the remaining unbound DNA, representing the DNA that is absent in the deletion mutant, is amplified by PCR.

The potential clinical applications are:

- **the detection of pathogen DNA in infected tissue**
- **the isolation of various disease genes**

Subtraction hybridization procedures use photoactivatable biotin, streptavidin binding, and organic extraction for solution hybridization for the analysis of mammalian genomic DNA as well as for **deletion cloning**.

1.2.2.4 ALLELE-SPECIFIC PCR (AS-PCR) (p. 21)

AS-PCR allows the direct determination of genotype by the presence or absence of an amplified fragment specific for the allele of interest. For example, using nonradioactive primers for the normal and sickle cell β-globin alleles in genomic DNA, the sickle cell mutation could be detected without hybridization, ligation, or restriction enzyme cleavage. The AS-PCR product can be captured on a streptavidin-agarose column or in titration plate wells, while the fluorescent label may be used for quantitating the amount of the amplified fragment.

Direct haplotyping is possible using AS-PCR, without the need for pedigree studies, when allelic sequences at the polymorphic priming site are known. Modifications of this procedure subsequently were reported employing **single-molecule dilution (SMD)** and the so-called "**booster**" **PCR**.

A recent elaboration of this approach offering a faster procedure utilizes pairs of allele-specific PCR primers to amplify each haplotype differentially. In this procedure, termed **double-PCR amplification of specific alleles (PASA)**, the detection of mutant alleles is based on the finding that primers mismatched within 2 bases of their 3′ ends are capable of unequivocally distinguishing between two different alleles by the presence or absence of the amplification products in agarose gel electrophoresis. If the desired allele is present, the allele-specific primer gives rise to an abundance of the amplified product, whereas **if a different (mutant) allele is present there is no amplification of the segment being tested.** The PASA approach has been tested in the detection of mutations responsible for hemophilia B and phenylketonuria.

A significant increase in the sensitivity of detection of mutations can be achieved using the technique of **whole-genome PCR**. In this technique, total human genomic DNA is cleaved into fragments only several hundred base pairs long, and ligated to the so-called "**catch-linkers**". These serve as primers for PCR amplification that produce DNA fragments for the subsequent **specific enrichment** and **selection** of the sequences of interest.

1.2.2.5 COMPETITIVE OLIGONUCLEOTIDE PRIMING (COP) PCR (p. 22)

COP uses a **mixture of primers**, which compete for the binding site on the template sequence so that the **highest degree of complementarity would exclude the mismatched primer**. The **COP-PCR** is used for specific allele recognition, in which either competing radiolabeled or fluorescently tagged primers identify the tested mutation.

1.2.2.6 OLIGONUCLEOTIDE LIGATION ASSAY (OLA) (p. 22)

The **OLA** technique requires knowledge of the target sequence. Two synthetic oligonucleotides are joined head-to-tail over the target sequence by the enzyme **DNA ligase. If a mismatch occurs due to alteration (mutation) of nucleotides in the target sequence at the junction region, the ligation fails to occur.**

1.2.2.7 LIGATION AMPLIFICATION REACTION (LAR) (p. 23)

This method of **allele-specific detection** utilizes the ligation of oligonucleotide pairs that are complementary to adjacent sites on the template DNA. The bacteriophage T4 DNA ligase is capable of joining adjacent oligonucleotide pairs annealed to a complementary DNA sequence in a specific fashion. The presence of an altered base in the template DNA **at the point of juncture** of the two oligonucleotides markedly reduces the efficiency of ligation by this enzyme.

The **ligation amplification reaction (LAR)** has been developed based on this property of the T4 DNA ligase. When a **single pair of oligonucleotides** is used the amplification increases in a **linear fashion**. An **exponen-**

tial increase in the LAR product is obtained when **two pairs of oligonucleotides**, one complementary to the upper strand of a target sequence and one to the lower strand, are present. The products of LAR serve as templates for subsequent rounds of ligation, feeding into the exponential growth of the product yield much like that in PCR.

1.2.2.8 LIGASE DETECTION REACTION (LDR) AND LIGASE CHAIN REACTION (LCR) (p. 23)

Another addition to amplification methods for DNA analysis is the use of a **thermostable DNA ligase,** which both amplifies DNA and discriminates a **single-base substitution**. The reaction is composed of two parts — one is a **ligase detection reaction (LDR)**, the product of which is then further amplified in a **ligase chain reaction (LCR)**, in which both strands of genomic DNA are used as targets for oligonucleotide hybridization. Similar to PCR, the ligation products from one round of ligation become templates for another round, and the amount of product exponentially increases when repeated thermal cycling is maintained.

The diagnostic utility of this assay lies in the fact that **the ligation/amplification cycle** is interrupted by the occurrence of even a single-base mismatch, as was demonstrated in comparison of normal β^A- and sickle β^S-globin genotypes from 10-μl blood samples.

1.2.2.9 TRANSCRIPTION-BASED AMPLIFICATION SYSTEM (TAS) (p. 23)

TAS is primary designed to produce **RNA copies** of a target **DNA** or **RNA** sequence. First, short nucleotide sequences — the **polymerase-binding sequences (PBSs)** — are recognized and positioned on the 3′ side of the target sequence by a DNA-dependent RNA polymerase. To achieve this, an oligonucleotide primer containing two domains is used. Depending on the target sequences to be amplified either a reverse transcriptase or a DNA polymerase is employed in a **primer-extension reaction**. A second oligonucleotide primer following a primer extension reaction complements the product of the first primer-extension reaction to form a double-stranded PBS-containing cDNA copy of the target sequence. A number of **RNA polymerases (T7, T3**, or **SP6**) can be used to recognize the PBS domains.

TAS can produce from 10 to 1000 RNA copies per DNA template. **A large-scale amplification** (10^6 copies produced in 4 TAS cycles) **of target nucleic acid sequences can be accomplished by a *relatively few cycles* of amplification** because the RNA produced during the first TAS cycle serves as target sequences for subsequent TAS cycles, similar to the situation in PCR.

The product of TAS is in part **single-stranded RNA**, which does not have to be denatured prior to detection by hybridization. The TAS protocol has been adapted to the detection of human immunodeficiency type 1 (HIV-1) RNA with the identification of the reaction products by a **bead-based sandwich hybridization**.

1.2.2.10 SELF-SUSTAINED SEQUENCE REPLICATION (3SR)

(p. 24)

While the TAS system requires temperature shifts for optimal reaction rates at the DNA synthesis and RNA transcription steps, a modification of this system using **three different enzymes** is an **isothermal process**. The **3SR system** in essence reproduces the strategy employed by retroviral replication. The three enzymes carrying out the cDNA synthesis and RNA transcription simultaneously are as follow:

- **avian myeloblastosis virus (AMV) reverse transcriptase**
- *Escherichia coli* **RNase H**, dissociating RNA-DNA duplexes
- **T7 RNA polymerase**

In the 3SR system, each priming nucleotide contains three domains in its sequence:

- a region for T7 RNA polymerase binding
- an area for the preferred transcriptional initiation, and
- a sequence complementary to the target sequence.

In contrast to both PCR and TAS, the 3SR reaction **operates at 37°C** — consequently, denaturation of double-stranded DNA cannot be achieved at this temperature and, therefore, cannot serve as the target template. The **main application of 3SR is in the specific detection and amplification of RNA sequences, and also in the evaluation of transcriptional activity of specific genes**.

The **efficiency of 3SR** appears to be **higher than that of PCR** because a 10^5-fold amplification has been observed within 15 min of 3SR that is theoretically attainable by PCR only within 85 min under ideal reaction conditions.

Yet another technique for the **amplification of broad classes of cDNA** using T7 RNA polymerase has been described.

1.2.2.11 QB AMPLIFICATION

(p. 26)

This is one of the first systems introduced for **amplification of the probe** rather than the target. The probe contains RNA bacteriophage replication sequences and an enzyme "**replicase**", the RNA-dependent RNA polymerase from the **bacteriophage QB**.

QB probes are constructed to contain two components:

- a portion of the QB phage RNA (**MDV RNA**) joined to
- a sequence complementary to target RNA or target DNA.

Following hybridization of QB probes to the target, the formed hybrids are purified from the unhybridized probe and QB replicase is added. QB amplification achieves a **billionfold** increase of the probe reporter in 30 min **without temperature cycling**.

The **specificity of QB amplification cannot be determined from the amplified product**, however, and confirmation of positive results must be performed. One suggested combination for clinical use would be **screening by QB amplification** followed by **confirmation with PCR**.

1.2.3 DETECTION OF UNDEFINED DNA ALTERATIONS: SCANNING FOR NEW MUTATIONS

1.2.3.1 RNase PROTECTION ASSAY (RPA) (p. 27)

While ASO and OLA **require the knowledge of expected alteration** in the targeted nucleic acid sequence, other methods can analyze **unknown single-base alterations** in heteroduplexes produced by point mutations by cleaving DNA with the **single-strand specific S1 nuclease**.

Single-base mismatches in **RNase protection assays** have been recognized by using **RNase A** to cleave RNA at the point of mismatch in RNA-DNA heteroduplexes, or in RNA-RNA duplexes. The RPA can detect only about one third of all possible single-base mutations. Its efficiency can be almost doubled by reversing the procedure and labeling the mutant RNA instead of the wild-type RNA or DNA.

1.2.3.2 CHEMICAL CLEAVAGE METHODS (p. 27)

Single-base pair mismatches in DNA-DNA heteroduplexes can be cleaved by chemicals such as **carbodiimide**, **hydroxylamine**, and **osmium tetroxide**. In the latter method, **heteroduplex DNA** containing mismatches is first incubated with osmium tetroxide to detect T and C mismatches, or with hydroxylamine to detect C mismatches, and then they are exposed to **piperidine** to cleave the DNA at the modified mismatched base. This methodology is **suitable for screening all possible mutational changes**. Known examples of its application include detection of **point mutations** in:

- **hemophilia B**
- **the *p53* gene in lung cancer**
- **human ornithine transcarbamylase gene**

1.2.3.3 NUCLEIC ACID ALTERATIONS DETECTED BY ELECTROPHORETIC MOBILITY

1.2.3.3.1 Denaturing Gradient Gel Electrophoresis (DGGE) (p. 27)

Generally, the **conformation of a DNA molecule** is determined by the relative abundance of specific nucleotides. The **strength of bonding** of GC pairs is greater than that of AT and, **under the same denaturing conditions** (e.g., concentrations of urea and formamide, or temperature level), **duplexes enriched in GC will retain their conformation longer than AT-rich regions**.

Heteroduplexes are formed when **one mutant** and **one wild-type** DNA (or RNA) strand associate in a duplex, whereas **homoduplexes** are formed by **two wild-type strands**. DGGE is based on **differential melting** of heteroduplexes compared with that of homoduplexes and is capable of resolving DNA molecules differing by even a single base.

Dissociation of a double helix under denaturing conditions such as in a gradient of denaturants in electrophoretic gels will result in an **abrupt**

reduction of migration. Mismatches or substitutions found in heteroduplexes reduce the strength of complementary strand bonding. The presence of weakened interstrand bonding can be detected by **comparing the electrophoretic mobility of unknown DNA samples under denaturing conditions to that of the wild-type DNA**. This powerful technique has been very widely used. One example, identification of polymorphisms useful in genetic linkage analysis of the proximal region of chromosome 21.

DNA strands separate first in the low-melting domains, thus altering the migration of the entire molecule, which may contain a mutation in the high-melting domains. Loss of detection of DNA alteration in the high-melting domains can occur due to interference from the low-melting domains. The addition of GC sequences known as **GC clamps** by PCR, using primers carrying a GC-rich sequence at their 5′ ends, induces a more gradual and complete melting of the target sequences. In this modification, variations in the melting profile of the entire sequence can be evaluated.

Temperature gradient gel electrophoresis can also be used to reveal variations in the mobility of nucleic acid duplexes. At this time the utility of this technique for clinical diagnostic applications appears to be limited.

1.2.3.3.2 Amplifications and Single-Strand Conformation Polymorphism (PCR-SSCP) Analysis (p. 28)

The electrophoretic mobility of single-stranded nucleic acids in **nondenaturing polyacrylamide gels** is affected not only by size but also by the sequence characteristics, being sensitive to even single-base substitutions. Under nondenaturing conditions the single-stranded DNA conformation is stabilized by interstrand interactions that influence DNA mobility in the gel. Therefore, the **conformation characteristics reflected in altered mobility are determined, in turn, by changes in the sequence**, justifying the term **single-strand conformation polymorphism (SSCP)**.

A combination of SSCP with PCR amplification of a target sequence (**PCR-SSCP**) allows the rapid and sensitive detection of most sequence changes, including single-base substitutions, without the need for restriction enzyme digestion, blotting, and hybridization to probes.

PCR-SSCP requires that the target sequences be known in order to design primers. It can be used for

- **detection of oncogene activation**
- **prenatal evaluation of the presence or absence of particular alleles**
- **linkage analysis for a known sequence**

Spurious results, however, do occur. Unfortunately, the effect of sequence changes on electrophoretic mobility **cannot be predicted**. Moreover, some sequence alterations may not appreciably affect the mobility. Nevertheless, in certain instances PCR-SSCP allows rapid and simple screening of mutations.

The **combination of PCR-SSCP with the Pharmacia Phastgel system** is a simplified, nonradioactive method for the detection of single-base mutations, as demonstrated in the identification of a rare variant of Tay-Sachs disease.

A **multistage procedure** involving the PCR-SSCP can be used to **specifically identify a mutated sequence in a mixture of amplified fragments.** This approach was successfully applied to the detection, in a surgical specimen of lung cancer, of a point mutation in exon 1 of the Ki-*ras*-2 gene that could not be identified by the conventional method of direct sequencing of genomic DNA. Likewise, PCR-SSCP analysis of the *APC* gene mutations in *FAP* and sporadic colon carcinomas yields clinically useful information.

1.2.3.4 A COLOR COMPLEMENTATION ASSAY (p. 29)

The **color complementation assay** is rapid, relatively simple, uses a nonradioactive PCR-based procedure, and has great potential as a routine diagnostic application for detecting infectious agents and various mutations in human DNA.

The simultaneous amplification of two or more DNA segments of specific genomic DNA can be performed using **fluorescent oligonucleotide primers.** Up to five **products of different colors** are generated. The fluorescence of each dye, corresponding to its amplified DNA locus, is evaluated on a fluorimeter. The main advantage of this system is that the identification of the PCR products is enormously simplified. Examples of successful application of the color complementation assay include the detection of

- the **α-globin deletion** in hydrops fetalis
- the **t(14;18) translocation** in follicular lymphoma
- cytomegalovirus (**CMV**)
- a **4-bp deletion** or a **single-base pair** substitution in β-thalassemia
- the **3-bp deletion** in cystic fibrosis

1.2.3.5 EXON-SCANNING TECHNIQUE (p. 29)

An exon-scanning technique uses RNA probes derived from cDNA templates (cRNAs) to identify lesions in suspect genes. The cRNA probe forms a heteroduplex with the ***exons of the target gene, but not with the introns***. The sites of cRNA probes not hybridized to the exon DNA are cleaved by **RNase A.**

The method is used for screening a large number of long stretches of DNA or RNA for **unknown sequence alterations** in the linkage analysis and in the isolation of "candidate genes". It has advantages over Southern and Northern blotting, which may fail to detect small genetic lesions, whereas ASOs are helpful in identifying **known** polymorphisms and cannot be used to screen large genes or mRNAs for unknown sequence alterations. Direct DNA sequencing is not a method of choice in screening for genetic alterations in disorders of unknown etiology.

The basic **difference from the RNase protection assay** is in using cRNA probes derived from cDNA templates, rather than colinear RNA transcribed from cloned genomic DNA. In the **exon-scanning system**, the **entire coding region composed of exons scattered over large lengths of DNA can be screened for point mutations** encountered only in the exons.

Some of the **deficiencies** include

- erroneous cleavage of the cRNA probe at perfectly matched base pairs
- cleavage at intron/exon junctions
- RNase nibbling at the ends of cRNA fragments

Some of the **advantages** of the exon-scanning technique are:

- Only a **small amount of DNA** is needed (which can be obtained from lymphocytes or an amniotic fluid sample).
- It can use **both sense and antisense cRNA probes** to evaluate both DNA strands.
- **Many DNA samples** can be analyzed on a single gel.
- Because it **screens only exons**, the polymorphisms occurring in introns may not complicate the analysis.
- This technique **can essentially map the implicated exon** with a sequence alteration for further characterization by PCR amplification and sequencing.

This approach finds **greatest use** in

- the **evaluation of candidate genes** in heterogeneous disorders
- the **characterization of expressed genes** identified by linkage analysis
- **clinical** and **prenatal diagnosis**

1.2.3.6 MULTIPLEX PCR (p. 30)

Several target DNA sequences can be amplified by PCR simultaneously in a system termed "**multiplex PCR**". The first examples of its application included

- analysis of nine regions of the dystrophin gene for deletions
- an eight-fragment multiplex PCR amplification of the human hypoxanthine guanine phosphoribosyltransferase (*HPRT*) gene in Lesch-Nyhan syndrome

Since its introduction multiplex PCR has become most widely accepted for a variety of applications.

1.2.3.7 AMPLIFICATION AND SEQUENCING (p. 30)

Amplification of the targeted DNA sequences by PCR **followed by direct sequencing** of the amplified products is another widely used approach. Among the many reported applications are the characterization of

- c-Ki-*ras* oncogene alleles
- β-thalassemia mutations
- mitochondrial DNA
- the HLA-DQA locus
- mutations leading to the Lesch-Nyhan syndrome
- α_1-antitrypsin deficiency
- the analysis of DNA polymorphisms

A 1000-fold increase in the sensitivity of the method is achieved using a modified primer in the PCR, thereby **introducing an *artificial restriction site* into the amplified product**. This creates an RFLP, indicative of a given mutation, that can be used for screening purposes or for monitoring the effects of therapy in oncology.

Significantly more definitive information on the polymorphisms is obtained when the **amplified DNA segments are directly sequenced**. Immobilization of the PCR products via biotinylated ASO primers and steptavidin-coated magnetic beads drastically improves this methodology. The **solid-phase DNA sequencing following PCR amplification** allows a rapid search of polymorphisms. There is no need for DNA cloning of genomic or PCR-amplified DNA prior to sequencing as required in earlier protocols.

Significant **improvements** are being continuously reported in simplifying PCR-sequencing procedures universally directed toward **automation** and analyzing DNA sequences from a single cell for evaluation of genetic disease characteristics. Even the unsurpassed resolution of the **tunneling microscope** is contemplated for use in DNA sequencing.

1.2.3.8 QUANTITATIVE AMPLIFICATIONS (p. 32)

Amplification procedures directed at either the target or the probe are **qualitative** methods primarily designed to establish the presence or absence of a nucleic acid characteristic in a condition under study. **Quantitation of the initial amount** of the template DNA or RNA significantly enhances the informative content of an amplification scheme. Quantitation of amplification procedures has been attempted by

- **coamplification**
- **restriction digestion of the amplified product** and **size analysis** on agarose gels
- **competitive PCR**
- hybridization techniques followed by **isotopic or immunoenzymatic estimates of the generated product**

A technique termed **DIANA** — for detection of immobilized, amplified nucleic acids — uses a qualitative colorimetric assay of *in vitro*-amplified DNA capturing the PCR products on **streptavidin-coated magnetic beads**. By introducing **competitive titration** with an *in vitro*-cloned sequence of the target into which the *lac* operator sequence is inserted, quantitation of specific DNA can be performed in clinical samples. This assay allowed the detection of *Plasmodium falciparum* DNA in the range of 20 to 50,000 parasites per sample, making the diagnosis of malaria infection possible not only in the acute but also in chronic phase.

1.2.3.9 AUTOMATED AMPLIFICATION ASSAYS (p. 32)

Although **thermocyclers** were introduced in 1987, efforts are continually being made to achieve a maximally automated system for a wide range of specific uses. The **GeneAmp PCR System 9600** (Perkin-Elmer Cetus) ensures better temperature control, eliminates the need for a mineral oil overlay to control evaporation, allows for compatibility with multichannel liquid-handling devices, and minimizes contamination among many other features described by the manufacturer.

Because **temperature control** of the PCR is critical for the specificity and speed of the reaction, a system capable of rapid temperature changes achieved by **hot air** instead of water circulation is particularly attractive. In addition, the typical reaction time can be reduced to 15 min or less.

Detection of specific DNA sequences by PCR coupled with the discrimination of allelic sequence variants by a colorimetric **oligonucleotide ligation assay (OLA)** has been automated. The system has been demonstrated in the detection of

- sickle cell hemoglobin
- cystic fibrosis (F508) mutation
- gene segments of the human T cell receptor β-chain locus

The **automated PCR/OLA system** eliminates the need to measure DNA fragment sizes, the use of radioisotopes, as well as the requirement for high-quality DNA, particularly when polymorphic human STSs become available for automated forensic typing. The entire ELISA-based assay can be performed in **microtiter plates**, and no centrifugation or electrophoresis is needed. **Digoxigenin-tagged oligonucleotides** serve as reporter molecules in the ELISA of the duplexes captured via their biotinylated oligonucleotides.

1.2.4 SELECTED CHROMOSOME ANALYSIS TECHNIQUES APPLICABLE TO GENE LEVEL DIAGNOSIS IN CLINICAL PRACTICE (p. 34)

Specific assignment of defined DNA sequences to **chromosomal locations** (chromosomal mapping) using *in situ* hybridization (ISH) techniques can, at the present stage of technology, directly place genetic sites within or outside of bands (5 Mb), or at distances as close as 100 kb apart.

The **top-down approach** identifies and makes assignments of progressively smaller fragments of DNA from a given chromosome. The **bottom-up strategy** identifies and aligns, in a continuum of sequences, overlapping sets of DNA clones (**contigs**). Techniques known as **chromosome walking, chromosome jumping**, and **probe walking** are used in these strategies.

Some of the genome **mapping techniques** may find their way into clinical gene level diagnostics. Among these are

- **ISH** and its variants, which are already helpful in diagnostic protocols
- **pulsed-field gel electrophoresis (PFGE)**, which is capable of resolving large fragments of DNA (up to 10 Mb)
- **controlled partial digestion** of DNA using, for example, *Not*I and *Mlu*I, which recognize **CpG islands**

- **YAC (yeast artificial chromosomes) vectors** containing large fragments of DNA up to 1 Mb in size
- **cell hybridization** and **microcell-mediated gene transfer**
- **PCR**

1.2.4.1 ANALYSIS OF DNA FROM SPECIFIC CHROMOSOMAL REGIONS (p. 34)

Genetic linkage analysis is capable of determining the physical location of a mutant allele. The isolation of some genes, however, may not be possible using the current cloning strategies due to limitations on the size of the restriction fragments that can be cloned in appropriate vectors. A novel approach of **microcloning region-specific chromosomal DNA** has been developed. In an experimental system, **single bands** are dissected from polytene chromosomes and digested with *Sau*3A, followed by ligation of oligonucleotide adapters to these fragments. The latter provide convenient primers for PCR in this "**microamplification**" approach.

An alternative approach uses the PCR amplification of restriction fragments obtained from **chromosome dissection** at a specific region. This rather complicated micromanipulation procedure is combined with **sequence-independent amplification of DNA**. A collection of DNA molecules representing unique and repetitive sequences can be obtained in this manner.

Isolation of expressed sequences from defined human chromosomal regions has been achieved using **somatic cell hybrids** in which the complexity of the genomic region under study is reduced. This approach, for example, allowed the isolation of transcribed sequences from the distal region of the X chromosome long arm.

Direct amplification of a microdissected Giemsa-banded chromosomal segment has been described in application to a cloned human brain sodium channel (α subunit) gene sequence from chromosome 2q22-q23. This technique, termed **chromosomal microdissection PCR (CM-PCR)**, provides a relatively simple, precise, and direct approach to gene mapping. Advantages of CM-PCR:

- the assay is **rapid**
- **rehydrated archival tissue** can be used
- CM-PCR produces a **higher level of resolution** than that obtained with ISH with radioactive probes

Although CM-PCR is hardly applicable in clinical diagnostic practice at this time, it can offer a **direct approach** to the analysis of chromosomal aberrations (deletions, duplications, translocations, inversions) encountered in various human malignancies. Also, CM-PCR can be used in linkage analysis by simultaneously amplifying loci of interest and identifying their allelic states.

Sequence data from the **GenBank** can be used to construct PCR primers, for example, in multiplexing PCR amplifications for a rapid and detailed analysis of human chromosomes. **Lists of PCR primers** for each chromosome that have very little overlap have been described, providing a basis for the comprehensive analysis of human chromosomes. This information, containing **sequence-tagged sites (STSs)** or **expressed sequence tags (ESTs)** for all the human chromosomes, allows the detection of most chromosomal aberrations applicable to molecular cytogenetic diagnosis.

An alternative approach is based on **two complete sets of small inserts**, and **complete digest DNA libraries** for each of the 24 human chromosomal types constructed at the **National Laboratory Gene Library Project**. The available information is used

- in **genome mapping**
- in the **search for RFLP markers** linked to various genetic diseases
- as a source of unique sequence probes for "**chromosome painting**", by **fluorescent ISH (FISH)**

The potential diagnostic utility of these probes is high for the detection of chromosome aberrations in metaphase and interphase nuclei. They can be used as **highly specific cytochemical stains for individual chromosomes**. Chromosome paints for all human chromosomes are converted into commercial products that are widely available (e.g., Life Technologies, Bethesda, MD; Oncor, Gaithersburg, MD; Bios Laboratories, New Haven, CT).

1.2.4.2 AUTOMATED KARYOTYPING (p. 36)

An example of attempts to standardize and automate the analysis of human chromosomes can be seen in the advances of the **Athena semiautomated karyotyping system**, which analyzes metaphase spreads. It provides

- **automated segmentation** of metaphase images into individual chromosomes
- **automated measurements** of each banded chromosome

The entire assay is complete within 1 to 2 min and provides a specific description of the karyotype according to the standard Paris convention.

Alternatively, individual chromosomes can be identified using **multiple dye-DNA interactions** followed by high resolution of the total and peak intensities of the dyes. Patterns are produced that are translated into karyotype images **without the use of banding**.

Representation of chromosomes in stylized images resembling bar codes is adaptable to automated karyotype analysis. Although this approach needs further refinement, it appears to offer a workable diagnostic system for wide use in clinical practice.

1.2.4.3 ISH IN CHROMOSOME ANALYSIS (p. 37)

Chromosomal aberrations such as deletions, translocations, the formation of isochromosomes, and other structural rearrangements can be identified by **high-resolution banding** techniques, which are continually being improved to the level appropriate for use in routine clinical laboratories.

The chromosomal and/or gene level evaluation of diseases by ISH allows

visualization of the target site relative to high-resolution banding. In cancer cytogenetics, in particular, this approach may be integrated into the clinical management of patients not only with hematological malignancies but with solid tumors as well.

The presence of chromosomal abnormalities such as a balanced chromosomal translocation can be confirmed by **quantitative hybridization dosage** studies. Examples of combined use of ISH and Southern blotting, which complement the high-resolution banding technique, are seen in the establishment of the exact molecular cause of gonadal dysgenesis, or the presence of aberrations in sexual development of patients with Y;autosome translocations even when no detectable genetic defects could be demonstrated by conventional cytogenetic methods.

The new technique of "**interphase cytogenetics**" allows the **recognition of specific chromosomal regions in interphase tumor cells** by ISH; this not only allows a better resolution by ISH, but also expands the range of diagnostic possibilities of ISH in clinical laboratories. In addition, it removes the need to culture and synchronize the cells, which are cumbersome and time-consuming procedures.

The practical utility of ISH for clinical gene level diagnosis is enormously enhanced by

- the introduction of **nonradioactive** and **multiple labels**
- the use of **shorter oligonucleotide probes**, both **sense** and **antisense RNA probes**
- the **extension of hybridization strategies** previously employed in solid support or soluble hybridization procedures **to tissue sections** and **chromosomes**
- the **combination of microdissection** of target chromosomal regions **with amplification**
- promising efforts to **automate** ISH

Emerging combinations of ISH with flow cytometry (FCM), electron microscopy (EM), and confocal microscopy (see below) may eventually evolve into diagnostic modalities of practical utility as well. Specific **"custom-designed" probes** tagged by easily and rapidly quantifiable nonradioactive labels, such as digoxigenin, allow the application of **ISH even to frozen sections.**

PCR amplification from **formalin-fixed, paraffin-embedded tissues** adds a new dimension to tissue analysis for infectious and genetic diseases. Examples of successful **amplification of target sequences** have been described in

- **formalin-** or **B-5-fixed** bone marrow specimens
- **boiled** clinical specimens using fast multiplex PCR amplification of human papillomavirus (HPV) sequences
- **formalin-fixed**, **paraffin-embedded** liver tissue for the analysis of gene expression and the detection of RNA viruses

The **tissue fixatives best suited** for optimizing subsequent PCR analysis were found to be

- acetone
- buffered neutral formalin (10%)

Amplification efficiency was **compromised** by

- Zamboni's
- Clarke's
- paraformaldehyde
- formalin-alcohol-acetic acid
- methacarn

The **worst results** were observed with tissues processed with

- Carnoy's
- Zenker's
- Bouin's fixatives

Although all of diagnostic pathology deals with establishing the distinction between a benign and malignant process, some attempts have been made to develop an **objective criterion** defining a population of cells characterized by proliferative activity usually associated with "deviant" cellular behavior. At least two assays have been reported for

- **DNA polymerase α** activity in paraformaldehyde-postfixed frozen sections (an immunohistochemical method informative even in precursor lesions)
- the detection by ISH of the expression of the **histone 3 (*H3*) gene,** which is absent in resting cells; its expression unambiguously identifies cycling cells

1.2.4.3.1 Nonradioactive ISH (p. 38)

The **main advantage** of using radioactive labels in ISH is the high level of sensitivity that can be attained, making it capable of identifying DNA sequences a few hundred base pairs in length. The **disadvantages** are associated with

- handling of radioactive materials
- need for autoradiography and lengthy exposure times
- limited shelf life of radioisotopes
- frequently extensive background

Therefore, **alternative reporter molecules** are much more attractive for diagnostic laboratories. Some of these include

- chemical modification of nucleic acids with **acetylaminofluorene (AAF)**, with the detection of AAF by poly- or monoclonal antibodies
- **mercuriation** of nucleic acids performed after ISH is completed, using haptens with SH groups that bind the mercury
- reporter molecules introduced as haptens, e.g., **trinitrophenyl, biotinyl,** and **fluorescent** groups
- **dimerized thymidine** probes
- DNA **modification at the cytosine** residues by bisulfite catalyzed **transamination**

- **5-bromouridinylated oligonucleotides** for analysis of DNA and RNA both for ISH and membrane hybridization assays
- **acridinium ester**-labeled DNA oligonucleotide probes and oligonucleotides covalently linked to **porphyrins**
- **oligo-[α]-thymidylate** covalently linked to an azidoproflavine derivative
- **sulfonylation** of DNA directly in tissue sections followed by immunocytochemical detection of modified DNA; this shows a high specificity for DNA without cross-reactivity with RNA and superior staining qualities compared to Feulgen staining of DNA
- **bioluminescent** systems based on luciferin derivatives; these systems display a high sensitivity and are amenable to quantitation
- **chemoluminescent** systems using derivatives of **1,2-dioxetane** in combination with various enzymatic triggering combinations, allowing multiplex reactions to be monitored simultaneously
- incorporation of the nucleotide analog **digoxigenin-11-dUTP**, a derivative of a plant steroid found in digitalis

A **double-labeling ISH** approach using various combinations of radioactive and nonradioactive reporters at the same time has also been described in the analysis of mRNAs, chromosomal DNA sequences, and proteins.

Biotinylated nucleic acid probes are useful in identifying the highly repetitive, middle repetitive, and large single-copy sequences (15 to 40 kb long). For example, biotinylated alphoid repeat probes have been used in the clinical diagnostic laboratory to detect the centromeres of chromosomes 13, 21, 14, and 22 by ISH. In some cases the biotin label has been found to reduce hybridization sensitivity. **Photochemical biotinylation** has also been described.

The **immunogold-silver** detection of biotinylated probes applicable to formalin-fixed, paraffin-embedded tissue dramatically increases ISH sensitivity. The **streptavidin-gold detection system**, however, failed to reveal the same degree of sensitivity, apparently due to **steric hindrance** of the bulky streptavidin-gold complex.

Fluorochromes have long been used to label nucleic acid and oligonucleotide probes for ISH, and now **multicolor fluorescence** has become commonplace for the simultaneous detection of different nucleic acid sequences.

Double- and **triple-hybridization** procedures have been described. A triple-hybridization protocol using three differently haptenized probes (AAF, mercuriated, and biotinylated chromosome-specific repetitive probes) combined with three fluorochromes (blue-amino methyl coumarin acetic acid (AMCA), green-FITC, and red-TRITC) was applied to study chromosomal abnormalities in human solid tumors using interphase nuclei of tumor cells. By combining double-haptenized probes and three fluorochromes **up to seven specific targets** can be evaluated simultaneously. Combined with the **digitized image evaluation**, which can be automated, this approach may lead to a standardized, rapid tool for molecular cytogenetic evaluation of patient specimens in clinical practice.

A hybridization signal detection system based on the **digoxigenin-anti-digoxigenin** interaction avoids **two major disadvantages** of the biotin-streptavidin system, which are

- the **ubiquitous presence of biotin** (vitamin H) in eukaryotic as well as prokaryotic cells, which contributes to the occurrence of nonspecific signals
- the **nonspecific binding of streptavidin to immobilizing supports**, such as nitrocellulose or nylon, used in hybridization. Digoxigenin reporter molecules allow for reduced background noise in filter hybridization and ISH

Digoxigeninated **simple repeats** $\mathbf{(CAC)_5}$ used as probes on human chromosomes reveal a pattern of banding resembling that of R bands. Likewise, digoxigenin-labeled probes have been used for ISH of the M13 **minisatellite tandem repeat sequences** that show an R bandlike pattern on human metaphase chromosomes. Digoxigenin labeling of oligonucleotides has also been used for **fingerprint analysis**.

PCR has been used to produce single-stranded or double-stranded cDNA probes for ISH, as well as for generating vector-free digoxigenin-dUTP- and biotin-11-dUTP-labeled probes.

Ingenious new nonradioactive detection methods for nucleic acids keep appearing. For example, the "**universal probe system**" has been developed, using the principle of **sandwich hybridization**: the primary probe is without a label and the second probe, complementary to a portion of the primary probe, is tagged with biotin or any other nonradioactive reporter molecule.

1.2.4.3.2 Banding and ISH (p. 41)

Until recently, a sequential protocol of ISH and then banding, with the **two images subsequently superimposed**, has been employed whenever ISH signal had to be assigned to a specific chromosomal location. The **simultaneous visualization of R-banded chromosomes and hybridization signal** generated by probes as small as 1 kb has been described in 23 unique DNA segments mapped by ISH to the long arm of chromosome 11.

Larger cosmid probes designed for marked chromosomal sites (e.g., 11q13 and 11q23 regions) associated with various disease or growth control genes (hematological neoplasias, Ewing's sarcoma, and ataxia-telangiectasia) can be used in **competitive suppression of hybridization to repetitive sequences**. This procedure combines biotin-labeled probes and staining of chromosomes with both **propidium iodide** and **4′,6-diamidino-2-phenylindole (DAPI)**, producing **R** and **Q banding** patterns, respectively. It allows unambiguous chromosome identification of various markers useful in practical analysis of chromosomal alterations.

A potentially clinically useful technique, **replicational banding**, can identify genes that are actively expressed; it offers a particular advantage in prenatal diagnosis of developmental abnormalities.

1.2.4.3.3 Repetitive DNA-ISH (p. 41)

Chromosome-specific probes containing only **repetitive (satellite) DNA** hybridize predominantly to the **centromeric** regions of the chromosome. Hybridization probes constructed from **chromosome-specific libraries** specifically stain individual chromosomes. In these protocols, total genomic DNA is used as a competitor to exclude the dispersed repetitive sequences from participation in ISH. Various chromosomal aberrations can thus be identified as demonstrated by the diagnosis of trisomy 18 with the probe L1.84. Likewise, the pericentric region of human chromosome 17 has been analyzed by a cDNA clone containing the entire alphoid repeat. Using a monomer of the higher order repeat, the chromosomal specificity of ISH was improved. **Shorter probes** can further improve the chromosome specificity, allowing the precise identification of individual chromosomes using alphoid repeats. Biotinylated DNA probes can be used for **high-resolution mapping of satellite DNA**.

1.2.4.3.4 Single-Copy ISH (p. 42)

Chromosome-specific "**painting**" in interphase nuclei, using biotinylated DNA library probes, pinpoints fine chromosomal aberrations associated with various pathological conditions. In one such study, DNA probes complementary to single-copy and repetitive sequences were used for the characterization of chromosomes 1, 4, 7, 18, and 22.

Single-copy genes can be detected by ISH, and a combination of ISH with routine histology and DNA analysis, in tissues as well as in cytospins prepared from cerebrospinal fluid, pleural fluid, bone marrow, peripheral blood, and cell lines.

Simultaneous hybridization with different probes combined with **digital image analysis** allows the visualization of complete chromosomes, deletions, and translocated segments of chromosomes in complex karyotypes. Enhancement of the detection sensitivity by ISH of single-copy DNA sequences using biotinylated probes is possible by **intensified-fluorescence digital imaging microscopy**.

1.2.4.3.5 Single-Copy RNA-ISH (p. 42)

By **combining microfluorometry** and **ISH**, the simultaneous evaluation of specific gene expression (mRNA level) and the DNA content in individual cells becomes possible without distorting the morphological picture. Using **riboprobes** to study the expression of colony-stimulating factor, granulocyte-macrophage, and interleukin 3 mRNA, a sensitivity comparable to that of northern blots can be achieved.

In some cases, ISH demonstrates a sensitivity even higher than that of northern blotting. **mRNA-ISH** has been used for two- and three-dimensional visualization of specific gene expression in a variety of cell types.

1.2.4.3.6 Quantitative ISH (p. 43)

Evaluation of the **magnitude of expression** of target genes significantly enhances the diagnostic utility of ISH. Using synthetic oligonucleotide probes constructed for different regions of a messenger RNA markedly enhances the sensitivity of quantitative ISH.

A quantitative ISH procedure using biotinylated probes has been developed and demonstrated in the measurement by computerized image analysis of the relative levels of cellular mRNA for proopiomelanocortin in the anterior lobe of the pituitary. Although this procedure does not provide absolute values for mRNA levels, the relative quantitation can be informative in evaluating specific gene expression.

1.2.4.3.7 Competitive ISH (CISH) — "Chromosome Painting" (p. 43)

In general, the **competitive hybridization** approach is used to enhance the specificity in detecting target sequences. "**Chromosome painting**" uses large pools of cloned genomic sequences from a single human chromosome as probe and, by performing a preannealing step in the presence of an excess of sonicated total human DNA, complete staining of a given chromosome can be achieved in metaphase and interphase nuclei. When applied to tissues, CISH can be helpful in reducing the contribution of the signal from repetitive DNA through the addition of competitor DNA, by analogy with the technique used in Southern hybridizations. A marked increase in the ratio of specific to nonspecific hybridization signal can be achieved under optimal preannealing conditions.

A variant of competitive ISH, called **chromosomal *in situ* suppression (CISS)-hybridization**, using biotinylated phage DNA library inserts from sorted human chromosomes, has been applied, for example, to detect aberrations of chromosomes in irradiated peripheral lymphocytes and hybrid cells.

1.2.4.3.8 Confocal Microscopy and ISH (p. 43)

A study of the **spatial organization** of chromosomes within the nucleus has been markedly advanced by the technique of optical sectioning of intact cells and stepwise movement of the microscope focus through successive planes of the DAPI-stained nucleus. Combining FISH with confocal scanning laser microscopy, a **three-dimensional topography** of specific nucleic acid sequences and their transcripts can be analyzed in suspended cells.

1.2.4.3.9 EM-ISH (p. 44)

Like confocal microscopy-ISH, the **EM-ISH combination** at this time is primarily utilized in basic research. Even single-base mutations can be detected in large DNA fragments by immuno-EM. The simultaneous determination of at least two sequences can be achieved by a modification of ISH at the EM level. The location of a sequence is expected to provide insights into its properties and function by **high-resolution sequence mapping**.

RNA probes appear to be more sensitive and produce lower background than DNA probes.

1.2.4.3.10 Flow Cytometry-ISH (FCM-FISH) (p. 44)

Using rare-cutting restriction endonucleases, such as *Not*I, **large DNA fragments** can be obtained, sorted by FCM, and subjected to appropriate analysis. The main obstacle is the **loss of morphological integrity** of ISH-processed cells, which interferes with informative FCM analysis. This drawback was markedly reduced when cellular morphology could be well retained in cells of erythroid lineage following new procedures developed for FCM-FISH.

The wide use of FCM for clinical diagnostic purposes may offer a suitable background for introducing its combination with ISH as a method of gene level testing in clinical practice.

1.2.4.3.11 Automation of ISH (p. 44)

So far, efforts to automate ISH have been reported either in the **processing of glass slides**, or in the development of **automated evaluation** of the final results of ISH, for example, using the high-resolution image analysis system IPS KONTRON.

When an X centromere probe that recognizes the alphoid satellite sequences is used for ISH, the automated detection of fragile X chromosomes is possible. This procedure accomplishes a fully automatic metaphase identification, digitization at high resolution, and segmentation analysis, offering an almost fully automated diagnostic ISH system. New, **triplet-specific probes** for the unstable region can be used in the same mode.

1.2.5 DNA POLYMORPHISMS

1.2.5.1 SITE POLYMORPHISMS (p. 45)

Molecular analysis of chromosomal DNA using restriction endonuclease digestion revealed DNA polymorphisms, the **restriction fragment length polymorphisms (RFLPs)** that can be used as molecular markers. RFLPs are due to base substitutions, microdeletions, or insertions that lead to the presence or absence of a recognition site for the enzyme used. Although the overall **variability of DNA** is low, and the variable regions are not uniformly distributed, the **informative potential** of DNA polymorphisms by far exceeds that of **protein polymorphisms**. While the coding sequences for proteins occupy only a **small portion of the genome**, DNA polymorphisms represent variability both in the coding and noncoding regions distributed throughout the **entire genome**. Furthermore, certain noncoding regions display high variability.

Judicious choice of restriction enzymes allows identification of **highly mutable sites** (e.g., at **CpG islands**). When combined with the use of long probes that cover multiple restriction sites of a given locus, RFLPs of virtually unlimited variety can be disclosed. Conventional methodology calls for separation of the DNA fragments subsequent to enzyme digestion. Newer techniques (e.g., using PCR) markedly increase the resolving power, distinguishing fragments with a single base difference.

A marked improvement in resolution of DNA fragments can be achieved with **high-performance capillary electrophoresis (HPCE),** which combines high resolving power at the single-nucleotide level and speed, the possibility of automation, and a subattomole level of detection. Gel slab electrophoretic systems allow resolution of fragments above 400 bp in length, whereas HPCE separates fragments from under 100 bp up to 12,000 bp; this is important in clinical, diagnostic, and forensic applications of RFLPs, PCR product analysis, and in genome mapping.

1.2.5.2 HYPERVARIABLE REGIONS (HVRs) (p. 46)

The first report of a **highly variable locus**, in which DNA polymorphism was due to **DNA rearrangements rather than base substitutions or modifications**, identified an **arbitrarily selected region** showing a significant frequency of **DNA sequence variation**. Regions displaying numerous DNA polymorphisms, known as **hypervariable regions (HVRs)**, have been subsequently identified in various parts of the genome.

In **site polymorphisms**, changes in the DNA sequence are produced by base modifications, insertions, or deletions that eliminate the site cut by a restriction enzyme. In contrast, the **different lengths of HVRs** are due to a **variable number of tandemly repeated sequences**. These repeats, called "**minisatellites**", can be identified in an unknown mixture of DNA fragments by hybridization of a **short unit sequence** to the DNA of interest. HVRs have also become known as **VNTRs (variable number of tandem repeats)**; they show extremely high variability in length, and the frequency of heterozygotes may be on the order of 80 to 100%

Some of the earlier **minisatellite probes** were

- **33.15**, consisting of 29 repeats of a 16-bp core sequence
- a related probe **(33.6)** that produced a different pattern of hybridization due to differences in the **core units**

The resulting hybridization patterns can be viewed as individual-specific DNA "**fingerprints**", in which each band (fragment) represents a different locus. **Such a characteristic, unique, and reproducible pattern can be used for individual identification**. Furthermore, because the majority of the bands in fingerprints are not genetically linked to each other, and are inherited in a Mendelian fashion with **heterogeneity of almost 100%**, these patterns can be used for **linkage analysis of genetic disorders**.

Population-related characteristics, procedural variability, and statistical aspects of DNA fingerprinting analysis for individual identification are hotly debated areas at present.

1.2.5.3 RFLP PROBES (p. 47)

A specific pattern can also be used to generate probes from the fragments (bands) displaying an informative association (linkage) with a given genetic trait. PCR amplification directly from agarose gels makes this approach a highly practical and fast route to molecular genetic probing.

1.2.5.4 SINGLE-LOCUS PROBES (p. 47)

High-stringency conditions, which allow the formation of hybrids with only the highest degree of complementarity of the probe to the target sequence, **are used to detect restriction fragments from a single genetic locus**. Higher stringency hybridization is used to reveal RFLPs that show **two alleles (fragments) from both chromosomes**, if a person is heterozygous for the probe.

Single-locus probes can be produced by cloning DNA of interest. They have been used in the early stages of human genome mapping. Likewise, oligonucleotide probes corresponding to the **consensus sequences** of the tandem repeats of various hypervariable loci have been constructed. Such VNTR probes, each derived from a single locus, have been used to identify a large number of polymorphisms spanning the entire human genome.

A **combination of RFLP analysis with PCR amplification of target sequences** is an approach receiving much attention. The simplicity of this approach for analysis of the inheritance of alleles within a pedigree was demonstrated in the case of the interphotoreceptor retinoid-binding protein (*IRBP*) gene. In contrast to Southern blotting, this combination of **PCR coupled to restriction digestion (PCR-RD)** does not require probing of the products following digestion, reducing the labor required by a factor of 10; this makes linkage analysis acceptable for a small clinical laboratory.

DNA **fingerprinting from single cells** can be performed when PCR technology is applied. Some of the useful repeats include the **$(dC\text{-}dA)_n$ repeats**, the allelic variations of which can also be easily detected with PCR. The wide distribution of this class of repeats, estimated to number up to 100,000 in the human genome, justified the efforts to develop an automated system for genotyping individuals with respect to $(dC\text{-}dA)_2$ polymorphisms.

Automation of the analysis of DNA polymorphisms offers a desirable modality not only for large-scale mapping efforts such as are involved in the HGP, but can offer a better controlled, faster, and reproducible tool for individual identification in forensic practice or parentage testing.

1.2.5.5 MULTIPLE-LOCI PROBES (p. 48)

Hybridization patterns generated under **less stringent conditions** allow the probes to form hybrids with a lower degree of complementarity to the DNA being tested, and **the RFLPs produced reveal fragments derived from multiple chromosome loci**.

The two related HVR probes, **33.15** and **33.6**, derived from the **"core" consensus repeat sequences**, were used to generate RFLP patterns. Scores of bands (alleles) derived from as many as 30 to 50 independent loci are produced under **reduced stringency** conditions in such patterns of *Hin*fI-digested chromosomal DNA. Another multilocus probe, **pAC365,** detects loci on 16 human chromosomes, producing a multiple-locus RFLP pattern under conditions of **high stringency**.

Because **each band represents a different locus**, the DNA "**finger-prints**" produced by multiple-locus probes can trace a number of loci simultaneously. The **probability** that all fragments (bands) detected in one individual are present in the RFLP pattern from another individual using probes 33.6 and 33.15 has been calculated to be less than 5×10^{-10}. Thus, with the **exception of identical twins**, who can share a large number of alleles, these DNA fingerprints are **totally individual specific** even within one family.

1.2.5.6 ANALYSIS OF DNA POLYMORPHISMS USING PCR (p. 48)

Analysis of DNA sequences is now possible at a **single-cell level** and, given the strategem for **inferring haplotypes from PCR-amplified samples**, the convenience of PCR-based evaluations in clinical diagnostic practice is readily apparent, particularly when noninvasive **sampling from buccal mucosa** has been shown to be sufficient for direct genetic analysis.

A general strategy for **rapid screening of genomic DNA for sequence variation** from many individuals has been devised as a two-step procedure using thermostable *Taq* DNA polymerase. First, sequences flanking the coding region of a target gene are amplified. Then sequencing primers are synthesized based on the cDNA sequence of the gene. The same *Taq* DNA polymerase is then used for sequence analysis of the amplified, linear, double-stranded DNA (dsDNA). This feature allows easy automation of the entire cycle, offering an attractive tool for large-scale screening of DNA variations applicable to clinical diagnosis.

Direct PCR amplification can be performed on **lyophilized** tissues and cells, markedly simplifying the process of genotyping in diagnostic laboratories.

Multiple polymorphic loci can be amplified simultaneously in a procedure using **amplified sequence polymorphisms (ASPs)**. Another modification of the PCR technique, based on the sharing of sequence motifs of **tRNA genes** (which occur in multiple copies dispersed throughout the genome in most species), can be used to **identify species** and **genera**. This version of **PCR-generated fingerprints** was termed **tDNA-PCR**. The method can be of use in determining the epidemiology of human diseases. Yet another version of PCR fingerprinting has been used to select HLA-matched, unrelated bone marrow donors.

PCR amplification to detect polymorphisms at VNTR loci allows the use of these highly specific polymorphic markers in a highly simplified format for typing **allelic variations at any hypervariable region**. This approach has another advantage over Southern blot analysis (traditionally used for this purpose), in that it can distinguish small differences (11 to 70 bp) between large DNA fragments. The method is suitable for a variety of **medical** and **forensic** applications as well as in **paternity determinations**, with the results available within 2 days.

Other polymorphic loci, such as the **HLA-DQα locus**, have been studied by PCR amplification with subsequent probing of the immobilized, amplified DNA for allelic diversity with nonradioactively labeled oligonucleotides in a **dot blot** format. Alternatively, the probes have been immobilized on a filter and hybridized to biotin-labeled, PCR-amplified DNA in a **reverse dot blot** technique. This approach allows rapid and relatively simple typing of a large number of samples.

Compared to the VNTR system defined by RFLPs, an **allele-specific typing system**, such as for the DQα locus, in addition to identifying discrete traits transmitted in a Mendelian fashion allows better discrimination of alleles than in a gel electrophoresis system.

Although the **discriminatory power** of the DQα markers is less than that attainable with some VNTR systems, this PCR-based typing can rapidly generate information starting from minute amounts of material. The convenience of the reverse blot format, in which all the probes come fixed to a typing strip, has been translated into a commercial product. This nonradioactive system has been widely used in forensic casework.

1.2.5.7 INTERSPERSED REPETITIVE SEQUENCES — PCR (IRS-PCR) (p. 50)

The presence of repetitive elements in the human genome distinguishes it from that of nonmammalian species. Combined biochemical, cytological, recombinant DNA approaches together with computational analysis indicate that the distribution of **interspersed repetitive DNA sequences** in the human genome can be best described by models assuming a **random distribution**.

Some sequences, such as those of the ***Alu* family**, display varying density. *Alu* elements represent the major family of **short interspersed repeats (SINEs)** in mammalian genomes and, characteristically, there are approximately 10^6 copies of a 300-bp sequence scattered throughout the human genome approximately every 3 to 4 kb. *Alu* sequences

- are **frequently interspersed** in the human genome
- **flank anonymous DNA segments** harboring yet undisclosed polymorphisms
- are **at least 50-fold less abundant in centromeric heterochromatic regions**, whereas other repetitive, tandemly arranged sequences are found there

Members of the other family of repetitive sequences, **L1**, have **an inverse relationship** with *Alu* repeats with respect to their cytologically identifiable locations on chromosomes. Abundance measurements for the repetitive sequences of the L1 and **GT/AC** families indicate that these should be located, on average, every 30 to 60 kb throughout the human genome.

Application of **interspersed repetitive sequences PCR (IRS-PCR)** directed toward both *Alu* and L1 sequences allows a **chromosome-specific pattern** of amplification products to be easily analyzed on agarose gels run with ethidium bromide. This "**PCR-karyotype**" approach displaying a characteristic, chromosome-specific banding pattern of PCR-karyotype products in agarose electrophoresis offers a convenient alternative to conventional cytogenetic procedures. Predigestion of template DNAs with restriction enzymes prior to *Alu*-PCR ("**restricted *Alu*-PCR**") allows the isolation of specific markers from a given chromosomal region.

When sequence information is available, the method of **amplified sequence polymorphisms (ASPs)** can be applied to analyze differences in the genome patterns being compared. Alternatively, when prior knowledge of the DNA sequence of each polymorphic locus to be amplified is not required, **polymorphisms flanked by *Alu* sequences** can be evaluated. There are approximately 100,000 *Alu*-flanked, relatively short (1 kb) DNA segments ("**alumorphs**") that can be amplified using *Alu*-specific primers. Granted that **alumorphs are less informative than the RFLP or VNTR markers**, they are **more uniformly scattered** throughout the genome than minisatellites or VNTR markers. This feature makes alumorphs attractive markers for gene mapping, as well as helpful adjuncts in forensic investigations and population studies.

By analogy with the amplification of *Alu* repeats, members of the other major class of IRSs — the **L1 elements** — have also been used for PCR-based analysis. These **long interspersed repetitive sequences (LINEs)** are present at 10^4 to 10^5 copies per genome in mammals. The human L1 element has also been known as *Kpn*I sequences.

Thus, primers constructed to a variety of IRS types can be used for the identification and amplification of defined chromosomal regions by IRS-PCR, as well as for cytogenetic studies on human chromosomes by FISH. They can also be used to serve as **STSs** to identify reference points in the physical mapping of the genome. A large number of repetitive sequences have been defined. Just two representative examples are

- the **telomeric repetitive sequences** featuring the **$(TTAGGG)_n$** motif
- a new class of DNA length polymorphisms associated with *Alu* sequences (***Alu* sequence-related polymorphisms**) with a **$(TTA)_n$** repeat motif in the 3-hydroxy-3-methylglutaryl-coenzyme A (HMG-CoA) reductase gene

1.2.5.8 AMPLIFICATION USING ARBITRARY PRIMERS (p. 51)

DNA polymorphisms can be characterized using **amplification with arbitrary primers** that, in distinction from conventional PCR, **does not require specific sequence information**, because it uses primers of arbitrary nucleotide sequences. These primers detect polymorphisms that are inherited in a Mendelian fashion, and can be used to construct genetic maps for DNA fingerprinting and other applications. The polymorphisms generated in this manner are termed **random amplified polymorphic DNA (RAPD)** markers. A significant advantage over other traditional methods is that nucleotide sequencing, hybridizations, and specific probes are not necessary when RAPD markers are used.

The genomic analysis that can be readily automated uses a **universal set of primers** and transfer of the information among various collaborators can be easily accomplished. Genetic maps using RAPD markers apparently can be generated more efficiently and with a greater marker density than by RFLP or targeted PCR-based techniques.

An essentially similar approach using arbitrary primers for PCR amplification, **arbitrary primed PCR (AP-PCR)**, to generate fingerprints **requires no prior sequence information**. Two cycles of low-stringency amplification are followed by PCR at higher stringency, generating polymorphic fingerprints. This method was demonstrated on a number of *Staphylococcus* and *Streptococcus* strains, and on three varieties of mice, and proved to be species and subspecies specific. The **method can be applied to virtually any species, including humans**.

1.2.5.9 MINISATELLITE VARIABLE REPEATS (MVRs) (p. 52)

Amplification of hypervariable loci by PCR significantly increases the sensitivity of single-locus minisatellite probe analysis. These **minisatellite variable repeats (MVRs)** may have different patterns of variation in minisatellite alleles, and so far little is known about the extent of its allelic variation.

A further refinement of the analysis of minisatellites can be achieved by **mapping variant repeat units within amplified alleles**. This approach dramatically enhances the number of different alleles that can be distinguished in a population. For example, in the human hypervariable locus D1S8 this PCR-based technique can distinguish over 10^{70} allelic states. Two rounds of PCR amplification of genomic DNA are performed: first, two parental alleles are isolated by PCR, and then much shorter mutant alleles produced by internal deletions **within** a variable repeat unit are amplified. A combination of two restriction digest patterns is used. Partial digestion with *Hin*fI, followed by electrophoretic separation of the fragments, allows the determination of the number of minisatellite repeat units. Comparison of the partial digest patterns produced by *Hae*III and *Hin*fI enables each repeat unit to be evaluated for the presence of a mutation reflected in the presence of a restriction site for *Hae*III.

The **MVR map** can be encoded in a **binary code**, in which restriction fragments cut by *Hae*III are designated 1 and the noncleaved repeats as 0, or in a two-letter code such as *a* for repeat cut, and *t* for the repeat not cut. This allows the entire ladder of bands in an electrophoretic gel to be expressed in a highly objective manner, and the coded information can be stored and manipulated as if it were a DNA sequence. **MVR-PCR** and even **multiplex MVR-PCR** can be used on a number of minisatellites for forensic and paternity determinations, among other applications, with a **high degree of objectivity**.

1.2.5.10 DNA FINGERPRINTING TO DETECT SOMATIC DNA CHANGES IN TUMORS (p. 53)

Changes in the genomic DNA in mammalian tumors have been traditionally implicated by cytogenetic studies and by the detection of tumor-associated chromosomal aberrations, such as loss of heterozygosity, using RFLPs.

Alternatively, the DNA fingerprinting technique, utilizing a variety of probes such as wild-type **M13 phage DNA** to detect repetitive sequences, proved to be helpful as a universal marker for DNA fingerprinting in humans, animals, plants, and microorganisms.

Human minisatellite and microsatellite probes have also proved useful in the analysis of genomic differences between tumors and unaffected tissues. These are reflected in the intensity of hybridization bands and in the emergence of novel bands detectable in tumor material, these differences appearing to be **tumor specific** rather than tissue specific.

Using an **oligonucleotide probe, $(GTG)_5$**, DNA fingerprints have been generated from intracranial tumor tissues removed surgically that demonstrated tumor-specific chromosomal aberrations not identifiable in corresponding unaffected tissues. Further characterization was achieved by additional hybridization with the **$(GT)_8$** and **$(GATA)_4$** probes. In many gliomas, the amplification unit contained **two simple repetitive DNA fingerprint loci, $(CAC/GTG)_n$** and **$(CA/GT)_n$**, in addition to the epidermal growth factor receptor (*EGFR*) gene. This approach provides unique data for the **definitive identification of certain tumors** that can be used in the clinical diagnostic laboratory. Restriction analysis can also be helpful in **establishing histogenesis** in some poorly differentiated carcinomas, as shown in a series of midline carcinomas of uncertain histology.

Advantages of the characterization of tumors by DNA fingerprinting are:

- it does not require metaphase chromosomes
- a **reproducible and specific pattern that can be associated with a given type of neoplasia** provides a relatively simple and rapid diagnostic tool for clinical practice

Analysis of DNA amplification that **is rare in normal human cells**, combined with DNA fingerprinting, may find its way into clinical practice in the diagnostic evaluation of tumors, and offers a relatively simple and sensitive tool for **monitoring the course of tumor progression** in addition to **assessing tumor clonality**.

Understandably, the efficiency of this approach wholly depends on the **reproducibility of the restriction enzyme digestion** of the tumor tissue. This implies, in turn, that the restriction enzymes cleave DNA in tumor and unaffected cells with the same specificity and efficiency. This premise, however, may not always hold. In one study, three enzymes, *Hin*dIII, *Kpn*I, and *Xba*I, consistently **failed to digest** the tumor DNA completely and, importantly, ethidium bromide staining could not reveal this failure. Such deficiencies should always be considered when comparing restriction digests of normal and tumor DNAs.

1.2.5.11 DNA POLYMORPHISMS IN IDENTITY TESTING

1.2.5.11.1 Forensic Applications (p.54)

Since the introduction of HVR probes by Jeffreys and colleagues they have been widely applied to **identity determinations.**

Analysis of nucleic acid material in forensic practice may have to be performed on specimens presented as stains (blood, semen, body fluids), tissue fragments (skin squames, hair), vaginal swabs, aspirates, and even bone. Frequent complications are:

- the **amount of material is often inadequate** for sufficient DNA or RNA to be extracted for "conventional" processing by gel electrophoresis.
- if **degraded DNA** is predominantly represented by fragments less than 2 kb in size it is considered to be too far degraded for forensic RFLP analysis.
- **contamination by bacterial DNA**: to ensure that the DNA recovered from the presented specimen is human DNA it is probed by *Alu* repeat sequences characteristic of mammalian origin.

DNA is then digested, usually by *Pst*I, and tested with DNA probes specific for the Y and X chromosomes. Southern blots are hybridized with a number of highly polymorphic DNA probes: e.g., **pAC255**, **pAC256**, **pAC225**, **pAC254**, **pAC061**, and **pAC404**. Finally, the membrane is challenged with a **bacterial ribosomal gene probe** to reveal the possible bacterial contamination of the specimen. This is the procedure reportedly employed at **Lifecodes Corporation** (Valhalla, New York).

Interpretation of the patterns obtained with the polymorphic probes is carried out within the framework of the known **frequency distributions of specific DNA fragments** for a given population (American blacks, Caucasians, Hispanics). Following that, the frequency of the evolving pattern is determined using the Hardy-Weinberg equation, and the patterns generated from various sources are used for comparison.

Another major U.S. laboratory providing DNA support to forensic investigations is **Cellmark Diagnostics** (Germantown, Maryland). This facility reportedly used four highly polymorphic **locus-specific minisatellite probes**, **MS1**, **MS31**, **MS43**, and **g3**. The probability of two unrelated individuals possessing the same genotype with these probes is calculated to be 3.4×10^{-12}. The **virtual uniqueness** of these patterns obviates the need for sizing alleles or determining specific loci.

At the present time, forensic laboratories predominantly use Southern analysis with the minisatellite and VNTR single-locus probes, and population databases for these probes are being developed.

An assessment of the use of DNA technology in forensic practice (Genetic Witness. Forensic uses of DNA tests, 1990) concluded that

methodological errors, including poor sample handling, incomplete DNA digestion, incomplete Southern transfer of DNA to a membrane, poor probe labeling, and inadequate autoradiography, may fail to produce expected results, **but will not lead to an error in identification. Validation studies** at the FBI also indicated that adverse exposure of the specimen and chemical and bacterial degradation also **lead to no result, rather than to an error in identification**.

The Office of Technology Assessment (OTA) finds that, when properly performed, DNA technologies per se are **reliable** with **acceptable reproducibility** of results within one laboratory and among different laboratories. According to the OTA, **forensic single-locus DNA probes** should

1. Contain DNA sequences that detect only **one chromosomal locus** under a reasonable range of hybridization conditions
2. Produce **well-characterized patterns** (such as defined size range, number of bands, and relative band intensities) so that unexpected patterns can be recognized
3. Detect a **single polymorphic fragment per allele** so that each person's test yields either one or two fragments, depending on whether an individual is homozygous or heterozygous, respectively
4. **Be avoided if they detect alleles of varying number or intensity** on Southern blots, in order to minimize the problem of identifying the true bands that comprise an RFLP pattern
5. Be selected such as to ensure that the **alleles** revealed **are independently segregating**
6. Be **sufficiently widely used** and available to other scientists to confirm their alleged properties
7. Be characterized with respect to its chromosomal assignment filed with the Human Gene Mapping Workshop

Two-dimensional electrophoresis of restricted DNA fragments probed by minisatellite core sequence fragments has been able to resolve up to 625 separate spots per probe per human individual. This technique offers a powerful tool for generating individualized fingerprints helpful not only in identity determinations, but also in the studies of disease associations, mutations, and other processes related to cancer and aging.

In-gel hybridization with nonradioactive **$(CAC)_5$** oligonucleotide probes markedly simplifies the process of fingerprinting.

Amplification of simple repeat DNA sequences by PCR followed by hybridization with **$(CAC)_5/(GTG)_5$**, **$(CT)_8$**, and **$(GACA)_4$** probes can be used for the analysis of forensic stains using a specific probe combination, depending on the amount of the recovered material and the informativeness of the results obtained with a given probe.

An important aspect in forensic casework is the ability to declare that two or more DNA patterns match. **Mixing experiments**, although capable of revealing minor discrepancies in the patterns being compared, are not routinely performed by forensic laboratories due to frequently limited amounts of the available DNA and difficulty in quantitating it when preparing an informative mix; however, many scientists believe that mixing tests are the "gold standard" for DNA typing of forensic specimens.

As an **alternative to mixing**, identity is inferred by comparing the **positions of bands**, and, thereby, their size in two separate lanes. Minor displacements in the positions of bands in different lanes of electrophoresis gels may lead to "**band shifts**". Therefore the best estimate of the size of an allele is obtained by comparing it to a set of **internal lane controls** — polymorphic markers. Scientists disagree as to the interpretation of matches based on the analysis of bands. The important issue is how close the two alleles have to be in size (because identical measurements are not achievable) to be declared a match based on the "molecular rule" reference. It is generally agreed that fragment lengths should be within a specified, and always observed, **range of deviation** observed empirically under a given set of reproducible experimental conditions when known forensic samples are repeatedly tested.

Multilocus probes, initially used in an **immigration case** in the United Kingdom in 1985, require more sample than single-locus probes and, in some cases, the large number of bands produced may complicate forensic analysis. **The current consensus is that multilocus probe analysis is *not* the most appropriate technique for criminal casework.**

Although even degraded specimens can be used for the PCR of samples containing picogram quantities of DNA that are inadequate for Southern analysis, some restraint in the use of PCR for criminal casework is justified:

- **contamination** can be produced by the reaction itself, and proper controls and precautions must be observed (see below)
- the **minuscule amount of starting material**, as may frequently be the case in forensic practice, can potentially present a problem if misincorporation occurs early in the amplification reaction
- **preferential amplification** of one allele over the other by PCR due to the condition of a forensic sample may result in spurious identification of the person as homozygous for a trait

Currently, the only genetic system generally employed in forensic casework using PCR is the HLA-DQα1 system, although it is not as discriminating as RFLP analysis.

DNA technology is developing at a rapid rate, to say the least, and the limitations of the present-day methods of identity testing may be obsolete in the nearest future.

The **standardization** of DNA genotyping methodology with the introduction of **minisatellite variable repeats (MVRs)** allows more secure identity testing, while objective interpretation of the restriction patterns in this system offers unparalleled and reproducible "**bar coding**" of an individual. Because no subjective steps such as size measurement of fragments are required, the reading of patterns and conversion of these into a digitalized format offers an exchangeable, reproducible, and unique identification system particularly attractive for forensic practice. The **digitalized transformant** of a given pattern can be stored and objectively compared to any set of similarly presented patterns for computerized criminal casework as if these were DNA sequences themselves with their inherent individual uniqueness. (See also Section 1.2.5.9.)

Computerized assessment of hypervariable DNA profiles has been developed and the initial results suggest that band matching should be performed only after analysis of the errors arising in the process of electrophoretic separation of fragments of various sizes.

A protocol termed **FINS** (forensically informative nucleotide sequencing) identifies the source of the material in forensic specimens as being of animal origin. PCR amplification is followed by DNA sequencing and computer-based **phylogenetic analysis** of the data.

Contrasting opinions have been voiced concerning quality control of DNA fingerprinting, the validity of the currently used statistical treatment of the data obtained, and the need for accumulating more elaborate and extensive racial and ethnic population databases. These controversies have been looked into by the National Academy of Sciences and panel recommendations were issued, upholding the use of DNA fingerprinting in forensics and stressing the need for maintenance and improvement of the highest quality control level in DNA typing laboratories. [*See Supplement 5.*]

1.2.5.11.2 Paternity Testing (p. 57)

The range of genetic markers composed of red cell antigens, red cell enzymes, serum protein, and HLA polymorphisms is considered by some to be fully adequate for efficient paternity testing service.

The Jeffreys probes 33.6 and 33.15 allow **unambiguous paternity testing** even when HLA typing cannot distinguish between tentative fathers. **Guidelines for standardization** of parentage testing laboratories have been issued by the American Association of Blood Banks.

As a rule, in parentage testing, the **high informative power** of **HVRs** is used. A large number of probes is available, combinations of which allow over 99% of men falsely accused in paternity suits to be excluded. Some of the widely used probes in paternity testing are those for

- the **D14S1 locus**
- the **flanking region of Ha-*ras*-1**
- the five probes for HVR loci (**D2S44, D14S1, D14S13, D17S79,** and **DXYS14**)
- **pa3′-HVR** for the **D16S85 locus**

Variability in the estimates of the length of allele fragments (usually within 0.6 to 1.0%) due to the **resolving power of agarose electrophoresis** used in separating the restricted fragments in single-locus probe testing must be entered into the statistical evaluation of the observed frequency of a restriction fragment in the population. An **averaging approach** that combines close alleles into the so-called "**frequency bins**" is used. Advantages of using DNA probes have been clearly demonstrated in a case of disputed paternity when seven probes had been used to produce the **cumulative paternity index** of 1.4×10^6 that was 316 times higher than that obtainable from the 23 standard blood group markers and HLA.

A **more informative pattern** is generated by probing Southern blots under reduced stringency conditions to reveal fragments derived from multiple chromosomal loci — **multilocus probing**. Save for the case of **identical twins**, such multilocus probes provide virtually unique individual identification.

In cases when either the putative father and/or mother is absent, single-locus probing involving relatives becomes much less informative, whereas DNA fingerprinting with multilocus probes is still effective.

Minisatellite probes have been successfully used on amniocytes or chorionic villi for **prenatal determination of familial relationships**.

An emerging approach for determining familial relationships is based on the maternal inheritance of the **mitochondrial genome**, and uses PCR amplification and sequencing of the D loop. So far, however, not enough population data are available to make this approach widely acceptable.

1.2.6 FLOW CYTOMETRY (FCM) AND IMAGE CYTOMETRY (ICM) ANALYSIS (p. 59)

FCM has been extensively used for

- the relatively **crude estimates of DNA content** in abnormal (benign and malignant) cells
- for the more sophisticated **experimental applications** in molecular genetics

In equivocal cytological cases, the additional benefit of FCM, or ICM, analysis is thought to be questionable, leaving cytological examination as the definitive technique. Until assessment of the specific gene expression becomes available on a routine basis for FCM and ICM techniques, these will remain essentially of limited diagnostic value in clinical practice.

Techniques have been devised to disintegrate even formalin-fixed, paraffin-embedded tissue to allow informative FCM measurements. Although **still lacking the necessary standardization** and support from extensive comparative clinical studies, this approach has been widely exploited to develop correlations between DNA ploidy and the biological behavior of tumors.

Comparison of FCM with ICM for DNA quantitation in solid tumors revealed that **ICM was able to identify some aneuploid populations where FCM failed.** On the other hand, selected cases could be detected by FCM, but not ICM. Thus, while FCM may fail because of cell loss during processing, **ICM shows a somewhat higher sensitivity** due to visual identification of target cells by the operator. FCM has been shown to offer

simultaneous assessment of the nuclear (e.g., DNA content), cytoplasmic, and cell surface parameters (e.g., cytoplasmic antigens or oncogenes). The **clinical utility** of such determinations is evident in the evaluation of the DNA index in breast cancer samples, which offers a distinct advantage over the biochemical assays of homogenates.

The frequently observed discrepancies in the assessment of DNA ploidy of tumors by FCM are in part related to the **wide scatter of FCM readings** on repeated aliquots of the same tissue sample due to the cellular heterogeneity of tumors. For example, **at least four samples are needed** for reliable determination of DNA ploidy of primary breast carcinomas tumors by FCM. (See Section 3.1.2.)

FCM of nucleic acids may, theoretically, offer a powerful tool for clinical diagnosis, provided more specific tumor markers can be determined in representative cell populations. Using the **relative abundance of single-stranded vs. double-stranded RNA** in solid nonhematopoietic neoplasms indicated that dsRNA content may be a useful parameter that complements DNA ploidy in the evaluation of solid tumors.

1.2.7 QUALITY CONTROL IN GENE LEVEL DIAGNOSIS

An increasing number of clinically useful molecular diagnostic tools are appearing due to the **user-friendly methodologies** the industry is developing. Molecular biological diagnostics will require a different level of facilities, equipment, and, most importantly, training of technologists and physicians.

1.2.7.1 FACILITIES (p. 60)

Organization of a molecular diagnostics laboratory requires that a higher level of precautions be considered against

- **interference** by the ever-present **nucleases**
- **cross-contamination of specimens** (e.g., by amplified products)
- **protection of personnel**
- **pollution of the environment**

Special attention should be given to creating the maximally affordable **dust-free environment** with relatively **low air flows** still adequate for appropriate air exchange. Work surfaces should be able to withstand strong cleansing solutions such as **bleach** and sterilization by **ultraviolet (UV) light** without creating cracks or other defects of the bench surface. Where possible **separate rooms** should be assigned to dedicated parts of certain procedures for handling specimens and/or nucleic acid preparations as in PCR assays.

Some procedures may still be expected to use **radioactive reporter molecules**. It is desirable to have a **liquid scintillation counter**, not only for certain testing protocols but also for monitoring the quality of reagents and supplies. Other essential equipment includes adequate **refrigeration** and

deep freezer (–70°C) space, preparative **high-speed centrifuges, table-top centrifuges**, a **vertical luminary flow hood** with dedicated exhaust conduit, UV-equipped table-top **hoods**, a **laser densitometer**, a **dark room**, a **water-purification system**, and a **glass washing** facility. A list of other laboratory equipment for a molecular biology laboratory can be found in a number of manuals.

Transfer of technology from basic research and biotechnology into clinical diagnostic laboratories is progressively accelerated with the wider **use of nucleic acid probes, automation,** and **chemo-** and **bioluminescence-based detection systems**. An example of the formerly research-oriented equipment that has great potential in the clinical diagnostic field is the **Phast System** (Pharmacia) for **electrophoresis** and **isoelectric focusing.** This programmable system offers an attractive modality to achieve operator-independent **standardization** and **reproducibility** of assays. Adaptation of **isoelectric focusing (IEF)** protocols for routine protein electrophoresis of clinical specimens in many cases allows one to avoid separate immunoelectrophoretic identification of immunoglobulins. IEF of hemoglobins by the Phast System is, in this author's opinion, superior to the methodologies currently accepted in routine clinical laboratories. The Phast System can also be used for high standardized, reproducible electrophoretic evaluation of amplification assays, restriction digestion, and in other frequently employed assays requiring electrophoretic separation.

1.2.7.2 QUALITY CONTROL IN FCM AND ICM (p. 61)

The quality of nucleic acids recovered from patient samples is of critical importance for informative gene-level determinations by FCM and ICM in clinical practice. Acceptability of the material retrieved by **fine-needle aspirates (FNAs)** from solid tumors for DNA FCM is adequate only if multiple (no fewer than three) aspirates are assayed.

Extracting **DNA from paraffin-embedded tissue for FCM** has gained widespread use in diagnostic pathology, although this approach has a **number of limitations** affecting

- informative **assessments** of the DNA content
- **reproducibility**
- variability in the DNA FCM of solid tumors arising from **analytical variations**, as well as from **differences in the detection rates** of near-diploid and tetraploid **DNA indexes** (DIs)
- **comparison** with either previous or later evaluations on similar material
- **interlaboratory comparisons**

Nevertheless, DIs offer a better criterion than histopathological grading for evaluating malignancy, giving a somewhat **lower intra-** and **interobserver variation** (83.9 and 82.2%, respectively) and **reproducibility** (65 and 57%, respectively).

Meaningful **interpretations of the DNA content** in isolated nuclei recovered from paraffin-embedded tissue must take into account

- the substantial decrease in **propidium iodide fluorescence** (the most commonly used stain for this purpose) produced by formalin fixation
- **cross-linking of the chromatin proteins** during fixation reduces accessibility of DNA to the dye
- the dependence of the DNA signal on the state of chromation **dispersion** or **condensation**

Importantly, the **degree of decrease** in FCM fluorescence varies in specimens with differing ploidy whereas ICM evaluation does not suffer from this artifact. Other artifactual phenomena are associated with

- **ion-dependent alterations** in the chromatin structure produced by changes in tonicity, in part accounting for variable state of chromatin dispersion
- the marked variability of the **staining characteristics of the nuclei** isolated from deparaffinized, rehydrated, and disaggregated tissue for **Feulgen staining. Disaggregation of tissue** significantly contributes to variations in the Feulgen stainability that can reach 300%
- the **enzymatic digestion step**, in the course of disaggregation, that leads to low levels of fluorescent signals

The evaluation of DNA ploidy by FCM usually makes use of the relative DNA content in the form of a **DI** calculated with respect to **calibrators**, or as the ratio of fluorescence intensity from the abnormal to normal nuclei.

Appropriate **internal standard** cells are preferred, whereas **external standard** cells appear to be unacceptable. When DNA FCM is performed on paraffin-embedded material, a tissue section cut from a paraffin block of the series should be run together with the disaggregated preparations in one batch as a **staining control**.

When Feulgen staining is used for the nuclear DNA quantitation by ICM the nature of fixative agents is critical. The **Riguad fixative** appears to be the best for automatic cytophotometry of Feulgen-stained nuclei.

Without the use of specific chromosomal paints, DNA FCM is not an acceptable technique for the **assessment of chromosomal abnormalities** (failing in up to 65% of patients with solid tumors) that are detectable by cytogenetic methods. A combination of DNA FCM and ICM is suggested, particularly when aneuploid populations are present.

1.2.7.3 QUALITY CONTROL IN ISH (p. 63)

Maintaining optimal procedural parameters cannot be overemphasized, because

- even minor deviations in the **salt concentration** or **temperature** may alter the stringency of hybridization either way, thus giving information, if at all, on the wrong set of genetic determinants than those being tested for
- minor mistakes in the **gel preparation**, or conditions of the **electrophoretic runs**, may fail to reveal the sought-for presence of informative alleles, and so on

The specificity of identification of a given nucleic acid target in a tissue largely rests on the **controls** used. **General test procedures** aimed at ensuring the specificity of ISH under varying conditions of pretreatment, hybridization, and probe labeling conditions include

- **preincubation** of some tissue with "**cold" probe** prior to addition of labeled probes to have a "competition" or "blocking" control
- hybridization to a **heterologous probe** and other tissue not expected to have the same target nucleic acids
- inclusion of a "**positive control"** with overabundance of the target
- nuclease (**RNase** and **DNase**) digestion
- hybridization with the **vector** without the probe
- tests for "**positive chemography"** (signals generated not by the probe but by other chemicals used in the procedure)
- tests for "**negative chemography"** (loss of signal due to the chemicals used)
- the selection, when possible, of the **"antisense" strands** when using cRNA probes to compare these to a control probe (**"sense" strand**)

A partial list of procedures helpful in optimizing ISH deals with a number of **frequently encountered problems**, including

- variations in **tissue fixation** that may lead to **excessive cross-linking** and thereby block accessibility of the nucleic acid target to the probe
- excessive **proteolytic treatment**, which although opening the target may lead to **topological distortions** affecting the ISH results
- false-positive signals caused by **lipofuscin** interacting with a variety of DNA probes. Appropriate nuclease treatment and **controls without any probes** help to resolve the problem
- **nonspecific binding** of DNA to eosinophils e.g., in the identification of viral nucleic acids (such as HIV or CMV) in peripheral blood; this can be effectively eliminated by pretreatment of slides with **carbol chromotrope**
- **strong background** staining, which may be encountered using **biotin-labeled probes**, particularly on liver sections. Biotin-blocking procedures should be employed, and **nonspecific signals from glycogen** can be removed by **amylase** pretreatment of the liver sections
- low specificity; **increased specificity** can be achieved through the use of **multiple and competitive probes**
- the **possible nonenzymatic hydrolysis of ssRNA** in aqueous solutions at neutral pH and elevated temperatures; **temperature control** is of paramount importance in hybridization, especially when **ssRNA** or **synthetic oligonucleotides** are used as probes

1.2.7.4 QUALITY CONTROL IN RESTRICTION ENDONUCLEASE ANALYSIS (REA) (p. 64)

When **comparing tumors** and **normal tissues** by REA, the DNA extracted from tumors (but not normal tissues) was shown to be incompletely digested by three of the nine commonly used restriction enzymes (*Hin*dIII, *Kpn*I, and *Xba*I). Ethidium bromide staining of the gels failed to reveal this difference. Specific **restriction enzyme inhibitors** in the tumors appear to be responsible.

Other artifacts in REA may arise from

- **bacterial contamination**, particularly in fingerprint studies using vector sequences for probing DNA polymorphisms
- **failure of some restriction enzymes** (e.g., *Bam*HI and *Hin*dIII) **to cut tissues exposed to prolonged** (24 h) **formalin fixation**

A detailed set of standards and controls for **quality assurance of DNA fingerprinting** for identity testing has been developed by the Technical Working Group on DNA Analysis Methods (TWGDAM), and by the American Association of Blood Banks. Among the standards are the following:

- DNA loci should to be specified for **parentage testing** using DNA polymorphisms
- The **chromosomal location** of the polymorphic loci should be traceable to the Yale Gene Library, or to the International Human Gene Mapping Workshop
- The use of appropriate restriction enzymes and probes as well as the conditions of hybridization and sizes of alleles should be **documented in the literature**
- The **type of polymorphism** being detected should be defined.
- Assurances must be made with respect to
  - the **completeness of endonuclease digestion** of DNA
  - the **adequacy of size markers**
  - the presence of **appropriate human DNA controls**

Optimal **restriction enzyme/probe combinations** can be assessed by computer programs that predict the ability of particular restriction enzyme to produce the most efficient digestion of a particular DNA molecule.

1.2.7.5 QUALITY CONTROL IN THE ISOLATION OF NUCLEIC ACIDS (p. 65)

Southern blot analysis in its various modifications may still be used for clinical molecular genetic diagnosis for some time to come, despite the revolutionary developments in the amplification of diagnostic targets.

In the isolation of **nucleic acids from blood samples**, the **acid citrate dextrose (ACD)** anticoagulant appears to be superior to ethylenediaminetetraacetic acid (**EDTA**) and **heparin** in

- **not affecting the yield of DNA**, even from frozen blood kept in ACD for 5 days at 23°C
- **allowing unaffected restriction patterns** to be obtained with three enzymes (*Eco*RI, *Hin*dIII, and *Xba*I)

Procedures have been developed to yield undegraded nucleic acids from a **large number of small samples** virtually **at the same time**, without the use of conventional phenol extraction.

Isolation of nucleic acids from **gram-positive bacteria** can be accomplished by the **thermal shock technique** (–196°C/+100°C) followed by 3% SDS lysis. It fails in mycobacteria, from which DNA can be isolated using **microwave oven thermal treatment**; the yield of high molecular weight DNA depends on the length of thermal treatment. Convenient modifications of DNA analysis for RFLPs particularly suitable for clinical diagnostic laboratories allow the **isolation** of DNA from cells, **digestion, separation** of the restricted fragments, and subsequent **blotting** for Southern analysis — **all in the agarose gel.**

Nucleic acids can be isolated even from formalin-fixed, paraffin-embedded tissues in a condition suitable for various molecular biological procedures.

1.2.7.6 NUCLEIC ACID MEASUREMENTS (p. 66)

Errors are frequently made in the determination of **nucleic acid concentrations**. Reproducible linear methods for DNA and RNA quantitation at picogram and even attomole levels include: **spectroscopic**, **dye-binding**, **radioactive**, **chemoluminescent digoxigenin-based probe**, and **bioluminescent** techniques.

Errors in estimates of DNA concentration based on **ethidium bromide fluorescence** in one- or two-dimensional gels may be large, therefore the following conditions should be observed:

- the **reproducibility** of the electrophoretic runs must be ensured
- the films must be calibrated by an **internal standard**
- appropriate **background controls** must be selected to determine the **baseline for integration**
- meticulous attention be given to the **photographic process**, and the limitations of the **film used**
- scanning using **high-resolution densitometry** should be used to obtain reliable quantitation of DNA from agarose gels

When samples are relatively free of contaminating nucleic acids, DNA-DNA hybridization with **sulfonated probes** or the **direct sulfonation procedure** can be used.

The DNA concentration determinations in multiple small-volume samples particularly prone to error due to **viscosity of DNA** can be improved by an adaptation of a **fluorescent mircotiter plate reader** that gives a linear response over the range of 250 to 2000 ng.

Transfer of nucleic acids from gels onto various membranes for Southern or Northern analysis is critical for standardization of these techniques. **Nylon membranes** display greater sensitivity than **nitrocellulose membranes.**

The **mode of retention** and **capture** of nucleic acid fragments drastically influence the sensitivity of hybridization methods. A **quantitative hybridization** has been developed using retention of nucleic acids by **dA-tailed capture probes**.

1.2.7.7 QUALITY CONTROL IN AMPLIFICATION SYSTEMS (p. 66)

The use of PCR in clinical diagnostic laboratories is gradually becoming less restricted due to patent licensing arrangements. [*See Supplement 3.*] Some potentially troublesome aspects of PCR assays, and the need to develop more standardized and reproducible protocols for clinical laboratories, contribute to delays in the wider introduction of PCR in diagnostic laboratories.

The basic quality control parameters of PCR have been addressed repeatedly. Some of these include

- the **amplification efficiency** of *Taq* polymerase, which may vary significantly, depending on the commercial sources
- the **primer concentration** — the wrong choice may completely block the amplification and higher primer concentrations are not always optimal
- the effect of **primer length** on the specificity and efficiency of PCR
- the **conformation of the template DNA** (supercoiled DNA supports lower amplification rates than does linearized DNA)
- the optimum range of **elongation temperature** and its effect on the size of the amplified product
- the effect of **hybridization temperature** on the efficiency of amplification with different primer sets
- the **number of cycles** (20 to 30)
- the **primer sequences** — these are of critical importance, not only in ensuring the **specificity** of PCR amplification, but also for **allowing the reaction to proceed**. Particularly critical are even single mismatches at the 3′ end, whereas **mismatches** at the 5′ end are generally without major negative consequences. Parallel reactions with degenerate primers or mixtures of 3′ degenerate primers are recommended to prevent failure of amplification due to 3′ mismatches
- the absolute requirement of *Taq* polymerases for **magnesium ions**, which appears to display two maxima at low (below 5 mmol/l) and high (10 to 12 mmol/l) concentrations. (Enzyme batches that require the higher Mg^{2+} concentration usually display lower activity; a high Mg^{2+} requirement may represent the presence of **contaminating chelating compounds** in the enzyme preparation)
- the **fidelity** of the DNA polymerase used in the PCR — this is of critical importance for the specificity of the amplification. At low concentrations of starting material (templates, targets) the **probability of an error in a PCR amplification is around 1%**
- the **variability in efficiency** of PCR from paraffin-embedded tissues,

which is associated with the particular **fixatives** used and the length of fixation. **Ethanol** and **Omnifix** do not affect amplification of the β-actin gene, whereas all specimens fixed in **Bouin's** or **B-5** are negative. Fixation in **formalin** and **Zenker's** solution leads to variability in the PCR product. **With almost all fixatives tested the *PCR signal for RNA was less affected* by fixation than that for DNA**

- the **extraneous contamination** by positive control material, as well as by the very products of PCR amplification in the so-called "**carryover**", that can yield false-positive results. The **carryover problem** presents a particular concern in clinical diagnostic laboratories when similar amplifications are routinely repeated

Sources of error in PCR amplification include the following:

- **false-negative results** obviously may occur when primers fail to anneal, whereas **nonspecific amplifications** may occur with primers also annealing to nontarget sequences. **Multiple primers** to various sequences within the DNA area of interest and strict monitoring of the **expected band size** can help with this problem
- **smearing patterns** and **wrong band positions** on the gels may suggest nonspecific priming
- the **formation of chimeric cDNA clones** due to the reverse transcriptase activity of the *Taq* polymerase may be another source of error in PCR amplifications
- an **intrinsic source of contamination** of the PCR process has been related to the presence of DNA in *Taq* polymerase preparations thought to arise during the purification process, or enters with the reagents added to the enzyme. This problem can be resolved by running a **no template control** and choosing other primers

Measures to reduce the **possibility of contamination** include

- **physical isolation** of PCR preparatory stages from the products of amplification
- **autoclaving** water and buffer solutions that do not deteriorate during autoclaving
- use of **disposable labware** — pipette tips and microcentifuge tubes
- allocating into **single doses** all of the **reagents already premixed** that are consumed in a single run
- use of **gloves at all stages** of the preparation and performance of the reaction
- use of **positive displacement pipettes**
- adding **DNA negative controls**
- treatment of reusable equipment (gel apparatus, etc.) with **1 *M* HCl** to depurinate any residual DNA. However, simple sterilization or denaturation is not adequate to prevent the amplified PCR products ("**amplicons**") from contaminating subsequent reactions because a single strand of nucleic acid, no matter how short, can be readily amplified and **can withstand sterilizing temperatures**

- a proposed "**post-PCR sterilization**" approach, based on the blockage of *Taq* DNA polymerase in the amplification reaction by a **photochemically modified base** in the amplified polynucleotide strand. This is achieved by the introduction of **isopsoralen derivatives** that form cyclobutane adducts with pyrimidine bases photoactivated after amplification
- **inactivation by ultraviolet light**; this is more efficient with larger DNA segments (>700 bp) compared to the shorter ones (<250 bp), and it is less efficient in eliminating dried DNA
- **omission** of **positive controls** in some applications, to avoid the possible introduction of the target sequences. **Southern analysis**, or **sequencing**, is then used for quality control, in addition to the identification of the product on **sizing gels**. In this context, males may be excluded from even performing PCR assays for **Y-specific sequences**
- use of **wipe tests**, similar to those used to monitor radioactive contaminations, to combat false-positive amplifications. The **laboratory locations** found to harbor DNA contributing to false-positive results include **freezer door handles**, **freezer shelf**, and **room door handle**. Replacement of contaminated objects and cleaning with **1 *M* HCl** is effective
- use of **anticontamination primers** is effective in avoiding false positive results in PCR

An alternative amplification approach, the **Q beta replicase system**, has received wide publicity. This method relies on the fidelity of hybridization of a given probe inserted into an RNA that is then efficiently amplified by a dedicated enzyme. The limit of detection is currently about 10,000 target molecules.

Chapter 1.3

Prenatal Diagnosis

1.3.1 INTRODUCTION (p. 71)

Inherited diseases are diagnosed prenatally or early in childhood using REA, DNA and oligonucleotide probes, and the PCR. The majority of the diagnostic achievements in this area remain, however, in the realm of specialized laboratories. **Simplification of the procedural aspects** of molecular diagnostics (e.g., construction of specific nucleic acid probes with **nonradioactive labels**, introduction of **PCR combined with automation**, and the development of **new** colorimetric and fluorescent **detection systems**) makes gene level diagnostics a potentially viable alternative to conventional methods.

Most cases may require a **combination of different methods**. Some typical examples include

- a specific 21-OH DNA probe should be combined with genotyping of the fetus by HLA DNA probes on chorionic villus samples for early prenatal diagnosis of **21-hydroxylase deficiency**
- the diagnosis of **familial amyloid polyneuropathy** relies on REA of white blood cell DNA followed by hybridization of a cloned prealbumin DNA to a Southern blot of the digests
- REA combined with specific oligonucleotide 20-mer probes has been applied to the diagnosis of **neuronopathic Gaucher's disease**
- prenatal diagnosis of **α-antitrypsin deficiency** can be accomplished either by RFLP analysis or by hybridization using M- and Z-specific oligonucleotides

In many cases, a combination of traditional methods of prenatal diagnosis (ultrasonography, amniotic fluid cells, and chorionic villus sampling) with molecular biological techniques is being used. Analysis of fetal DNA from the circulating nucleated erythrocytes in maternal blood has been described. The identification of a **carrier status** or **fetal infection** is performed on a routine basis. Even **pre-implantation** and **pre-conceptual** diagnosis has now become possible (see Section 1.3.2.3). Given adequate molecular tools are devised, the prenatal evaluation of **predisposition** to various diseases will also find widespread use.

A wide range of new **ethical**, **moral**, and **legal issues** is raised by the availability of molecular diagnosis that reduces the level of uncertainty in predicting unfavorable developments later in life. The availability of prenatal diagnosis per se does not necessarily ensure this modality is to be universally accepted even by populations at risk, as exemplified by the parental attitude to neonatal and prenatal screening for cystic fibrosis (CF).

Specific probes for the Y chromosome (**p75/79, Y84, Y190**) and the X chromosome (**pSV2X5**) allow the fine analysis of **Turner's syndrome** patients. Sex-specific probes (e.g., **pY3.4** for the Y chromosome) have been used to monitor the presence of host cells in the peripheral blood of bone marrow recipients for the development of **graft-vs-host disease** or relapse after transplantation.

1.3.2 FETAL SEXING (p. 72)

Definitive sex identification at various stages in human development can be important in cases of **sex-linked inherited conditions** for which an option to discontinue a pregnancy can be considered if the trait is going to be expressed. In the practice of ***in vitro* fertilization**, molecular diagnostics allow sex determination at the eight-cell stage.

The first probes specific for **repeat sequences of the heterochromatic region of the long arm of chromosome Y** have been used for dot-blot and *in situ* hybridization (ISH) in sex diagnosis. Currently, a large number of **DNA probes for unique** and **repetitive sequences of sex chromosomes are available**. Examples include:

- **pS4** detects repeated sequences on Yq12 with a sensitivity of 5 ng
- By combining pS4 with the probe **4B-2**, analysis of defective gametogenesis associated with translocated Y chromosome has been performed
- With a combination of two probes, **λ KC8**, derived from a unique sequence localized on the short arm of the X chromosome, and **L1.28**, also located on the short arm of the X chromosome, the absence of maternal contamination of chorionic villi can be demonstrated
- A **commercially available probe, pHY2.1** (Amersham) recognizes 2000 copies on the Y chromosome and 100 copies elsewhere in the gene. This probe and a similar probe (**pY431a**) have been used for sex determination using fluorescent hybridization on uncultured amniotic fluid cells and even on degraded DNA from old blood stains. They can also be used for ISH evaluation of Y;autosome translocations in metaphase spreads or interphase nuclei

1.3.2.1 FETAL SEX DETERMINATION BY PCR-BASED ASSAYS (p. 73)

The next generation of tests for fetal sex determination is based on **PCR amplification of selected sequences of either X or Y chromosomes**. One rapid fetal sex determination procedure can be used on any specimen, even those collected as dried blood samples (Guthrie cards). It is based on the 5.5-kb *Eco*RI sequences of the **alphoid (α) satellite family**, which is located in the pericentromeric regions of the Y chromosome; these sequences are highly characteristic of the **Y alphoid repeats**.

Another PCR-based assay for fetal sexing identifies a single and discrete 530-bp fragment using two oligonucleotide primers specific for the ***ZFY* gene** cloned from a region of the human Y chromosome. While these procedures are definitely attractive for a number of clinical and forensic uses when sex determination is sought, their application to maternal blood, although possible, has not yet been reported.

A method for sex determination of the fetus by testing the **maternal blood** at 9 to 11 weeks and at 32 to 41 weeks gestation has been described based on amplification of a Y-specific repeat sequence. To minimize the chance of contamination DNA was extracted under **class II containment conditions**, the reagents were **predigested**, no DNA was extracted from men during these experiments, and all blood samples were taken and handled by women. Despite these rigorous measures, amplification **beyond 20 cycles** in the second nest consistently yielded false-positive results. One of the sources of erroneous results appears to be the extension of the fragmented DNA in a reaction known as "**jumping PCR**". Neither predigestion of reagents with *Eco*RI nor autoclaving could eliminate this interference.

Using a similar approach, other investigators encountered problems with **reproducibility**. Choosing a different probe for PCR amplification, **27A**, derived from a unique Y-specific sequence, the defection limit of male cells after Southern hybridization was 10 pg male DNA, corresponding to one or two male cells. Weak false positivity persisted in spite of isolating DNA under class II containment conditions and predigestion of all reagents. The only measure, although impractical, appears to be **ultrafiltration of all reagents in addition to class II containment** and **predigestion** with restriction enzymes. Of major concern is **variability of the amplification efficiency** combined with the **variability of fetal cell concentration** in the maternal circulation, which may compromise the reproducibility of the assay to an unacceptable level.

Further refinement of **PCR-based fetal sex determination** using a **maternal blood assay** came with targeting a single-copy sequence that is part of a gene expressed in testicular tissue. **Persisting male cells in maternal circulation** from a previous pregnancy are thought to account for the "internal" contamination. Other causes include autosomal cross-reactivity of Y primers, and the "vanishing" male twin of a female sibling. **In practice, however, *contamination may be the main problem*.**

PCR-based identification of the Y-specific DNA combined with ISH using chromosome-specific probes allows noninvasive identification of such conditions as trisomy 21 and trisomy 18 in fetal nucleated erythrocytes.

1.3.2.2 SEX DETERMINATION GENES (p. 74)

Sex identification of a **fetus** or **embryo**, in particular, would be unequivocal if the very gene(s) responsible for sex determination could be the targets of evaluation. So far, all the reported sex evaluation studies have relied on the identification of repeated or unique sequences of the Y or X chromosomes, rather than the genes now considered to be the **candidate sex-determinant genes**.

The gene encoding **testis determining factor (TDF)** has been mapped to a short segment of the Y chromosome adjacent to the pseudoautosomal boundary within an area located 35 kb distal to the boundary. It is likely that more than one gene is required for male or female sex determination. Attempts to identify and clone TDF are underway.

An **open reading frame (ORF)** that is part of a conserved, Y-specific gene, named ***SRY*** — for **sex-determining region Y** — has been identified. It encodes a testis-specific transcript.

Genetic studies of XY females with **gonadal dysgenesis**, as well as of normal males, identified mutations within the *SRY* gene responsible for sex reversal in the affected individuals. **Amplification of *SRY* sequences** from the DNA of XY females or normal males by PCR revealed that although SRY is required for male sex determination, this is not the "ultimate" *TDF* gene. A **specific DNA-binding activity encoded by *SRY* appears to be essential for sex determination.**

1.3.2.3 PREIMPLANTATION ANALYSIS AND SEX DETERMINATION OF HUMAN EMBRYOS (p. 75)

Biotinylated Y-specific sequences (**pHY 2.1**), available commercially (Amersham), have been used as probes for ISH to allow the undelayed processing of female embryos for transfer to the mother's uterus when X-linked disorders are to be avoided. The entire procedure can be accomplished within 48 h. In male interphase nuclei, this probe (**pHY 2.1**) has an efficiency of 66% in detecting the Y body, which is comparable to that of sex determination by fluorescent staining of metaphase chromosomes.

A **PCR-based sex determination technique** has also been developed. Human embryos at the 6- to 10-cell cleavage stage, 3 days following *in vitro* fertilization, were manipulated and a single cell was removed. Oligonucleotide primers selected from the published 3.4-repeat sequences on the Y chromosome were selected. A cell was judged male by the presence on electrophoresis gels of a 149-bp fragment of the Y-specific repeat, and female if the band was absent or if it was very faint. The entire procedure took 5 h.

The **micromanipulation of preimplantation embryos** has been successfully performed to allow nondestructive embryo biopsy. ISH with probes specific for chromosome X and chromosome 3 was subsequently performed on the biopsy sample.

The fast pace of accumulated experience in these micromanipulation techniques combined with ISH and/or PCR approaches makes preimplantation genetic diagnosis suitable for clinical application when medical indications justify the risks for the woman and her fetus associated with these diagnostic procedures.

1.3.3 PRENATAL DETERMINATION OF PATERNITY (p. 76)

The **parentage of a fetus** may be requested in pregnancies resulting from **sexual assault**, or when the mother has a history of **multiple sexual partners**. One of the major drawbacks in evaluating polymorphic protein markers, such as HLA antigens on **amniotic cells**, or **chorionic villus** samples as well as **red cell antigens**, and **serum enzymes** lies in the poor expression of some of the protein markers in fetal tissues.

In contrast, **genotypic evaluation of fetal material** provides substantially better evidence of highly specific polymorphisms helpful in human identity testing. The most widely used approach to genotyping has been **molecular genetic fingerprinting** that relies primarily on the restriction patterns generated with probes for short, tandemly repeated sequences found in the hypervariable regions (see Section 1.2.5).

The fingerprinting method is a simple, reliable, and accurate method for prenatal paternity testing when performed on material routinely obtained at **amniocentesis**, or on **chorionic villus samples**. Undoubtedly, application of PCR to this task will become routine.

Yet another use of REA in prenatal diagnosis is the **analysis of zygosity** in a twin tubal gestation.

1.3.4 DEVELOPMENTS IN THE PRENATAL GENE LEVEL DIAGNOSIS OF HEMOGLOBINOPATHIES

In addition to amniocytes usually retrieved during gestation week 16, fetal blood sampling (gestation weeks 18 to 20), and chorionic villus sampling (gestation weeks 8 to 12) the more recent studies use cordocentesis.

1.3.4.1 SICKLE CELL ANEMIA (p. 77)

Direct and indirect identification of **β-globin point mutations** by linkage analysis has long been performed during the first trimester on the material obtained by amniocentesis or chorionic villus biopsy. These earlier methods, based on REA of the β-globin gene, are much more labor intensive than the newer assays based on PCR.

One such PCR assay identifies **Hb** β^A, β^S, as well as **Hb** β^C alleles. Following amplification, the genomic DNA sequences are hybridized with radiolabeled oligonucleotide probes, and the annealed hybrids are then sequentially digested by two restriction enzymes, *Dde*I and *Hin*fI. The **GTG sickle mutation** of the normal **GAG codon** introduces a single base-pair mismatch that inhibits cleavage by the restriction enzyme. Comparison of the cleavage products generated by the two enzymes digesting two different kinds of duplexes leads to the diagnosis.

A **simplified PCR** for the sickle cell mutation uses amplification of a 294-bp segment of the β-globin gene followed by digestion of the amplified products by *Oxa*NI. The identification of an abnormal band by ethidium bromide or silver staining without radioactive reagents allows a diagnosis to be established within 3 to 4 h.

1.3.4.2 THALASSEMIA SYNDROMES

1.3.4.2.1 Prenatal Diagnosis of β-Thalassemias (p. 77)

In contrast to sickle cell anemia, which is the consequence of a single mutation, β-thalassemia is a collection of different mutations with similar phenotypic characteristics. **Direct identification** of such mutations by REA would require the mutation under study to affect the restriction site. Only a few mutations can be recognized in this way. The most common ones cannot be identified by REA alone. By combining REA with hybridization using short synthetic oligonucleotides labeled to a high specific activity, prenatal identification of the mutations can be performed under appropriate stringency conditions.

Strict adherence to appropriate hybridization and washing conditions is mandatory, as is the knowledge of the mutations to be tested, since **the probes designed for assays in Mediterranean populations will fail in analysis of Asians due to clustering of specific mutations** in different areas of the genome in particular ethnic groups.

The indirect method of detecting β-thalassemia mutations relies, like other indirect DNA assays, on **linked polymorphisms.**

One strategy for prenatal detection of β-thalassemia mutations in prospective parents utilizes **amplification of DNA followed by dot blot analysis with a set of allele-specific oligonucleotide (ASO) probes** complementary to the most common mutations in a given population.

Noninvasive (with respect to the fetus) **prenatal testing** for a hemoglobinopathy has been successfully demonstrated on maternal blood or, more precisely, the **nucleated fetal cells in maternal circulation**. At this time, one limitation of this procedure is that **only paternal alleles would be detected** while a maternal allele inherited by the fetus would not be identified.

Certain types of β-globin mutations in various parts of the world **cluster into haplotypes** that provide a basis for prenatal diagnosis of thalassemia in a clinical setting. A combination of synthetic oligonucleotide probe hybridization, PCR amplification of the genomic DNA, followed by direct sequencing as well as cloning and sequencing of the β-globin genes has been successful in identifying the β-globin mutation in 97% of 116 β-thalassemia genes evaluated.

At least in an **experimental system**, PCR-based analysis at the **preimplantation stage**, done by sampling the embryo directly, has successfully identified the normal β-globin gene, allowing a diagnosis of β-thalassemia to be excluded.

1.3.4.2.2 Prenatal Diagnosis of α-Thalassemias (p. 79)

α-Thalassemia, a heterogeneous group of inherited microcytic anemias, results from a decrease in α-globin chain synthesis. **Hydrops fetalis** occurs when all four α-globin genes are either deleted or inactive. Intrauterine fetal demise occurs in the third trimester with fetal red cells carrying nonfunctional hemoglobin composed of γ-globin tetramers (**Hb Bart's**). A prenatal identification of the predominant haplotype found in Southeast Asians (α^0-thal-1) has been developed.

The new approach using **PCR** and **dual REA** offers a significant advantage over the red blood cell (RBC) analysis, globin chain electrophoresis, and single REA formerly used for the diagnosis of α-thalassemia. Combinations of the restriction enzymes and the expected restriction patterns in some of the most important α-globin gene mutations have been published. The PCR tests for α-thalassemia are not capable of identifying all fetal genotypes. They offer, nevertheless, a fast and reliable approach to distinguishing the most prevalent mutations.

1.3.5 COAGULATION DISORDERS

1.3.5.1 HEMOPHILIA A (p. 79)

This sex-linked bleeding disorder, largely related to the defects of the **factor VIII (FVIII)** cofactor activity, is (unlike single-mutation disorders) caused by a **variety of mutations** at the *FVIII* locus. The *FVIII* gene is located at the telomere of the long arm of X chromosome at Xq28. The recognition of female carriers and prenatal diagnosis relies on RFLPs in and near the *FVIII* locus.

These **restriction site polymorphisms** can be analyzed with Southern blots, whereas **sequence polymorphisms** can be identified by ASO hybridization to genomic Southern blots under high-stringency conditions. Both types of polymorphisms can also be analyzed by **PCR**. The polymorphism evaluation and fetal sex determination by PCR in hemophilia A cases provide useful information in about 70% of hemophilia pedigrees.

Using intragenic *Hin*dIII polymorphism of exons 79 and 20 of the *FVIII* gene, a prenatal diagnosis of hemophilia A can be established in the first trimester of pregnancy.

1.3.5.2 HEMOPHILIA B (p. 80)

This X-linked recessive bleeding disorder clinically resembling hemophilia A is produced by decreased factor IX (FIX) activity in plasma due to a variety of genetic lesions causing the disease. Up to one third of cases of hemophilia B are due to **new mutations**. In a given pedigree these must be precisely identified for the prenatal diagnosis to be reliable. In the Caucasian population, the status of 89% of the potential carrier population has been determined with 99.9% certainty.

The molecular genetic approach to various coagulation factor deficiencies for prenatal diagnosis in clinical practice appears to be of limited use at present.

1.3.5.3 VON WILLEBRAND DISEASE (vWD) (p. 80)

von Willebrand factor (vWF) is a multimeric glycoprotein synthesized and secreted into plasma by endothelial cells and accumulated in platelet α granules. A highly informative (heterozygosity rate, 75%) VNTR sequence has been identified in intron 40 of the ***vWF*** gene. Based on this information a **PCR** assay of leukocyte DNA from the cord blood was able to show that the infant was homozygous for the **vWF** VNTR marker. Others successfully established a prenatal diagnosis of a severe case of vWD by amplification of the ***vWF*** gene segment containing **TCTA repeats**.

1.3.6 PLATELET DISORDERS (p. 80)

One platelet disorder is **neonatal alloimmune thrombocytopenia** — a serious bleeding disorder with a risk of intracranial bleeding *in utero* or during delivery. The alloantigen most frequently responsible for the condition is **HPA-la (Zw[a] or P1A)**. The alloantigens can be tested on platelets, but when these are not available the amplification of a noncoding DNA sequence directly upstream of the **HPA-1a** site can be used for analysis.

1.3.7 PRENATAL DIAGNOSIS OF INFECTIOUS DISEASES

1.3.7.1 CYTOMEGALOVIRUS INFECTION (CMV) (p. 81)

Congenital CMV infection affects up to 1% of live births. The significance of **intrauterine diagnosis** of CMV transmission acquires additional importance because there are many genetic variants of CMV circulating in the general population.

The experience in prenatal diagnosis of congenital CMV infection accumulated over the past 20 years emphasizes the **usefulness of amniotic fluid analysis** as an adjunct procedure. Although predominantly immunological in nature, the detection methods undoubtedly will soon be complemented by gene probes for CMV, as demonstrated by DNA ISH (see Section 6.3.1.4).

1.3.7.2 HUMAN IMMUNODEFICIENCY VIRUS (HIV) (p. 81)

The significance of prenatal diagnosis of HIV-1 infection of the fetus is emphasized by the documented predominant route of transmission of the virus **transplacentally**. In which period of pregnancy (i.e., **intrauterine**, **intrapartum**, or **postpartum**) the infection of the fetus/newborn occurs has not been definitively established. **Intrapartum transmission** is a possibility; **postnatal infection via breast milk** has already been documented.

The **conventional methods** of viral diagnosis (e.g., culturing HIV-1, and detection of the p24, p31, p55, gp160 antigens and other immunologically identifiable markers) are largely unreliable for a variety of reasons, including the differences in the clinical course of HIV-1 infection in the pediatric population.

PCR amplification of the HIV-1 sequences **in the fetus** appears to hold the greatest promise due to its inherent sensitivity and specificity. HIV-1 DNA in **newborns** and **children** has been identified by PCR in uncultured peripheral blood mononuclear cells (PBMCs) in children of HIV-1-infected mothers in the absence of HIV-1 antigenemia.

Efforts to maximize the **specificity of PCR detection of HIV-1 infection** involved the use of **multiple primers** for *gag, pol, env, tat, vif,* and *ref* regions of the HIV-1 genome in a **multistage PCR**. The positive amplified products of PCR can be tested in more detail in **yet one more PCR** using two sets of primers that span the hypervariable region of the *env* gene. Products of amplification of this reaction are then run on polyacrylamide gels to create a **pattern of amplified DNA length variation**. These patterns were found to be **characteristic of each child** and persisted for at least 3 to 7 months.

We are not aware of any reports at the time of this writing on the **prenatal identification of HILV-1 in the fetus** by PCR or by other techniques with continuous monitoring through pregnancy, birth, and postnatally. Several problems inherent in such approaches can be recognized:

- **contaminating nucleic acid material** may be amplified
- in the case of fetuses of infected mothers, the very process of **invasive retrieval** of a sample for amplification may introduce maternal tissue, and thus lead to false-positive results
- **false-positive** results may stem from amplification of **non-HIV sequences**; however, this interference can be largely diminished
- **false-negative** results may result from inadequate sensitivity of the assay due simply to the **vanishingly low viral load** in the fetus as well as to **sampling errors**
- **problems of interpretation** of the results of sensitive techniques such as PCR must be emphasized

Interestingly, in the case of human T cell leukemia/lymphoma virus type 1 (**HTLV-1**), analysis of cord blood from 20 seropositive women by PCR revealed no evidence of HTLV-1 infection, suggesting among other things that no significant contamination of the cord blood sample by maternal blood occurs that may interfere with PCR amplification.

1.3.7.3 VARICELLA ZOSTER VIRUS (VZV) INFECTION (p. 83)

In utero diagnosis of VZV infection can be established by probing chorionic villi by PCR with specific primers. This, however, may not correlate with the clinical manifestations of fetal disease.

1.3.8 PRENATAL MOLECULAR GENETIC DIAGNOSIS OF CHROMOSOME INSTABILITY SYNDROMES (p. 83)

These genetic disorders, of differing etiologies and manifestations, are united by a common feature — **cellular hypersensitivity to mutagenic chemicals**. This property is used for prenatal diagnosis. The principal diseases in this group include **Fanconi anemia** (FA), **ataxia-telangiectasia** (A-T), **Bloom syndrome** (BS), **xeroderma pigmentosum** (XP), and **Cockayne's syndrome** (CS). These disorders are associated with a variety of developmental defects, from growth retardation to immunodeficiency and premature aging.

The common approach to the diagnosis of these disorders has been the demonstration of mutagen hypersensitivity in cultured amniotic cells or cultured trophoblast cells from chorionic villi. **Failure to repair DNA damage** in irradiated cells from patients is also used for diagnosis (see Section 1.5.7).

1.3.9 OTHER EXAMPLES OF DISORDERS FOR WHICH THE DISEASE LOCUS HAS BEEN IDENTIFIED

1.3.9.1 α_1-ANTITRYPSIN (AAT) DEFICIENCY (p. 83)

AAT deficiency leads to early-onset **emphysema** and **fatal cirrhosis** of the liver. **Prenatal diagnosis** of AAT deficiency has long been performed by **PI typing of fetal blood** samples retrieved at fetoscopy. **Chorionic villus** biopsy is the preferred source of DNA for prenatal diagnosis of the AAT deficiency.

An **oligonucleotide hybridization** technique to detect the point mutation in the *M* and *Z* alleles, and subsequently in the *S* allele, has been developed as well. Comparison of the results of an RFLP analysis in and around the *AAT* gene, particularly using *Ava*II digestion, with those obtained by oligonucleotide hybridization suggests that the latter approach is preferable in families in which no siblings are available for RFLP analysis.

In informative kindreds, the **linkage analysis** appears as accurate and reliable as oligonucleotide hybridization and is technically easier. It is applicable to various types of AAT deficiency, and is an acceptable method for a routine diagnostic laboratory.

In prenatal identification of *AAT* mutations by **PCR** amplification less than 1 μg of fetal tissue is required and the result can be available within 72 h. In addition to the *Ava*II polymorphism, two other enzymes are used (*Mae*III and *Bst*EII) to increase the frequency of informative families. By amplifying the sequence containing the *Bst*EII polymorphic site followed by REA of the PCR amplification products, a rapid and relatively simple RFLP analysis is performed.

1.3.9.2 ADULT POLYCYSTIC KIDNEY DISEASE (APKD) (p. 84)

This autosomal dominant **monogenic** disorder has a prevalence of about 1:1000, with a highly variable age of onset and high expressivity in adult life. The ***APKD* gene** has been assigned to **chromosome 16 (16p13.3-p13.1**) and a number of RFLPs in and near the ***PKD*1 locus** have been identified and used for prenatal diagnosis. Another locus yet to be mapped **(*PKD*2)** has also been identified.

Prenatal diagnosis of APKD by RFLP analysis has been performed on DNA obtained from chorionic villus biopsy as early as gestation weeks 9 and 10.

1.3.9.3 TAY-SACHS DISEASE (TSD) (p. 84)

Until recently the only approach to prenatal diagnosis of TSD was the measurement of **hexosaminidase A activity** in cultured amniocytes or chorionic villus samples. The identification of **HEXA mutations**, accounting for the majority of TSD alleles in Ashkenazi Jews, made a direct DNA-based diagnosis possible.

Prenatal diagnosis would be indicated particularly in couples who either already had an affected child, or have a positive family history. In one study, DNA from peripheral blood lymphocytes was restricted with *Hae*III and a 4-bp insertion mutation in exon 11 was found in both parents. At gestation week 10, chorionic villi were obtained and fetal DNA was restricted with *Hae*III. Comparison with the parental pattern revealed that the fetus was heterozygous for the 4-bp insertion allele, the pregnancy was allowed to proceed, and the infant proved to be healthy.

A combination of PCR amplification of mutant alleles followed by cleavage of a *Dde*I restriction site generated by the mutation can be used to identify carriers of the disease.

Other frequently encountered mutations defined in Ashkenazi Jews lead to defects in **sphingomyelin metabolism**, resulting in **Niemann-Pick disease**, and in the **glucocerebrosidase gene** responsible for **Gaucher's disease.**

1.3.9.4 ORNITHINE TRANSCARBAMYLASE (OTC) DEFICIENCY (p. 85)

OTC is the second enzyme of the urea cycle that is involved in **detoxification** of the products of **nitrogen metabolism**, primarily in the liver, to yield an excretable product, **urea**. OTC deficiency is a X-linked disorder with a severe, and often fatal, **hyperammonia intoxication** in hemizygous affected males. Prenatal diagnosis of this disorder by biochemical techniques is difficult because the enzyme is not expressed in amniocytes, and the metabolic products of OTC deficiency are not detectable in the amniotic fluid.

Probes generated with the restriction enzymes *Msp*I, *Bam*HI, *Taq*I, as well as *Pst*I allow the prediction of affected fetuses. Two RFLPs produced by *Msp*I and *Bam*HI are informative in 74% of known female carriers, whereas by using *Taq*I site alterations the mutation can be detected directly in 17% of affected males or heterozygous females, suggesting that all families should also be screened for *Taq*I alterations.

In cases in which new mutations may account for the failure of the known RFLPs to provide diagnostic information, then biochemical testing appears to provide the most information. In such cases, DNA analysis of antecedents in an affected family may prove useful.

The characterization of the *OTC* gene and some of the mutations associated with the disease allows not only the screening of female carriers, but also the performance of prenatal diagnosis of OTC deficiency. The technique of **chemical mismatch cleavage** in combination with **PCR** has also been used in the prenatal diagnosis of OTC deficiency.

1.3.9.5 HEREDITARY AMYLOIDOSIS (HA) (p. 85)

HA is an autosomal dominant disorder (**familial amyloidotic polyneuropathy type II**) manifested as a late-onset carpal tunnel syndrome, progressive peripheral neuropathy, vitreous opacities, gastrointestinal dysfunction, and eventual death usually following cardiomyopathy. This is one of a number of familial amyloidotic polyneuropathies related to mutations of the plasma protein **transthyretin (TTR)**.

DNA restriction analysis has been used for the identification of presymptomatic carriers and prenatal determination of heterozygosity. **PCR amplification** of an exon 3 region containing a fragment of the *TTR* gene can be performed on DNA extracted from amniocytes of a fetus at risk of carrying the serine-84 prealbumin (*TTR*) gene. **ASO** hybridization and **REA** of the PCR-amplified products have been used (for the first time) to determine prenatally the carrier status of a fetus for the *HA* gene.

1.3.9.6 SEVERE COMBINED IMMUNODEFICIENCY (SCID) AND DIGEORGE SYNDROME (p. 86)

SCID comprises a number of sex-linked and autosomal genetic defects in cellular and humoral immunity. In over 80% of cases the genetic cause is unknown, and prenatal diagnosis of SCID has been relying on lymphocyte functional studies in fetal blood samples.

Progress in the prenatal diagnosis of SCID came from studies on the **inactivation pattern of the X chromosome** in a woman's T cells; the pattern appears to be nonrandom in cases of the X-linked form of SCID. Furthermore, genetic linkage studies identified the X-linked *SCID* locus on the proximal long arm of the **X chromosome (Xq13.1-21.1)**. So far, no specific genetic alteration responsible for the X-linked form of SCID has been identified, but **polymorphic DNA markers have proved their usefulness in prenatal diagnosis**.

In **DiGeorge syndrome**, there are three deletions consistently found at loci D22S75, D22S66, and D22S259 in **chromosome 22q11**. The prenatal diagnostic use of RFLPs and linkage analysis is markedly improved by FISH for the deleted region.

Chapter 1.4
Aging

1.4.1 INTRODUCTION (p. 87)

Evaluation of **pathological** conditions at the gene level is inevitably affected by the ongoing **physiological** alterations in cells and tissues involved in the disease process related to **DNA repair**, **aging**, **drug resistance**, and **environmental effects** on the genome.

Some of the changes occurring in the normal process of aging will (hopefully) be distinguished from aberrations associated with disease. So far, **no unifying theory of aging has been developed**, and the plausibility of attaining a single theory of aging is far from certain.

1.4.2 GENETIC INSTABILITY (p. 87)

The main **theories of aging** consider at least three alternatives:

- first, it had been postulated that aging is directly beneficial at the species or group level to **prevent overcrowding**, among other things, and specific genes may have evolved to **control life span**. Support for this view has not been strong
- second, aging may reflect the **late expression of mutations accumulated in the germ line**, and **selection** may exert little pressure on these potentially deleterious genetic alterations. Experimental support for that view is also lacking
- a third category of genes invoked in the mechanism of aging includes those that offer some **short-term benefit** in a **trade-off for the potential for long-term survival**

An extension of the "**disposable soma**" **theory** is the prediction that

- first, aging is due to an accumulation of **unrepaired somatic defects**. **Mutation interactions**, **transposable elements**, and **epimutations** involving **demethylation** (loss of 5-methylcytosine, in particular) appear to have a role. **Experimental** verification of the assumptions of the **somatic mutation theory** fails, however, to provide strong support for it
- second, the **rate of aging** can be modulated by altering the level of important maintenance and repair processes. The **homeostasis of maintenance** and **repair functions** is thought to be the point of application of the damage and defective repair. **Genetic instability is viewed as a major cause of aging**

The importance of **efficient DNA repair** contributing to longevity lends support to the notion that genetic instability is involved in aging.

Candidate genes determining aging and longevity in humans include genes controlling

- **protein synthesis**
- **free radical scavenging**
- **mitochondrial functions**
- **reparatory systems**, particularly DNA repair

DNA damage can be expressed in various **epigenetic effects** such as induction of the protooncogene c-*fos*, the activity of which can lead to **poly-ADP-ribosylation** of chromosomal proteins, resulting in modulation of gene expression. Additional mechanisms of genetic instability appear to involve

- **DNA recombination**
- **transposon-related inactivation** of genes
- **DNA amplification**

1.4.3 PROGRAMMED CELL DEATH (p. 88)

The role of **programmed cell death** in the process of aging is not unequivocally established as one of cardinal causes of senescence. Although intermediate forms of cell death between the extremes of **necrosis** (resulting from **acute trauma** to the cell) and **apoptosis (physiological cell death)** certainly exist, the phenomenon of "programmed cell death" is seen as a definable sequence of events.

An **active genetic death program** may be responsible for cell death, justifying the search for genes controlling this process. Some of these include **growth regulatory genes**, such as the gene producing **transforming growth factor β_1 (TGF-β_1)**, that can inhibit cell proliferation and lead to increased apoptosis.

The ***ced* (cell death) mutation** in the nematode *Caenorhabditis elegans* prevents almost all of the programmed cell deaths. Reconstitution experiments confirm that the **death program is an active process** involving specific protein synthetic patterns.

Not all tissues seem to require active protein synthesis for dying, however. The activity of **lysosomal enzymes** and the **calcium-activated endonuclease (CaN)**, in particular, have been linked to cell death in a variety of tissues. A host of other gene products become expressed in the dying cell [**c-*myc*, c-*fos*, heat shock protein 70, testosterone-repressed prostate message 2 (TRPM-2), transglutaminases**, etc.]. Programmed cell death is operational in neurons in response to numerous influences that include changed anatomical, hormonal, growth factor, and energy environments.

The products of the **major histocompatibility complex (MHC)** in their interaction with nonimmune systems affect **mixed function oxidases, reproductive senescence, DNA repair**, and **free radical-scavenging enzymes.**

1.4.4 ROLE OF CARBOHYDRATE METABOLISM IN AGING (p. 88)

A decrease in diverse physiological functions of the cell has been linked to the age-related **slowing** and **impairment of induction of enzymes of**

carbohydrate metabolism. The intracellular reduction in carbohydrate metabolism is believed to accompany various pathologies associated with the aging process, such as **glucose intolerance** and **diabetes**.

1.4.5 THE OXIDATIVE STRESS HYPOTHESIS AND THE FREE RADICAL THEORY OF AGING (p. 89)

Another theory, or hypothesis, visualizes **oxidative damage** resulting from **peroxidation** of cellular membrane lipids, inactivation of enzymes, and DNA damage as the central cause of the multifactorial, deleterious process of aging. **Aging constitutes the terminal stage of development**, and the level of oxidative stress (which is the balance between prooxidants and antioxidants) increases during cellular differentiation and aging so that **aging is not a genetically programmed phenomenon**, but results from the consequences of oxidative stress on genetic programs.

The effects of oxidative stress on aging, as mediated through its effects on gene expression, are **not conclusive** and numerous **alternative interpretations** of the data are possible. The role of **direct oxidative DNA damage** under the influence of environmental agents (e.g., mutagens and carcinogens) has also been considered.

The **free radical theory** of aging is yet another actively pursued line of research on aging. This theory includes cancer and atherosclerosis among the growing number of **free radical diseases**, as well as other common degenerative diseases that can be counteracted, to a degree, by the administration of antioxidants.

1.4.6 ROLE OF DNA METHYLATION IN AGING (p. 90)

Most organisms appear to exhibit a loss of **5-methyldeoxycytosine (5mC)** during aging, and the extent of this loss inversely correlates with the life span potential of a particular species.

A pattern of age-related changes in DNA methylation, particularly **CpG dinucleotides** (predominantly methylated in high eukaryotes), can be established. **Age-related patterns of increase in DNA methylation with age** have been observed in

- **rRNA genes** on chromosome 16 in aging mice
- the **c-*myc* gene**; these patterns were found to be more related to **biological aging** than to chronological aging

1.4.7 MITOCHONDRIAL DNA (mtDNA) IN AGING (p. 90)

Somatic mtDNA mutations have been observed in tissues of aging organisms and implicated in the reduced functional capacity of tissues characteristic of advanced age. PCR amplification of mtDNA from adult mitochondria from the heart and brain lends support to the idea that **accumulation of specific mtDNA deletions** may be associated with aging.

At the **cellular level**, a deregulation of the normal biosynthetic process in the mitochondria brought about in lymphocytes by the **t(14;18) translo-**

cation leads to **prolonged cell survival**. It appears that overexpression of the *bcl*-2 gene **blocks the apoptotic death** of the cells. Furthermore, induction of *bcl*-2 expression in human B lymphocytes by **Epstein-Barr virus (EBV) latent membrane protein 1 (LPM 1)**, which specifically **up regulates** the cellular oncogene *bcl*-2 in mediating its effect on apoptosis, protects these cells from programmed cell death.

Other mechanisms are being studied that lead to **cellular immortalization**, including the effects of the DNA tumor viruses SV40, HPV 16 and 18 and type 5 adenovirus, and of the tumor suppressor genes *RB* and *p53* and their interaction with E1A, E1B, and *ras*.

Chapter 1.5
DNA Repair

1.5.1 INTRODUCTION (p. 93)

Altered homeostasis, either transient or permanent, affects a number of physiological functions at different levels, ranging from molecular to organismal. The **damage** underlying a disease process, be it an infection, a heritable mutation, or a physical-chemical injury, activates mechanisms aimed at restoring homeostasis. The cellular and molecular mechanisms immediately involved in the **repair** of the primary focus of injury have been extensively studied. In human pathology, the study of such responses has gained the most dramatic success in unraveling the **reparatory processes at the DNA level**. The model systems that have been studied in greatest detail use **ultraviolet (UV)** and **X irradiation**, **chemical mutagens**, and **chemotherapeutic agents**.

Although the assessment of DNA repair in pathological conditions at this time remains essentially a basic research endeavor, this area of molecular pathology will in the near future most likely become increasingly important for evaluating disease processes in a clinical setting.

1.5.2 DNA DAMAGE (p. 93)

Damage of **genomic DNA** occurs as a result of application of physical (e.g., UV or X irradiation, heat), chemical (e.g., known antigens, chemotherapeutics), or biological (e.g., viruses) agents. All of these can **damage DNA indirectly**, usually through the **secondary changes** produced in the intracellular milieu, which could be of a specific or nonspecific nature. Depending on the **nature** and **dose** of an agent, a **specific cell response**, including that originating at the genome level, can be elicited by a largely **nonspecific external stimulus** such as heat shock.

Transduction of the external stimuli originating at the cell periphery through a network of intracellular structural and chemical mediators results in specific responses of various biosynthetic apparatuses of the cell.

The concept of **organ specificity** of chemical damage accommodates the **multiplicity of stages, factors,** and **target genes** participating in the process of carcinogenesis. Four significant processes appear to be involved in DNA damage:

- **oxidation**
- **methylation**

- **deamination**
- **depurination**

Highly specific DNA repair enzyme systems have been demonstrated. The major type of DNA damage appears to be related to oxidation.

1.5.2.1 ENVIRONMENTAL EFFECTS ON THE HUMAN GENOME (p. 94)

The study of **environmental mutagenic effects** is progressively expanding. Some of the gene level studies address **human genome-environment interactions**, which establish links between **genetic susceptibility** and **risk factors** for disease, on the one hand, and the environment, on the other. **Epidemiological methods** have produced plausible models of these interactions. Examples of these are:

- the interaction between the recessive gene for phenylketonuria (**PKU**), blood levels of phenylalanine, and mental retardation
- the avoidance of sun exposure on the part of individuals with **xeroderma pigmentosum** to reduce the genetically determined predisposition to skin cancer
- the enhancing effect of **barbiturates** on the manifestations and consequences of **porphyria variegata**
- **hemolytic episodes** provoked by **fava bean** consumption in glucose-6-phosphate dehydrogenase (**G6PD**)-deficient individuals
- the predisposition of **α_1-antitrypsin-deficient** persons to **emphysema**; in these people **smoking** significantly increases the risk of developing the disease
- the possible induction of inherited defects in humans by exposure to environmental mutagens, such as in the long-term administration of **antischistosomal drugs**, which manifest genotoxicity and carcinogenicity
- the complex influences of many disease-related risk factors associated with **familial** environment linked to cultural inheritance, and genetic predisposition

Direct assessment of effects of the environment on the human genome can be performed by demonstrating the formation of structural chromosome aberrations — **chromatid** and **isochromatid breaks** and "**minutes**". These and other chromosome aberrations have been reported, for example, in

- the lymphocytes of persons exposed to **polychlorinated biphenyl (PCB)**
- the effects of **dioxins**
- the effects of **agent orange**
- the peripheral blood lymphocytes of nurses handling **cytostatic drugs** without a safety cover
- the **nuclear industry**, in which chromosomal aberration analysis has long been used to monitor exposure to **external radiation**
- blood lymphocyte cultures from workers occupationally exposed to **uranium**

It appears that at present a surprisingly minor effort is being given to the assessment of environmental effects on the human genome and their interaction with the disease process in clinical practice.

1.5.2.2 RADIATION DAMAGE (p. 96)

In molecular terms, UV-induced damage eliminated via the **nucleotide excision repair** process and recovery from **chemical adducts** is the best characterized reparatory process. Excision repair has been dissected into consecutive steps:

1. **Preincision recognition** of the damaged area of DNA
2. **Incision of the damaged DNA strand** in the vicinity of the site of damage
3. **Excision of the defective site** with concomitant or **sequential degradation** of the excised product
4. **Repair replication**, filling in the excised area with normal nucleotides
5. **Ligation of the repaired segment** at its 3′ end to the undegraded stretch of DNA

UV irradiation (254 nm) has been shown most commonly to covalently link neighboring bases into **pyrimidine dimers**, and less commonly to produce a 6–4 linkage between adjacent pyrimidines (**6–4 photoproduct**). The UV-induced (6–4) photoproducts and their repair in XP-A cells can be quantitatively evaluated by **laser imaging microspectrofluorimetry** or by **autoradiography** with the help of a monoclonal antibody raised against (6–4) photoproducts.

More than one type of damage usually occurs as a result of action of radiation or carcinogens. Some of the other common types of DNA damage in eukaryotic cells include

- **DNA-protein cross-links (DPCs)**
- **base damage**
- **single-** and **double-strand breaks**
- **bulky lesions**

Efficiency of DNA excision repair is affected by

- **nuclear structure**
- **accessibility of chromatin to endonuclease attack** reflected, for example, in the altered sensitivity of chromatin DNA in malignant cells to DNase I hydrolysis
- the size of the available **pool of DNA precursors**

1.5.2.3 ALKYLATING AGENTS (p. 97)

Alkylating agents are efficient mutagens that produce a number of lesions in DNA, such as alkylation at the N^7 position of guanine, the O^6 position of guanine, and the N^3 position of adenine. The action of ***O*^6^-alkylguanine-DNA alkyltransferase (AGT)**, which transfers the alkyl groups to itself and performs excision repair, is responsible for protection against cell killing and the accumulation of mutations.

This large group of chemicals includes:

- **carcinogens** such as *N*-methyl-*N′*-nitro-*N*-nitrosoguanidine (**MNNG**) and ethylnitrosourea (**ENU**)
- many widely used chemotherapeutic drugs (e.g., **cyclophosphamide, melphalan, busulfan, chlorambucil**)

Monofunctional alkylating agents, such as ***O*^{6}-methylguanine (m^6-Gua)**, produce methylation of bases in DNA, and a specific **repair enzyme**, **m^6-Gua-DNA methyltransferase**, affords protection to most human cell lines. Alkylating agents produce **mismatch base pairing**, which is corrected by proteins binding specifically to DNA mismatches. A rapid **genotoxicity test** measures DNA strand breaks by unwinding DNA strands in alkaline solutions and evaluating the dissociated strands by **hydroxyapatite elution chromatography**; its potential clinical usefulness has been demonstrated in mouse lymphoma cells treated *in vitro* with some 78 compounds.

1.5.2.4 HEAVY METAL COMPOUNDS (p. 98)

Cis(II)-platinum diamminedichloride (**cisplatin**) and **carboplatinum** are widely used antitumor agents that interfere with DNA-related functions through the formation of **intra**- and **interstrand cross-links**. The details of cisplatin action have not yet been finalized.

1.5.2.5 ANTIBIOTICS (p. 98)

A variety of **antitumor antibiotics** targeted to DNA have been introduced into clinical practice. The specific effects on DNA include

- **single-** and **double-strand breaks** (e.g., bleomycin, daunomycin, and Adriamycin)
- **intercalation** (mitomycin C)
- **inhibition** of **transcription** and **replication** (actinomycin D)
- **interaction** with **topoisomerases**, leading to strand breaks [epipodophyllotoxins, etoposide (VP16), and teniposide (VM26)]

1.5.2.6 FRAME-SHIFT MUTAGENESIS (p. 98)

Acridines represent a model class of compounds known to form DNA-acridine complexes by **reversible binding**. Through **insertion** or **deletion** of a base, a **shift out of frame** occurs in the particular genetic message due to a distortion of a previously uninterrupted, meaningful sequence of triplets.

Cytotoxicity of most antitumor **intercalating agents** is due to the stabilization of the cleavable complexes between mammalian **topoisomerase II** and DNA. This prevents the passage of other double strands through the cleavage site during replication. The specific cellular repair mechanisms involved in counteracting the effects of acridines appear to be different from those of the enzymes repairing damage produced by radiation and chemical agents.

1.5.3 DNA REPAIR GENES (p. 99)

The enzymatic factors involved in eliminating the consequences of various types of damage have been defined and the corresponding genes are known as **DNA repair genes**. Significant progress has been made in cloning and characterizing DNA repair genes such as ***ERCC*-1**, the ***den* V** gene, the gene coding for **DNA ligase** I (defective in Bloom syndrome), and the human repair gene coding for **O^6-methylguanine-DNA methyltransferase**.

1.5.4 PREFERENTIAL DNA REPAIR (p. 100)

An important concept with solid experimental support describes the **heterogeneity of DNA repair** as reflected in the preferential repair of active genes. Repair primarily occurs in the transcription of DNA strands and certain genomic regions rather than being determined by the accessibility of the chromatin structure. The **level of methylation** of selected areas in the genome may be coordinated with the preferential DNA repair activity. Preferential repair of **actively transcribed genes** has been shown in the c-*abl* and c-*mos* protooncogenes (the former being much more actively transcribed), and in the **metallothionein gene**.

Because the repair is much more efficient when the gene is activated, an **interaction between transcription and DNA repair is apparent**. Preferential repair activity is also found in the areas of a gene involved **in more frequent structural rearrangements**, as is the case with the c-*myc* gene. Preferential repair in certain regions of the gene varies with the inherent resistance of the organism.

The preferential repair of the actively transcribed genes has been demonstrated in UV-irradiated fibroblasts from the XP complementation group C.

1.5.5 DNA REPAIR AND DRUG RESISTANCE (p. 101)

The phenomenon of multidrug resistance accounts for a large proportion of chemotherapeutic failures. Different types of chemotherapeutic agents display a selective preference for interaction with coding or noncoding areas of the genome as well as for active vs. nonactive genes.

1.5.5.1 TOPOISOMERASES (p. 101)

DNA topoisomerases are involved in **DNA replication** and **transcription** in mammalian cells. Topoisomerases regulate the **superhelical configuration** of cellular DNA. They relax the supercoils in the chromatin by passing DNA strands through one another: single-stranded DNA (ssDNA) by **topoisomerase I (TI)**, double-stranded DNA (dsDNA) in the case of **topoisomerase II (TII)**.

TII induces dsDNA cleavage, the formation of an **intermediate bridging complex** of the enzyme and the **double-strand breaks (DSBs),** and, eventually, **resealing** of the DNA strand breaks. In the presence of antitumor drugs that inhibit TII, such intermediate cleavable complexes accumulate in the genome. There is a **direct correlation** between the failure of the cell to repair DSBs induced by radiation and their sensitivity to drugs that stabilize the cleavable complexes.

TI, which is capable of producing **single-strand breaks (SSBs)**, is specifically inhibited by the antitumor agent **camptothecin**, which traps TI-DNA covalent complexes in which one enzyme molecule is covalently bound to the 3′ terminus of an SSB. The mechanism of cell death produced by camptothecin is believed to be due to the collision between moving replicating forks and the camptothecin-induced and stabilized cleavable complexes between DNA and TI.

It is already evident that in the near future clinical patient management will include the evaluation of cellular resistance to specific chemical or physical agents at the gene-specific level. In the meantime, **quantitative relationships between the cytotoxicity due to DNA strand breaks and DNA immunoreactivity to anti-DNA monoclonal antibodies can be used to predict the sensitivity of the individual patient to alkylating agents at the genome level.**

1.5.6 ASSESSMENT OF DNA DAMAGE AND REPAIR (p. 102)

The detection of DNA lesions and evaluation of the cellular reparatory efforts are accomplished by chemical, physical, enzymatic, and immunological techniques. Antibodies have been produced to quantitate the **DNA adducts** produced by various carcinogens (**benzo[*a*]pyrene, aflatoxin B_1, ethylnitrosourea**) as well as **thymidine dimers** and **thymine glycols** produced by radiation. DNA lesions can be converted to a break in the DNA and then identified by **sequencing**.

A number of techniques can be used on patient DNA samples, including **alkaline elution** or **alkaline sucrose gradient** analysis. This may give information on the **efficiency of DNA ligation** in the process of DNA repair, for example, in suspected Bloom syndrome. Quantitation of DNA DSBs produced by various agents can now be measured by **pulsed-field gel electrophoresis** and its modifications.

To analyze molecular details of DNA damage and repair, methods have been devised for **molecular dosimetry of DNA damage** at the level of individual cells or tissue sections employing, for example, **quantitative immunofluorescence microscopy**.

A PCR restriction site mutation method has been introduced for the study of mutations in mammalian cells. Some of the potential problems of this approach are

- the **variable efficiency** of nonmutant DNA cleavage
- the potentially **inadequate fidelity** of PCR amplification
- the **need for removal of restricted sequences** so as not to contaminate the mutant DNA being amplified

One ingenious method **eliminates unmutated sequences** using biotinylated oligonucleotides that bind to nonmutated DNA sequences, and the formed duplexes are then separated by capturing these on streptavidin-coated magnetic beads. [*See also Supplement 1.*]

A number of **human repair genes** [*ERCC*-1, *ERCC*-2, X-ray repair cross-complementing 1 (*XRCC*-1)] have been identified by **complementing studies**, in which the **candidate genes** were able to correct the repair deficiency of cultured cells and were assigned to human **chromosome 19**. An elaborate technique using a **cisplatin cross-linked plasmid shuttle vector** has been devised for use as a specific probe to assess the DNA repair capacity of XP cells.

A number of *in vivo* and *in vitro* assay systems have been devised for the study of **purified repair proteins** introduced into repair-deficient cells or subcellular systems. The effects of *in vivo* exposure of a patient to antineoplastic drugs can be evaluated by analyzing the DNA repair efficiency and replicative synthesis in peripheral blood lymphocytes following their exposure *in vitro* to UV irradiation.

1.5.7 DNA REPAIR IN HUMAN DISORDERS (p. 103)

Table 1 lists **disorders involving DNA repair deficiency**, conditions **suspected** to involve DNA repair deficiency, conditions with **possible** DNA repair defects, and diseases with **demonstrated hypersensitivity** to DNA damage, in which DNA repair may be important.

1.5.7.1 XERODERMA PIGMENTOSUM (XP) (p. 103)

XP is a rare autosomal recessive disorder with deficient excision of **UV-induced DNA photoproducts**. It is manifested by excessive solar damage to the skin with increased frequency of skin cancers (basal cell carcinoma and malignant melanoma) and internal malignancies affecting the lung and brain. Pathogenesis of XP-associated malignancies underscores the fact that carcinogenesis involves DNA damage, and that DNA repair plays a major role in maintaining cell homeostasis.

Eight **complementation groups** (A through H) and a variant (XP-V) have been identified, the four most common being A, C, D, and V. The predominant biochemical defect is in the **removal of pyrimidine dimers**, whereas XP-V is deficient in **postreplication repair** (**PRR**).

The **failure of endonucleases** in XP-A cells to excise the damaged areas of DNA early in the repair process does not appear to reside in aberrations of the enzymes themselves. It appears that XP cells fail to accomplish the **initial incision**, but are capable of carrying out later steps of repair. The broad spectrum of clinical symptoms and cellular phenotypes in repair-deficient cells can be explained by a **combination of single-gene defects** involving DNA repair. Some of the repair genes for XP-A are located on **chromosome 9**. A constant tendency toward chromosome instability in XP mutation carriers is observed.

The **opposing notions** suggest that **multiple gene loci** operate simultaneously in determining XP and the disorder may be, in fact, composed of several different clinical abnormalities. Recently, an immunological defect

TABLE 1
DNA Damage-Sensitive Human Hereditary Disorders with Established or Potential DNA Repair Deficiency

Disorder	Risk of cancer	Clinical features	Characteristics
DNA repair deficiency			
Xeroderma pigmentosum	++++	Photosensitivity Neurological abnormalities Mental impairment	Hypersensitive to UV[a] Defective excision repair Nine complementation groups Increased chromosome breakage and sister chromatid exchanges after UV
Cockayne's syndrome	None	Dwarfism Neurological abnormalities Mental retardation Photosensitivity	Hypersensitivity to UV Lack of preferential DNA repair Impaired post-UV DNA and RNA synthesis
Bloom syndrome	++++	Telangiectasia Photosensitivity Immunodeficiency	Ligase I deficiency Spontaneous chromosome breaks, aberrations and sister chromatid exchanges. Altered glycosylase
Probable defect			
Fanconi's anemia	++++	Skeletal abnormalities Bone marrow hypofunction Mental deficiency Leukemia	Spontaneous chromosome damage Hypersensitivity to cross-linking agents Possible deficiency in cross-link repair
Ataxia telangiectasia	+++	Telangiectasia Cerebellar ataxia Immunodeficiency Neurologic abnormalities	Hypersensitivity to X-irradiation and some carcinogens Spontaneous chromosome rearrangements

			X-ray-resistant DNA synthesis Possible deficiency in repair of strand breaks
Suspected DNA repair defect			
Tricothiodystrophy (PIBID)	None	Telangiectasia Sulfur deficient hair Neurological abnormalities Mental impairment	Possible defective excision repair
Dyskeratosis congenita	++	Hyperpigmentation Telangiectasis Leukoplakia	Possible defective cross-link repair
Nevoid basal cell carcinoma	+++	Skeletal abnormalities Mental impairment	Increased rate of spontaneous chromosome breaks and sister chromatid exchanges Possible deficiency in UV repair
Gardner's syndrome	++++	Polyps in colon Osteomas Dental deformities	Increased rate of sister chromatid exchanges after UV
No known repair defect			
Dysplastic nevus syndrome	++++	Dysplastic nevi	Hypersensitive to UV Increased sister chromatid exchanges Chromosomal rearrangements after UV
Retinoblastoma	++++	Hypersensitive to ionizing irradiation Ocular malignancy Dysmorphic features	Hypersensitivity to ionizing irradiation Spontaneous chromosomal deletion

[a] UV, Ultraviolet irradiation.

From Bohr, V. A. et al. (1989). Lab. Invest. *61*:143–161. © by The U.S. and Canadian Academy of Pathology, Inc. With permission.

has been suggested to play a crucial role in determining dermatological symptoms of XP patients.

The molecular **diagnosis of XP** is based on the hypersensitivity of cultured cells to UV radiation due to **defective repair of UV-induced pyrimidine dimers** in DNA.

Prenatal diagnosis relies on the **analysis of DNA repair** of cultured amniotic fluid or trophoblast cells following UV irradiation. The **unscheduled DNA synthesis** is evaluated by autoradiography of [^{3}H]thymidine incorporation into nonreplicating cells.

A more complex situation arises when the proband has a **variant form** of XP. Then the PRR defect is detected by measuring the **conversion** of low to high molecular weight DNA after irradiation of cells. Fractionation of DNA on **alkaline sucrose gradients** following pulse labeling of DNA is a relatively common research procedure that is somewhat complicated to perform in a clinical laboratory setting.

1.5.7.2 BLOOM SYNDROME (BS) (p. 105)

Bloom syndrome is manifested in low birth weight, stunted growth, normally proportioned but dwarfed body, photosensitivity, butterfly telangiectasia of the face, and high susceptibility to infection resulting from moderate to severe immunodeficiency. Chromosomal abnormalities include

- **sister chromatid exchanges (SCEs)**
- **mitotic crossing over**
- **nonhomologous translocations**

Fibroblasts from BS patients display a **decreased UV resistance**, pointing to a defect in DNA repair, which has been traced to **DNA ligase I** and **glycosylase deficiency**.

Prenatal diagnosis utilizes the characteristic property of BS cells to form **SCEs** in cultured amniocytes.

1.5.7.3 COCKAYNE'S SYNDROME (CS) (p. 106)

CS is transmitted in an autosomal recessive way and manifested in growth and developmental arrest leading to dwarfism, photosensitivity, deafness, intracranial calcifications, mental deficiency, sunken eyes, disproportionately long arms and legs, and a general appearance of **premature aging**. Progressive neurological degeneration leads to death at an early age; however, **an increase in the incidence of malignancies is not observed.** Similar to XP, but to a lesser extent, CS fibroblasts and lymphocytes show **hypersensitivity to UV**.

The **diagnostic procedure** is simple, evaluating the level of [^{3}H]uridine incorporation into RNA in amniotic fluid cells by scintillation counting or autoradiography.

1.5.7.4 FANCONI'S ANEMIA (FA) (p. 106)

Fanconi's anemia, a **pancytopenic** disorder, is apparently transmitted in an autosomal recessive mode and is accompanied by congenital and developmental abnormalities and a **threefold increased risk of malignancy. Chromosomal instability** in FA leads to an increased frequency of spontaneous chromatid aberrations and abnormal susceptibility to the action of DNA cross-linking agents.

Variability of the phenotypic expression of FA precludes accurate clinical diagnosis and parents of affected children prefer **prenatal diagnosis** in subsequent pregnancies.

Prenatal diagnosis has been based on the hypersensitivity of FA cells to the blastogenic effect of **DNA cross-linking agents**, such as

- **diepoxybutane (DEB)**
- **mitomycin C**
- **cyclophosphamide**
- **nitrogen mustard**
- **psoralens followed by long-wavelength UV irradiation** (PUVA)
- **cisplatin**

DEB-induced chromosomal breakage can be detected in cultured amniotic fluid cells and trophoblast cells from fetuses with FA when compared to control cells, thereby providing a reliable method of prenatal diagnosis of FA.

1.5.7.5 ATAXIA-TELANGIECTASIA (A-T) (p. 106)

A-T occurs with a frequency of 1 in 40,000 live births, affects the nervous and immune systems and the skin, and is transmitted in an autosomal recessive mode of inheritance. Clinically, the dominating manifestation is a progressive ataxia due to severe muscular discoordination and progressive mental retardation. An immune dysfunction leads to increased susceptibility to infections.

As in other chromosomal breakage syndromes (BS, FA, XP, and Werner's syndrome), A-T patients have a markedly elevated incidence of malignancy (~10%). A distinguishing chromosomal abnormality is a stable, **clonal translocation** involving **chromosomes 7** and **14**. A-T lymphocytes and cultured fibroblasts are exceedingly sensitive to the lethal effects of ionizing radiation and various other DNA-damaging agents causing DNA strand breaks.

Interestingly, despite the **hypersensitivity of A-T cells to X irradiation**, A-T cells display **radioresistant DNA replication initiation** and **chain elongation**. The exact deficiency in the DNA repair system in A-T has not been identified so far.

The radioresistant DNA synthesis can be demonstrated in fetuses with A-T using autoradiographic detection of incorporated tritiated thymidine **in cultured chorionic villus cells**. This approach offers a better approach for the **prenatal diagnosis** of A-T than conventional cytogenetic procedures.

1.5.7.6 BASAL CELL CARCINOMA (BCC) AND OTHER CONDITIONS (p. 107)

Although there is a 25% reduction in maximal DNA repair in the multiple BCC lesions of the **nevoid BCC syndrome**, as reflected in unscheduled

DNA synthesis, the exact lesion has not been identified. Individuals who develop BCC in sun-exposed areas have been shown to accumulate a higher number of **UV-induced pyrimidine dimers**, and to require more time for their removal, than do normal individuals.

Other dermatological conditions such as **dyskeratosis congenita, dysplastic nevus syndrome, trichothiodystrophy, and mer-phenotype**, although presenting symptoms consistent with a deficiency in DNA repair, have not yet been defined in specific molecular terms.

Defective DNA repair systems appear to be responsible for various clinical manifestations in a number of **primary neuronal degenerations** of children (**Friedreich's ataxia, familial dysautonomia, and Usher's syndrome**) as well as of adults (**Huntington's disease, Alzheimer's disease**, and **Parkinson's disease**).

1.5.8 DNA REPAIR AND CANCER (p. 107)

The role of DNA repair in carcinogenesis is more than intuitively relevant. Given the involvement of diverse, and often precisely targeted, repair systems in the restoration of normal DNA and cellular functions, the probability of malignization as a result of defective DNA repair is certainly high. Furthermore, location of damage sites in functional or structural association with **oncogenes** offers one of the plausible mechanisms of tumorigenesis. In fact, the **methylating agent *N*-methyl-*N*′-nitrosourea** (**MNU**) has a surprisingly specific effect on the Ha-*ras* protooncogene in animal systems, producing the **G-to-A base change**.

Chemical carcinogenesis studies provide the following support for the **existence of interaction between two repair-related mechanisms** operating in carcinogenesis: (1) presentation of particular base positions as easy targets to mutagens, and (2) inaccessibility of some DNA sequences to reparatory activity of the cell that allows the lesions to persist. Preferential repair (see also Section 1.5.4) of the UV-induced damage of the more active c-*abl* protooncogene, compared to the c-*mos* gene, underscores the importance of a detailed understanding of the **reparability of protooncogenes**.

The deficiency of **methyltransferase** activities involved in the repair of O^6-methyl-G produced by alkylating agents correlates with the high incidence of tumors in such cells. A **lowered DNA reparatory capacity**, as manifested in the decreased repair of DNA strand breaks, appears to play an important role in the development of some of the most commonly occurring human malignancies, for example, in patients with secondary leukemia.

1.5.9 DNA REPAIR AND AGING (p. 108)

Several theories linking DNA damage and repair with aging have been proposed, implicating

- **oxidative damage** of DNA
- the efficiency of **nucleotide excision**
- **accumulation of mutation defects**, particularly **SSBs** and **DSBs**

Applying a new **single-cell gel assay** methodology capable of evaluating SSBs in individual cells following X-irradiation of human lymphocytes, a distinctly **lower reparatory capacity of the cells from older persons can be demonstrated**. Cells derived from a **progeria (Hutchinson-Gilford syndrome)** patient, characterized by premature aging, display a reduced capacity to repair UV-excision damage.

Chapter 1.6

Drug Resistance

1.6.1 INTRODUCTION (p. 111)

The development of optimal treatment protocols for bacterial infections requires a knowledge of the susceptibility to a chosen drug(s). Testing for this parameter has become routine in clinical microbiological laboratories and is based predominantly on registering the phenotypic response of bacteria to a panel of antimicrobial drugs.

Newer testing methods are being introduced, such as an assay for erythromycin resistance using PCR amplification of the sequences of the *erm* **genes encoding rRNA methylases**, that appear to be universal for the detection of *erm* genes in gram-positive cocci. In *Escherichia coli*, biotin-labeled DNA probes for the **dihydrofolate reductase (DHRF) gene** can assess the plasmid-encoded **trimethoprim resistance**.

In the case of tumors, failure of chemotherapy may become apparent in a clinical situation mostly on a trial-and-error basis. A substantially more efficient and specific identification of potential failures of a chosen type of chemotherapy can be obtained using molecular biological tools.

1.6.2 MDR GENES (p. 111)

The **multidrug resistance (*mdr*) genes** control intrinsic or acquired resistance of tumor cells to a wide range of natural products and hydrophobic drugs such as **vincristine, actinomycin D, and adriamycin**.

The human **P-glycoprotein gene** family is composed of two genes, ***MDR*-1** and ***MDR*-3**, also called ***MDR*-2**, which are similar in structure. The possibility of modulating the expression of the *MDR*-1 gene for clinical purposes by affecting the efficiency of the *MDR*-1 proximal promoter has been demonstrated. Interestingly, the extent of **hypomethylation** of the **CpG-rich sequences** at the 5′ ends of both genes roughly correlates with their transcriptional activity.

Only the *MDR*-1 gene has been unequivocally demonstrated to be responsible for the production of a 170,000-Da cell surface glycoprotein, called **P-glycoprotein**, responsible for the energy-dependent **drug efflux** of various hydrophobic chemotherapeutic agents from tumor cells.

Both *MDR*-1 and *MDR*-2, are located on **chromosome 7q21-1** and encode mRNAs of 4.5 and 4.1 kb, respectively. The MDR family of transmembrane transporters is related to the major histocompatibility (MHC) class II region, linking the presentation of metabolized drugs at the cell surface to the immune system.

Anticancer drug resistance is not solely determined by the action of the "**efflux pump**". The **enhanced DNA repair** resembles the more efficient DNA/chromosome break repair in radioresistant tumor cell lines. The expression of the **c-*myc*** gene has been implicated in the acquisition of drug resistance in transfected NIH 3T3 cells.

The P-glycoprotein gene is **amplified** and **overexpressed** in multidrug-resistant cell lines and the **level of P-glycoprotein mRNA** correlates with the degree of resistance. The **degree of overexpression** of the gene, however, is not necessarily associated with a proportionate increase of the gene copy numbers. The levels of expression of the *MDR*-1 gene and the *MDR*-1 mRNA were consistent with the extent of tumor resistance to daunomycin and/or with the eventual relapses following chemotherapy. This regularity was observed in cancers of the kidney, liver, colon, adrenal gland, and pancreas as well as in neuroblastoma, lymphomas, leukemias, and ovarian cancer.

Expression of the *MDR*-1 gene is observed in most **acute myeloid leukemia (AML)** patients who die early in the course of the disease, indicating that *MDR*-1 gene expression is an unfavorable prognostic factor and that multidrug resistance plays a role in AML.

In **gastric adenocarcinoma**, the expression of *MDR*-1 was heterogeneous in individual cells and its expression was also noted in the intestinal metaplastic and dysplastic cells.

***MDR*-1 transcripts** can be quantitated by slot blot hybridization with *MDR*-1 mRNA as a probe. The finding of **overexpression of the *MDR*-1 gene correlates with cellular resistance to cytotoxic drugs** and corresponds to that found in human **gastric** and **colorectal cell lines, metastatic undifferentiated sarcoma of the liver**, and **human ovarian cancer** cells. However, increased expression of the *MDR*-1 gene is **not necessarily** associated with increased cytotoxic drug resistance. The relationship between the *MDR*-1 gene expression and drug resistance is rather complex.

Flow karyotyping has been used to determine the DNA content of a six-gene amplicon in the multidrug-resistant Chinese hamster cell line CH^rB3. This technique estimates chromosomal aberrations by FCM analysis of propidium iodide-stained chromosome suspensions. The **size of** the amplified DNA visible as a large, **homogeneously staining region (HSR) correlates with the degree of relative drug resistance** of the cells. Most of the human cells lines expressing *mdr* genes, however, do not show amplification of the *MDR*-1 gene, suggesting that **overexpression of the gene occurs by mechanisms other than amplification**.

In the meantime, the characterization of the *MDR*-1 gene product allows the specific chemical circumvention or reversal of P-glycoprotein activity by **verapamil** and related drugs. **Verapamil enhances the retention of antitumor agents** in the cells by competing for the binding sites on the P-glycoprotein transporter.

1.6.3 TOPOISOMERASES IN DRUG RESISTANCE (p. 113)

Defective **topoisomerases I** and **II** (TI and TII) account for some aspects of the resistance of tumor cells to a group of chemotherapeutic drugs targeted to these enzymes. This resistance can be traced to the **rearrangements** in the topoisomerase gene and **hypermethylation** that resulted in defective transcription of the gene. Genetic changes other than affecting chromosome 17, to which the *TII* locus is assigned, may also account for the resistance of TII to the inhibiting actions of the **specific TII-DNA intercalator, m-AMSA**.

1.6.4 CISPLATIN RESISTANCE (p. 113)

Cisplatin is widely used alone or in combinational chemotherapy of cancers. Its most significant limitation, however, is the development of resistance, the mechanism of which is complex. Cytotoxic effects of cisplatin on cells result from the formation of **cisplatin-DNA adducts**. The ability of the cell to repair this damage depends on the availability of repair enzymes and their substrates (deoxynucleotides). Cisplatin affects both **folate** and **methionine metabolism**, thus interfering with the supply of thymidine for repair of damage. The cells relying more on the endogenous folate metabolism are thus less affected by cisplatin toxicity.

The **detoxification role of glutathione (GSH)** in cisplatin resistance, although highly suggestive, remains to be defined, as is the contribution of **metallothionein** and **DNA polymerase β** in contributing to the **repair of cisplatin-induced DNA damage**.

The repair enzymes counteracting the effect of cisplatin-DNA complexes require deoxynucleotides. The **dTMP synthase** cycle supplying thymidine thus becomes rate limiting in DNA synthesis. **DNA repair enzymes are potentially involved in determining cisplatin resistance, but so far no evidence of gene amplification for these enzymes in human cells has been obtained**, although DNA polymerase β mRNA increases severalfold in cisplatin-resistant cell lines.

Amplification of oncogenes **c-*fos*** and **c-Ha-*ras*** has been observed in patients who developed resistance to cisplatin treatment. In ovarian and breast carcinoma cell lines, **c-*myc*** mRNA expression is selectively increased three- to fivefold in cisplatin-resistant cells.

1.6.5 DRUG RESISTANCE RELATED TO METABOLIC DEFECTS (p. 114)

The inherent ability to oxidize a test drug such as **debrisoquine**, **sparteine**, or **dextromethorphan**, and the ratio between the parent drug and its metabolites, can be used to classify individuals as **poor metabolizers**. Such persons are at risk for adverse drug reactions, the phenotype being inherited as an autosomal recessive trait in 5 to 10% of European and North American populations. Routine debrisoquine phenotyping is recommended particularly for psychiatric patients. A link between the **debrisoquine phenotype** and susceptibility to lung cancer has also been suggested.

The poor metabolizer phenotype is caused by the absence of a **cytochrome isozyme, P450IID6**, due to aberrant splicing of its pre-mRNA. Three mutant alleles of the **P450IID6 gene locus (*CYP2D*)** associated with the poor metabolizer phenotype have been identified by RFLPs.

The evaluation of oxidation phenotype **is not recommended for routine application** prior to drug therapy because the clinical importance of the sparteine-debrisoquine polymorphism is still not fully resolved.

1.6.6 CLINICALLY RELEVANT EVALUATION OF DRUG RESISTANCE AT THE GENE LEVEL (p. 115)

In vitro assays of synergistic or antagonistic effects of various cytotoxic drug combinations sometimes fail to reproduce or predict their clinical efficacy. A monoclonal antibody assay for the improved *in vitro* and *in vivo* detection of DNA modifications induced by cisplatin and carboplatin has been described. Analysis of mutations in the **dihydrofolate reductase gene (*DHFR*)** possibly accounting for **methotrexate resistance** in human tumors can be useful.

The evaluation of **P-glycoprotein** clearly has practical significance for the appropriate modification of chemotherapeutic protocols as demonstrated in lymphoma, leukemia, neuroblastoma, and childhood soft-tissue sarcomas. **The *MDR*-1 gene, fortunately, is not activated in all the cancers tested so far.**

A PCR-based assay for the analysis of **cisplatin resistance** is founded on the ability to specifically identify potentially responsible genes such as those encoding **dTMP synthase, dihydrofolate reductase**, and **thymidine kinase;** the assay is highly sensitive, requiring as little as 2 ng of RNA (in the case of the DNA polymerase β gene assay).

PCR assays complementing REA can identify over 95% of resistant individuals. Low levels of *MDR*-1 mRNA have been detected by PCR in the many tumors tested. Yet another PCR assay for measuring the levels of *MDR*-1 mRNA in clinical specimens demonstrated low levels of *MDR*-1 expression that are undetectable by conventional assays in most solid tumors and leukemias tested before chemotherapy.

The **clinical significance** of the low level of expression of *MDR*-1 mRNA (one molecule per cell or lower) detectable only by PCR is still unclear. Even at these low levels, however, the extent of the *MDR*-1 gene expression correlates with the proportion of acute nonlymphocytic leukemia patients who fail to respond to chemotherapy. It remains to be determined, however, what is the minimal level of the *MDR*-1 gene expression capable of conferring clinically observed resistance to chemotherapeutics.

An important finding links the phenomenon of multidrug resistance to **c-Ha-*ras*-1** and ***p53***. The promoter of the human *MDR*-1 gene is a target for the products of both c-Ha-*ras*-1 and *p53*, both of which are associated with tumor progression. Mutant p53 stimulates the *MDR*-1 promoter, whereas wild-type p53 specifically represses its expression.

Evaluation of the *MDR*-1 gene expression can be important for patient management.

Chapter 1.7

Metastasis

1.7.1 PHENOMENOLOGY OF METASTASIS (p. 117)

The study of molecular events involved in the metastatic process is relatively young, although emerging facts deserve attention as potential candidates for laboratory assays in the comprehensive management of oncological patients.

A set of diverse events related to **spread of tumor cells to distant sites from the primary focus** as well as to their **ability to establish secondary tumor growths** independent of the primary tumor comprises the metastatic process.

Biological **heterogeneity** of primary and metastatic tumors is reflected in receptor status, antigenic profiles, sensitivity to chemotherapeutic agents, and many other phenotypic characteristics. The specificity of **intercellular interactions** in the process of metastasis is responsible for targeted metastatic spread occurring only when the transplants are placed in specific organ locations, supporting the so-called **"seed and soil" hypothesis of Paget**. Selective organ- and intraorgan-specific **homing of metastasizing cells** of various tumors cannot be explained solely on the basis of blood vessel and lymphatic topography.

1.7.2 CELLULAR AND TISSUE ELEMENTS IN METASTASIS (p. 118)

A number of conditions contribute to site-specific tumor progression and metastasis:

- the **action of inhibitors** of cellular growth and specific growth factors
- the relative **abundance of mitogenic factors** in specific organ sites

Neovascularization that develops in neoplasias larger than only a few millimeters is supported by the **angiogenic factors** produced by the tumor cells as well as by activated lymphocytes and macrophages. The **passive transport of tumor cells** to distant sites alone is not sufficient for the development of metastatic lesions.

Interaction of tumor cells with **extracellular matrix (ECM)** is an active process in which the coordinated action of various degradative enzymes (e.g., collagenase type IV, plasminogen activator, glycosidases, cathepsin B) facilitates the **invasion**, **intravasation**, and later **extravasation** by tumor cells. In addition to coordinated proteolytic activities, metastasizing tumor cells display **motility** and **tumor cell colony formation**, eventually resulting in the establishment of secondary growths.

The elucidation of specific markers that are closely linked with the metastatic spread of the tumor presents particular interest in patient management.

1.7.2.1 CELL ADHESION IN METASTASIS (p. 118)

The process of **cell adhesion** is one of the critical determinants of site-specific metastasis. **Metastatic cells preferentially adhere to endothelial cells from their target organs** as well as to other ECM elements (**fibronectin, laminin,** and **collagen types IV** and **V**) displaying fine characteristics of organ specificity.

Adhesion of metastatic tumor cells is mediated via a limited number of **specific peptide sequences.** Screening patient specimens for genetic determinants in tumor cells responsible for such determinants may reveal a propensity to metastasis early in this process and, eventually, serve to develop specific antimetastatic tools. In fact, expression of the genes for **elastase** and **type IV collagenase** appears to correlate significantly with the potential for pulmonary metastasis. A receptor for **elastin** on the surface of tumor cells has been identified and its expression appears to correlate with organ-specific tumor metastasis in the lung.

1.7.2.1.1 Integrins (p. 118)

The attachment of tumor cells to the **basement membrane** surface is mediated by integrin and nonintegrin-type **receptors** on the tumor cell surface. Integrins are a family of related **integral membrane glycoproteins** acting as the **primary mediators of cell-extracellular matrix adhesion,** and also as one of the many types of **molecular determinants in cell-cell adhesion.**

The **bridging** of the intracellular cytoskeleton with the extracellular matrix occurs through integrins and is cell type specific. **Extracellular matrix ligands** for integrins include

- **fibronectin**
- **fibrin** and **fibrinogen**
- **laminin**
- **collagens**
- **entactin**
- **tenascin**
- **thrombospondin**
- **von Willebrand factor**
- **vitronectin**

Intercellular adhesion proteins such as **ICAM-1, ICAM-2,** and **VCAM-1** also act as ligands for integrins.

The **integrin cDNAs** have been cloned and the amino acid sequences of many subunits have been derived. The highly active **tripeptide Arg-Gly-Asp (RGD)** peptides interfere with specific adhesion functions. Such proteins have been termed "**disintegrins**".

Correlations between the **level of expression of specific integrins** and phenotypic behavior of various tumors are being established. Cytokines and growth factors play a role in modulating the expression of integrins.

1.7.3 PROTEOLYTIC ACTIVITY OF TUMORS (p. 119)

The coordinated, highly **localized**, and **multicomponent proteolytic activity** of invading tumor cells is the intrinsic characteristic of metastatic tumors and is currently a focus of intense investigations. Of particular interest is the regulation of expression of the **matrix metalloproteinase gene family**, which includes three categories of enzymes:

- **interstitial collagenases**
- **type IV collagenases (gelatinases)**
- **stromelysins**

Positive correlation between type IV collagenase activity and tumor cell invasion acquires **practical significance**: an increase in this enzyme activity, particularly associated with the appearance of **72-kDa type IV collagenase mRNA** species, reflects the genetic induction of a metastatic phenotype. Its overexpression can be interpreted as a marker of many invasive and metastatic human cancers.

The opposite effect on the invasive phenotype can be seen in the expression of **proteinase inhibitors**. A group of **tissue inhibitors of metalloproteinases (TIMPs)** may function as **metastasis suppressors** and the degree of their expression can also characterize the metastatic potential of a given tumor.

1.7.4 ROLE OF ONCOGENES, GROWTH FACTORS, AND TUMOR SUPPRESSOR GENES IN METASTASIS (p. 119)

Oncogenes (such as *ras* and *myc*), capable of conferring a **malignant phenotype**, do not necessarily lead to the appearance of a **metastatic phenotype**. Invasion and metastasis require the activation of genes distinct from those accounting for uncontrolled growth of tumors.

Among those are

- **c-*fos***, apparently capable of suppressing metastasis indirectly through the histocompatibility system
- genes encoding epidermal growth factor receptor (**EGFR**) and platelet-derived growth factor (**PDGF**), the enhanced expression of which is correlated with invasiveness
- a number of **oncogenes other than *ras***, although probably acting via a common *ras*-dependent pathway (***mos***, ***raf***, ***src***, ***fes***, ***fms***)

1.7.4.1 METASTASIS SUPPRESSOR GENES (p. 119)

Recent somatic cell hybridization and DNA transfection experiments as well as the isolation of cDNA clones by subtractive hybridization point to the existence of **metastasis suppressor genes** that inhibit invasion and metastasis.

Particularly interesting is the differential expression of the ***nm*23 gene** in several experimental metastatic models. The level of expression of the *nm*23 gene in highly metastasizing cells (cultured and in infiltrating ductal breast carcinomas from patients with lymph node metastases) is reduced by an order of magnitude compared to its level in cells with low metastatic propensity. **The loss of *nm*23 mRNA has been associated with poor survival in breast cancer patients.**

Transfection of melanoma cells with *nm*23 leads to a reduced incidence of primary tumor formation, marked reduction in tumor metastatic potential, and altered responses to the cytokine **transforming growth factor β_1**, suggesting that ***nm*23, which plays a role in signal transduction in cell-cell communication, is capable of suppressing several aspects of carcinogenesis including metastasis.**

At least two cellular functions known to be associated with cancer and development, **microtubule assembly/disassembly** and **signal transduction through G proteins**, can be affected by defective **nucleoside diphosphate** (NDP) kinases. It is believed that members of the NDP kinase family, one type of which may be *nm*23, could act as positive or negative regulators of oncogenic, developmental, or metastatic processes.

Part 2
Gene Level Evaluation of Selected Disorders of the Heart, Lung, Gastrointestinal Tract and Tumors of the Head and Neck

Chapter 2.1

Atherosclerosis and Coronary Heart Disease

2.1.1 INTRODUCTION (p. 163)

The key role of **lipids** and **lipoprotein disorders** in the development of atherosclerosis and associated vascular abnormalities has been supported by epidemiological evidence. Other contributing factors — genetic, male gender, hypertension, obesity, family history of CHD, etc. — account for the **multifactorial nature** of atherosclerosis. The role of lipids and **sex hormones** in modulating the effects of elevated lipid levels contributing to atherosclerosis still remains poorly defined in **women**. Much is to be learned about the importance of normalization of cholesterol level in **children** and the **elderly**, and about the contribution of various diseases to hyperlipidemias. On the other hand, **stabilization** and even **regression of advanced atherosclerotic lesions** as a result of lipid-lowering treatments have been reported in animal models and humans.

The diagnostic armament of practicing physicians in the near future will likely extend beyond evaluation of lipids, to lipoproteins, including Lp(a), apolipoproteins A, B, and E, and in selected cases also to the analysis of the *apo* A, B, and E gene polymorphisms, and of lipoprotein receptor gene defects.

2.1.2 SYNOPSIS OF MOLECULAR BIOLOGY OF DYSLIPOPROTEINEMIAS (p. 163)

Lipid homeostasis in the body is controlled by proteins classified into three major categories:

- **apolipoproteins**, the integral proteins that determine the structure of lipid particles and their specific interaction with other molecules and tissues; examples are **A-I, A-II, A-IV, C-I, C-II, C-III, D, E**, and **apo(a)**
- **lipid-modifying proteins** and **enzymes**, affecting primarily the lipid moiety of lipid particles; for example, lipoprotein lipase (**LPL**), hepatic triglyceride lipase (**HTGL**), lecithin-cholesterol acyltransferase (**LCAT**), cholesterol ester transfer protein (**CETP**)
- proteins and enzymes of cellular lipid metabolism involved in **metabolic, catabolic**, and **transport** functions. Examples include the lipoprotein receptors [low-density lipoprotein (**LDL**) **receptor, chylomi-**

cron remnant receptor, and **scavenger receptor**, **lipoxygenases**, **phospholipases**, etc.

The **genetic component** is important and, in many situations, is predominantly responsible for the observed lipid disorders.

2.1.3 GENETIC DEFECTS IN EXOGENOUS LIPID TRANSPORT (p. 164)

Exogenous lipid transport includes the breakdown of dietary fats in the intestine and the formation of triglyceride-rich cholesterol-containing particles called **chylomicrons**. The proteins of chylomicrons include the **intestinal form** of apo B (**B-48**), **A-I**, and **A-IV** located at the surface of chylomicrons. Chylomicrons acquire **apo C-II** and **apo E**, gradually cleaving the lipids into **chylomicron remnants** that are catabolized in the liver. Genetic defects have been recognized in chylomicron secretion, processing, and clearance of the remnants.

Chylomicron secretion is defective in two rare conditions, **abetalipoproteinemia** and **homozygous hypobetalipoproteinemia**, occurring at a frequency of less than 1:1 million. Clinical manifestations in both conditions are related to fat malabsorption and consequent fat-soluble vitamin A, D, E, and K deficiencies leading to ataxic neuropathy, retinitis pigmentosa, and acanthocytosis of the red blood cells (RBCs).

The absence of apo B precludes the formation of chylomicrons and very low-density lipoprotein (VLDL) and LDL particles. Genetic linkage studies identified defects in the *apo* B gene in **homozygous hypobetalipoproteinemia**. In the other condition, **abetalipoproteinemia**, no defect of the *apo* B gene has been found and the genetic basis of the disease is not understood, and a posttranslational defect or a mutation in other gene(s) encoding factors involved in lipoprotein assembly or secretion appear to be at fault. A form of **familial hypobetalipoproteinemia** has also been described, in which a dominantly transmitted mutation in a gene other than the *apo* B gene is responsible for the low plasma cholesterol and apo B-carrying lipoproteins.

Processing of chylomicrons is defective in **LPL deficiency** and **apo C-II deficiency** — two other rare conditions occurring at a frequency of less than 1:1 million. *LPL* gene polymorphisms appear to be associated with high-density lipoprotein (HDL) and total cholesterol levels. A number of specific mutations in the *LPL* gene result in the synthesis of enzymatically inactive proteins. The **accumulation of large chylomicrons** is not associated with increased risk of atherosclerosis, but rather leads to **pancreatitis** and **eruptive xanthomas**.

2.1.3.1 APOLIPOPROTEIN E (p. 165)

Apo E is a 299 amino acid polypeptide with the receptor-binding region in chylomicrons; VLDL and HDL particles participating in receptor-mediated clearance of these lipoprotein particles. Mutations of the *apo* E gene affect receptor recognition and underlie **type III hyperlipoproteinemia**.

Three common alleles of the *apo* E gene have been recognized, **E4**, **E3**, and **E2**, which can produce six phenotypes. The **most common allele is E3** (77% frequency). The product of the allele E2 fails to bind to receptors. The **most common phenotype is E3/3** (60% of the population).

The three isoforms of the *apo* E gene product, recognizable by isoelectric focusing, differ by **cysteine-arginine interchanges** at residues 112 and 158 in the protein. Apo E2 contains cysteine at both positions, thought to contribute to the poor recognition of the specific receptor.

Although type III hyperlipidemia found in the E2/2 phenotype, occurs in 1 in 5000 individuals, which characteristically have elevated fasting levels of cholesterol and triglycerides, these patients may have either low, normal, or elevated LDL cholesterol levels. These people have an increased incidence of CHD or peripheral vascular disease and xanthomas.

Another variant of defective chylomicron clearance does not present with fasting hyperlipidemia, the so-called **normolipidemic E2/2** individuals. They constitute up to 1% of the population and although their chylomicron level is elevated their LDL level is decreased, and they neither develop xanthomas, nor have an elevated risk of CHD.

And, finally, the last group of people with defective chylomicron remnant clearance includes **E2 heterozygotes** (e.g., E4/2 and E3/2). These individuals have a milder elevation of chylomicron remnants and intermediate-density lipoprotein (IDL) than that found in E2 homozygotes and do not have an increased risk of CHD. The Apo E phenotype affects the LDL cholesterol level in a number of ways, in which an **interplay of E4 and E3 gene expression accounts for a wide spectrum of biochemical and clinical phenotypes**.

The **E7** and **E5** mutations have been identified in Japanese, not Caucasians, and appear more frequently in individuals with hyperlipidemia and/ or artherosclerosis.

2.1.3.2 APOLIPOPROTEIN E POLYMORPHISM: POPULATION STUDIES (p. 166)

A higher frequency of the E4 allele and a lower frequency of the E2 allele are observed in the Finnish population than in others. Total cholesterol, LDL-cholesterol, and apo B levels are markedly higher in the E4/4 homozygotes, and tend to be lower in the E2 heterozygotes.

The frequency of the E3 allele is highest among the Japanese and frequency of the E2 allele is highest in New Zealand. In the Netherlands the *apo* E gene frequency is similar to that in Europe and the U.S. A strikingly high frequency of the E4 allele has been observed among blacks in the U.S. and Nigeria that is almost twice as high as that reported in other populations. The effect of the E4 allele is to lower apo E levels, and the average effect of the E2 allele is to raise apo E levels by 0.95 mg/dl, lower apo B levels by 9.46 mg/dl, and lower total cholesterol levels by 14.2 mg/dl.

2.1.3.3 *apo* E GENE EXPRESSION, ATHEROSCLEROSIS, AND CHD (p. 167)

Apo E affects the level of plasma lipid and lipoprotein levels in normal and hyperlipidemic individuals. **Apo E polymorphism** plays a role in the **predisposition to atherosclerosis**. The **practical value** of determining apo E polymorphism is also supported by the observed influence of this polymorphism on the **serum cholesterol response to dietary intervention**.

Type III hyperlipidemia with the **E2/2 phenotype was overrepresented in the group of patients with myocardial infarction (MI)**, even though the frequency of the E2/2 phenotype was the same in MI survivors and controls. The **ε2 allele** was thought to **protect against atherosclerosis** in the absence of hyperlipoproteinemia. In contrast, the **ε4 allele may confer an increased risk of MI at an earlier age**; however, a host of other factors may modulate the effects of apo E polymorphism. The evidence, however, is inconclusive in patients undergoing **coronary angiography** for documented or suspected CHD, where the apo E phenotype distribution appears to be essentially similar in patients with and without CHD.

Nevertheless, a study in Finnish patients with angiographically confirmed CHD demonstrated that the frequency of the ε4 allele was 1.4-fold higher than that in the normal Finnish population and twice as high as in other Caucasian populations. Among Japanese patients with CHD the frequencies of ε2 and ε4 alleles were significantly higher than in controls.

One of the proposed reasons for the apo E polymorphism-related differences in plasma and LDL cholesterol concentrations is believed to be due to **differential responses to dietary lipids**, suggesting that the **effect of the *apo* E genotype on plasma cholesterol may be modulated by dietary lipids**.

At the tissue level, **cholesterol feeding was found to markedly increase *apo* E gene expression** in atherosclerotically changed aortas.

2.1.3.4 *apo* E GENOTYPING METHODS (p. 168)

The three major apo E isoforms, E2, E3, and E4, and structural variants of apo E can be evaluated by **isoelectric focusing (IEF)** of VLDL proteins, IEF using **thrombin inhibitors**, or by **two-dimensional gel electrophoresis**. Separation of isoforms is **based on charge differences** related to the presence of **cysteine** and **arginine** at specific sites in the apo E molecule. These methods are quite laborious and have certain limitations.

2.1.3.4.1 PCR in Apo E Assays (p. 168)

The analysis of *apo* E alleles at the DNA level allowed the **demonstration of structural alterations** affecting apo E isoform **receptor-binding capacities**. The routine detection of *apo* E alleles is possible using allele-specific oligonucleotides (**ASOs**) in hybridization and **PCR** amplification. The discrimination of ε3-ε4 and ε2-ε3 alleles by the ASO hybridization approach is based on the **difference in thermal stability** between the hybrids formed by fully matched 19- or 20-mer oligonucleotides and duplexes with a minor G-T mismatch or duplexes with a more destabilizing mismatch. **PCR amplification** produces much stronger signals that are less affected by nonhomologous hybridization. **Combining PCR with Southern blotting** allowed the identification of *apo* E isoforms and could be modified to recognize any mutation in the gene of interest.

An improved PCR procedure followed by ASO hybridization in a **slot-blot** format identifies only a **single amplified apo E-specific product**, rather than a number of different-sized bands, most of which may be unrelated to apo E.

The benefits of combining DNA amplification with ASOs for the determination of *apo* E genotype are numerous:

- it is **less labor intensive** than IEF
- it can be performed on a **much smaller sample**
- a **large number of samples** can be assayed
- **unequivocal genotype assignments** can be made

apo E genotyping using **automated sequence analysis of PCR products** has also been used. PCR amplification of the *apo* E gene region harboring the two variable nucleotides giving rise to **Arg → Cys interchange** in apo E isoforms has been combined with restriction digestion by *Hha*I and ASO hybridization. A **nonradioactive** PCR-based method for *apo* E genotyping using biotin label, as well as a **one-step primer extension reaction** have also been developed.

A technique described as **restriction isotyping** (**restriction enzyme isoform genotyping**) avoids the use of costly and labor-intensive hybridization and sequence procedures and offers a simpler and faster method for typing the common *apo* E isoforms. Each genotype can be easily distinguished by a **unique pattern of fragment sizes** in all homozygous and heterozygous patients. **Visualization** of these patterns in stained gels does not required radioactive labels as in the case of ASO probes. Although this approach cannot detect rare unknown mutations, its simplicity and high efficiency make it a **convenient method for laboratories performing evaluation of *apo* E gene effects**. On the other hand, automated sequencing of PCR amplification products is suitable for the **identification of any mutations** in the target DNA sequences.

A **novel PCR technique** has been successfully applied to identify all possible combinations of the three *apo* E alleles. It first amplifies the DNA fragment of interest with a primer biotinylated at its 5′ end, and the amplified product is then immobilized on an avidin matrix and double strands are separated. Then, **a one-step primer extension** is performed to identify the variable nucleotide using a single nucleoside triphosphate (radioactively or nonradioactively labeled).

Several **other PCR-based protocols** have been described for the analysis of *apo* E genotypes that do not use radioactive labels and can be performed in diagnostic clinical laboratories.

2.1.4 GENETIC DEFECTS IN ENDOGENOUS LIPID TRANSPORT (p. 169)

The **endogenous lipid transport** starts with the formation of VLDL in the liver, followed by hydrolysis of its triglycerides by LPL, yielding IDLs. These are then either taken up by the liver, where they are recognized by **hepatic LDL receptors** that bind apo E on the IDL surface, or further processed by HTGL to LDLs that are cleared from the plasma by the LDL receptor. VLDL contains the liver form of apo B (**apo B-100**), apo C, and apo E associated with triglycerides and cholesterol. Seventy percent of LDL particles enriched in cholesterol are cleared from the plasma by LDL receptors on hepatocytes and extrahepatic cells.

2.1.4.1 STRUCTURE, FUNCTION, AND MUTATIONS OF APO B (p. 169)

Apo B is essential for the assembly of VLDL particles and constitutes the protein component (ca. 25% by weight) of LDL particles. It is situated at the outer shell of a lipoprotein particle. Human apo B is a glycoprotein existing in two forms, **hepatic** (**B-100**) and **intestinal** (**B-48**). B-100 is a simple polypeptide (apparent M_r 550,000 Da) encoded by a single-copy gene and produced primarily in the liver. B-48 is approximately one half that molecular mass and is synthesized primarily in the small intestine.

In contrast to other apolipoproteins such as apo A-I, A-II, and E, which transfer freely between lipoprotein particles, **apo B-100 remains associated with the same particle**. Apo B-100 appears to interact with lipids in a manner resembling that of intramembrane proteins.

Modification of LDL particles through oxidation and acetylation and the association of apo B-containing lipoproteins with acidic glycosaminoglycans or specific proteoglycans contribute to the atherogenicity of these particles.

The **sequencing** of apo B-100 DNA allowed the determination of **variation in both the quantity** and **quality of *apo* B genes**. Five pairs of allelic variants designated Ag(c)/g, (x)/(y), (t)/(z), (a1)/(d), and (h/i) have been immunologically defined that result from variations in the apo B protein. One of these determinants, **Ag(x)**, has been linked to elevated cholesterol and triglyceride levels. RFLPs of *apo* B-100 for *Xba*I have been associated with myocardial infarct.

An immunochemical polymorphism, Ag(c/g), in a Finnish population is associated with elevated cholesterol levels. A comparison between Ag and DNA polymorphisms may be used to search for significant correlations between certain *apo* B alleles and lipid levels, atherosclerosis, and CHD.

Informative RFLPs of the *apo* B gene have been found in the **tandem repeat sequences in hypervariable minisatellites** in the *apo* B gene. Several more polymorphisms of the *apo* B gene have been identified (*Ava*II and *Bal*I), and the molecular basis and frequencies of the two previously reported RFLPs (*Hinc*II and *Pvu*II) were characterized. These may be useful in establishing disease associations and in linkage analysis studies revealing the contribution of this locus to atherosclerosis susceptibility.

While some studies demonstrated a **positive correlation** between *apo* B polymorphisms and CHD, other researchers **failed to observe** an association in European and Asian populations. Importantly, variation in the *apo* B gene has been found to express itself **differently in different populations.**

It appears that the **weight of the accumulated evidence favors an association between *apo* B gene polymorphism and CHD**. Although a correlation between the polymorphisms and CHD risk has been repeatedly established, the single-site polymorphisms observed could hardly provide more information about the CHD risk of a patient than conventional lipid parameters.

A high-resolution method using **PCR amplification of hypervariable alleles of a minisatellite region** close to the *apo* B gene has been used to demonstrate that alleles containing 38, 44, 46, or 48 hypervariable elements displayed an association with CHD, elevated cholesterol and apo B among patients, and with elevated total triglycerides in the controls.

Thus, **although the *apo* B gene polymorphisms do appear to be associated with CHD and, in some cases, with elevated levels of atherogenic lipids, *their clinical utility at this time is not clear*. So far, this approach appears to offer information essentially comparable to that gained in serum lipid and apolipoprotein studies.**

2.1.4.2 HETEROZYGOUS HYPOBETALIPOPROTEINEMIA AND ABETALIPOPROTEINEMIA (p. 171)

The most important genetic disorders resulting in clinically important dyslipidemias have been recognized to affect **VLDL secretion** or **formation**, and **LDL clearance**.

Defective VLDL secretion linked to the specific mutations in the *apo* B gene occurs in **heterozygous hypobetalipoproteinemia**, a condition occurring with a frequency of 1 in 1000 in the general population. Affected individuals have a lowered LDL cholesterol level (30 to 50% of normal) and, in effect, have a **lower risk of atherosclerosis**. Selective deficiencies of either hepatic (apo B-100) or intestinal (apo B-48) forms of apo B have also been described.

Abetalipoproteinemia, characterized by low or undetectable concentrations of apo B and of the apo B-containing lipoproteins, is associated with severe fat malabsorption, acanthocytosis, progressive spinocerebellar degeneration, and retinopathy. The **molecular defect causing this condition has not been defined yet**. Apparently a defect in another gene involved in the synthesis or secretion of apo B-containing lipoproteins is responsible for this disorder.

In a related condition, **Anderson's disease**, the affected individuals suffer from fat malabsorption, reduced plasma levels of cholesterol, triglycerides, and apo B, A-I, and C. The results of amplification of genomic DNA by PCR followed by restriction endonuclease analysis (REA) of the amplified products derived from **regions flanking the *apo* B gene 3′ hypervariable locus** were incompatible with involvement of the *apo* B gene. A **posttranslational defect** or **a mutation in another gene** encoding a protein apparently involved in the processing, assembly, or secretion of apo B-containing lipoprotein particles is suggested.

2.1.4.3 FAMILIAL DEFECTIVE APO B-100 (FDB) (p. 172)

Another category of conditions with defective endogenous lipid transport includes two major disorders — **familial defective apo B-100 (FDB)** and **familial hypercholesterolemia (FH)**. Both disorders are characterized by defective LDL clearance. The first condition occurs at a frequency of 1 in 500 to 1 in 1000 in the general population and 1 in 50 to 1 in 100 in persons with upper-decile LDL cholesterol. **No homozygotes for FDB have been found so far**. At a frequency of 1 in 500 in the general population, similar to that of FH, **FDB is one of the most common single-gene mutations leading to a clinical abnormality**.

Unlike the situation with FH, where no single mutation is responsible for the observed phenotype, in FDB the **mutation at codon 3500** is by far the most common one.

2.1.4.4 PCR IN THE ANALYSIS OF *apo* B DNA POLYMORPHISM (p. 173)

A general method has been described for rapidly and accurately typing the 3′-flanking region of the *apo* B gene, composed of a variable number of tandemly repeated short A+T-rich DNA sequences (VNTRs). Typing different regions of the *apo* B gene-related DNA sequences by a combination of PCR amplification and hybridization with ASOs has been performed to analyze the association between the *apo* B gene polymorphisms and the apo B Ag immunological epitopes. The PCR approach is substantially faster and less expensive than laborious Southern analyses, which are poorly suitable for large-scale population studies. Other PCR-based identifications of *apo* B-100 mutations use sequence-modifying primers creating recognizable RFLPs in the amplified product, or the **amplification refractory mutation system (ARMS)** with allele-specific primers exactly matching the target sequence.

2.1.4.5 FAMILIAL HYPERCHOLESTEROLEMIA (FH) (p. 173)

Defects in the **LDL receptor** also account for the elevation of circulating LDL particles in another condition inherited as a dominant trait — **FH** (frequency, 1 in 500 in the general population). Patients with FH develop xanthomas and an increased incidence of **heart attacks at younger ages**. **FH heterozygotes** with one normal gene and one defective gene have long been shown to have one half of functional LDL receptors on the surface of their cells. **FH homozygotes** had virtually no LDL receptors and are found at a frequency of 1 in 1 million.

The **LDL receptor precursor** (120,000 Da) is transported to the Golgi apparatus, where maturation of the carbohydrate chains occurs to increase the molecular weight of the protein to 160,000 Da. This **mature receptor** then migrates to specialized regions of the plasma membrane called **coated pits**. The receptor with its ligand enter the cell by **receptor-mediated endocytosis**, the receptor dissociates from the ligand in an endosome, and **recycles** back to the cell surface, capable of a new round of LDL internalization.

The cholesterol released from the LDL particles is used for the synthesis of cellular membranes, and it influences the expression of several genes involved in cholesterol metabolism, including the LDL receptor itself. **Six domains** have been discovered in the LDL receptor protein.

At least **four major classes of mutations** are responsible for the **FH phenotype**:

- **Class 1 mutations** preclude the synthesis of the LDL receptor
- **Class 2 mutations** encode LDL receptors inefficiently transported from the endoplasmic reticulum (ER) to the cell surface
- **Class 3 mutations** affect the ability of LDL receptor proteins to bind LDL particles
- **Class 4 mutations** produce LDL receptors that fail to cluster in coated pits on the cell surface and fail to internalize bound LDL

The **gene for the LDL receptor** has been cloned, a number of mutations have been identified in it, and their structural basis defined. The LDL receptor mRNA has an 3′-untranslated region of 2 to 5 kb, containing three copies of the middle-repetitive *Alu* family of DNA sequences; this is unusual for exons.

The LDL receptor gene has been assigned by *in situ* hybridization (ISH) to bands p13.1-13.3 of **chromosome 19**. In individuals with the heterozygous and homozygous forms of FH up to 16 mutations are recognized, which include large deletions and insertions as well as single-base changes resulting in nonsense of missense codons.

Although **variability of mutation** of the LDL receptor gene resulting in the FH phenotype has been observed, some **predominance of one type of mutation does occur**, such as the so-called **French Canadian deletion**. Other common LDL receptor mutations were identified as **missense mutations** found in 14% of affected individuals. Mutations of the LDL receptor gene among **Japanese kindreds with homozygous FH** were found to affect all four biochemical classes of mutations (synthesis, processing, ligand-binding activity, as well as internalization of the LDL receptor).

It appears that a large number of the LDL receptor gene mutations will be discovered that are associated with FH. Therefore, in a genetically heterogeneous population it may be difficult to achieve a diagnosis of FH at the DNA level. In fact, the potential **usefulness of RFLPs for diagnosis of FH** has been addressed, and no difference in the relative allele frequency between the group of FH patients and controls for any of the polymorphisms has been detected. In that study, defective LDL receptor genes occurred on six different chromosomes determined by the four RFLPs, suggesting that at least six different genetic defects could cause FH in the patients tested. Using linkage analysis, however, appropriate diagnosis could be established in over 90% of informative families using these four RFLPs.

The **phenotypic expression of FH in homozygotes** appears to be dominated by the consequences of the genotypic variation at the LDL receptor locus. In the case of **heterozygous FH**, however, the consequences of the underlying mutation are compounded by other influences such as gender, diet, and other forms of genetic polymorphism, including the apo E phenotype, which determine the phenotypic characteristics. In fact, **female FH patients** with the apo E3/3 phenotype were also found to respond better to **simvastin treatment** than male FH patients with the same apo E3/3 phenotype, as judged by LDL cholesterol level. Furthermore, the existence of a **suppressor gene** that ameliorated or suppressed the hypercholesterolemic effect of the LDL receptor mutation has been suggested.

PCR assays can be used to detect specific mutations in the LDL receptor gene accounting for the FH phenotype (see Section 2.1.4.4).

2.1.4.6 Lp(a) (p. 175)

In **Lp(a)** a large glycoprotein, **apo(a)**, is **covalently bound** to the apo B-100 moiety of LDL. An **association of Lp(a) with atherosclerosis and CHD** has been established in **epidemiological studies**, although the correlation between Lp(a) levels and lipid and lipoprotein levels is weak or absent. The role of Lp(a) heterogeneity and its relationship to atherogenicity is not yet clear.

The **lipid** and **protein composition of Lp(a)** is quite similar to that of LDL. The Lp(a) particle varies widely in size among individuals, ranging between 400 and 700 kDa, and at least six different Lp(a) protein species have been identified, using SDS-PAGE and immunoblotting, that are inherited in a codominant manner. The **apo(a) size polymorphism** displays a varying association with the Lp(a) concentration in different ethnic groups. Although **no effects of diet on Lp(a) level** can be demonstrated, improved glycemic control in insulin-dependent diabetes mellitus decreases Lp(a) level. **Decrease in Lp(a) concentration in the blood is not mediated via its removal by LDL receptors.**

2.1.4.6.1 Lp(a) and Plasminogen (p. 176)

Both the gene for Lp(a) and the protein have a striking resemblance to plasminogen. The primary amino acid and DNA sequences of apo(a) show the presence of one kringle V-like domain, a serine protease domain, and between 15 and 40 tandem repeats resembling the kringle IV domain of plasminogen. The structural gene for the Lp(a) glycoprotein [apo(a)] is closely linked to the plasminogen locus on chromosome 6. Assessment of

the kringle IV:kringle V ratio is proposed as an independent DNA marker for atherosclerosis risk associated with Lp(a).

Based on the structural similarity of Lp(a) to plasminogen, it has been proposed that Lp(a) might have both **proatherosclerotic** and **prothrombotic** properties. Similar to the higher atherogenicity of acetylated or oxidized LDL particles, the uptake of **modified Lp(a)** particles by cells is more efficient. It appears that Lp(a) may cross the endothelium of blood vessels by **nonreceptor-mediated mechanisms**, possibly due to a direct action of Lp(a) on endothelial functions.

The **potential thrombotic role of Lp(a) in atherogenesis** has been suggested by

- its **structural characteristics**
- the ability of Lp(a) to compete *in vitro* with the **binding of plasminogen** to fibrinogen on fibrin monomers
- the ability of Lp(a) to compete with the **streptokinase-mediated activation of human plasminogen**
- the ability of Lp(a) to compete with the **tissue plasminogen activator-mediated lysis of fibrin clots**
- the reversible **binding of Lp(a)** to **tetranectin**, a plasma protein that itself binds reversibly to plasminogen and enhances plasminogen activation by tPA

In clinical practice, however, even drastically elevated levels of Lp(a) were found not to affect the recanalization of coronary vessels in MI patients treated by **pro-urokinase** or **alteplase (rt-PA)** thrombolytic therapy. With respect to Lp(a) interaction with cells, Lp(a) is able to compete for the binding of plasminogen to the plasminogen receptor on endothelial cells and macrophages at Lp(a) concentrations comparable to those found in the circulation. Lp(a) has been convincingly shown to regulate the endothelial synthesis of a major fibrinolysis involved protein, **plasminogen activator inhibitor 1 (PAI-1)**: by increasing PAI-1 expression, Lp(a) may support a specific prothrombotic endothelial cell phenotype.

One of the unresolved properties of Lp(a) is the **independence of its plasma concentration of the influence of environmental factors** such as diet and drugs, which are known to markedly affect other lipoproteins. At Lp(a) levels of 0.25 to 0.4 mg/ml, which are associated *in vivo* with an increased risk of CHD, approximately 16 to 24% of plasminogen-binding sites can be occupied by Lp(a). **By blocking plasminogen binding to vascular endothelial cells, Lp(a) apparently inhibits fibrinolysis and promotes surface coagulation**. Although isolated reports of Lp(a) expressing protease activity appear, at this point **the coincidental nature of the similarity of apo(a) and plasminogen cannot be fully excluded**.

2.1.4.7 Lp(a) IN CHD (p. 177)

The **frequencies of specific apo(a) phenotypes** determined by immunoblotting with monoclonal antibodies are different between patients with and without CHD. **The allele** Lp^{S2}**, which is associated with high**

Lp(a) levels, is almost three times as frequent among patients with CHD. On the contrary, the allele Lp^{S4}, which is associated with low Lp(a) levels, is almost twice as frequent among patients without CHD. Likewise, apo(a) levels in FH patients with CHD are higher than in patients without CHD.

Thus, it appears that **the level of Lp(a) and the phenotypic Lp(a) pattern are *inherited characteristics predisposing an individual to CHD* and are not affected by race, sex, diet, or age**.

Attempts to find effective **pharmacological agents** capable of reducing the Lp(a) level in plasma are being made. Limited success has been reported with **neomycin** and **niacin**, which interfere with Lp(a) synthesis. A marked reduction of the Lp(a) level has been achieved in patients receiving ***N*-acetylcysteine** (NAC) to an extent not seen with any other treatment. Interference of NAC in the measurement of Lp(a) has been suggested, however, and caution in considering NAC an efficient Lp(a)-lowering agent was advocated.

In a clinical setting, the **measurement of Lp(a)** is offered by selected laboratories. The heterogeneity of Lp(a) combined with the scarcity of diagnostic methods at present, most of which rely on immunological detection of the lipoprotein complex, require significant efforts to be directed toward **rigorous standardization of the procedure.**

2.1.5 GENETIC DEFECTS IN REVERSE CHOLESTEROL TRANSPORT (p. 178)

The main lipoprotein particle involved in transfer of cholesterol from peripheral tissues back to the liver for excretion is **HDL**. The principal protein of the nascent HDL is **Apo A-I** which attracts free cholesterol from extrahepatic cells. **LCAT** esterifies this cholesterol in plasma, and apo A-I acts as a cofactor to the enzyme.

The recognized **genetic defects** of reverse cholesterol transport are few in number. A homozygous form of apo A-I deficiency resulting in defective HDL production leads to severe premature CHD and corneal opacities. Heterozygotes have half-normal HDL levels, and **three types of mutations** are associated with this condition:

- in **type I (apo A-I/C-III)**, both apo A-I and C-III are deficient
- deletion of the entire locus leads to the **type II mutation**, in which apo A-I, C-III, and C-IV are all deficient
- in **type III**, apo A-I is deficient due to a small insertion in the *apo* A-I gene leading to premature termination of translation of the *apo* A-I mRNA

Other mutations have also been described resulting in **defective HDL metabolism, a complete absence of HDL**, or altered **processing of HDL particles**. Among the latter is **Tangier disease** with rapid clearance of HDL from circulation, however, the exact underlying genetic alteration has not been established so far. **Frame-shift** and **nonsense mutations** leading to

apo A-I deficiency and premature atherosclerotic changes have been described in the *apo* A-I gene.

A **variant allele (*S2*)** of the *apo* A-I/C-III/A-IV complex is associated with high plasma HDL levels in some populations, particularly in patient groups with premature coronary heart disease.

In patients with **peripheral vascular disease**, within-individual variations in serum cholesterol seem to be associated with DNA polymorphisms at the *apo* B and A-I/C-III/A-IV loci.

Polymorphisms affecting regulatory elements of the *apo* A-I/C-III/A-IV region do account for **hyperalphalipoproteinemia**. The strong **antiatherosclerotic potential of this type of HDL alteration** has been subsequently demonstrated. Polymorphisms in the gene coding for CETP, particularly the *Taq*I, appear to be related to plasma HDL-cholesterol and transfer protein activity.

Modulation of lipoprotein gene expression by **environmental influences** has been addressed at the gene level, particularly as related to *apo* C-III genotyping.

A more intimate role in HDL metabolism appears to emerge for a group of **steroid hormones** since the finding of regulation of the *apo* A-I gene by a novel member of the steroid receptor superfamily, protein **ARP-1**, which is capable of down regulating *apo* A-I gene expression.

At this time, **only general trends toward RFLP allele—lipid level association can be observed in population studies**. Some studies suggest that a **predisposition to the development of CHD may be enhanced by variants in the *apo* A-I/C-III/A-IV gene complex or in the associated regulators**, affecting, for example, the apo A-I catabolic rate. Others, however, **find no such association.**

Significant variability of the above-discussed observations is due, in part, to

- a marked **variation of the allele frequency** of many of the RFLPs so far examined in different populations with significant racial and ethnic heterogeneity
- the **multifactorial nature of atherosclerosis and CHD**, which may be masked by effects of allele frequencies characteristic of contributing subpopulations, races, and ethnicities
- **nonuniformity in clinical definitions of CHD**, the **criteria** used in sample selections, and **methodologies** used

All these factors in many cases may contribute to the inconclusive nature and limited clinical usefulness of the molecular genetic evidence accumulated so far.

2.1.6 OPTIONS IN PREDICTIVE MOLECULAR GENETIC TESTING FOR CHD (p. 181)

At this time, the use of predictive testing at the gene level may be limited to some RFLP displaying strong associations with CHD, especially in family studies. Although testing for Lp(a) and apo E isoforms appears to provide a useful addition to the evaluation of total, HDL, and LDL cholesterol, triglycerides, and apo A and B, **it is not clear at present whether more clinically useful information on an individual's risk of CHD can be obtained by evaluating the respective genes**.

Undoubtedly the emerging **role of regulatory factors**, the development of **simpler molecular tools**, and the **accumulation of relevant genetic data** on lipoprotein disorders will make gene probe methodologies a helpful adjunct in wide predictive testing for CHD. Discussions on the genetic aspects of predictive CHD testing have strengthened the importance of the "**candidate gene**" approach and the role of "**variability genes**" in determining an individual's risk of CHD.

Chapter 2.2

Gastrointestinal Diseases

2.2.1 INTRODUCTION (p. 183)

Molecular cytogenetic approaches hold the promise of developing the least invasive means of objective assessment of gastrointestinal diseases. **Family studies** of patients with colorectal cancer indicate that in over 90% of patients, a first-degree relative is affected.

Several **models of colorectal tumorigenesis** based on the evidence accumulated in molecular genetics have been proposed, suggesting practical approaches to patient management. With respect to **gastric malignancies,** however, no unified conceptual scheme has been proposed so far.

2.2.2 DNA PLOIDY IN COLORECTAL TUMORS (p. 183)

Two characteristics — **the mean nuclear cytoplasmic (*N*/*C*) ratio** and **the coefficient of variation of the nucleus-to-cell apex distance** — appear to provide discriminatory, informative variables in colorectal tumors. The *N*/*C* ratio was almost twice as high in adenocarcinoma cells compared to normal epithelium (39.7 ± 7.0 vs. 20.4 ± 2.0) and was accompanied by eccentricity of nuclei in adenocarcinoma cells (47.8 ± 9.1 vs. 19.2 ± 7.5).

The same set of criteria fails, however, to distinguish between **regenerative changes** and low-grade **dysplasia** as well as between these two groups and the normal and tumor tissues.

DNA content has been proposed as an objective criterion for differential diagnosis and/or prognosis of colorectal neoplasias. In longstanding **ulcerative colitis** (UC), low-grade dysplasia was associated with **hypertetraploid** cells (27% of cases), and areas of high-grade dysplasia were associated with **aneuploid** cell populations (62.5% of cases). In **adenocarcinomas** histopathological evaluation suggested that DNA content can be used for prognostic purposes: a better prognosis could be entertained for cases of **polyploid DNA content** of carcinomas compared to those with an aneuploid pattern. However, the finding of aneuploidy per se should not be overinterpreted to indicate or predict invasion, or to distinguish between low- and high-grade dysplastic lesions, or between low- and high-grade malignant tumors.

Total nuclear DNA content as measured by FCM or ICM is clearly a gross measure of the eventual accumulation of changes occurring at the chromosomal level, but is **insensitive to fine alterations at the gene level.** Several parameters correlated with ploidy subdivisions of colorectal tumors based on the **DNA index** are:

- **histological grade**
- vascular and serosal **invasion**
- the expression of **secretory component** (SC) as a marker of epithelial differentiation
- the **carcinoembryonic antigen** (**CEA**) level
- the expression of the major histocompatibility (**MHC**) **class II antigens**

However, **reproducibility** on replicate samples **was poor** (discordance of up to 20%). It resulted from either **sampling imperfection** or the **intrinsic heterogeneity** of the tumors, and no clearcut correlations of **practical value** could be established except to emphasize the heterogeneous nature of diploid and aneuploid nuclear DNA populations. **Multiple sampling** is strongly advocated to reduce the effects of marked heterogeneity of ploidy characteristics in individual samples from colorectal carcinomas.

Adenomas are considered to be precursors of carcinoma of the large intestine, and the characterization of transitional stages can be of value for patient management. **Increased proliferative activity** in adenomas is correlated with the degree of **cellular atypia** arbitrarily defined as low, moderate, and high. Regardless of atypia, the proliferative activity was higher in **adenomas adjacent to carcinomas**. Similar to normal epithelium, however, neither polyploidy nor aneuploidy could be detected in the adenomas. The identification of aneuploidy in the otherwise benign-appearing lesions is, therefore, considered as suggestive of colorectal carcinoma.

A correlation of ploidy with prognosis of patients with colorectal carcinomas has been repeatedly demonstrated in earlier studies, particularly in cases with **Dukes' C stage** cancers. The FCM data on DNA ploidy and cell proliferation may be, however, **no better than Dukes' staging**, especially in **rectal** cancer. In fact, in a large prospective study DNA and RNA indices as well as the proliferative characteristics derived from FCM analysis **failed to be of independent prognostic value**.

Extensive studies of colorectal adenocarcinomas by **FCM of fresh whole-cell suspensions**, or by **ICM of fresh samples**, **emphasize the lack of correlation between ploidy and prognosis** within individual Dukes' stages. Ploidy as well as Dukes' stage and the degree of differentiation were, however, significantly **correlated with overall and disease-free survival** in patients who had had surgery.

The prognostic indicators in **colonic adenocarcinoma**, in particular DNA ploidy, seem to differ from those of **adenocarcinoma of the rectum**. Aneuploidy is a significant predictor of poor overall and disease-free survival in patients with rectal malignancy, but not in those with colonic tumors. The **variability in DNA measurements** in multiple tumor samples of colonic adenocarcinoma may contribute to the discrepant prognostic relationships.

Substantially wider areas of involvement of **mucosa adjacent to carci-**

noma of the large bowel than were previously thought to exist have been suggested by the finding of a high frequency of aneuploidy in mucosa up to 10 cm from a colorectal cancer. **Ploidy patterns did not show any correlation with histological abnormalities in the adjacent mucosa.**

The DNA ploidy pattern of the **metastatic lesions** did not correlate with the metastasis period, the number of lesions, or extent of their spread; however, overall prognosis in diploid tumors was much better than in aneuploid tumors.

Another approach correlated the degree of **susceptibility of chromatin DNA to acid denaturation** with different stages of colorectal tumors. Indeed, **DNA denaturability *in situ***, assessed by FCM, was considered to be a sensitive probe for discriminating **tumors of differing aggressiveness**. Substantially more specific gene level probes are available, however, making evaluation of the gross physicochemical characteristics of chromatin DNA unlikely as a practical, useful approach to managing patients with colorectal and, for that matter, other types of neoplasias.

2.2.3 DNA PLOIDY IN GASTRIC TUMORS (p. 186)

Only a loose correlation could be observed between the nuclear DNA content of gastric cancers (**papillary adenocarcinoma**, **tubular adenocarcinoma** of varying degrees of **differentiation**, and **mucinous adenocarcinomas**) and the clinical and histological parameters. A significant positive correlation was noted, however, between polyploidy and metastases to lymph nodes or the liver.

The DNA ploidy pattern determined by FCM also appears to display a significant correlation with survival and tumor grade in **gastric smooth muscle tumors** (leiomyomas, leiomyosarcomas, and leiomyoblastomas). Although DNA ploidy cannot be used to differentiate between benign and malignant gastric smooth muscle tumors, it apparently can be used as a prognostic tool.

The wide heterogeneity of ploidy in the primary lesions becomes markedly reduced in metastatic tumors, and the tumor cells with a lower DNA ploidy tend to metastasize to lymph nodes, particularly in cases of a differentiated carcinoma.

Multiple samples from a tumor specimen must be evaluated because of **the intratumoral regional heterogeneity of DNA ploidy** parameters, which is much higher in gastric carcinomas (40% of cases) than in colorectal carcinomas (7.4% of cases).

DNA ploidy assessment has some prognostic value in gastric carcinomas, but only in advanced tumors with lymph node metastases.

2.2.4 CHROMOSOME ABERRATIONS (p. 187)

A loss of specific chromosomal regions in colorectal adenomas and carcinomas has been well documented and related to alterations in **tumor suppressor genes** and **oncogenes**. The gene responsible for the inherited predisposition to the development of multiple colorectal adenoma, **familial adenomatous polyposis (FAP)**, has been identified and assigned to chromosome 5q21-22. Although **loss of heterozygosity (LOH)** of 5q in colorectal adenomas is rare, it is observed in up to 50% of colorectal carcinomas.

The original probe used to establish linkage to the ***FAP* locus (C11p11)** had limited utility for family studies because of low heterozygosity and distance from the *FAP* gene. The other probes useful in linkage analysis, **cKK5.33** and **pi227**, substantially enhance risk calculation, allowing predictive testing in presymptomatic FAP kindred, surveillance, genetic counseling, and development of treatment regimens.

A gene showing tumor-specific LOH in at least 50% of informative cases of colorectal carcinoma encodes the **colony-stimulating factor 1 receptor**, and has been mapped to **5q21-22**, close to the *FAP* locus. Alleles of the ***APC* locus** (responsible for FAP) may be associated with susceptibility to colon cancer not necessarily accompanied by polyposis.

In **benign adenomas**, specific genetic changes that seem to involve the **D5S43 locus on chromosome 5** appear to have a role in malignization of the epithelium. In fact, the C11p11 probe as well as two other closely linked polymorphic DNA probes, **pi227** and **YN5.48**, which flank the *FAP* locus, allow the establishment of presymptomatic diagnosis of FAP and can be used for prenatal diagnosis with over 99.9% reliability in the majority of families.

Another chromosome loss, at **17p**, infrequent in adenomas, is also present in colorectal carcinomas and is involved in the progression of an adenoma to malignancy. The ***p53* gene** located on 17p may be lost due to point mutations in this gene, and loss of the wild-type allele is considered to impart a selective growth advantage to affected tissues, leading to tumor progression.

Chromosome 18q is lost in more than 70% of colorectal carcinomas and in up to 50% of late adenomas. The work of Vogelstein's group had identified a gene, termed ***DCC*** (for **d**eleted in **c**olon **c**ancer), assigned to this region. The gene encodes a protein that displays a significant homology to the **adhesion family of molecules**; it is expressed in normal colonic mucosa, and its expression is reduced or absent in the majority of colorectal carcinomas.

Losses of alleles from other chromosomes have also been observed in colorectal carcinomas: **1q, 4p, 6p, 6q, 8p, 9q,** and **22q** are lost in 25 to 50% of tumors, and band **1p35** deletions, fragile **8q22**, loss of **12q**, and clonal allele loss of **17p** also occur.

In patients with FAP, the following **progression of events** in the **adenoma-carcinoma sequence** is suggested: the heterozygous mutant/wild-type condition at the ***APC* gene** causes development of mild to moderate adenomas, while the loss of the normal allele in the *APC* gene accounts for a transition from moderate to severe adenoma, as does the Ki-*ras* mutation. Conversion of the latter to intramucosal carcinoma occurs following the LOH on chromosome 17p, and further progression to invasiveness involves additional LOH on **chromosomes 18** and **22q**.

A **reduction in length of the telomeric repeat sequences** has been noted in chromosomes from colorectal carcinomas compared to that of normal mucosa from the same patient. Although the role of **terminal repeat arrays** (**TRAs**) is not known, the length of TRAs in adenomas is reduced to the same extent as that found in the corresponding carcinomas, suggesting that TRA reduction may be involved in malignization.

In gastric and esophageal adenocarcinomas, a highly specific chromosomal region at **11p13-15** harboring nonrandom rearrangements has been identified that could prove to be of diagnostic value.

2.2.5 *ras* FAMILY ONCOGENES IN COLORECTAL AND GASTRIC NEOPLASIAS (p. 189)

Mutations in all three members of the *ras* family of protooncogenes (Ha-*ras*, Ki-*ras*, and N-*ras*) have been studied with respect to their prevalence, temporal occurrence, association with specific histopathological characteristics, and their prognostic value in gastric and colorectal neoplasias.

Mutations of the *ras* gene, predominantly in the **c-Ki-*ras*** gene, mostly at **codon 12**, have been identified using an amplification-ASO assay in colorectal cancers. Importantly, the same mutation could be frequently identified in both the benign (adenomatous) and the malignant (carcinomatous) portions of the same tumor, implying at least a temporal relationship of the mutations to malignization. No correlation could be established between the *ras* oncogene mutations and the degree of invasiveness or progression of the tumors, or the clinicopathological grade (location, invasion, and degree of differentiation). Although this oncogene could not be directly implicated in transformation or tumorigenicity, there was a greater than a fivefold increase in Ha-*ras*, and a somewhat smaller increase in Ki-*ras*, protooncogene expression in the **mucin-secreting** cells.

ras mutations are **not the initial event** in colorectal tumorigenesis, but appear to **occur early** in the malignization process before changes in ploidy take place.

The expression of the **Ha-*ras* product, $p21^{Ha\text{-}ras}$**, is stronger in **deeply invasive** colorectal carcinomas than in superficially invasive tumors, suggesting a role for $p21^{Ha\text{-}ras}$ in tumor proliferation, invasion, as well as in promoting metastases. The degree of p21 expression correlates with the degree of epithelial dysplasia and the size of adenoma, but not the histological type. Adenocarcinomas show an even greater presence of the $p21^{ras}$ (77.1%), more intense staining, and staining of a greater number of cells than that found in adenomas.

The level of $p21^{ras}$ in Dukes' B and C **rectal adenocarcinomas** as determined by the RAP-5 monoclonal antibody **correlates with survival** independent of Dukes' stage. **Measuring $p21^{ras}$ may be recommended to identify patients at high risk for recurrent disease.**

2.2.6 c-*myc* (p. 191)

In normal gastric and colonic mucosa, the level of expression of the c-*myc* gene is relatively low. Increased levels become detectable in chronic inflammatory bowel conditions — **Crohn's disease** and **ulcerative colitis**, and in advanced carcinomas. Positivity for **p62**$^{c\text{-}myc}$ was correlated with a better prognosis.

Although in the majority of cases the enhanced expression did not involve aberrations of the c-*myc* gene, marked **amplification** of the c-*myc* gene has been found in mucinous (53.8% of cases) and poorly differentiated tumors (42.8% of cases). Cell lines established from human **signet ring cell gastric carcinoma**, known for the worst clinical prognosis of all types of gastric cancer, demonstrated multiple chromosomal aberrations, enhanced c-*myc* expression, and a 16- to 32-fold amplification of the c-*myc* oncogene. Overexpression of c-*myc* mRNA significantly correlates with chromosome 5q deletions.

Elevated expression of the c-*myc* gene appears to correlate with the **degree of differentiation** of colorectal neoplasms, declining in more differentiated tissues. It seems to play an important role in the progression of colorectal carcinomas.

The other *myc* protooncogene, **L-*myc*,** failed to show any correlation with conventional characteristics of colorectal cancers (Dukes' stage, metastases, or the degree of differentiation), in contrast to the finding that L-*myc* RFLPs are related to the progression of human colorectal and gastric cancer. A low level of **N-*myc*-** and L-*myc*-specific transcripts can be detected in **normal** colonic mucosa. These genes are frequently, if only modestly, overexpressed in colorectal polyps and carcinomas. **L-*myc* overexpression occurs more frequently in adenomatous polyps than in tumors.**

2.2.7 *erb*A AND *erb*B GENES (p. 192)

One *erb*A-β- and two *erb*A-α-related transcripts of thyroid hormone receptors are expressed in normal human colonic mucosa. A **marked and selective decrease** of *erb*A-β, but not of *erb*A-α, transcripts is observed in colon carcinoma and viewed as a novel marker of malignant transformation in the colon.

Coamplification of c-*erb*B-1, the **EGF receptor gene**, and c-*myc* has been seen in a human gastric cancer and in human oral squamous cell carcinoma cell lines, representing simultaneous amplification of protooncogenes encoding the growth factor receptor (c-*erb*B-1) and the nuclear proteins (c-*myc*).

The other protooncogene of the *erb* family, **c-*erb*B-2**, also known as ***neu*** (***HER*-2**), encodes a transmembrane glycoprotein with **tyrosine-specific protein kinase** activity. Prominent expression of the ***neu* gene product, p185**^{neu}, has been observed only on the luminal surface (with no expression in the crypts) in mucosal epithelium of stomach, small intestine, and colon, as well as in liver parenchyma, the exocrine and endocrine pancreas, and the salivary gland. This protein is more abundant in adenomatous and preneoplastic polyps than in adjacent normal colonic epithelium, or areas of malignant degeneration. Correlations between *neu* gene alterations, p185^{neu} product expression, and colorectal tumor characteristics remain to be fully defined.

The expression of an **aberrant HER-2/Neu polypeptide** with a molecular weight of 190,000 (exceeding that of the normal HER-2/neu protein by 5000) has been observed in colon adenocarcinomas.

In **gastric cell lines**, the level of *neu* expression is comparable to that in colorectal carcinoma cell lines. A more than fivefold amplification of the *neu* gene has been observed in 21% of gastric and esophageal adenocarcinomas. The ***erb*B-2 gene is amplified** in 25% of metastatic, but not early primary, **gastric tumors** (regardless of histological type) with occasional coamplification of ***hst*-1** and ***int*-2 genes**.

In gastric cancer, erbB-2 protein expression is not associated with histological type. It is, however, significantly associated with serosal, but not venous, invasion, lymphatic invasion, and the high number of lymph node metastases. Patients with *erb*B-2/*neu*-positive tumors had a fivefold greater relative risk of death compared to *neu*-negative tumors, suggesting that **erbB-2/neu protein expression can serve as an independent prognostic factor in gastric carcinomas.**

2.2.8 TUMOR SUPPRESSOR GENES (p. 193)

2.2.8.1 *p53*

The frequent loss of **chromosome 17p** alleles and the location of the *p53* gene in that area suggested the involvement of this gene in colorectal carcinomas. Frequent point mutations in one allele of the *p53* gene are associated with the loss of the wild-type allele in gastric and colorectal tumors, similar to that in lung, breast, and brain tumors. Importantly, in addition to deletion of one allele of the *p53* gene that is present in 75% of colon cancers, mutation of the remaining *p53* allele almost always occurs as well. In adenoma cell lines, the transforming activity appears on joint expression of mutant *ras* and *p53* genes.

The **frequency of *p53* gene mutations detected by sequencing cloned PCR products in adenomas is low irrespective of the size and the fact that the adenomas occur in FAP patients.** The *p53* gene mutations usually occur near the time of transition from benign to malignant morphology, pointing to the causal role of the *p53* gene in malignization. In HCC the pattern of *p53* mutations is somewhat different than in the lung, colon, esophagus, and breast.

Using an **RNase protection assay** and northern and Southern analysis to study abnormalities in the *p53* gene and its expression, the incidence of *p53* gene abnormalities was found to be similar in gastric and colorectal carcinoma cell lines. It was present in one third of human esophageal tumors as well. Amplification by PCR combined with direct sequencing revealed that the *p53* gene mutations are also common genetic events in the pathogenesis of **squamous cell carcinomas of the esophagus.**

Asymmetric PCR-based assays can be used as diagnostic and prognostic tools to rapidly and reliably detect *p53* mutations. Assessment of **p53**

protein expression by immunohistochemistry showed positive tumors were significantly more frequent in the distal colon, although no correlation with tumor grade, Dukes' stage, DNA ploidy, or patient survival has been established.

2.2.8.2 RETINOBLASTOMA (RB) (p. 194)

The majority of colorectal carcinomas have been found to **concomitantly overexpress retinoblastoma (*RB*) and *p53* mRNA** compared to adjacent normal mucosa apparently associated with the increased number of cycling cells in tumor tissue.

2.2.8.3 DCC (p. 194)

The other tumor suppressor gene, termed ***DCC*** (for **d**eleted in **c**olon **c**ancer), has been identified at **chromosome 18q**. *DCC* is expressed in normal colonic mucosa, and its expression is reduced or absent in the majority of colorectal carcinomas. The homology of the ***DCC* gene product** to the **cell adhesion family of molecules**, and a decrease in its expression in cancer, are believed to be related to the altered adhesion properties of mucosal cells and derangement of growth-restricting controls.

A strong confirmation for the role of *DCC* and ***APC* genes** as TSGs in colonic tumors comes from the finding that introduction of normal human **chromosome 5** (with the *APC* gene) and **18** (with the *DCC* gene) into a human colon carcinoma cell line, COKFu, through microcell hybridization, completely suppresses the tumorigenicity of the hybrid cells and normalizes their growth in culture. Similar results have been obtained on introduction of chromosome 5 into DT cells.

Yet another putative TSG related to colon cancer appears to be the **protein kinase C (*PKC*) gene**, when it is expressed at high levels.

2.2.8.4 MCC (p. 195)

The recently identified ***MCC* gene** (**m**utated in **c**olorectal **c**ancer) is located at chromosome **5q21** and encodes an 829-amino acid protein showing a similarity to the G protein-coupled m3 muscarinic acetylcholine receptor. The presence of rearrangements and somatically acquired point mutations in *MCC* in sporadic colorectal carcinomas makes this a candidate for the putative colorectal tumor suppressor gene. *MCC* is present in normal colonic mucosa and, like the *APC* gene, is also expressed in a wide variety of tissues. **Both *MCC* and *APC* genes are somatically altered in tumors from colorectal cancer patients,** and it appears that **mutations in the *APC* gene may contribute to both FAP** and **Gardner's syndrome.**

2.2.9 OTHER GENES ASSOCIATED WITH TUMORS OF THE DIGESTIVE TRACT (p. 195)

Class I HLA gene expression is completely lost in some cases of colorectal, gastric, and laryngeal carcinomas, whereas a selective loss of HLA-B antigens and a different pattern of expression of class I antigens is found in primary tumors and metastases. The search for possible molecular genetic markers helpful in the earliest detection of tissue changes predicting subsequent neoplasias continues and a number of cloned DNA sequences are being tested for possible nonrandom association with specific abnormal phenotypes. The higher level of *mdr*-1 gene expression correlates with a higher degree of differentiation of colorectal and gastric carcinomas. In precursor lesions of gastric carcinoma, yet another oncogenic rearrangement (*tpr-met*) has been detected.

2.2.9.1 ORNITHINE DECARBOXYLASE (ODC) (p. 195)

ODC is a rate-limiting enzyme catalyzing the conversion of ornithine to putrescine, a member of the family of low molecular weight polyamine bases essential for cellular proliferation and differentiation.

Two genes encoding ODC, *odc*-1 and *odc*-2, have been localized to **chromosomes 2p25** and **7q31-qter**, respectively. Although the activity of ODC is elevated in colorectal polyps and carcinomas, no amplification of the loci could be detected, suggesting that the **regulation of ODC in colorectal neoplasia occurs at the posttranscriptional level.**

2.2.9.2 LAMININ (p. 195)

Laminin is one of the major components of basement membranes and has been implicated in numerous cellular functions, including attachment, spreading, migration, cell growth, differentiation, mitogenesis, and binding to extracellular matrix components. Using a cDNA clone encoding a human **laminin-binding protein**, a significant increase in the corresponding mRNA expression has been observed in colorectal carcinomas that can serve as a marker for tumor progression and aggressiveness in clinical practice. Moreover, the **increased expression of laminin receptor is specifically associated with cancer cells and displays a correlation with the Dukes' classification and the invasive phenotype of colon carcinoma.**

2.2.9.3 c-*src* (p. 195)

The 10- to 30-fold increased activity of the product of the *c-src* protooncogene, **pp60**$^{\textbf{c-}src}$, is observed in cultured human colon carcinoma cell lines and is apparently related to activated protein kinases rather than increased synthesis of the c-*src*-encoded protein.

2.2.9.4 *trk* (p. 195)

Yet another protooncogene, ***trk*** **(tropomyosin receptor kinase)**, which encodes a hybrid molecule that contains parts resembling nonmuscle tropomyosin combined with tyrosine kinase (similar to that found in retroviral transforming genes), has been identified in colon carcinomas. This protein has features characteristic of cell surface receptors. The product of the human *trk* protooncogene is a 140,000-Da glycoprotein, designated **gp140**$^{\textbf{proto-}trk}$.

The hybrid nature of gene products in colon carcinomas is not limited to gp140$^{proto\text{-}trk}$, and is also observed in a fusion product of **ubiquitin** and the **ribosomal protein S27a**, encoded by a novel **ubiquitin hybrid protein gene**. This novel gene is overexpressed in tumor tissues compared to adjacent normal mucosa, and a correlation between the level of its expression and the clinical tumor staging has been reported.

2.2.9.5 HUMAN PAPILLOMAVIRUS (HPV) (p. 196)

The presence of and a putative role for HPV type 6 have been described in the progression of **laryngeal papilloma** to carcinoma. An intriguing, although still unexplained, association of **HPV** infection and **colon carcinomas** has also been demonstrated by immunohistochemistry and ISH. A correlation appears to exist between the viral burden and the degree of malignant transformation of the glandular epithelium. Chromatin changes suggest the presence of a lytic virus. Further studies are certainly warranted to confirm these preliminary observations by more rigorous molecular biological techniques.

2.2.9.6 CEA (p. 197)

The well-known tumor marker **carcinoembryonic antigen (CEA)** is a cell surface glycoprotein whose expression greatly increases in human colon carcinomas. CEA has a strong similarity to members of the immunoglobin superfamily that includes **intercellular adhesion molecules (ICAMs)**. CEA mRNA expression is much stronger in carcinomas than in adenomas, and is not present in noninvaded tissue. The possibility of CEA having a role in carcinogenesis and in metastasis through its involvement in tissue architecture is intriguing. In an experimental system, **exogenously added CEA induced metastases, suggesting a direct or indirect role for CEA in the metastatic process.**

2.2.9.7 Ki-*sam* (p. 197)

The **Ki-*sam* gene**, encoding a **heparin-binding growth factor receptor with tyrosin kinase activity**, has been isolated from gastric cancer cells, where it is markedly amplified. This amplification, however, is restricted to **poorly differentiated** types of stomach cancer. The biological significance of this gene remains to be established.

2.2.9.8 *nm*23 (p. 197)

The *nm*23 gene encodes a potential **tumor metastasis suppressor**; it is expressed in normal colonic mucosa and its expression rises early in colon carcinogenesis and remains increased in metastatic colon cancers, suggesting that, in the colon, **a dissociation of *nm*23 gene expression from loss of tumor metastatic competence occurs in a tissue-specific manner**.

2.2.10 GROWTH FACTORS (p. 197)

It appears that evaluation of protooncogenes may eventually be of practical significance in patient management. Less is known about the utility of assessing **growth factor (GF) genes** and their products for diagnostic purposes. By analyzing the *nm*23 allelic deletions with a specific probe (*nm*23-H1), a prognostic evaluation of colorectal carcinoma with respect to distant metastases becomes possible.

Tumor suppressor genes, known to be consistently altered or deleted in various neoplasias, including those of the gastrointestinal tract, also encode products essential for the maintenance of normal **growth controls**. The aberrant expression of GFs and their receptors is, in part, responsible for autonomic growth of tumors through **autocrine stimulation**.

2.2.10.1 ESOPHAGUS (p. 198)

Coamplification of the ***hst*-1** and ***int*-2** genes occurs in approximately 50% of the primary **squamous carcinomas of the esophagus** and in 100% of the metastatic tumors. The *hst*-1 gene product is a heparin-binding GF for mouse fibroblasts and human vascular endothelial cells. In spite of their amplification, **these genes do not appear to be expressed** in esophageal carcinomas. A positive correlation has been observed, however, between coamplification of the *hst*-1 and *int*-2 genes and the **clinical stage** or presence of **metastases**.

Amplification and overexpression of the **EGFR gene** occurs in 71% of SCC of the esophagus, and **a good correlation exists between EGF expression and the depth of tumor invasion and prognosis.**

2.2.10.2 STOMACH (p. 198)

Normal human gastrointestinal mucosa contains **T cell growth factor α (TGF-α)** and **urogastrone epidermal growth factor (URO-EGF)**. Elevated expression of genes for TGF-α and **EGFR** (c-*erb*B-1) is present in **gastric carcinomas**. In one third of cases, the gene encoding EGF is also expressed. A positive correlation exists between the depth of tumor invasion, poor prognosis, and coexpression of the genes for **EGF** and **EGFR**. Yet another correlation has been found between the degree of malignancy or the presence of metastases of gastric carcinomas and coexpression of **TGF-α** and $p21^{c\text{-}Ha\text{-}ras}$.

About 10% of gastric carcinomas show coamplification of the **EGFR** (c-*erb*B-1), **c-*myc*, c-*yes***, and **c-Ki-*ras*** genes. **c-*erb*B-2** is frequently found in adenocarcinoma of the stomach, and even more frequently in metastatic lesions of any histological type.

Other GFs expressed in gastric carcinomas are **PDGF** and **TGF-β**, frequently simultaneously, suggesting a posssible role for PDGF in the production of **scirrhous changes**. In addition to TGF-β, ***sam* gene** amplification is present only in **scirrhous carcinoma** and **poorly differentiated gastric adenocarcinoma**.

2.2.10.3 COLORECTAL TUMORS (p. 198)

In **colorectal carcinomas**, genes for **TGF-α, EGFR, EGF, PDGF-A, PDGF-B, and PDGF-R** are expressed, some simultaneously. Almost 50% of colorectal cancers show the presence of mRNA for tumor necrosis factor (**TNF**); its expression, however, does not correlate with the stage of disease, degree of lymphocytic infiltration, or tumor necrosis, and is present predominantly in stromal cells.

The determination of **EGF** in the urine (**urogastrone**) although revealing a statistically significant difference between patients and control, could not unequivocally demonstrate its clinical utility.

Gastrin promotes the growth of gastrointestinal epithelial cells, and thereby acts as a GF. Gastrin positivity is found in 17% of **gastric adenocarcinomas**, and in 50% of **scirrhous gastric carcinomas**. High-affinity **gastrin receptors** present in 65% of colorectal carcinomas correlate with a better prognosis, however, the clinical usefulness of these observations for patient management at this time is not clear.

2.2.11 DNA METHYLATION (p. 199)

Changes in gene expression associated with the degree of **methylation of cytosines** at the 5′ position by a methyltransferase enzyme specific for the **CpG** sequence have been recognized as an important characteristic of neoplasia. In carcinomas either **total genomic hypomethylation** or **selective hypomethylation** of specific genes is predominantly observed. **Hypermethylation** patterns have also been seen. While hypermethylation appears to decrease gene expression consistently, hypomethylation has a variable effect. The consistency of hypermethylation patterns in colon carcinoma may be associated with progression through various stages of malignization.

In lung cancers, lymphomas, and acute myeloid malignancies, extensive hypermethylation of several genes on the short arm of **chromosome 11**, including the **calcitonin gene**, has been observed. The novel pattern of calcitonin gene methylation is a feature of the majority of **colonic adenomas** and established colon carcinoma cell lines, **but not of colon carcinomas**.

2.2.12 GENOMIC IMPRINTING (p. 199)

The phenomenon of **genomic (or genetic) imprinting**, implying the preferential expression of maternally, or paternally, derived genes, may in part be mediated through variations in DNA methylation (see Section 1.1.6). Differences in the degree of methylation of specific alleles have been described in the Ha-*ras* gene. Using REA, electrophoresis, and recovery of the bands from gels on ion-exchange paper can simplify and make the analysis of specific oncogene methylation patterns in clinical material much easier and reproducible. It remains to be seen whether the phenomenon of genomic imprinting, which frequently amounts to the functional suppression of expression of certain genes, plays a role in the inheritance pattern of gastrointestinal neoplasias.

2.2.13 CURRENT HYPOTHETIC MODELS OF COLORECTAL TUMORIGENESIS (p. 199)

Confronted with a problem as widespread and complex as the development of colorectal tumors in humans, one is tempted to organize the accumulated evidence in the hope of finding certain reproducible features in the etiology and molecular mechanisms that would allow coherent analysis of clinical cases. Only a few such attempts, of different research groups, that are relevant to events occurring at the gene level will be discussed here.

The **pros and cons** of five hypotheses on the origin of colon cancer that are probably farthest removed from the level of genetically operated mechanisms include

1. The role of **fecapentaenes** as potent mutagens produced in the colonic lumen, and acting as initiators of colon cancer, is not supported by the current findings;
2. The effect of **3-ketosteroids** as cytotoxic and genotoxic substances initiating or promoting colorectal tumors, although tentatively implicated in *in vitro* studies, still remains far from established *in vivo*;
3. The significance of nuclear aberrations allegedly produced by **pyrolysis products** in food is far from clear, although theoretically possible;
4. The importance of **calcium deficiency** in enhancing the presence of free **bile** and **fatty acids**, which are toxic to colonic epithelium, is not supported by experimental data, which fail to demonstrate the protective effect of calcium;
5. The impact of **high fecal pH**, leading to epithelial proliferation and to an increased sensitivity to carcinogens, could not be demonstrated unequivocally.

In spite of this, some investigators pursue the elucidation of the role of diet in colonic tumorigenesis.

Turning to molecular genetics data, the role of oncogenes and tumor suppressor factors has been incorporated into a model for colon carcinogenesis. At least two forms of the disease are suggested:

- One is characterized by the loss of *FAP* gene function and deregulation of c-*myc* expression.
- The other involves an unidentified TSG locus that predisposes to cancer in the proximal (right sided) colon and does not involve c-*myc*.
- In both forms, the activation of the *ras* oncogene by point mutations plays a significant role.

Essentially two coupled events are seen as responsible and necessary for the neoplastic transformation of colonic epithelium:

- **disruption of the cell cycle** control by deregulated c-*myc*

- **activation of the signal transduction** as a result of point mutations in *ras*

Multiple, and largely independent, genetic events, **occurring essentially at random** without any particular order, are thought by some to be involved in tumor formation in the colon.

Others argue that, in the case of colorectal carcinomas, the accumulated evidence points to losses involving multiple chromosomes, and multiple mutational events are likely to have occurred involving both recessive and "dominant" oncogenes by the time the diagnosis is made.

Yet others discount the likelihood of a major role for nuclear oncogene(s) (N-*myc*, L-*myc*, and *p53*) structural alterations and/or amplification in induction or progression of human colorectal malignancies.

A **model of colorectal tumorigenesis** developed by the Vogelstein group, and incorporating their own discoveries and molecular genetics studies as well as those of others, appears to offer the most comprehensive synthesis of the currently available evidence. Their model comprises the following concepts and findings:

- **accumulated changes** associated with mutations of at least four to six genes result in tumorigenesis
- mutations mostly affect ***p53*** and ***DCC*** **(tumor suppressor genes)**, located at loci 17p and 18p, respectively
- the only **consistently amplified** oncogene in colorectal carcinomas is *ras*
- **hypomethylation** of DNA occurs early and apparently leads to aberrant segregation, resulting in the loss of wild-type tumor suppressor genes
- the ***FAP*** **gene**, mapped to 5q, although only infrequently lost in the adenomas of FAP patients in sporadic adenomas, is affected in over a third of cases
- a certain **order of genetic changes** is visualized: hypomethylation occurs early, followed by *ras* gene mutations and 5q allelic losses, which precede 18q allelic losses and the 17p allelic deletions, completing the sequence. The critical point, however, is the **total accumulation of changes rather than the order of changes**

Based on this coherent model, the practical uses of the molecular genetic findings are envisaged:

- **defective gene products** in the blood or feces of patients known to be predisposed to FAP or the Lynch syndrome **should be identified** (even by immunological methods)
- individuals who do not have the affected genes can be selected early on, sparing them regular colorectal examinations
- irrespective of the histopathological features of the tumor at the time of evaluation, **analysis of the total accumulated allelic deletions in an individual patient may suggest the likelihood of potential metastases**. This may provide helpful information to clinicians for individualizing therapeutic regimens

Chapter 2.3

Hepatocellular Carcinoma

2.3.1 ANIMAL MODELS OF CHEMICALLY INDUCED HCC

2.3.1.1 PLOIDY ASSESSMENT (p. 203)

The earliest morphological events produced in the liver of rats in the course of **chemically induced carcinogenesis** are the appearance of enzyme-altered foci (EAF) and altered hepatic foci (AHF), marked by enzyme elevations. The **γ-glutamyl transpeptidase-positive** preneoplastic foci contain aneuploid cell populations with the 2C, 4C, and 8C ploidy patterns, whereas in AHF the ploidy is predominantly diploid, tetraploid, or heterogeneous. **Aneuploidy may not be the necessary condition for an irreversible commitment to malignancy**. It constitutes an increased risk factor for carcinogenesis reflecting **genomic instability** as a stage in the multistep process of carcinogenesis.

2.3.1.2 ONCOGENE STUDIES (p. 203)

Studies of the course of chemical carcinogenesis alterations of several oncogenes led to the following conclusions:

- spontaneous mutation of **Ha-*ras*** at the first and second positions of **codon 61** is one of the early events
- the "**hot spots**" at **codons 12** and **61** of Ha-*ras* play a role in the susceptibility to chemical carcinogenesis; however, truncation of the 5′ exon of the Ha-*ras* gene, rather than hot spots, is thought by others to be involved
- the **causal relationship of oncogene activation to carcinogenesis** is complicated by the observed increase in Ha-*ras*, **Ki-*ras***, **c-*myc***, and **c-*fos*** protooncogene expression due to cell proliferation in the course of regenerative growth of affected liver parenchyma
- studies by the Pitot laboratory established that increased expression of Ha-*ras* and c-*myc* is absent in the majority of preneoplastic foci; however, Ha-*ras* appears to be activated at the stage of progression
- activation of protooncogenes in the rat model is observed at the stage of progression, whereas in the mouse some protooncogenes appear to participate in carcinogenesis at earlier stages

- alteration of the mRNA levels of **c-*raf***, and the **gap junction protein**, appear to be involved in the malignant development of preneoplastic foci
- some of the early, and possibly critical, events in chemical hepatocarcinogenesis include transient elevation of **c-*fos***, c-*myc*, and Ha-*ras*
- the **α-fetoprotein** (**AFP**) genes are expressed only 5 weeks after the initiation of hepatocarcinogenesis; at later stages hepatocytes are always negative for AFP
- it appears that activation of **γ-glutamyl transpeptidase** and **glutathione transferase P** is induced by activated *ras* protooncogene expression

2.3.2 VIRALLY-INDUCED HCC IN ANIMAL MODELS (p. 204)

Enhanced expression of the c-*myc* oncogene is seen in **woodchuck hepatitis virus (WHV)-induced HCC**. A high frequency of integration of WHV DNA into N-*myc* genes has been established.

2.3.3 HUMAN HCC

2.3.3.1 PLOIDY OF HUMAN HCC (p. 204)

The larger the tumor, the lower the percentage of diploid cells detected in HCC. Those over 5 cm in diameter typically have 82% of cells featuring the nondiploid pattern. Although the DNA pattern **does correlate** with the morphological grading and age of patients, **it fails to be of use in the prognosis** of patients with HCC.

2.3.3.2 CHROMOSOME ABERRATIONS (p. 205)

Allele loss on **chromosome 16q (16q22.1** and **16q22.3-q23.2)** is a common genetic alteration in **human HCC** detectable in poorly differentiated larger tumors, and in those with metastases, rather than at the earliest stage. These chromosome alterations are not associated with hepatitis B virus (HBV) DNA integration or **hepatitis C virus (HCV)** infection.

Loss of heterozygosity (LOH) has also frequently been observed on **chromosome 4**, **11p**, **13q**, and **17q**. Losses possibly involving **TSGs** located on chromosomes **5q, 10q, 11p, 16q,** and **17p** occur frequently. High levels of allele loss were also observed on **chromosome 16p** (83%) and **8q** (44% of cases), suggesting the presence of additional genes, the loss of which may contribute to the development of HCC.

Consistent **rearrangements of DNA** occurring in about 10% of primary HCC have been confirmed by the isolation of a unique cellular DNA sequence adjacent to an HBV integration site.

In **hepatoblastomas**, malignant hepatic tumors of childhood, the pattern of chromosomal aberrations (**trisomy 2q, trisomy 20**, and **LOH at 11p**) is similar to that observed in **pediatric embryonal rhabdomyosarcomas**, suggesting a certain parallelism in the genetic pathways.

2.3.3.3 ONCOGENES IN HUMAN HCC (p. 205)

An elevated expression of ***erb*B, *erb*A** and ***erb*B, Ha-*ras, fms, fos*,** and ***myc*** has been observed in human hepatoma. On the other hand, **Ki-*ras, rel, myb, sis, mos, src*,** and ***bas*** did not show any clear changes in their expression either in fetal liver during development, or in HCC.

c-*myc* expression appears to be higher in HCC than in cirrhotic liver tissue. In contrast to the earlier findings, no significant differences in mRNA levels of c-*fos*, N-*myc*, N-*ras*, Ha-*ras*, c-*erb*A, c-*erb*B, and c-*abl* could be detected between patients with HCC, cirrhosis, and those in the normal-chronic hepatitis groups.

Growth factors and TSGs have been also implicated in liver cancer. Expression of **insulin-like growth factor II (IGF-II)** appears to serve as a marker of liver cell differentiation, and it is markedly enhanced in primary liver cancer. Loss of ***p53*** expression in HCC is associated with deletions or rearrangements, or aberrations of its product. Specific point mutations in the *p53* gene in HCC differ from those found in other human cancers. The development of HCC related to HBV integration is associated with the loss of one allele of the ***p53*** gene (chromosome 17p13) in 60% of patients with HCC. **A mutation at codon 249 of the *p53* gene, characteristic of HCC associated with high risk of exposure to aflatoxins, which can be used as a specific marker.**

2.3.3.4 HBV DNA IN HCC

2.3.3.4.1 Epidemiological Evidence (p. 206)

A strong association between **persistent HBV infection** reflected in the prevalence of the hepatitis B surface antigen (HBsAg) carrier status and HCC has long been established. The relative risk of developing HCC in carriers has been estimated to be over 200-fold compared to noncarrier control populations. Although the presence of HBV DNA always precedes the development of HCC, other factors, such as **aflatoxins**, **tobacco smoking**, **alcohol**, and **oral contraceptives**, are involved in the onset of malignization. Again, oncogenicity of HBV may be limited to specific tissues, because a number of extrahepatic sites of HBV DNA have been identified that are not prone to malignant transformation. **Chronicity of HBV infection also does not appear to be a prerequisite for the development of HCC**, as shown in pediatric cases.

2.3.3.4.2 Oncogenesis Associated with HBV (p. 206)

The potential mechanisms of hepatocarcinogenesis related to HBV infection range from the predisposition to malignancy associated with **cirrhosis** to **integration of HBV DNA** leading to dramatic changes in hepatocyte gene expression. **Chromosomal rearrangements, insertional mutagenesis**, as well as **trans-activation** of cellular genes by the **viral protein X** have been implicated.

In recent years, the molecular aspects of **viral DNA replication, gene**

expression, and **integration** have been deciphered. The presence of **direct 11-nucleotide repeat sequences** (DR1 and DR2), their role in HBV integration, and the functions of the four **open reading frames** (ORFs) in the course of infection and transformation have been studied in detail.

It appears that HBV can replicate and integrate into the genome of nonhepatic cells as well, and that the X protein of HBV is a trans-activating factor for viral enhancers possibly involved in hepatocarcinogenesis. Another likely HBV gene sequence, that of the **truncated pre-S/S region**, has been identified as a transcriptional trans-activator possibly associated with HBV-associated oncogenesis. **HBV integration** appears to be a **random event** with an apparently specific predilection for the viral sequences near the direct repeats **DR1** and **DR2 as integration sites**. The possibility is entertained that HBV DNA integration within an intron of the **cyclin A gene**, identified at an early stage in human primary liver cancer cells, contributes to hepatocarcinogenesis.

The loss of **TSGs** has been invoked as one of the plausible modes of tumorigenesis by analogy with other tumors (retinoblastoma, colorectal cancer, breast cancer, etc.).

Other observations in hepatocarcinogenesis include:

- activation of a **steroid hormone receptor-related gene, *hap***, resembling the **c-*erb*A** protooncogene, and inappropriate expression of the *hap* gene product in the HCC is thought to contribute to hepatocellular carcinogenesis
- activation of ***c-myc***, and a **retinoic acid receptor** have been identified in HCC
- the transforming oncogene, ***hst*-1**, has been detected in HCC with integrated HBV genomes. One integration occurred in proximity to the *hst*-1 gene on **chromosome 11q13**
- the *hst*-1 and ***int*-2** genes as well as the HBV DNA can be coamplified, but not rearranged, in tumor cells. The **fibroblast growth factor** properties of the *hst*-1 and *int*-2 homologs are known to induce transformation
- the **reverse transcriptase activity of HBV** has long been recognized. The **92-kDa gene product** of HBV ORF P is believed to be the precursor of the reverse transcriptase of HBV

2.3.4 CLINICAL RELEVANCE OF THE HBV DNA IDENTIFICATION IN LIVER DISEASE (p. 208)

The identification of antibodies to HBsAg by immunohistochemistry revealed superior qualities of **ISH**, even in routine clinical practice using biotinylated, commercially available DNA probes. ISH can be efficiently performed, even in an automated fashion on formalin-fixed, paraffin-embedded material, and various methodological improvements aimed at increasing the sensitivity of nonradioactive probes have been reported.

Because HBV DNA can be detected in the liver of patients negative for conventional serum markers, the sensitivity of HBV DNA detection may prove to be of critical importance. The **Southern blot** analysis has proved to be comparable in sensitivity to the detection of HBsAg in serum.

A much more sensitive detection is offered by **PCR**, which detects HBV DNA and RNA sequences in primary liver carcinomas from patients negative for HBsAg. Clearly, the issue of technical differences in the detection methods (e.g., Southern blotting vs. PCR, or the use of a different set of primers in the PCR amplification) acquires major importance in establishing the etiology of liver pathology. [*See Supplements 1 and 2.*] Using primers corresponding to the HBV genome regions coding for the **pre-S**, **S**, **C**, and **X genes**, two different patterns of HBV DNA molecules were identified in the tumor cells of HBsAg-negative patients with HCC: those with deletions and without deletions. **A reduction in the expression of the viral genome seems to occur in the course of hepatocarcinogenesis**, and could be paralleled by the masking of HBsAg in immune complexes. HBV infection may induce an initial event leading to transformation, followed by deletion of the viral sequences in some cases.

Although the etiological link between HBV and HCC is strongly suggested, the accumulated findings do not unequivocally establish the causal relationship. The high **genomic variability of HBV** is well documented and its role in transformation of hepatocytes by HBV alone or in association with other viruses or cellular factors remains to be established. Furthermore, it is not clear whether **HCV**, which may act as a cofactor, plays a role in hepatocarcinogenesis related to HBV. As discussed in Section 6.2.3, HCV has been detected in up to 60 to 70% of patients with HCC.

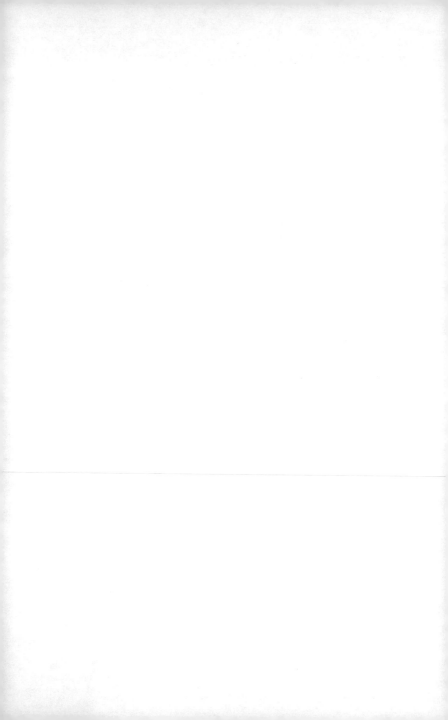

Chapter 2.4

Pancreas

2.4.1 IMMUNOHISTOCHEMICAL MARKERS (p. 211)

A controversy still exists concerning the type of precursor cell undergoing neoplastic transformation into a pancreatic carcinoma. Flow cytometry (FCM) allows the separatation of **endocrine pancreas** cells into individual cell populations. Immunohistochemical markers can also be used. Antibodies against **chromogranin A** and **somatostatin** identify islet cells, secretory component, and carbonic anhydrase II; **pancreatic cancer mucin Span-1** can be used to reveal ductal cells; and **trypsin** and **lipase** identify acinar cells. It appears that the original transformed cell type in many **exocrine** pancreatic carcinomas belongs to the **endodermal "stem cells"** capable of differentiating along multiple cell lineage pathways. This is in contrast to the often held view, based on morphological characteristics, that the majority of pancreatic cancers derive from differentiated ductal epithelium cells.

The **pattern of expression of chromogranin A can be diagnostic**: it is homogeneous in islet cell tumors and heterogeneous in exocrine pancreatic adenocarcinomas. The well-documented correlation of **serum tumor markers** of pancreatic tumors — **CA 19-9, CA-50, DU-PAN-2, CEA** — with malignancy and tumor load **cannot be used for diagnosis or screening** beyond providing a tentative indication of tumor or for monitoring tumor recurrence.

2.4.2 DNA PLOIDY (p. 211)

The fraction of cells with aneuploid DNA content in **ductal adenocarcinomas** of the pancreas and **adenocarcinomas of the ampulla of Vater**, although being relatively low (15% of the tumors), was still significantly higher in carcinomas than in nonneoplastic tissue. While **ploidy** proved to be of **no prognostic value** for survival, aneuploidy was found predominantly in advanced tumors with lymph node metastases and in higher-grade tumors. The combined fraction of cells in **$S + G_2M$** phases has an **independent prognostic value**: a high proportion of cells in $S + G_2M$ phases of the cell cycle portends a shorter survival time. These cells are also more **responsive to chemotherapy**, as can be expected for rapidly dividing cells in general. **The value of ploidy determinations, which do show meaningful correlations, is limited, however, because *diploid pancreatic cancer is also rapidly fatal*.**

A tentative and weak correlation of aneuploidy of **pancreatic neuroendocrine tumors** with a poor prognosis has been observed by FCM. Ploidy data for a particular type of tumor should be extrapolated to diagnostic and prognostic conclusions with caution. In the case of **insulinoma**, for example, half of the tumors studied displayed aneuploidy, despite that most of the patients were still alive and disease free 2 to 5 years postresection. A similar survival pattern was observed in patients with normal DNA profiles, suggesting that **DNA ploidy should not be used for prognostic purposes in patients with insulinomas**.

2.4.3 CHROMOSOME STUDIES (p. 212)

Numerical or structural abnormalities of **chromosome 7** have been found in human pancreatic carcinoma cell lines; however, no amplification of the **EGF-R gene** assigned to chromosome 7, could be observed.

The gene for **MEN-1 syndrome** has been assigned to **chromosome 11**, and a loss of alleles in **insulinomas** suggests that unmasking of a recessive mutation at the *MEN*-1 locus contributes to carcinogenesis. Subsequently three unrelated cases of sporadic insulinoma were found to have an allele loss on both **11p** and **11q**.

2.4.4 SPECIFIC GENE EXPRESSION IN PANCREATIC TUMORS (p. 212)

Studies in **cultured pancreatic endocrine tumor cells** have identified the genes encoding various **DNA-binding proteins** specifically interacting with DNA regulatory elements in controlling the function of islet cell hormones.

So far no explanation is given for the role of **parathyroid hormone-like peptide (PLP) gene** expression in normal and neoplastic **islets of Langerhans**, where its abundant expression was not associated with hypercalcemia.

The expression of the **transthyretin (TTR) gene**, normally identifiable in the pancreatic islet cells at a low level, is specifically enhanced in endocrine pancreatic tumors (**glucagonomas** and **insulinomas**) as well as in intestinal **carcinoids**. Interestingly, other endocrine tumors — **pheochromocytoma** and **paraganglioma** as well as **adenocarcinoma of the pancreas** — are TTR mRNA negative.

The genes for other pancreatic carcinoma markers, such as **DU-PAN-2** and **pancreatic apomucin**, are being studied; however, the range of expression of these genes in the context of clinical presentation of pancreatic tumors remains to be defined.

2.4.4.1 ONCOGENES (p. 213)

Pancreatic β cell replication is markedly stimulated in cultured cells into which the oncogenes *src* or a combination of *myc* and *ras* are introduced. Ki-*ras* mutations in codon 12 have been detected by PCR in the majority of patients with pancreatic adenocarcinoma, but not in benign aspiration cytologies of the pancreas. Although the frequency of these mutations in **atypical cytologies** is much lower (25%) than that found in fine-needle aspirates (FNAs) displaying malignant features, **no significant correlation with survival could be detected**.

A combination of FNA technique and PCR amplification can distinguish between normal and mutant *ras* oncogene (or other genes)-containing tumors, and the detection of such changes can be used for clinically useful diagnostic or therapeutic decisions. However, the evaluation of c-Ki-*ras* oncogene mutations at codon 12 in pancreatic aspirates **cannot be considered a reliable diagnostic criterion of pancreatic adenocarcinoma, nor can it be seen as an indicator of a benign disease.** Opposing findings have also been reported. On the other hand, in **endocrine** pancreatic tumors the level of **overexpression** of **Ha-*ras*** and **Ki-*ras* is associated with tumor prognosis.**

The **histological type** of pancreatic adenocarcinomas has been shown to vary depending on the **initiating transforming influence**. The most predominant (ductal) type of pancreatic adenocarcinoma appears to derive from transformed acinar cells specifically modified by the oncogenic action of c-*myc*, but not by the tumor antigen of **SV40** or *ras*. Recently the expression of **c-*erb*B-2** has also been demonstrated in pancreatic adenocarcinomas.

Because there appears to be a higher incidence of pancreatic cancers in families with a **history** of this disease (6.7% of cases vs. 0.7% of controls), molecular genetic evaluations may also offer a powerful tool for the identification of **persons at increased risk** for developing this highly lethal malignancy.

It remains to be seen whether evaluation of specific changes of oncogenes or other genes reported to be characteristic of endocrine or exocrine pancreatic tumors may provide a clinically useful test.

Chapter 2.5

Cystic Fibrosis

2.5.1 INTRODUCTION (p. 215)

Cystic fibrosis (CF) is the most frequent, fatal autosomal recessive disease, affecting Caucasians with an incidence of 1:2000–2500. It is caused by a dysfunction of the fluid and electrolyte transport in exocrine epithelia. The increased (up to fivefold the normal level) concentration of **sweat NaCl** of CF patients has long served as a diagnostic marker for this disorder. The inherent defect of Cl^- transport increasing the viscosity of secretions leads to severe pulmonary infections, eventually causing death in 90% of CF homozygotes.

A set of **consensus guidelines** for CF patient management has been issued, and the diagnosis of CF at the gene level is believed to "contribute to confirmation of the diagnosis of CF in the future" (The Cystic Fibrosis Foundation Center Committee and Guidelines Subcommittee, 1990).

The disturbances in electrolyte transport and its regulation have been linked to the *CF* gene product, the **cystic fibrosis transmembrane conductance regulator (CFTR).**

2.5.2 THE *CF* GENE AND ITS PRODUCTS (p. 215)

The ***CF* locus** has been linked to a polymorphic locus controlling the activity of serum **aryl esterase paraoxanase (PON)** subsequently assigned to **chromosome 7**. Other linkage associations were established between the *CF* locus and the DNA marker **DOCRI-917** on 7q, an anonymous DNA probe, **pJ3.11**, and the *met* oncogene locus, which placed the *CF* locus in the middle third of the long arm of chromosome 7, between bands q21 and q31.

Physical mapping of the *CF* region, using **pulsed-field gel electrophoresis, saturation cloning, chromosome jumping**, and **chromosome walking**, helped to accelerate cloning of DNA from the target region. Eventually, cDNA clones in sweat gland and tracheal cDNA libraries were identified, and a **6.5-kb mRNA transcript** was detected in tissues from CF patients.

The gene size was estimated to be 250 kb, and the presence of a **3-bp deletion** affecting **phenylalanine at position 508 (ΔF508)** from the coding region identified the gene as the *CF* gene. This deletion is the **predominant mutation** in Caucasians, affects about 75% of CF patients, and is present only in affected persons.

Other mutations have already been identified. The following are examples:

- in **exon 11**, three mutations cause **amino acid substitutions** and one produces a termination codon
- two **frame-shift mutations** have been described in **exon 7** of the *CFTR* gene, one caused by a two-nucleotide insertion and the other caused by a deletion of one nucleotide, which introduce a termination codon UAA (Ochre) at residues 369 and 368, respectively
- a mutation causing **a two-nucleotide insertion** in the *CF* allele has been described

A compendium of the results of a worldwide survey of the most common mutations (ΔF508) shows a marked variation in the prevalence of ΔF508 in different geographic populations.

The *CF* gene product, called **CFTR**, is 1480 amino acids long with a molecular mass of 168,138 Da. The predicted CFTR structure shows two amino acid sequence domains, resembling consensus nucleotide (ATP)-binding folds, and two repeated sequence motifs, suggesting structures characteristic of membrane proteins. Although the molecular function of CFTR is still not fully defined, the protein resembles the mammalian **multidrug resistance P-glycoprotein**. This so-called **ATP-binding cassette (ABC)** superfamily of transport systems includes, in addition to CFTR, the **periplasmic binding protein-dependent uptake systems** of eukaryotes, **bacterial exporters**, and the P-glycoprotein product of the ***MDR*-1 gene** among others.

Based on the presumed localization of CFTR in the epithelial membrane it is speculated that CFTR operates as an **unidirectional pump**. It adjusts the intracellular level of chloride channel inhibitors by regulating their transport across the cell membrane. **The CFTR function should not be equated, however, with that of chloride channel activity**.

Analysis of the effects of CFTR mutations on ion channel selectivity indicates that **CFTR is a cAMP-regulated chloride channel**. At the intracellular level, defective CFTR function results in defective glycoprotein processing and ligand transport. The mutant versions of CFTR are recognized as abnormal, are incompletely processed in the endoplasmic reticulum, and are subsequently degraded there. It is the **absence of mature, completely glycosylated CFTR** at the correct cellular location that **accounts for the molecular defects in most cases of CF**. Defective incorporation of the CFTR protein into the membrane leads to abnormal chloride ion transport.

The CF protein, found at elevated concentrations in the blood of CF carriers, is immunologically identical to a complex of two proteins, **MRP-8** and **MRP-14** — **calcium-binding proteins** expressed in the myeloid cells. This suggests additional molecular functions of the *CF* gene product.

2.5.3 TESTING FOR CF CARRIERS (p. 217)

An **estimated carrier frequency** of CF in a Caucasian population of northern European ancestry is high: 1 in 25. Parents carrying the CF mutation have a 1 in 4 chance of having either a normal or an affected child, and there is a 50% chance that the child will be a carrier. The risk of CF in each subsequent child is 1 in 4. The child of a sibling of a CF patient has a 1 in 120 risk of CF, compared to a 1 in 1600 to 2000 risk of CF in the child of a couple having no relatives with CF. Relative risk calculations have been determined in various countries for the most common CF mutation (ΔF508), and its frequency is heterogeneous in different populations, ranging from 30.3% to above 70% in Ashkenazi Jewish families.

Indirect assessment of **CF carrier status** is possible by **linkage analysis** using DNA markers known to be linked to the *CF* locus. A number of probes have been developed that detect polymorphisms in strong linkage disequilibrium with the *CF* locus: **J3.11, met H, met D, 7C22, KM.19, XV2C, B79a,** and **CS7**.

Not all asymptomatic persons with borderline or elevated **sweat chloride** concentrations and a positive family history are CF carriers. Whether such individuals are at greater risk for carrying a *CF* mutation can be answered only with a **direct carrier test for the *CF* gene**, rather than by linkage studies. Direct gene diagnosis of CF is now a possibility.

The appropriateness of **wide population screening** has been debated, and a view was expressed suggesting the **postponement of massive population screening until probes specific for all *CF* mutations become available** in order to avoid a significant proportion of false-positive results. At present, it is recommended that testing of CF status be provided to those with a family history of CF and to spouses of CF carriers or those affected with the disease.

Analysis of **carrier screening** efficiency at the present state of CF testing technology (only around 70% for CF carriers) suggests that the risk from **false-negative tests** in couples without previous children, who may be more interested in prenatal testing, can be more important than or equal to the mean risk from false-negative tests in the general population.

Using **PCR protocols**, the frequency of the most common *CF* mutation, ΔF508, has been evaluated to occur in 72% of affected chromosomes. Similar findings have been reported by others who observed the 96% frequency of occurrence of haplotype B on *CF* chromosomes carrying the ΔF508 mutations, whereas the same haplotype is encountered on 54% of chromosomes with non-ΔF508 mutations.

Screening of unaffected persons by PCR for a **4-bp tandem repeat (GATT)** at the 3′ end of **intron 6** in the *CF* gene revealed that this repeat existed only in two polymorphic allelic forms, either as a hexamer or a heptamer, with a predicted heterozygote frequency of 41%. The **GATT motif** appears to be a valuable DNA marker for haplotype analysis, and it significantly increases the informative nature of the XV2c and KM19 probes, allowing prenatal diagnosis to be informative when XV2c and KM19 alone fail.

2.5.4 CF WITH AND WITHOUT PANCREATIC INVOLVEMENT (p. 220)

Most CF patients develop **progressive pulmonary disease** and **pancreatic insufficiency with malabsorption**; however, approximately 10 to

15% of CF patients, who are **pancreas sufficient (PS)**, have only mild pulmonary involvement. Higher genetic heterogeneity is found among PS patients than among **pancreas insufficient (PI)** patients. In patients who are compound heterozygotes for the ΔF508 mutation (individuals carry the ΔF508 and other types of mutations), the **as yet unidentified mutations influence the severity of lung involvement**. The variability of the phenotype is apparently due to the influence of genes outside the *CF* locus.

The role of genetic factors in determining the **degree of pancreatic involvement** and the **rate of its progression** is better understood. The variable clinical course is at least partly attributable to specific genotypes of the *CF* gene. The genetic factors have a dominant role over nongenetic factors in determining the severity of pancreatic and lung disease. Testing for the presence of four other mutations (**G551D, R553 Stop, S549N**, and **R347P**) did not establish a correlation between the phenotype and the specific mutation. Caution should be exercised in basing the presumptive diagnosis of pancreatic functional status, and the decision to use enzyme therapy, on molecular genetic analysis.

This view is strengthened by a correlation observed between the severity of CF disease and the genotypes of the **more distant flanking marker loci**. Furthemore, different degrees of severity of the disease are associated with the **HLA haplotypes** — the C and D haplotypes are associated with lower age-corrected sweat sodium levels.

2.5.5 PRENATAL DIAGNOSIS OF CF (p. 221)

As soon as the first probes demonstrating tight linkage with the *CF* locus became available (J3.11 and *met*), they were used for **prenatal diagnosis** on **chorionic villus DNA**. Understandably, **linkage analysis** was not able to resolve all cases unequivocally. Comparison of **microvillar enzyme** and DNA analysis in the prenatal diagnosis of CF showed the enzyme results are **falsely negative** in over 10% of cases.

One of the most useful probes for prenatal diagnosis by linkage analysis, giving an adequate diagnosis in about 70% of families at risk for CF, is **KM-19**, which was used to develop a **PCR-based prenatal diagnostic test**. Another prenatal test for CF based on PCR has been reported that tests trophoblast tissue with oligonucleotide primers specific for a sequence identifying the **CS-7 locus**. Only about 50% of high-risk CF families were fully informative with CS-7 alone, necessitating the use of additional markers.

A combination of closely linked DNA markers (**XV2c, KM-19, *met* H,** and **J3.11**) can be used for prenatal diagnosis of CF.

In certain cases, the informativeness of direct detection is inferior to an indirect RFLP approach. When screening pregnancies at risk for CF in **Italy** and **Spain**, KM-19, XV2c, and **MP6d-9** probes used in RFLP analysis were informative in 86% of the families, whereas direct analysis was fully informative only in 30% of the cases, and partially informative in 34% of the cases. The limited effectiveness of using the ΔF508 deletion for prenatal diagnosis of CF and carrier screening in families of **South European ancestry** demonstrated **regional variations in the predominance of major *CF* gene mutations.**

ΔF508 is much more informative (in up 75% of cases) in **North American** and **North European populations**. In the **French population**, the informativeness of the XV2c and KM-19 RFLPs was the same as in Caucasian populations in North America.

In **Italian populations**, PCR analysis of the gene regions detected by the KM-19 and CS-7 probes for first-trimester prenatal diagnosis is informative in 86% of cases, and the results were confirmed by Southern blotting with a panel of probes.

In the **Belgian population**, prenatal diagnosis by RFLP analysis, using **seven DNA markers** in four families, allowed the accuracy of prediction to exceed 99%.

A number of **other non-ΔF508 mutations** have been described in various ethnic groups and geographic locales. Oligonucleotide primers have been described for the amplification of individual exons and the immediate flanking sequences in the introns for more efficient detection of CF-causing mutations.

A definite **advantage of DNA probes** over the frequently inconclusive nature of enzyme analysis of amniotic fluid has been reported. The **convenience of PCR amplification** allows the performance of assays on DNA extracted from Guthrie blood spots that had been in storage at room temperature for up to 7 years.

Prenatal testing for couples carrying the ΔF508 mutation has been extended to **preconception** and **preimplantation diagnosis** the usefulness of which has already been demonstrated in successful full-term pregnancies.

2.5.6 OTHER ASPECTS OF PCR ASSAYS FOR CF (p. 222)

PCR assay for the known DNA sequences characteristic of the ΔF508 deletion directly and unambiguously identifies affected individuals in about 70% of cases. **This test cannot be used for the direct determination of carrier status, because other as yet unidentified mutations will not be identified without the use of mutation-specific oligonucleotide primers.** In such cases, linkage analysis within families will still have to be used.

A novel approach, capable of detecting known DNA sequence variations down to a single nucleotide, uses the **primer extension strategy**. It is based on the fidelity of the DNA polymerase and allows the identification of a mutant product in gel electrophoresis and autoradiography of the extended primer. This general testing strategy, called **"single nucleotide primer extension" (SNuPE)**, is also suggested for any genetic disease where sequence variation is known.

To detect *CF* mutations when they do not create a useful restriction site, **such a site can be generated by PCR and specially chosen primers.** Likewise, a **PCR-mediated site-directed mutagenesis** has been used to create new allele-specific restriction sites, arising from a **single-base mismatch specifically introduced into a PCR primer** that abuts the mutation under study.

A **simplified PCR protocol** has been described for the detection of the ΔF508 mutation. Instead of the usual three-temperature cycling, the new procedure uses a **two-temperature PCR**, which combines annealing and extension steps performed for 30 sec at 72°C alternating with a 60-s denaturation step at 94°C. This method favorably compared to the more labor-intensive traditional approach that calls for hybridization of allele-specific oligonucleotide (ASO) probes to Southern blots prepared from PCR products amplified by a three-temperature method. A promising adaptation of minisequencing of the PCR-amplified DNA using biotinylated primers and capture of the product on solid support allows this assay to be automated for large-scale quantitative detection of point mutations and deletions.

A **sample pooling strategy** has been proposed in order to reduce the number of PCR tests in a mass screening program for the ΔF508 mutation. For a German population, with an ΔF508 heterozygosity incidence of about 1 in 35, the optimum number of samples to be pooled in one assay is calculated to be 24. This approach allowed the achievement of substantial economy in analyzing large numbers of DNA samples, **reducing the number of PCR assays by as much as 77%, without reducing the efficiency of screening for this particular mutation.**

Chapter 2.6
Lung

2.6.1 SMALL CELL LUNG CARCINOMA (SCLC) (p. 225)

SCLC comprises up to 20% of all lung cancers, occurring predominantly in male smokers, and can be divided into three types: **oat cell**, **fusiform**, and **polygonal cell type**. Ultrastructural studies may be needed for definitive diagnosis. Other characteristics can be used to establish the type of lung cancer:

- **nuclear chromatin** and **nucleolar patterns**
- the presence of **dense-core neurosecretory-type granules**
- immunohistochemical markers such as for **neuron-specific enolase** (NSE), **neurofilaments, Leu-7, synaptophysin, chromogranin A**, **L-dopa decarboxylase**, **pituitary polypeptide** (**7B2**), antigens of the **membrane cluster-5** and **cluster-5A**, and **p21**^{ras}, which are identifiable by the **monoclonal antibody RAP-5**

2.6.1.1 CHROMOSOME ABERRATIONS (p. 225)

Structural abnormalities have been found in virtually all chromosomes in cell cultures established from **small cell** and **non-small cell lung cancers** (**SCLC** and **NSCLC**). The most consistent aberration is a deletion of the short arm of **chromosome 3**, particularly in SCLC (93%). The **thyroid hormone receptor**, assigned to **3p24**, may have a specific role in lung cancer. The pattern of loss of 3p alleles was found to be different for NSCLC and SCLC.

The analysis of SCLC is of particular interest because the ***RB*** **gene** (chromosomal region **13q14.1**) and the ***p53*** **gene** (chromosomal region **17p13**) — the two genes with established tumor suppressor functions — appear to be involved in the development of this malignancy.

2.6.1.2 PLOIDY IN SCLC (p. 226)

SCLC features a significantly higher DNA content than **typical carcinoids**, whereas **atypical carcinoids** show intermediate values. Computer-assisted ICM of cytospin preparations differentiated between typical and atypical carcinoids and SCLC based on DNA content of cells. Nevertheless, the clinical usefulness of DNA ploidy determinations is not certain as an independent criterion of the malignant potential of carcinoid tumors and **atypical adenomatous hyperplasia** of the lung.

2.6.1.3 ONCOGENES IN SCLC

2.6.1.3.1 The Retinoblastoma (*RB*) Gene (p. 226)

Structural abnormalities of the ***RB*** gene, aberrant *RB* mRNA expression, and absent or dysfunctional RB protein have all been observed in the majority of SCLC and only in 10 to 20% of NSCLC. The *RB* cDNA clones **pRB4.5** and **p6NRO.5** can be used to assess whether tumor cells lost an allele of the *RB* gene.

No clear understanding has so far been gained of the **specific functions** of the *RB* gene product, the M_r **110,000–116,000 phosphoprotein**, expressed in more than 50% of SCLC tumor cell lines, and to what extent the degree of phosphorylation of this protein plays a role in SCLC. It appears, however, that a point mutation in exon 21 of the *RB* gene is sufficient to alter its functional integrity drastically.

2.6.1.3.2 *p53* (p. 227)

A role for *p53* in the natural history of lung cancer is suggested by abnormalities of **chromosome 17p**, the frequent mutations of ***p53*** in lung cancer, and assignment of the single copy of the *p53* gene to chromosome region **17p13**. In addition to DNA abnormalities, which include a **homozygous deletion**, and **point** or **small mutations** occurring in the **open reading frame** of *p53*, abnormalities of *p53* mRNA have also been detected in the tumor tissue of patients. Using a **chemical mismatch cleavage technique** combined with preliminary **amplification** of targeted genomic DNAs by **PCR**, single-base mutations in the *p53* gene have been efficiently identified. The **wild-type p53 proteins** have a short half-life, whereas the mutant proteins have a much longer half-life (4 to 8 h vs. 6 to 20 min); this change accounts for the frequent finding of p53 proteins in cancer cells. **The very fact of identification of the protein may be viewed as diagnostic of mutation**. Usually, p53 appears to be involved in the later stages of tumor development, interacting with other oncogenes such as c-*myc* and Ha-*ras* and playing a role in SCLC **tumor progression** rather than initiation. PCR-sequencing protocol allows one to detect mutations of *p53* in exons 5 to 9 that are characteristic of exposure to radon in uranium miners.

2.6.1.3.3 The *myc* Gene (p. 227)

In SCLC cell lines amplification and expression of **c-*myc*, N-*myc***, and **L-*myc*** protooncogenes occur. Using RNA-RNA *in situ* hybridization (ISH) the amplification of c-*myc* observed in cell lines could also be confirmed in surgical biopsy and autopsy material from patients with SCLC.

Amplification of the N-*myc* protooncogene has been readily demonstrated directly in tumor specimens taken from patients. **Restriction endonuclease mapping** and **DNA sequence analysis** allowed the isolation of L-*myc* first defined in a lung tumor cell live. A correlation between amplification of c-*myc* or N-*myc* with a shorter survival time and the more malignant behavior of the variant (as opposed to classic) histology of SCLC has been reported. On the other hand, no morphological difference between cells expressing the c-*myc*, N-*myc*, or L-*myc* oncogenes could be observed. Somewhat discrepant observations are reported by other investigators on the amplification of various members of the *myc* gene family in SCLC. No difference in gene frequency of the L-*myc* RFLP is found between lung cancer patients and normal controls.

c-*myc* amplification appears to play a more significant role in **malignant progression**, rather than in development of SCLC. **Chemotherapeutic regimens** may affect the reported involvement of *myc* protooncogenes in lung cancer.

It does not appear at this point that assessment of alterations in the amplification and/or expression of the *myc* gene family may significantly contribute to management of patients with SCLC.

2.6.1.3.4 The *jun* Gene (p. 228)

The *jun* gene has a role as a mediator of growth factor or tumor promoter action on transcription. In about 25% of cell lines derived from different types of lung cancer, the level of ***jun*A** expression has been found markedly elevated. A related protooncogene, ***jun*B**, is expressed to high levels in all the tumor cell lines studied, and, like *jun*A, it is expressed to high levels in normal lung. The high level of expression is due to transcriptional activation.

Although deregulated expression of the *jun* family transcription factors is thought to contribute to the development of lung cancer, **it does not appear practical to consider evaluation of the *jun* family transcription factors for diagnostic or prognostic purposes at this time**; the more so because similar levels of c-*jun* expression, with large variations between patient cases, have also been found by ISH in matched pairs of NSCLC and normal lung cells.

2.6.1.3.5 *erb*B, *ras*, and Growth Factors in SCLC (p. 229)

The ***myc***, ***ras***, and ***erb*B** family genes constitute about 90% of the amplified protooncogenes studied in primary human lung carcinomas. The majority of amplified *ras* genes were present in advanced stages of lung cancer. The ***int*-2** gene was amplified in two squamous cell carcinomas. Another gene, ***src***, has been correlated with neuroendocrine differentiation, and reported in SCLC. More information is definitely needed to make practical use of the observed alterations for patient management.

In this context, the differential expression of c-*erb*B-2 mRNA in SCLC is of interest. The highest level of c-*erb*B-2/*neu* expression has been found in four of four adenocarcinomas, with the expression of this protooncogene being different between NSCLC and SCLC. The *neu* gene was not amplified. No expression of the *neu* gene, or only minimal at best, could be detected in SCLC, whereas 38% of NSCLC specimens, in contrast, showed high *neu* expression both at early stages and in advanced tumors.

While **p21^{ras} oncoprotein** is overexpressed in adenocarcinomas compared to squamous cell carcinomas, overexpression of the c-*myc* gene product, **p62**, preferentially occurs in **poorly differentiated squamous cell carcinomas** compared to the **well-** and **moderately differentiated** tumors. Interaction of N-*myc* and v-Ha-*ras* induces a specific change in cultured SCLC cells, including the expression of **platelet-derived growth factor B chain**, **transforming growth factor α**, and **epidermal growth factor receptor**. Apparently the Ha-*ras*-1 gene is responsible, to a certain degree, for the development of the NSCLC phenotype. The definite association of activated Ha-*ras*-1 gene with lung cancer can be shown by RFLP analysis with *Msp*I/*Hpa*II restriction digestion.

Patients with Ki-*ras* mutations have **significantly worse survival** than lung cancer patients without an activation. *ras* oncogenes, but not the *L-myc* gene, are implicated in the development of **metastases** in lung adenocarcinomas.

Transforming growth factor α (TGF-α) levels in lung carcinomas failed to correlate with histological type, stage, grade, or degree of desmoplasia in lung cancer. Nevertheless, overexpression of the **growth factor receptors, EGFR** (epidermal growth factor receptor), and **HER-2/Neu**, is observed in tumors and normal bronchial epithelium of patients with lung cancer.

Preliminary correlations begin to emerge between the **level of expression** of various growth factors (e.g., TGF-α and EGF) and growth factor receptors (e.g., EGFR) and **survival of lung cancer patients**. The progression of SCLC to the NSCLC phenotype may be influenced by these factors.

A role for activated protooncogenes and growth factors in lung carcinogenesis is certainly suggested by the above-mentioned observations, although the specific functions of individual genes remain to be defined. The importance of these studies is underscored by the effective and selective inhibition of Ki-*ras* expression and associated tumorigenicity of lung cancer cells by **antisense RNA**. The practicality of evaluating the expression of various growth factor genes in lung cancers still awaits further investigation.

2.6.2 NON-SMALL CELL LUNG CARCINOMA (NSCLC)

NSCLC includes a diverse group of lung carcinomas comprising 75% of lung neoplasms. Its distinction from SCLC is, in part, justified by the difference in the response of these malignancies to treatment. While SCLC is predominantly treated by chemotherapy, surgery is used in NSCLC.

2.6.2.1 CHROMOSOME ABERRATIONS (p. 230)

The most frequent chromosome alteration is the loss of DNA from the short arm of **chromosome 17**, occurring more in **squamous cell carcinomas (SCC)** than in **adenocarcinomas**. Loss of DNA sequences from **chromosome 11** accompanied the majority of chromosome 17 losses in SCC. In almost 50% of cases losses were also noted in **chromosome 3**.

Finally, loss of heterozygosity (LOH) at the **13q** locus was observed in some cases. **The pattern of DNA sequence losses is different not only between SCLC and NSCLC, but also between SCC and adenocarcinomas, the latter showing similarity to NSCLC.**

All NSCLC tumors (which included adenocarcinomas, SCC, and large cell carcinomas) appear to be aneuploid and display a **complex pattern of karyotypes** with multiple structural and numerical abnormalities. Loss of materials from **chromosome 9p** occurred in 90% of cases. Other **nonrandom alterations** involved **chromosomes 1p, 3p, 5q, 6p, 6q, 7p, 8p, 11q, 13p, 14p, 15p, 17p,** and **19p**.

2.6.2.2 ONCOGENES AND GROWTH FACTORS (p. 231)

In addition to the above-mentioned deletions from 11p, 3p, and 17p, evaluation of **protooncogenes** and **growth factors** in NSCLC revealed alterations of **c-*myc*, N-*myc*, *ras*, *erb*B,** and ***neu***. At this time, a still-inconclusive picture emerges concerning the consistency of alterations of particular protooncogenes. Changes in **c-*erb*B-1** and **c-*erb*B-2** were found to correlate with the histological type of tumor and were more common in advanced tumors. Loss of c-*ras*-Ha or c-*myc* was frequent in primary tumors with more aggressive behavior. In primary giant cell carcinoma of the lung a specific rearrangement of the c-*myc* gene has been identified.

Amplification of protooncogenes, in particular c-*myc* and Ha-*ras*, appears to be a relatively rare event in NSCLC, although the correlation with relapses appears quite strong. The amplification of c-*erb*B-1, which encodes the EGFR, is commonly detected in squamous and large cell undifferentiated cancers. EGFR expression is seen more frequently in well-differentiated tumors. Opinions differ on the **frequency of protooncogene amplification**, ranging from 56% to rare, and **no consensus exists on the utility of these measurements for diagnosis or prognosis**.

Mutations of Ki-*ras*, leading to its activation and 20-fold amplification, are thought to be among the early events in the development of NSCLC. Moreover, **the detection of Ki-*ras* point mutations in codon 12, found by PCR analysis to be more frequent in lung adenocarcinomas from smokers, appears to provide a useful indicator defining a subgroup of lung adenocarcinoma patients with very poor prognosis and relapses**, occurring despite a small tumor load and radical surgery. Tobacco smoke contributes to the appearance of point mutations in Ki-*ras* in human lung adenocarcinomas.

The role of the **TSGs** in NSCLC is suggested by the finding of functional and structural abnormalities in the ***RB* gene**, ranging from the absence or alteration of *RB* mRNA to deletions of the *RB* gene. Correlations with the clinical behavior are still being investigated. ***p53***, which is known to be involved in the genesis of SCLC (see above), also plays a role in the pathogenesis of early-stage NSCLC.

The cellular and molecular mechanisms underlying **transition between**

lung cancer phenotypes is an active field of research undoubtedly bound to influence patient management strategies in the near future. Evidence on the involvement of various genes in lung tumors also includes the expression of

- the **atrial natriuretic factor gene** in SCLC
- **endothelin mRNA**
- the expression of **blood-group antigen A** in NSCLC tumor cells, which can be used as an important favorable prognostic factor

Evaluation of a **combination of molecular markers** is certainly preferable to reliance on a single prognosticator when devising treatment protocols.

2.6.3 PLEURAL MESOTHELIOMA AND PULMONARY FIBROSIS

2.6.3.1 CHROMOSOME ABERRATIONS (p. 232)

Although **clonal chromosomal abnormalities** have been detected, **no chromosomal aberration specific to mesothelioma could be demonstrated**. However, the number of copies of the short arm of **chromosome 7** proved to be **inversely correlated with survival**, suggesting this numerical chromosome change may be used as a possible prognostic factor.

2.6.3.2 ONCOGENES AND GROWTH FACTORS (p. 232)

Several correlations have been established between nontumorous pulmonary disease and oncogene(s) and growth factor(s) expression. Some examples include

- an association between the **expression of c-*sis* protooncogene**, encoding the B chain of platelet-derived growth factor (PDGF), and **pulmonary fibrosis**
- the release of four times more **PDGF** by alveolar macrophages recovered from the lungs of patients with **idiopathic pulmonary fibrosis** than by alveolar macrophages from normal persons
- the linkage of pulmonary fibrosis induced by **silica** to **tumor necrosis factor α (TNF-α)** in laboratory animals
- the expression of **intercellular adhesion molecule 1 (ICAM-1)**, which has been strongly implicated in airway inflammation associated with **asthma, chronic bronchitis, emphysema**, and **idiopathic pulmonary fibrosis**
- by measuring the expression of **EGFR** no correlation could be established, however, between **malignant** and **benign mesothelial tissue** or between different histological subgroups of malignant mesothelioma and EGFR expression

2.6.4 NUCLEAR DNA CONTENT AND RESPONSE TO TREATMENT OF LUNG CANCERS (p. 233)

A significantly shorter survival was observed in patients with aneuploid tumors, as determined by flow cytometry (FCM) of archival paraffin blocks. Although ploidy did not show a reproducible correlation with the clinical and pathological characteristics (adenocarcinoma vs. SCC), it proved to be the most important determinant of survival. **Ploidy determinations are recommended for use in deciding whether adjunct therapy can be of benefit — as in the case of aneuploid tumors.**

The hyperdiploid DNA pattern was associated with a better response to combined **cytoxan + adriamycin + vincristine therapy** than was the near-diploid pattern. Others have noted no difference, however, in survival characteristics between patients with near-diploid and aneuploid tumors.

Multidrug resistance in lung cancer has been evaluated by the FCM analysis of **Hoechst 33342 dye uptake**. This dye accumulation parallels that of a number of cytotoxic drugs affected by the MDR phenotype. Measurement of the nuclear binding of this fluorescent compound by cells of different histological types of lung cancer (SCLC vs. NSCLC) suggested that different mechanisms of drug resistance operate in lung tumors of different tissue type.

Expression of the major polycyclic aromatic hydrocarbon-inducible cytochrome P4501A1 gene (**CYP1A1**), implicated in lung carcinogenesis through modification of procarcinogens in tobacco smoke into potent carcinogens by the cytochrome-dependent monooxygenase, has been shown to increase in smokers and lung cancers.

Chapter 2.7

Tumors of the Head and Neck

2.7.1 DNA PLOIDY (p. 235)

The high percentage of **aneuploidy** (48%) in **squamous cell carcinoma (SCC)** of the **oral cavity** correlates with the size of the tumor, low histological grade, and the presence of lymph node metastases. Likewise, polyploidy is higher in poorly differentiated tumors and in metastases. The **proliferative index** does not reach statistically significant correlation with tumor size or duration of the disease, although it is **inversely correlated** with the degree of differentiation.

In **nasopharyngeal cancer (NPC),** the **highest survival** following radiation therapy was observed in patients with **anaplastic carcinoma** (75%), followed by **lymphoepithelioma** (60%), and the worst survival was in patients with SCC (30%). There was no significant correlation between survival and the DNA index. The most important predictor of survival was the degree of local invasion.

The majority (62%) of **adenoid cystic carcinomas** of the submandibular gland appear to be diploid. Aneuploidy is more frequently encountered in advanced clinical stages, and it seems DNA ploidy measurements may be an effective adjunct to the clinicopathological assessment in this type of tumors.

Similar to NPC, the majority (57.1%) of SCCs of the **larynx** are aneuploid or tetraploid. Survival rates correlate with ploidy among tumors with a similar degree of differentiation, the lower survival rate being observed in nondiploid cases compared to diploid (27.7 vs. 41.7%, respectively) as well as in advanced cases, such as those entering chemotherapy for **end-stage disease**.

2.7.2 CHROMOSOME ABERRATIONS (p. 236)

Multiple unrelated structural clonal chromosome aberrations have been a consistent finding in the predominant type of cancer (SCC) of the head and neck region. A **clustering** of chromosomal **breakpoints, homogeneously staining regions**, and **double minutes** was found in some tumors. The **candidate genes** potentially affected in those regions were ***bcl*-1, *int*-2, *hst*-1,** and **N-*ras*.** In fact, all of these protooncogenes have been found to be amplified in head and neck carcinomas (see below).

2.7.3 ONCOGENES AND GROWTH FACTORS

2.7.3.1 *RAS* (p. 236)

PCR amplification combined with mutation-specific oligonucleotide probing revealed the presence of a **codon 61** mutation in Ki-*ras* in SCC of the oral cavity. Interestingly, the signal was stronger in the blood of an affected patient than in the DNA from the neoplasm or the salivary gland.

The expression of the Ha-*ras* protooncogene in SCC of the oral cavity has been localized to the tumor and is expressed more in areas with higher proliferative activity and invasion, suggesting that the evaluation of the pattern of Ha-*ras* expression may be helpful in predicting the development and progression of oral SCC.

Heterozygosity at the c-Ha-*ras* locus and LOH at **11p** may be related to the involvement of **tumor suppression genes** in oral SCC.

2.7.3.2 *MYC* (p. 236)

The expression of c-*myc* in SCC of the head and neck fails to correlate with various clinicopathological parameters, including histopathological differentiation, lymph node invasion, extracapsular extension of the tumor, TNM staging, and so on. On the other hand, the **survival** of patients with tumors having elevated levels of the c-Myc oncoprotein was **significantly shorter** than in those with lower levels of c-*myc* expression. It appears, therefore, that **measuring c-*myc* expression can be used as an effective prognostic indicator in head and neck SCC**.

Of the several protooncogenes evaluated (**c-*myc*, N-*myc*, L-*myc*, N-*ras*, Ha-*ras*, Ki-*ras*, *erb*B, *erb*B-2, *raf*,** and ***int*-2**), only amplification of *int*-2 and c-*myc* could be observed. Other oncogenes were not amplified in any of the tumors of the head and neck studied in that series.

The same RFLP pattern of L-*myc* gene expression can be demonstrated in **peripheral blood cells** in patients with oral SCC as in the tumor tissue. The presence, rather than the absence, of the smaller, 6.6-kb **S fragment** correlated with the likelihood of developing a poor to moderately differentiated tumor.

A **variable pattern** of oncogene amplification and deletions has been demonstrated in SCC cell lines and primary tumors of the head and neck. Amplification of c-*myc* and **EGFR** was observed in stage 3 tumors, whereas N-*myc* and ***neu*** amplifications as well as ***p53*** deletion correlated with recurrence. A strong positive correlation exists between a smoking and drinking history with the overexpression of the *p53* gene in head and neck SCC.

Although the expression of *neu* is observed in normal salivary glands, SCC of the oral and laryngeal mucosa did not display an enhanced *neu* transcription level. In contrast, parotid **pleomorphic adenomas** and a salivary gland adenocarcinoma featured enhanced *neu* expression. In pleomorphic adenoma, the degree of *neu* expression varies and mostly is not accompanied by amplification of the gene.

2.7.3.3 EGFR (p. 237)

Conflicting findings have been reported concerning the expression of the gene encoding the EGFR in head and neck SCC tumors. **Amplification** or **overexpression** of the gene for EGFR has been reported.

2.7.4 THE AMYLASE GENE (p. 237)

No evidence of structural rearrangements has been obtained by Southern analysis of the **amylase gene** in a **Warthin's tumor**, an **adenoid cystic carcinoma**, and a **mucoepidermoid carcinoma of the parotid gland**.

Pleiomorphic adenoma, although positive for the amylase protein, failed to show an elevation of gene expression. Warthin tumors were negative for both the amylase protein and amylase mRNA. One of the consistent findings in the salivary glands of **Sjögren's syndrome** patients is the presence of Epstein-Barr virus (EBV), detectable by immunocytochemistry, ISH, and PCR.

2.7.5 EPSTEIN-BARR VIRUS (p. 237)

EBV can be found in nasopharyngeal carcinoma by PCR in fine-needle aspirate material, and its identification is considered important in preventing the development of advanced lesions with squamous cell differentiation.

Part 3
Gene Level Evaluation of Selected Breast, Gynecologic and Urologic Tumors

Chapter 3.1

Breast

3.1.1 INTRODUCTION (p. 271)

Efforts to establish a systematic approach to **management** of **patients** with breast cancer led to the development of **quality assurance guidelines** for specifying the type of information and the format of its presentation in the pathologist's report.

TNM clinical and pathological staging (**T**, extent of primary tumor; **N**, absence or presence and extent of regional lymph node metastases; **M**, absence or presence of distant metastases) provides a reference framework for the assessment of cases, allowing uniform management strategies to be developed and compared on an intra- and intercase basis. The **new TNM** staging system of the American Joint Committee on Cancer/Union Internationale Centre le Cancer (AJCC/UICC) is given in Table 3.1 (Handbook p. 272).

The morphological analysis of breast tumors follows the currently accepted **WHO histological classification** (Table 3.2; Handbook p. 273).

3.1.2 PLOIDY ANALYSIS (p. 271)

The choice and administration of **systemic adjuvant therapy** in the treatment of breast cancer relies on the predictive clinical behavior of a given form of cancer and the evaluation of its histopathological parameters. A particular challenge is presented by **node-negative disease**. The **proliferative activity** of tumor cells and **DNA content** of individual cells as determined by **static cytomorphometry (ICM)** can complement and correlate with histopathological evaluation of the tumor by microscopy. The **S-phase fraction** (SPF) characteristically shows a significant correlation with **nuclear grading** and **mitotic index**.

The validity of measuring the **thymidine labeling index** as a prognostic indicator **cannot be confirmed.** On the other hand, **nuclear DNA content** of tumor cells measured by ICM **significantly correlates with prognosis** of breast adenocarcinomas and the histopathological grading of ductal carcinomas followed for up to 13 years. It does not correlate, however, with axillary node status. Nevertheless, in both node-negative and -positive patient groups DNA content can serve as an objective biological **marker of aggressiveness.**

The **predictive value of aneuploidy of primary tumors** decreases in more advanced disease. In general, ploidy proves not to have significant influence on **long-term survival** (minimum follow-up of 22 years) in patients with postsurgical stage T_{1-2} N_0 breast cancer.

Some workers include the following among parameters of independent prognostic value for long-term survival:

- **SPF**
- a **DNA index (DI)** of less than 1.2, which carries a more favorable prognosis than one above 1.2
- **lymphatic vessel invasion** of cancer cells — considered the most important independent prognostic variable

However, other investigators observe that in node-negative breast cancer following surgical resection, the most valuable prognostic information is afforded by an estimate of cells in S phase: a **high SPF is significantly associated with early relapse.**

In the case of **cystosarcoma phyllodes** of the breast, the DNA ploidy evaluation appears to offer a useful, objective adjunct to the clinicopathological assessment of the neoplasm.

Two-color multiparametric FCM analysis, in which the cells are labeled by **cytokeratin monoclonal antibody** and **propidium iodide**, enhances the informative prognostic value of ploidy and SPF estimates. Cytokeratin-gated DNA evaluation allows the exclusion of contaminating stromal and inflammatory cells. Even in this format, the **SPF appears to be of greater prognostic value than ploidy.**

The significantly better prognostic value of the combined evaluation of **ploidy** and **SPF** has been amply demonstrated. A **combination** of DI and SPF afforded the distinction of three types of DNA histograms and subdivided the SPF into **three prognostic categories** (<7%, 7 to 11.9%, and >12%) relative to the risk of death (eightfold increase in SPF >12%). However, aneuploidy and SPF estimates may be significantly influenced by the very treatment the patients receive.

The **extreme DNA heterogeneity** of breast cancer is thought to account for the disagreement in the findings on the prognostic value of DNA ploidy determinations. No less than **four samples** are now recommended for a reliable determination of the DNA ploidy status of primary breast tumors by FCM.

3.1.3 CHROMOSOMAL ABNORMALITIES IN BREAST CANCER (p. 275)

No single systematic study has been performed to define chromosomal alterations predisposing to the development of various forms of breast cancer (and/or benign lesions), aberrations associated with the progression of the tumor, and, finally, those contributing to metastatic spread.

The most frequently involved areas of cytogenetic alterations in primary breast tumors are **1p32-36**, **1q21**, and **1p13**. While 1p alteration may be associated with early stages of breast cancer, changes on **1q** can be related to **metastatic spread** of the disease.

Other chromosomal abnormalities in breast carcinomas include

- deletions of **17p** in the **p13-3 region**, **3p**
- abnormalities on **chromosomes 6**, **8**, **11**, and **13**

For the sake of simplicity, the genes associated with breast cancer (as well as other neoplasias) can be subdivided into those involved in the **initiation** of malignancy and those promoting **progression** of the disease.

Cytogenetic changes on 11p13-15 detected in pleural effusions suggested an association with **progression** of breast carcinoma, whereas changes on chromosomes 1 and 3 appear to be associated with an **early** event. Others believe, however, that chromosome 1p36 abnormalities are related to later stages of tumor progression, whereas deletion or rearrangements noted on **chromosome 16q22** may be important in the early stages of carcinogenesis. Although in breast cancer no common marker chromosome could be found several breakpoints involving, for example, 1q11, 3q11, 7p11, 9q11, and 13q11 could frequently be detected.

Linkage analysis of pedigrees with early-onset breast cancer places the responsible gene on **chromosome 17q**, in the region 17q12-q23 close to **D17S74**. The frequent loss of heterozygosity (LOH) on 17p has been defined in the p13-3 region and associated with overexpression of *p53* mRNA, pointing to a gene regulating *p53* expression in the majority of breast cancers.

3.1.4 PROTOONCOGENES AND TUMOR SUPPRESSOR GENES IN BREAST CANCER (p. 276)

In animal models, the expression of ***int*-1**, ***int*-2**, ***int*-3**, ***int*-4**, and ***hst*** cellular genes has been detected. The relatedness of these genes to the **fibroblast growth factor** (FGF) family of genes suggests that their abnormal expression may delay **senescence** of mammary epithelium or provide a stimulus for **angiogenesis** in the tumor.

Other genes, such as c-*myc* and Ha-*ras*, when inserted into transgenic mice do not interfere with normal mammary gland development, whereas in the presence of mutated, but not normal, *neu* oncogene normal mammary gland development does not occur. The chromosomal studies discussed above suggested that some of the well-known tumor **suppressor genes** (***RB*-1**, ***p53***, ***DCC***) may also be involved. In the following sections, pertinent findings will be summarized that **can potentially be used for clinical purposes.**

3.1.4.1 *erb*B-2/*HER*-2/*neu*

3.1.4.1.1 Amplification of the c-*erb*B-2/*HER*-2/*neu* Gene (p. 276)

The c-*erb*B-2 protooncogene encodes a membrane receptor with protein kinase activity; although related to **epidermal growth factor receptor** (EGFR), its amplification in human breast tumors can be identified and distinguished from a complex pattern of EGFR. Another member of the EGFR family, ***erb*B-3**, has been identified, mapped to **chromosome 12q13**, and found to be overexpressed in a subset of human mammary tumors.

The c-*erb*B-2 protooncogene, also termed ***neu*** (because it had been known to be activated in neuro- and glioblastomas), has 50% of its amino acid sequence identical to that of the EGFR.

A point mutation in the *neu* protooncogene converts it into an oncogene and the ***neu* gene product, p185^{neu}**, is oncogenic, whereas the product of c-*neu*, p185$^{c\text{-}neu}$, is not. In addition to point mutations in the kinase domain, which alone can impart sarcomagenic potential to the *neu* product, other modifications leading to activation of the transforming potential exist as well, including **carboxyl-terminal truncation** and **internal deletion**.

A 2- to over 20-fold **amplification** of the *neu* gene has been reported in 30% of human breast cancers, providing greater prognostic value than other prognostic factors, including hormonal status. This observation has been made in lymph node-positive cases. Since that finding a large number of reports have appeared that support and extend this observation to node-negative cases, refute the prognostic value of *neu* amplification entirely, or admit its usefulness with certain limitations.

Amplification of the *neu* oncogene (usually in 20 to 30% of cases) can be accompanied by **rearrangements** or **translocations** (7%) of the gene.

Amplification of the *neu* gene is thought to be an **early event** in breast carcinogenesis that is important in the **initiation of breast cancer**. The high proportion of cases with **amplified *neu*** (88%) are node positive, indicative of an **association between *neu* amplification and early metastatic spread** of the tumor.

In a retrospective analysis, the extent of *neu* amplification was found to be of less importance than the histological grade. Both parameters, however, were strongly associated with the **aggressiveness** of the tumor, rather than its **metastatic spread**. The *neu* gene amplification in node-negative tumors has been more frequently found in aneuploid tumors, and in those with poor nuclear grade.

In *neu*-negative patients, aneuploidy is associated with a significantly **shorter relapse-free survival**. In a large series of node-negative cases (704 patients) followed up to 16 years, 18% of the node-negative patients with amplified *neu* gene had relapsed compared to only 5% of the patients who had remained in remission.

In node-positive patients, DNA ploidy did not influence either survival or relapse-free survival. **Amplification** of the *neu* oncogene in tumors with aneuploidy, poor nuclear grade, and early metastatic spread suggests and, in fact, has been documented to predict **poor prognosis of node-positive patients**. The occurrence of ***neu* amplification proved to be a significant predictor of a short disease-free interval and of poor prognosis**. It is strongly suggested that amplification of *neu* may contribute to the pathogenesis of some forms of node-negative breast cancer, and can be used to identify a subset of high-risk patients.

That *neu* amplification can at least identify a subset of node-negative patients at higher risk for poor clinical outcome is acknowledged by investigators who tend to deemphasize the prognostic value of *neu* amplification. In a series of 362 tumors from patients with primary breast cancer (both node positive and node negative) the overall *neu* amplification rate was 33%. **Amplification of *neu* did not correlate with either disease-free or overall survival.** The absence of correlation was particularly unambiguous for patients with node-negative disease. **Analysis of *neu* alone is not considered to be of value**, because no significant correlation could be observed between its amplification and recurrence of the 157 primary and 14 metastatic breast cancers.

Coamplification of c-*erb*B-2 (*neu*) and c-*erb*A oncogenes in breast cancers has been considered to be either fortuitous, or to be indicative of lymph node metastasis.

More frequent early recurrence of tumors with detectable amplification of any of the five protooncogenes (**c-Ki-*ras*, c-*myc*, c-*myb*, *neu***, and ***int*-2**) was not considered to offer additional prognostic information.

With respect to **hormone receptor status**, *neu* amplification, although present in 14% of **estrogen-positive** (ER+) patients, is twice as frequent (28%) among **estrogen-negative** (ER–) patients. Similarly, it is more frequent in **progesterone-negative** (PR–) compared to **progesterone-positive** (PR+) patients (22 vs. 16%).

The degree of *neu* gene amplification is suggested as an additional parameter to be considered in the clinical and laboratory assessment of a specific subset of patients who can benefit from adjuvant therapy.

3.1.4.1.2 Expression of the *neu* Gene (p. 278)

A number of studies have evaluated the **degree of expression** of the *neu* gene, its association with specific histological characteristics of breast cancer, a correlation between **overexpression** and **amplification** of the gene, as well as its utility for prognosis. The ***neu* gene product**, a 185-kDa protein, can be readily identified in frozen sections and paraffin-embedded tissue, and routine antibody-staining techniques can be used to reveal overexpression of the protein.

An association of the amplification of the *neu* protooncogene with increased levels of expression of the Neu protein frequently occurs. The demonstration of overexpression of the gene accompanying its amplification is important because, for example, the *erb*A gene is amplified in breast cancer but not expressed. In a large series of 668 breast cancer specimens, *neu* DNA amplification was paralleled in all cases by overexpression of *neu* mRNA and Neu protein.

A further degree of complexity in this situation comes from the fact that **overexpression has not always been found to be linked to amplification of the *neu* gene**, although combined amplification and overexpression of the gene is significantly associated with poor prognosis. Others believe that it is the ***neu* amplification, but not *neu* mRNA expression**, that is correlated with poor prognosis and significantly shorter time to treatment failure.

Overexpression of the 185-kDa oncoprotein is related to aggressiveness of breast carcinoma, and only relatively rare positivity was detected in the other malignancies such as lung (1%), colon (4%), and bladder (2%). In prostate, skin, or SCLC, no positive staining could be observed at all. However, **only tumors of solid/comedo (large cell) type showed *neu* expression**.

No expression of the *neu* gene product could be observed in

- **atypical lobular** or **ductal hyperplasia**
- **lobular carcinoma** associated *in situ* with **ductal carcinoma *in situ*** **(DCIS)** or **papillary**, **cribriform**, or other *in situ* carcinomas

The **staining pattern of the *neu* gene product** can be detected with two types of monoclonal antibodies:

- **3B5**, raised again a synthetic peptide corresponding to a portion of the *neu* gene product
- **9G6**, raised against $p185^{neu}$ protein present on intact cells transfected with the human *neu* DNA.

Positive, strong staining along the cytoplasmic membrane is specific for malignancy. In contrast, a diffuse intracytoplasmic granular pattern of staining was observed in all the other tissues as well as in breast tissue that did not have strong cytoplasmic membrane staining. The intracytoplasmic staining reflects a **155-kDa** protein, whereas the membrane staining is due to the **185-kDa** Neu protein.

Positive staining for the Neu protein is seen in 33% of **invasive ductal carcinoma** and foci of *in situ* **ductal** and **lobular carcinoma**. Focal positive staining was also noted in normal breast epithelium and benign lesions of **fibrocystic disease** and **fibroadenoma**. The strongest staining was observed in all *in situ* comedo, solid, and papillary carcinomas, but not in the cribriform patterns or lobulated type.

Cross-reactivity of anti-Neu monoclonal antibodies with EGFR has been observed. When cross-reactivity is eliminated, 29% of breast carcinomas show amplified *neu* DNA and mRNA overexpression that closely correlates with the p185 expression. Weak staining of nonmalignant epithelium for p185 can be also noted.

neu gene expression is modulated by estrogens, but not by progestins. The **aggressiveness** of breast cancer, correlated with high levels of *neu* mRNA and protein, **is uncoupled from estrogen-stimulated proliferation**. Although positive staining for the *neu* gene product was also observed in **benign fibroadenomas**, it was not correlated with the expression of estrogen receptors.

***neu* mRNA overexpression is a significant predictor of early relapse,** as significant as ER negativity and ER positivity (R2). ER-negative patients with no overexpression of *neu* constitute a group with a relatively good prognosis.

It would be helpful if the degree of *neu* overexpression could be **quantified** and used in patient management. Morphological parameters seemed to offer better criteria in one such study. Quantitation can have value, however, as an independent risk factor in predicting patients at risk for **hematogenous spread** and in predicting unfavorable prognosis. Likewise, strong staining fully correlates with amplification of the *neu* gene, and is associated with poor overall and disease-free survival. Others argue that *neu* amplification is **not directly linked to the metastasizing potential** of breast cancer.

Clearly, as seen from the above-cited observations, a great deal of heterogeneity exists in the reported correlations between *neu* gene amplification/expression and prognosis. To add to the confusing picture, still others report a definite trend toward poorer prognosis in patients demonstrating *neu* mRNA **overexpression regardless of amplification.** One reason for the **apparent variability** of *neu* oncoprotein expression may be the failure to correlate it with cellular growth rate or proliferation index. It appears that **dysregulation of protooncogene expression**, rather than overexpression per se, should be regarded as the contributing factor in breast carcinogenesis.

3.1.4.2 *myc* (p. 281)

Similar controversy exists concerning the role of *myc* family protooncogenes in breast carcinoma. Estrogen stimulation of breast cancer cell growth induces c-*myc* oncogene expression at the transcription level, and leads to posttranscriptional modulation of its expression.

Tumor-specific rearrangement of one *myc* locus and **amplification** of the other *myc* locus have been detected in 56% of the **ductal adenocarcinomas.** These genomic alterations were **not** correlated with the aggressive behavior of breast tumors, but rather seemed to be associated with development of breast carcinoma.

Others link a rearrangement in the c-*myc* gene to **tumor aggressiveness.** Augmented *myc* expression may not necessarily induce more malignant patterns of growth as seen in other human cancers. **Amplification** of c-*myc* is significantly associated with PR-negative status, and is prevalent in high-grade tumors.

A correlation exists between **tumor stage** and c-*myc* expression. No correlation, however, could be observed between **survival** and elevated c-*myc* expression. High levels of $p62^{c\text{-}myc}$ protein are seen in well-differentiated tumors.

Amplification of ***L-myc*** has been related to a significantly shorter survival period after relapse.

3.1.4.3 *ras* PROTOONCOGENES AND c-*fos* (p. 281)

Activating *ras* mutations in Ki-, Ha-, and N-*ras* codons 12, 13, and 61 **appear to be only rarely involved in either initiation or metastatic progression of human breast cancer.** Likewise, the expression of the p21 product shows **no correlation** with tumor size, histological type, or grade. The **wide scatter of the observed $p21^{ras}$ values makes this parameter**

unsuitable for predicting the behavior of breast carcinomas. An additional influence of **transforming growth factor α (TGF-α)** was suggested as necessary for malignization and the enhanced growth rate in *ras*-transformed mammary epithelial cells.

p21^{ras} expression appears, however, to be quantitatively enhanced in a variety of tumors (stomach, lung, colon, and bladder) as well as in breast carcinomas, and in mammary and extramammary **Paget's disease**. Thus, although Ha-*ras* seems to be the predominantly transcribed gene of the *ras* gene family in breast cancer, **its diagnostic and prognostic value in patient management is low**.

Equal levels of expression of the **c-Myc** and **c-Fos** proteins are found in tumor and benign breast lesions. **c-*fos*** appears to be the most constantly and significantly expressed nuclear oncogene in surgical specimens of breast carcinomas according to some authors, while others note its expression is similar to the level found in benign controls. Even if c-*fos* is expressed to a higher level in carcinomas than in benign lesions or normal tissue, **no distinct relationship can be established with the degree of proliferation or differentiation.**

3.1.4.4 *int* (p. 282)

Implication of protooncogenes of the *int* family in human breast carcinoma stems from studies on the mechanism of oncogenesis produced by a retrovirus, the **mouse mammary tumor virus** (MMTV). ***int*-1** and ***int*-2** were identified and shown to map to different chromosomes. *int*-2 has been mapped to human **chromosome 11q13** in the region where another protooncogene, *hst*-1, is located. Both protooncogenes encode members of the fibroblast growth factor family and are located only 35 kb apart. *int*-2 has been detected in some human tumors, including breast tumors and squamous carcinomas of the head and neck. Frequent amplification of the chromosome 11q13 region harboring both protooncogene sites has also been reported in other malignancies: hepatocellular carcinoma, stomach, esophageal and bladder carcinoma, and melanomas.

Amplification of *hst*/*int*-2 is always accompanied by amplification of ***bcl*-1**, suggesting that there is a genetic element between *hst*/*int*-2 and *bcl*-1 that could be important in the development of a subset of breast tumors. **Amplification** of *hst*/*int*-2, found predominantly in the more differentiated low-grade carcinomas, usually on the order of 10 to 20% of cases, is frequently accompanied by an **alteration** of the *int*-2 gene. Amplification of *int*-2 is accompanied by the relatively low abundance of *int*-2 mRNA and their levels do not seem to correlate with the level of gene amplification.

Coamplification of the *int*-2/*hst*-1 genes in 9% of cases was not correlated with either tumor size, axillary lymph node status, patient age, menopausal status, or the stage of disease. Node-negative patients with *int*-2/*hst*-1 amplification had a significantly shorter **disease-free** survival rate. However, no significant correlation to overall survival was observed.

Amplification of *int*-2/*hst*-1 seems to identify a subset of aggressive tumors, quite different from the breast tumors identified by *her*-2/*neu* amplification.

The third gene of the *int* family, ***int*-3**, is also related to the gene encoding EGF, but **apparently only *int*-2 may have prognostic significance in human breast cancer.**

3.1.5 GROWTH FACTORS (p. 283)

Aside from c-*erb*B-2 (*neu*), which encodes a protein **related to** the EGFR, a number of growth factors involved in breast carcinogenesis have been described. The expression of c-*erb*B-2 and *EGFR* does not occur simultaneously, and follows different paths.

However, the close structural homology of the *EGFR* and c-*erb*B-2 gene products also extends to a similarity in their enzymatic activities as **autophosphorylating tyrosine kinases**. The c-ErbB-2/Neu protein is a substrate of EGFR, and the binding of EGF to its receptor (EGFR) causes a rapid increase in tyrosine phosphorylation of the c-ErbB-2/Neu protein. This suggests the possibility of **communication** between the c-ErbB-2/Neu protein and EGFR early in the signal transduction process.

No correlation between amplification of the *HER*-2/*neu* gene with EGFR and androgen receptor content has been observed. The expression of *EGFR*, like that of *neu*, is not correlated with tumor size and is reciprocally correlated with the expression of ER.

EGF receptors are modulated not only by their respective ligands but also by other growth factors such as **platelet-derived growth factor (PDGF), interferon α, and TGF-β**. The antiproliferative activity of TGF-β is modulated via its interaction with EGFR, and thus can amplify the EGF-induced inhibitory response of the cells.

In breast carcinoma biopsy samples, the level of EGF mRNA expression is **correlated with the ER/PR status** being higher in ER- and PR-positive tumors. **EGFR negativity is associated with good responses to endocrine therapy in ER-positive patients.** The prognostic value of EGF, EGFR, TGF-α, and TGF-β mRNA in breast carcinoma, however, still remains to be established.

The **protein marker Ki67** appears to influence the response to therapy in EGFR- and ER-positive cases.

Among the other growth factors expressed by cultured breast carcinoma cells are those typically related to the growth of mesenchymal cells:

- **PDGF A chain**
- **PDGF B chain**
- **TGF-α**
- **insulin-like growth factor II** (IGF-II) and **IGF-I**

IGF-I appears to be an independent prognostic factor and the **presence of IGF-I receptor is correlated with better overall and relapse-free survival.**

3.1.6 TUMOR SUPPRESSOR GENES

3.1.6.1 *p53* (p. 284)

The frequently observed LOH of a certain **chromosome 17p** region in breast carcinomas pointed to the possible involvement of the ***p53*** gene, manifested in an overexpression of *p53* mRNA. Frequent LOH was also found on

- **13q** (21% of cases)
- **16p** (45%)
- **17q** (56%)

The LOH most frequently noted in **lobular carcinoma** is on chromosome 22, and in **ductal carcinoma** on chromosome 17.

The **regulatory gene located 20 Mb telomeric to the *p53* gene is involved in the majority of breast cancers**.

Other tumor suppressor activity in breast carcinomas is associated with an LOH on **chromosome 18** between 18q21.3 and 18q23, where the candidate gene ***DCC*** (deleted in colorectal carcinomas) is assigned. While LOH on **17q** is correlated with ER-negative cancers, LOH on 18q is associated with histopathological grade III cancers.

LOH on both chromosomes 17q and 18q seems to be associated with more aggressive tumors, although **neither chromosomal change by itself is predictive of tumor stage or patient prognosis**.

It is hypothesized that different subsets of mutations may make comparable contributions to the malignant phenotype, and that **there are subsets of tumors characterized by the particular mutations they contain**. Over 50% of malignant breast lesions strongly express *p53*, whereas no expression is detectable in benign lesions. Because overexpression of p53 is associated with mutation of the *p53* gene, there is a possibility that alteration of this TSG plays a role in breast carcinogenesis.

The possibility that mutations in *p53* occur in **heritable breast cancer** has been explored. No structural abnormalities in the *p53* gene were noted in pedigree studies, discounting the possibility that structural abnormalities of the *p53* gene contribute to the heritable predisposition to the development of breast cancers.

In highly susceptible strains of rats, a **COP gene**, a new **mammary cancer suppressor gene**, has been identified that **has exclusive specificity for the mammary gland**.

3.1.6.2 *RB* GENE (p. 286)

The **retinoblastoma susceptibility gene (*RB*)**, is a recessive gene whose mutational inactivation has been implicated in the development of retinoblastoma tumors and other neoplasms. Inactivation of *RB* has been demonstrated in some human breast cancer cell lines. The *RB* gene product, **pp110^{RB}**, could not be detected immunologically with specific antibodies. In addition, linkage analysis for 14 loci on **chromosome 13q** and structural analysis of *RB*-1 indicate that changes in the *RB*-1 locus either occur by chance, or represent secondary alterations associated with progression of some tumors. Other TSGs are likely to be discovered in breast cancer. Thus, an association of frequent LOH on chromosome 7q31 with short, metastasis free, overall survival points to a possible TSG or metastasis suppressor gene.

3.1.7 OTHER LINKAGE STUDIES IN BREAST CANCER (p. 286)

Findings of a linkage between breast cancers and specific gene alterations suggest a role for deletions on chromosome 11p resulting in hemizygosity of **c-Ha-*ras***, **β-globin**, **parathyroid hormone (PTH)**, **calcitonin**, and **catalase** genes in the aggressiveness of breast tumors. c-Ha-*ras* RFLP data were not conclusive, but **c-*mos*** RFLPs could be detected only in patients with breast cancers or other types of tumors.

Testing of genetic linkage of breast cancer susceptibility to 9 oncogenes (**Ha-*ras*, Ki-*ras*, N-*ras*, *myc*, *myb*, *erb*A-2, *int*-2, *raf*-1**, and ***mos***) in 12 extended families strongly suggested that **these oncogenes are not the sites of primary alterations leading to breast cancer**. A significant association has been observed between ***int*-2** RFLPs and the presence of more than three positive lymph nodes. The low frequency of ***int*-1, *int*-2, *neu***, and **c-*myc*** amplification, however, limits their usefulness as clinical predictors of disease recurrence.

Fingerprinting analysis using *Hae*III enzyme digestion of tumor DNA can detect minimal DNA changes, which could allegedly identify the fine distinction between normal and malignant cells. It remains to be seen whether this direction of studies can offer a clinically useful discriminatory tool.

3.1.8 CATHEPSIN D (CD) (p. 286)

CD, a lysosomal protease, is one of the most recently evaluated prognostic factors in breast cancer. However, lack of uniformity in evaluation of CD leads to seemingly conflicting reports. Some find that overexpression of CD correlates with **aggressive tumor behavior** and, consequently, shortened disease-free and overall survival. In large screening studies, on the other hand, **positive CD staining is associated with significantly superior overall survival, particularly demonstrable in the subgroup of patients with lymph node metastases**. Furthermore, there is a correlation between the intensity of CD staining and overall survival.

In node-negative patients, however, a relationship between CD elevation and **worse survival** appears to be more apparent.

3.1.9 *nm*23, WDNM-1, WDNM-2, pGM-21, AND Gla PROTEINS (p. 287)

The ***nm*23** gene, isolated by differential hybridization from murine melanoma cell lines, **is expressed more in cells with low metastasis potential** than in clones with high metastasis potential. In the mouse, the *nm*23 gene product is an M_r 17,000 peptide and shares 96% homology with its counterpart in humans.

Reintroduction of the gene into a metastatic cell line results in reduction of spontaneous metastases, and those developed following intravenous injection. A low level of *nm*23 gene expression was found in one study in 86%, and in another in 73%, of lymph node-positive patients. **Better relapse-free survival and overall survival were significantly correlated with the higher levels of *nm*23 mRNA expression.** The higher levels of *nm*23 mRNA in adenomas and carcinomas from node-negative patients than node-positive cases is also confirmed by ISH. The usefulness of *nm*23 as an indicator of the metastatic potential of human cancers deserves further study.

A gene associated with **an effect apparently opposite to that of *nm*23** has been isolated by differential hybridization and termed ***pGM*-21**. The enhanced expression of *pGM*-21 mRNA occurs in mammary adenocarcinoma cell lines with high metastatic potential. In the highly metastatic DMBA8 ascites cell line, the *pGM*-21 gene overexpression was up to 25-fold higher than in the less metastatic cell lines.

Other **genes that are augmented in less metastatic** rat mammary cell lines are ***WDNM*-1** and ***WDNM*-2**. Their products, however, have not been characterized so far.

Specific expression of the gene encoding **matrix Gla protein**, which may be involved in breast cancer metastases, has been shown by differential cDNA hybridization. The matrix **Gla protein** gene appears to be specifically expressed in breast carcinoma cell lines and **may be involved in breast cancer metastasis.**

3.1.10 OTHER MARKERS OF BREAST CARCINOMA

3.1.10.1 *pS-2/pNR-2* (p. 287)

The human ***pS*-2/*pNR*-2 gene** was originally isolated from estrogen-dependent breast carcinoma cell lines and was absent from cell lines unresponsive to estrogen. The gene product, pS-2/pNR-2, has also been identified in normal stomach mucosa, although its function is not obvious. The expression of *pNR* mRNA occurs only in a subset of ER-positive tumors. The level of the *pNR*-2 gene product was **positively correlated with response to primary tamoxifen therapy**.

Significantly better survival of ER-positive patients with pS-2/pNR-2-positive tumors was particularly pronounced in patients also positive for PR. In the group of node-positive as well as node-negative patients, pS-2 was able to discriminate between a good and bad prognostic group. So far, no definite role for the protein has been established.

3.1.10.2 p24 AND p29 PROTEINS (p. 288)

These estrogen-regulated proteins have been described in human breast cancers that show some correlation with ER status:

- the **p24** protein is more frequent in ER-positive tumors

- the **p29** protein shows variable relation to ER status and no relationship to PR levels; no association with lymph node status has been observed

3.1.10.3 c-*myb* (p. 288)

c-*myb* is detected in 64% of breast carcinomas, and its expression is **associated with good prognosis**, suggesting its usefulness for the characterization of a new class of estrogen-dependent tumors.

3.1.10.4 STRESS RESPONSE PROTEIN (Srp-27) (p. 288)

Srp-27 is a 27,000-Da heat-shock or stress-response protein. It can be an independent prognostic indicator for disease-free survival only in patients having an advanced breast cancer with one to three positive lymph nodes. A significant correlation has been observed in such patients between Srp-27 overexpression and ER content, pS-2 protein expression, advanced stage of the disease, and lymphatic and vascular invasion.

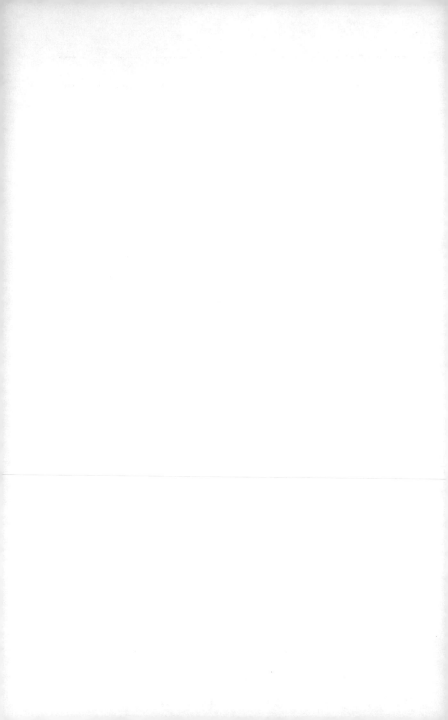

Chapter 3.2

Tumors of the Uterine Cervix and Endometrium

3.2.1 INTRODUCTION (p. 289)

The level of **estrogen** and **progesterone receptor** protein (ER and PR, respectively) in endometrial cancer is inversely proportional to the grade of the tumor, i.e., **the more differentiated tumors have higher levels of ER and PR protein**. This parallels the survival data — a favorable survival is expected with high levels of the sex steroid receptor protein.

Although the receptor status may offer a better prognostic tool than histological grading, an improvement of technical aspects of receptor protein assessment is needed before this parameter may be used clinically for prognosis, and appropriate selection of hormonal and/or chemotherapy.

3.2.2 DNA PLOIDY OF CERVICAL TUMORS (p. 289)

A significantly elevated percentage of dividing cells can be detected by FCM analysis of cervical cells from patients with **cervical intraepithelial neoplasia** (CIN) and HPV lesions. DNA aneuploidy, on the other hand, varies insignificantly between different stages of HPV lesions.

In **older women**, about 80% of CIN III lesions are aneuploid, irrespective of the coexisting invasive lesions. On the other hand, in **younger patients**, the majority (60%) of CIN III lesions with concomitant invasive cancer are diploid.

Progressive cytological signs of viral (HPV) effects and dysplasia positively correlate with the proportion of cells having aneuploid DNA content. Aneuploidy and the high percentage of cells in the **S phase** correlate with tumor size, histological grade, and clinical stage, confirming the earlier observations. Triploidy carries an especially poor prognosis, being associated with an aggressive disease. Thus, expectedly, **tumors with diploid DNA content and a low percentage of cells in the S phase have the best prognosis in cervical adenocarcinoma.**

3.2.3 CHROMOSOME ABERRATIONS IN CERVICAL CARCINOMA (p. 290)

In invasive cervical carcinoma, abnormalities of **chromosome 1** in multiple locations proved to be the most frequent chromosomal aberration.

Loss of heterozygosity (LOH) at the D3S2 locus is consistently observed on **chromosome 3p**, suggesting the **involvement of recessive genetic factors** in the development of cervical carcinomas.

3.2.4 DNA PLOIDY OF ENDOMETRIAL TUMORS (p. 291)

DNA FCM of **hyperplastic endometrium**, known to be the precursor of endometrial adenocarcinoma, showed a higher percentage of nondiploid cells (20%), comparable to that in well-differentiated adenocarcinoma (29%).

The majority (57%) of endometrial carcinomas have no differences compared to the ploidy characteristics of benign endometrium. Although DNA FCM is of no value for screening purposes, **the degree of aneuploidy correlates with histopathological findings**. DNA ploidy has a significantly better prognostic value for 2-year recurrence-free survival than the degree of tumor differentiation, range of myometrial invasion, or ER/PR concentrations.

In stage I PR-positive endometrial carcinoma, a combination of the **mean shortest nuclear axis, DNA index (DI)**, and **depth of myometrial invasion** offers the best prognosticator. In diploid tumors, the proliferative activity cannot predict the clinical outcome. It appears, therefore, that **DNA FCM of endometrial neoplasias can be of help in selecting a subgroup of patients who have low-stage, high-grade endometrial cancers, and who may benefit from adjuvant therapy.** DNA ploidy is suggested as an adjunct to other traditional modes of tumor assessment.

The pattern of infiltration (**contiguous vs. discontinuous growth**) shows a strong correlation with the tumor cell DNA content. A contiguous growth pattern appears to be a sign of a low-grade malignant (mostly diploid) tumor with delayed spread and fewer metastases before therapy is instituted. On the other hand, **tumors with a discontinuous pattern of growth had a predominantly aneuploid nuclear DNA pattern.**

The **range of myometrial invasion, however, is the strongest prognostic characteristic, superseding even the DNA parameters**.

3.2.5 CHROMOSOME ABNORMALITIES IN UTERINE CANCER (p. 291)

In addition to **heterogeneous chromosome profiles** a number of complex chromosomal aberrations such as minutes, rings, homogeneously stained regions, premature chromosome condensations, premature centromere division, and telomere joinings have been described in endometrial carcinomas. Chromosomal aberrations in endometrial carcinomas most frequently feature **chromosome 1** and **chromosome 10**.

The possible role of **tumor suppressor genes** in the development of endometrial adenocarcinoma is suggested by the complete suppression of the tumorigenicity of human endometrial carcinoma tumor cells (HHUA) on introduction via microcell hybrids of **chromosomes 1**, **6**, or **9**, and in some cases **chromosome 11**.

Smooth muscle tumors of the uterus (uterine leiomyomas, parametrial leiomyomas) show an identical abnormal stemline pattern in different tumors from the same patient, with each leiomyoma arising as a separate clone.

3.2.6 ONCOGENES IN CERVICAL TUMORS (p. 291)

With respect to cervical cancer the prognosis of the disease is not always predictable on the basis of involvement of the lymph nodes. Other noninvasive markers are being sought. The mutant oncogene **c-Ha-*ras*** detected in 40% of cervical carcinomas is accompanied by a deletion of the gene on the other allele. The loss of one c-Ha-*ras* allele in tumor cells from cervical cancers is not, however, correlated with the aggressiveness of cervical cancer.

The **c-*myc*** protooncogene is overexpressed (4 to 20 times and up to 60-fold the level in normal tissue) in stage I and II SCC of the uterine cervix, it is highly correlated with the eightfold higher incidence of early recurrences, and the 18-month relapse-free survival rates. A significantly higher rate of c-*myc* amplification is found in stage III and IV carcinomas (49%), compared with only 6% in stage I and II tumors, suggesting **a role for c-*myc* protooncogene in cervical cancer progression**. Overexpression of c-*myc* is correlated not only with the advanced stage cervical carcinoma, but shows a significant value as a prognosticator of the 3-year disease-free survival rate.

Concordant interaction of several oncogenes is seen, for example, in the activation of the c-*myc* protooncogene in all tumors that contained a mutated c-Ha-*ras* gene. Amplification of all three oncogenes — Ha-*ras*, c-*myc*, and *erb*B-2, has been found in CIN III and invasive cancer compared with normal cervix and CIN I. It appears, therefore, given these observations are confirmed in larger studies, that **evaluation of c-*myc* mRNA levels can serve as a practical clinical tool for identifying patients with cervical cancer at high risk for early recurrence, thus facilitating the customized therapeutic approach on an individual patient basis**.

3.2.7 ONCOGENES AND GROWTH FACTORS IN ENDOMETRIAL TUMORS (p. 292)

Little is known so far about the role of protooncogenes in endometrial neoplasias. In the rat uterus, estrogen is known to produce transient activation of **c-*fos*** and **c-*myc*** protooncogene expression, which precedes increases in DNA synthesis and cell proliferation. Changes of c-*fos* oncogene expression in uterine malignancies may offer a diagnostically useful tool.

In addition, **c-*jun*** gene transcription is increased three- to fourfold upon estrogen stimulation and constitutes a primary response to the hormone. The composite transcription factor Jun/AP-1 points to the possibility that c-*jun* as well as c-*fos*, which is regulated by the same growth-promoting factors, may be activated in human endometrial neoplasias.

The aggressive biological behavior of **serous papillary adenocarcinoma** of the endometrium may, to a certain extent, be related to c-*myc* amplification observed in the tumor itself (tenfold), and in its metastases to the omentum (fivefold), and may be associated with the particularly aggressive behavior of this type of carcinoma.

The similarity of c-*myc* amplification in endometrial and ovarian serous epithelial neoplasms appears to emphasize the **possible association of c-*myc* amplification with serous papillary morphology**. Estrogens appear to exert their control on the growth of all three uterine cell types (epithelial, stromal, and myometrial) through several oncogenes: N-*myc*, c-*myc*, c-Ha-*ras*, *erb*B, and c-*fos*, at least.

The activation of Ki-*ras* in endometrial cancer is frequently detected by PCR amplification. In contrast, cervical and ovarian tumors do not show *ras* activation.

The effects of **epidermal growth factor** (EGF) and **transforming growth factors α** and **β** (TGF-α and -β) on human endometrial carcinoma cells demonstrated a complex pattern of inhibitory and stimulatory activities. The inhibitory activity of TGF-β_1 appears to be inversely correlated with the level of expression of TGF-β_1 mRNA by the cells.

The role of EGF in endometrial growth control is implicated by the progressive decrease in the concentration of **EGF receptor** (EGFR) in endometrial cancers of increasing grade. While the role of EGFR in endometrial adenocarcinoma may be closely related to estrogen influences, this relationship is not as evident in sarcomas, which also demonstrate a 72% decrease in EGFR. At this time, **the diagnostic utility of these observations for endometrial cancer management is low** and further research and clinical correlations are needed.

The ectopic expression of the **placental alkaline phosphatase** gene has been demonstrated in a human endometrial adenocarcinoma cell line. Estrogen stimulates this activity at the level of transcription, and can be interrupted by a new antiestrogen, **ICI 164,384**. Much effort is going into unraveling the fine details of hormonally regulated cascades of mitogenic activity related to normal and neoplastic endometrial growth.

3.2.8 UTERINE SMOOTH MUSCLE TUMORS (p. 294)

The majority of **leiomyomas** display a DNA diploid pattern, whereas **leiomyosarcomas** exhibit a DNA tetraploid/polyploid or aneuploid DNA pattern. Although FCM evaluation of smooth muscle tumors is not recommended as a diagnostic tool, or for differentiation between benign and malignant tumors, its use for prognosis of the clinical behavior of these tumors may be of practical value. A combination of several histopathological characteristics usually improves the diagnostic accuracy and/or predictive value of FCM evaluations.

In addition to normal karyotypes in **leiomyomas**, clonal chromosomal abnormalities occur that include

- **del(7) (q21.2q31.2)**
- **trisomy 12**
- **t(12;14) (q14-15;q23-24)**
- multiple aberrations, including deletions, inversions, and translocations on **chromosome 6p**
- abnormalities in **chromosomes 2, 6, 9, 10, 11, 16, 17, 18, 21,** and **22**

The multitude of chromosome aberrations appears to suggest that **several genetic changes are necessary for the development of uterine leiomyomas**. Cytogenetic changes in these benign tumors cannot be used as discriminating diagnostic probes.

In uterine leiomyomas, the most frequent finding is **translocation t(12,14)**, although the specific gene rearrangement responsible for the tumor are not yet known. In contrast, significant variability exists from case to case in the complex rearrangements observed in the malignant smooth muscle tumor, **angioleiomyoma**. In yet another type of malignant smooth muscle tumor, **rhabdomyosarcoma**, the characteristic reciprocal **translocation t(2,13) (q37,q14)** is found in embryonal, undifferentiated, and alveolar histologic subtypes.

Leiomyosarcoma demonstrates extreme cytogenetic instability, in contrast to leiomyoma, which is characterized by cytogenetic stability. A grading system of practical diagnostic value is proposed for the evaluation of cytogenetic instability. Identification of **high-grade cytogenetic instability in uterine smooth muscle tumors is likely to predict the aggressive clinical behavior of the tumor.**

Specific molecular characteristics are sought to allow better **differentiation between histologically similar patterns**. In fact, a malignant smooth muscle tumor leiomyosarcoma is known, among other things, to express the gene encoding **insulin-like growth factor II** (IGF-II) at levels markedly higher than is observed in other, normal tissues known to express the *IGF-II* gene.

Chapter 3.3

Ovarian Neoplasms

3.3.1 DNA PLOIDY OF OVARIAN NEOPLASMS (p. 297)

Typing and staging of **primary ovarian carcinomas** of the **FIGO I stage** are prognostically important, but tend to be poorly reproducible. DNA ploidy and DNA cytometric studies potentially may offer a more objective approach. All the benign and semimalignant tumors as well as 48% of malignant tumors in one study proved to be diploid. Aneuploidy is more frequent in

- **advanced stages** of the disease
- tumors with a **low degree of differentiation**
- **older patients**

Ovarian tumors tend to display the following features:

- Tumors found in postmenopausal patients tend to have a higher percentage of cells in the **S phase**
- Ploidy is not associated with the **tumor type**
- Ploidy of **metastases** is the same as that of primary tumors
- The DNA index (**DI**) predicts mainly the **degree of differentiation** of ovarian tumors; a good correlation had been detected between the DI and FIGO stages. A **DI value over 1.3** proved to be the most effective parameter in predicting inferior prognosis. Although the DI may be correlated with prognosis in some cases, the overall consensus is that the **DI does not have good prognostic power**
- The majority of **low-grade tumors** are diploid
- The **worst prognosis** is reflected by a combination of grade 2 or 3, or a clinically advanced stage, with the G_1 cell fraction ≤85%
- **Intra-** and **interlesional heterogeneity** of cellular DNA content in ovarian carcinomas affects the sampling and averaging of the data obtained
- The **combined use of DNA flow cytometry (FCM) with DNA image cytometry** markedly improves the predictive power of either modality
- The **size** of tumor seems to be of prognostic significance. DNA diploid tumors are smaller in size, and contain a significantly lower percentage of cells in the S phase than DNA aneuploid tumors. The **S phase fraction (SPF)** increases the prognostic value of DNA ploidy evaluations, particularly in DNA diploid tumors. However, in a large study conducted over 10 years and on almost 1000 ovarian cancer patients the **residual tumor size** was found to be the major determinant of survival

- The majority (over 70%) of advanced (**FIGO stages III** and **IV**) malignant common epithelial ovarian cancers are aneuploid
- The type of **surgical intervention**, cytostatic treatment, histological tumor type, and the degree of differentiation had no significant effects on the length of survival
- Pathologists widely differ in their acceptance of DNA ploidy as a diagnostic aid in **borderline tumors**. Moreover, opinions vary on the definition of the histological criteria of borderline tumors, especially those spread beyond the ovary. Because **borderline diploid** tumors, despite spread to the perifoveal cavity, had an unusually good prognosis, defying conventional histological criteria, it is suggested that staging should be supplemented by DNA FCM when selecting therapeutic regimens
- In **granulosa cell tumor**, the prognostic value of DNA FCM prediction of outcome is highly questionable
- In **epithelial ovarian cancer**, DNA FCM did not significantly improve the prognostic value of stage, SPF, and the size of the largest metastases. DNA FCM can be helpful in the assessment of multiple, predominantly bilateral, ovarian tumors
- An interesting reversal of the situation with DNA ploidy-related prognosis is found in **partial hydatidiform moles**. All patients who developed gestational trophoblastic tumors arising from triploid partial moles achieved complete remission with one course of single-agent chemotherapy. In contrast, **diploid partial moles with persistent tumor were less sensitive to the same chemotherapy**
- A significant correlation exists between **the proliferative index (PI)** and **cyclic 3′,5′-guanosine monophosphate (cGMP)** excreted in the urine of patients with epithelial ovarian cancer. The clinical utility of this correlation has not been established
- No variable has been identified so far to predict ovarian cancer response to **cisplatin therapy**

3.3.2 CHROMOSOMAL ABNORMALITIES IN OVARIAN NEOPLASMS (p. 300)

Loss of heterozygosity (LOH) on **chromosome 6q** is observed in 64% of human ovarian carcinomas, and **appears to be specific to ovarian carcinomas**. Also noted are

- LOH on **chromosome 17p** at two loci
- an allelic loss at the Ha-*ras*-1 gene locus on **chromosome 11p**
- the **translocation t(15;20) (p11;q11)**
- **chromosome 1q heterochromatin asymmetry**
- an increased incidence of **C-band inversion**

***Pvu*II** digests probed with **variable number of terminal repeat (VNTR) probes (CMM86, LCN5A2,** and **pTHH59)** in familial breast-ovarian cancer reveal LOH at CMM86 in 65% of informative tumors, compared to only 9% in sporadic cases, suggesting the presence of a candidate gene near this locus.

Losses on 11p and 17p, also noted in other cancers, indicate that **loss of TSGs is involved in the genesis or progression of epithelial ovarian carcinomas.**

3.3.3 DNA FINGERPRINTING IN OVARIAN TUMORS (p. 300)

The most common trophoblastic tumor, **complete mole**, is highly prevalent, ranging from 1 per 100 to 1 per 1200 pregnancies.

Certainly **karyotyping** combined with chromosome marking can distinguish **complete mole, which usually has a 46,XX set of chromosomes, and occasionally 46XY**. The lengthy tissue culture procedure (several weeks) makes this approach unacceptable for resolving patient management problems.

Restriction fragment length polymorphisms (RFLPs) with the **minisatellite DNA probe 33.15 allow the determination of androgenesis of the complete mole**. Comparison of parental fingerprints with those of the moles can establish that **specific bands are exclusively inherited from the fathers.**

The **partial hydatidiform mole**, which is typically triploid (69,XXX), presents a complicated situation for DNA fingerprint analysis. Because partial mole results from the fertilization of a normal haploid ovum by two normal sperms, DNA fingerprints will show the polymorphic fragments transmitted from their parents as in normal pregnancy or hydropic abortus. However, because **partial mole is not associated with choriocarcinoma**, differentiation of partial mole from hydropic abortus, is of lesser clinical significance.

By evaluating the **quinacrine-fluorescent heteromorphism** and applying **RFLP analysis** to a **choriocarcinoma** sample, the origin of the tumor can be traced to the previous pregnancy. Restriction patterns must be compared between the spouses and the son (previous pregnancy) to elucidate whether maternal markers are present in the tumor.

DNA fingerprinting reveals specific DNA alterations in **epithelial ovarian tumors**. The satellite probes **33.15**, **228S**, and **216S** identify somatic changes in 70% of ovarian tumors, most commonly presented as a deletion or reduction in the intensity of a band, suggesting LOH.

3.3.4 PROTOONCOGENES IN OVARIAN NEOPLASMS (p. 301)

- Multiple oncogenes altered in approximately 20 to 30% of ovarian carcinomas include **c-Ki-*ras*, *sis*, *neu*, c-*jun*, c-*myc*, c-*fos*, *fes*, *fms*, *trk***, and ***PDGF-A***.
- Importantly, a correlation exists between protooncogene amplification and the serous type or some degree of **serous differentiation** of the

tumors, while none of the purely mucinous carcinomas showed any evidence of amplification.

- **Protooncogene, particularly c-*myc*, amplification apparently may be used as a marker of aggressiveness of epithelial tumors of the ovary**. Moreover, it may serve as an **earlier marker** of subsequent aggressive tumor behavior **than aneuploidy**.
- No correlation between ploidy and oncogene amplification is observed.
- **No protooncogene amplification** (**c-*myc*, *int*-2,** and **c-*erb*B-2**) is defected in **benign** ovarian neoplasms, or in ovarian carcinomas with **low malignant potential**.
- Conflicting observations have been made on the involvement of the **Ki-*ras*, Ha-*ras***, and c-*myc* genes in **serous cystadenoma** of the ovary.
- Levels of ***fes*** transcripts (usually observed in rapidly proliferating cells) were characteristic of aggressive neoplasms, distinguishing them from benign or borderline cases.
- The expression of the ***src*** family oncogenes ***erb*B, *neu*, *fms*,** and ***ros*** suggests that these kinases and their ligands play a role in the control of ovarian epithelial cell proliferation.
- Similar to the situation in breast cancer, **the expression of the gene encoding epidermal growth factor receptor (EGFR) and *neu*, in epithelial ovarian cancer correlates with poor prognosis.** Likewise, **overexpression** of *HER-2/neu* oncogenes in ovarian cancer is associated with resistance to several different cytotoxic influences.
- Homozygous **deletion of the *RB* gene** has been implicated in
    - the aggressive biological behavior of an **endometrioid tumor of the ovary of low malignant potential**
    - 30% of the informative cases of various ovarian neoplasms [**leiomyosarcoma, sex cord-stromal tumor,** and **epithelial carcinomas (mucinous, endometrioid, mixed Müllerian,** and **serous)** and their **metastases**]
- Two genes (***PSGG*A** and ***PSGG*B**) coding for the human **pregnancy-specific b glycoprotein** have been cloned. *PSGG*B encodes a 1.7-kb mRNA in **hydatidiform mole** tissues, and a 2.0-kb mRNA in **term placenta** tissues. The *PSGG*B-specific probe preferentially hybridizes to mRNA from trophoblastic tissue, suggesting a practical way to diagnose gestational trophoblastic disease at the gene level.
- The **multidrug-resistant protein P** accounts in part for the drug resistance of ovarian carcinoma cells. Its expression can be readily detected in **routinely processed tissues**, and helps in predicting failure to respond to standard chemotherapeutic treatments.

Chapter 3.4

Prostate Neoplasms

3.4.1 INTRODUCTION (p. 304)

Evaluation of prostate hyperplasia for possible foci of malignancy has been traditionally performed by light microscopy of tissue sections on the basis **of morphological characteristics** grouped into **diagnostic clusters** reflecting our understanding of the natural history of prostatic cancers. Histological grading is also used for **prognosis** of the disease.

These criteria, however, frequently fail because patients with the same pathological stage and histological grade of prostatic adenocarcinoma may experience a **different course** of the disease. Therefore an important challenge facing prostatic cancer evaluation is the development of **individualized prognosis** to supplement conventional assessment by grade and stage.

3.4.2 PLOIDY IN PROSTATE NEOPLASMS (p. 205)

DNA FCM of prostate adenocarcinoma at advanced metastasized stages (D1) reveals 42% of the tumors to be diploid, 45% tetraploid, and only 13% aneuploid. Differences in the tumor behavior are strongly correlated with ploidy: 75% of tetra- or aneuploid tumors subsequently progress locally or systemically. None of the patients with diploid tumors died during the period of observation, which lasted up to 19 years.

Even metastasized diploid tumors grow slowly and are unlikely to kill the patient. **Ploidy pattern, therefore, cannot serve as an index of metastasizing potential of prostate adenocarcinoma**, but rather may reflect the **rate of tumor progression.**

Thus, some workers believe that in patients with **stage A** or **B** prostatic cancer **nuclear DNA quantitation on needle biopsy specimens allows better estimates of prostatic cancer progression than that based on tumor grade alone**. Comparison of the advantages and drawbacks of evaluating prostatic carcinoma in smears and tissue sections indicates that tissue histograms are difficult to interpret and **cytological preparations should be preferred for DNA ICM evaluations.**

Again, initial ploidy evaluation correlates with the **cytological grade** (but not tumor stage, **prostatic acid phosphatase** activity, or patient age) and seems to have value even as a sole predictor.

A largely **contrasting conclusion** has been reached in later studies by urologists at Stanford University. Although aneuploidy was more predominant in advanced stages of the disease and in less differentiated tumors, diploid tumors were also found in this category, and **the relationship between aneuploidy vs. pathological stage and grade was not significant**. It suggested an association of aneuploidy with tumor progression, but not a requirement for progression to occur. On this basis, **DNA content analysis is discounted as an independent predictor of the clinical course of prostatic adenocarcinoma.**

Other workers also failed to significantly enhance the ability of standard histopathological grading to predict disease recurrence by adding DNA FCM and replication rate estimates. Numerous other studies also **question the validity of aneuploidy as a prognostic factor**.

It appears that techniques allowing finer evaluation of changes at the gene, rather than "bulk" DNA, level may provide clues to reliable, objective, and reproducible means of prostate cancer assessment on biopsy material.

3.4.3 CHROMOSOMAL ABERRATIONS IN PROSTATE CANCER (p. 307)

The following chromosomal aberrations in human prostatic adenocarcinoma cells have been detected by independent investigators:

- **chromosomes 1, 2, 3, 6, 7, 10,** and **15** are involved in **rearrangements** in over 50% of all prostatic carcinomas. Chromosomes 1 (90% of cases), 7 (80%), and 10 (70%) appear to be consistently involved in all the primary tumors studied
- the most common chromosomal changes include the **loss of chromosomes 1, 2, 5,** and **Y**, the **gain of chromosomes 7, 14, 20,** and **22**, and **rearrangements involving 2p, 7q,** and **10q**
- the role of **TSGs** in prostate cancer is suggested by the frequent LOH (30% of all tumors), predominantly affecting **chromosomes 16q** and **10q**

No single chromosomal aberration can be used, however, as a marker for early-stage prostate cancer

3.4.4 DNA FINGERPRINTING OF PROSTATIC CANCERS (p. 307)

When DNA probes for **hypervariable loci** (33.15 and 33.6) are used to generate DNA fingerprints after digestion with *Hae*III or *Hin*fI of matched tumor and blood DNAs, one finds **altered band intensities** (compared to constitutive DNA) in DNA fingerprints prepared from the majority of prostatic adenocarcinoma DNA as well as in almost half of BPH cases:

- **decreased band intensities** are significantly more common in adenocarcinoma than in BPH specimens
- **increased band intensities** are noted in both adenocarcinoma and BPH specimens

- the probe **33.15** proves more informative than **33.6** in revealing distinctions between adenocarcinoma and BPH
- the patterns are consistently correlated with the **type of neoplasia**
- changes in **methylation** of bases adjacent to the minisatellite regions can be **partially** responsible for changes in band intensities in BPH specimens

3.4.5 GROWTH FACTORS AND ONCOGENES IN PROSTATE TUMORS (p. 308)

A number of growth factors and their receptors have been identified in the prostate gland:

- **fibroblast growth factor (FGF)**
- **transforming growth factor β (TGF-β)** families
- **TGF-α**
- **epidermal growth factor (EGF)**

Within the FGF family of growth factors, overexpression of the ***int*-2** gene, **basic fibroblast growth factor** (bFGF) mRNA, and **acidic fibroblast growth factor** (aFGF) mRNA has been studied. Members of the TGF-β family appear to be expressed at higher levels in the poorly differentiated prostatic adenocarcinomas.

EGF receptor (EGFR) binding and its mRNA levels appear to be elevated in prostate cancer. The level of EGFR in BPH is higher than that in prostate cancer, and the well-differentiated tumors demonstrate a higher expression of EGFR.

Changes in **androgen** responsiveness do not appear to be associated with detectable alterations in the structure of the **androgen receptor gene**, at least in prostate carcinoma cell lines.

Elevated expression of ***myc***, and the expression of activated ***ras***, have also been detected in prostate cancer.

Activation of *ras* genes by point mutations has been amply documented in human prostate tumors. The frequency of *ras* gene mutations in prostate tumors, however, is low. It is believed they may play a role in the **progression** of prostate cancer or in the development of the rare ductal variant of prostatic adenocarcinoma.

p53, ***sis***, and ***fos*** are other protooncogenes the levels of which are elevated in prostate cancer.

Marked phenotypic changes of cultured human prostate carcinoma cells occurring upon introduction of a normal ***RB* gene** suggest this tumor suppressor gene may have a role in the genesis of prostate cancer.

A **unique mitogen**, an **M_r 26,000–30,000 basic protein** released by prostate adenocarcinoma cells, has been isolated that has a higher specificity for human osteoblasts than for human fibroblasts. It is believed that the production of this unique growth factor plays a role in the intense **osteoblast-specific stimulation** that is observed at the sites of prostatic adenocarcinoma bone metastases.

3.4.6 PROSTATE-SPECIFIC MARKERS: PSA AND hGK-I (p. 310)

Attempts to find objective molecular distinctions between BPH and adenocarcinoma (CA) have drawn attention to possible differences at the level of gene expression. One such effort analyzed the pattern of expression of two genes encoding closely related protein products — **prostate-specific antigen (PSA)** and **human glandular kallikrein-I (hGK-I)**.

PSA is a 33-kDa glycoprotein helpful in the evaluation of prostate hyperplasia and/or cancer. Most people do not exceed a certain low level of PSA, if the prostate is normal and has not been manipulated. The serum concentration of PSA has been found to be roughly **proportional to the mass** of either benign hyperplastic tissue or cancer, and it is especially useful in **monitoring the effects of specific therapy**, as well as in **screening** for prostate cancer in conjunction with rectal examination.

The determination of the *PSA* mRNA and *hGK*-I mRNA species in BPH and CA revealed that there were **no significant differences in the pattern of *PSA* gene expression between BPH and CA**. There was a threefold range of **intersample variation** in the *PSA* mRNA levels in CA patients.

The *hGK*-I mRNA variation in BPH samples was greater than that of the *PSA* mRNA. Differences are present in the expression of these two genes in the CA samples.

PSA has been found to occur in complex with **α_1-antitrypsin**, in addition to its existence in a monomeric form. Because the proportion of **PSA-α_1-antichymotrypsin complex** was higher in prostate cancer than in BPH patients, assaying for **this complex has a higher sensitivity for cancer than an assay for total PSA immunoreactivity alone**.

A combination of PSA with **α_2-macroglobulin** and **inter-α-trypsin inhibitor** has also been identified. The occurrence of these complexes, however, is similar in BPH and cancer patients.

High levels of **5′-nucleotidase** have been found in human prostates without cancer, and this enzyme activity decreases in prostatic carcinoma.

Thus it appears that, in spite of the existence of a number of specific markers correlated with pathological conditions of the human prostate, **at this time no clinically useful gene level probe can be reliably selected for patient management**.

The **specific expression of a gene per unit target tissue**, and related to the specific and independent epithelial and/or stromal markers, could reveal finer and more consistent distinctions we are trying to define between benign hyperplasia and cancer.

Chapter 3.5

Kidney and Urinary Bladder

3.5.1 RENAL NEOPLASMS (p. 313)

Renal neoplasms include

- renal cell carcinoma (adenocarcinoma)
- transitional cell carcinoma (TCC) and squamous cell carcinoma (SCC) of the renal pelvis
- sarcomas of the kidney
- nephroblastoma

3.5.1.1 RENAL CELL CARCINOMA (RCC) (ADENOCARCINOMA)

3.5.1.1.1 DNA Ploidy (p. 313)

- The **size criterion** (conventionally, tumors less than 3 mm in diameter are considered adenomas) frequently fails to differentiate between RCC and renal adenoma, and so does **eosinophilia** of tumors, the latter referred to as **oncocytomas** and **tubular adenomas.**
- Up to 75% of large RCCs contain aneuploid populations of cells. **Abnormal DNA content and heterogeneous cell populations begin to appear in tumors between 2.5 to 5.0 cm in diameter**. All tumors over 5.0 cm contained nondiploid cell populations, indicating more aggressive RCC behavior.
- **DNA ICM** data on adenomas and adenocarcinomas differs substantially from controls. A statistical correlation exists between the **DNA flow cytometry (FCM)** data and tumor grade. DNA FCM accurately predicts the favorable and poor outcomes of RCC. However, **differentiation between** renal adenomas and grade I adenocarcinomas, or between oncocytic and clear cell tumors of the same grade, cannot be made by FCM.
- In renal tumors of childhood, FCM has been viewed as a helpful supportive tool for determining **prognosis**; however, no correlation was observed in these tumors between ploidy and staging in any of the histological type of Wilms' tumor studied.
- **The summary evaluation of DNA content in RCC is not a sensitive criterion**. On the other hand, **high RNA content** is correlated with high nuclear grade, large tumor size, advanced clinical stage, and cytoplasmic granularity, and is helpful in the clinicopathological evaluation of RCC.

3.5.1.1.2 Molecular Cytogenetics of RCC (p. 314)

The four **histological categories** of renal cortical tumors are (1) clear cell, nonpapillary RCC, (2) nonclear cell, nonpapillary RCC, (3) papillary RCC, and (4) oncocytic tumors; **all four groups correlate with the cytogenetic and RFLP data**. These include

- **unbalanced translocations** and terminal or interstitial **deletions** on **chromosome 3p**
  chromosome 5
- trisomy or more copies of **chromosomes 17, 7,** and **12**
- **lack of association between loss of alleles of 17p and 18 and a particular histological type**
- an **increased dosage** of gene(s) on **chromosomes 5q** and **7**, suggesting these phenomena are primarily related to **progression** of the disease

The relatively high occurrence (30% of cases) of LOH at informative loci on six **chromosomes (5q, 6q, 10q, 11q, 17p,** and **19p**) suggests that more than one TSG appear to be involved in the development of RCC.

In RCC, the **chromosome 3** rearrangement (3p11-p21) has been repeatedly observed in all ***nonhereditary*** **RCCs** studied, and appears to be important in the **origin** or **progression** of RCC. The frequency of LOH on 3p is 53% in ***sporadic*** **RCC**, predominantly associated with clear cell RCC (in 75% of cases). Although the specific gene lost from that region is not known yet, the involvement of a recessive oncogene located on 3p is proposed.

A balanced **constitutional t(3;8) translocation** has been identified in eight of the family members who developed renal cancer. The breakpoint occurs either at 3p21 or 3p14.2 and can be identified using **ISH** by three probes (**pH3E4/D3S48**, **pHF12-32/D3S2**, and **pMS1-37/D3S3**). The first two probes are located in the 3p21 region, whereas the third probe hybridizes at 3p14.

The loss of the putative **"*RCC* gene"** located near the most frequently lost region on 3p, **D3F15S2**, either severely reduces the level of expression or leads to the complete absence of expression of another gene located near this site. This gene has been provisionally designated "**RIK**", and is found to be fully expressed in normal kidney.

The accumulated clinical and molecular genetic experience allows for the earliest recognition of **familial RCC**. The most frequent cause of inherited RCC is **von Hippel-Lindau (VHL)** disease, a rare autosomal dominant syndrome with a heterozygote prevalence of 1 in 50,000 individuals. Patients at risk for familial RCC should be **screened annually from age 20 years**, as in VHL disease, to ensure the earliest detection of the tumor that may significantly improve prognosis.

3.5.1.1.3 Papillary and Nonpapillary RCC (p. 315)

Hereditary, nonpapillary RCC features:

- **a breakpoint** on **chromosome 3** (3p13-14.2)
- **functional inactivation** of one allele of a putative TSG in each germline cell

Loss of the chromosome 3p segment is even proposed as a diagnostic marker for *nonpapillary* RCC.

- **Trisomy of 5q22-qter** is observed in about 50% of nonpapillary RCC and frequently accompanies the loss of 3p segment. **Growth factor** or **receptor genes** are localized to this segment, suggesting their involvement in proliferation of tubular cells.
- The next most common chromosomal aberration observed in about 50% of RCC cases is the **loss of chromosome 14**, which may point to the involvement of the c-*fos* protooncogene assigned to 14q.

In ***papillary*** **RCC**, in contrast to nonpapillary RCC, **no aberration of chromosome 3p or 5q is reported.**

- Characteristically, about 70% of **papillary RCCs** have **trisomy 17**, and frequently also **trisomy 7**. It is suggested that papillary tumors of the kidney parenchyma showing only trisomy 7 and 17 be considered benign renal adenomas.
- **Progression of renal adenomas to RCC** may involve c-*erb*B-1 (*HER*-1) (chromosome 7) and c-*erb*B-2 (*HER*-2) (chromosome 17).

3.5.1.1.4 Oncogenes and Growth Factors in RCC (p. 316)

The expression of three groups of protooncogenes in RCC can be demonstrated by northern blot:

- **nuclear oncogenes** c-*myc*, N-*myc*, and c-*fos*
- **G proteins** encoded by c-Ha-*ras*, c-Ki-*ras*, and N-*ras*
- **protein kinases** encoded by c-*abl*, c-*fes*, c-*fms*, c-*raf*, and c-*erb*B-2
- at least two protooncogenes show **consistent overexpression** — c-*myc* (73% of cases) and c-*erb*B-1 (47% of cases)

A correlation is observed

- between **the degree of c-*myc* expression** in the nuclei of cells in RCC, and **the grade of the tumor** reflected in nuclear pleomorphism
- between **the frequency of distant organ metastases** in RCC patients and the **RFLP pattern** of another protooncogene of the *myc* family, **L-*myc***. It appears that the **L-*myc* RFLP can be widely used as a molecular tool to predict prognosis in RCC patients**

The genes for the **precursors** of

- **epidermal growth factor** (pro-EGF)
- **transforming growth factor α** (pro-TGF-α)
- the **EGF receptor** (EGFR)

show differential expression in RCC and uninvolved kidney tissues so that

- the gene for a growth factor (pro-EGF) is **underexpressed** in RCC
- the genes for pro-TGF-α and EGFR are **overexpressed** in RCC
- a **reciprocal relationship** between the high expression of the *EGFR* and low expression of *HER-2/neu* genes is observed in RCC
- the **reverse holds true for normal kidney**: high expression of the *HER-2/neu* gene and low expression of the *EGFR* gene.

A correlation with the degree of differentiation of RCC has been seen in other protooncogenes: **c-*myc* expression is enhanced and c-*fos* oncogene expression is concomitantly decreased with a progressively malignant**

(poorly differentiated) phenotype.

The progressive decrease of c-*fos* expression with increasing malignancy in RCC can be related to the frequently observed loss of one or possibly both c-*fos* alleles from a region of **chromosome 14 [del(14)(q22-qter)]**.

LOH in the *p53* gene, as determined by PCR analysis of the **polymorphism in codon 72**, occurs in 60% of RCC and in 73% of bladder cancer specimens.

3.5.1.2 RENAL CORTICAL ADENOMAS (p. 317)

A consistent pattern of chromosome abnormalities, observed by **G banding** of metaphases in short-term cell cultures established from these tumors, includes

- a **monosomy 21**
- a **trisomy 12(+12)**
- a **trisomy** 17(+17)
- a **tetrasomy 7(+7,+7)**
- a **loss of Y chromosome**

A subgroup of renal cell tumors, **giant cortical adenomas**, are defined by the specific and unusual combination of **+7,+7,+17,** and **–Y**. On the basis of these criteria, a revision of some cases diagnosed as RCC may be warranted.

3.5.1.3 RENAL ONCOCYTOMA (p. 318)

In contrast to other renal tumors, renal oncocytomas typically have a distinctive mosaic pattern of cells with normal and aberrant karyotypes.

- **Ploidy patterns correlate with the stage of disease and predict overall prognosis**.

ICM helps to distinguish between the ploidy patterns of oncocytic and **oncocytic granular cell tumors** (both being diploid), which are not differentiated by a simple morphological evaluation. They may constitute **mitochondrial disease**, rather than being related to deletion of a TSG.

The **distinctive restriction pattern** that may be partially efficient as a **differential diagnostic tool** is observed only in oncocytomas, and is not encountered in papillary or nonpapillary RCC.

3.5.2 POLYCYSTIC KIDNEY DISEASE (p. 318)

Autosomal dominant polycystic kidney disease (PKD) is most commonly (95% of cases) caused by a mutation on **chromosome 16p13**. So far, attempts to isolate candidate genes from a number of CpG islands in regions expected to harbor the disease mutations suggested a **possible role for a mutated proton channel** in the pathogenesis of PDK.

Genetic markers on chromosome 16 can be used for prenatal and presymptomatic diagnosis of PKD. The limiting factor, however, is that some of the earlier markers display 5% recombination with the *PKD*-1 gene, and in some families genetic heterogeneity of PKD exists. Additional paired flanking markers have been described that increase the diagnostic utility of RFLP analysis in PKD.

In about 4% of affected families the disease is caused by mutations not occurring at the *PKD*-1 locus, and **linkage studies in that case are of little value**. A correlation between linkage data and clinical variation of the disease suggests that, in people with negative ultrasonography in early adult life, there is small likelihood of inheriting a *PKD*-1 mutation. In those who inherit a non-*PKD*-1 mutation for the disease, renal failure is likely to occur late in life.

A linkage map for the distal part of 16p has been constructed that incorporates newly defined RFLPs valuable for presymptomatic and prenatal diagnosis of *PKD*-1. In fact, a two-step procedure for the diagnosis of *PKD*-1 by family studies, utilizing these new markers (particularly **CMM 65**), has been described. In nonrecombinants, which comprise 90% of family members, **the accuracy of diagnosis using these new markers exceeds 99%**.

A novel, promising approach uses polymorphic variations in the length of **simple sequence repeats (SSR) — microsatellites**, such as those composed of **$(CA)_n$ repeats**, that can be detected by PCR and electrophoresis of amplified products. Some of these markers are

- **Sm 7**, for the **$(AC)_{19}$ repeat** at the D16S283 locus
- **Sm 5B**, for the **$(A)_7GTT(ATT)_5(ATTT)_2(GT)_{16}$** sequence at the D16S284 locus

These markers lie closer to the *PKD*-1 locus than any of those previously described.

Because autosomal dominant PKD is known to be genetically heterogeneous, the prenatal diagnosis of PKD on chorionic villi samples is recommended **only after completion of haplotyping** of the index family.

In laboratory animals, **elevated** and potentially **abnormal c-*myc* expression** was implicated in the pathogenesis of PKD.

3.5.3 BLADDER CANCER (p. 319)

A correlation has been established between the WHO grading of **transitional cell carcinoma** (TCC) of the bladder and the fraction of cells in S and G_2 phases as determined by high-resolution FCM measurement of cellular DNA content.

ICM of voided urine sediment is a sensitive technique for the recognition of abnormal DNA patterns.

FCM of bladder irrigation specimens allows the differentiation of urothelial cells from granulocytes and squamous cells. However, differential staining of blood and urothelial cells in FCM analysis of bladder irrigation specimens, using **propidium iodide staining** for DNA, can mimic aneuploidy or contribute to an increased hyperdiploid fraction.

Absolute nuclear fluorescence intensity of individual **acridine orange**-stained cells was able to detect recurrent and precancerous bladder lesions as well as kidney, ureter, and prostate lesions. The method is thought to be suitable for use in screening for early malignancy of the urinary tract.

Slit-scan enhancement of morphological features allows the enrichment of urothelial cells based on the ratio of nuclear diameter to cell diameter, resulting in significantly better discrimination, and an increase in the specificity of the method from 61 to 77%.

FCM combined with **T16** or **DU 83.21 monoclonal antibody** (MAb) offers enhanced discriminatory power. It is able to differentiate between normal, metaplastic, and neoplastic urothelium from the inflammatory cells, fibroblasts, smooth muscle cells, and other nonepithelial cells found in a bladder irrigation specimen.

The quantitative assessment of morphology appears to be of value as a predictor of tumor recurrence. DNA ploidy offers the best discriminator. None of the classic prognosticators, including histological grade or morphometry, enhance its prognostic value.

At the **electron microscopic** (EM) level reproducible characteristics of **nucleoli** have been tested as new morphological criteria for the classification of bladder cancers. These characteristics include

- **nucleolar volumes**
- **nucleolus:nucleus volume ratio**
- **volume densities of granular and fibrillar nucleolar components**
- **number of fibrillar centers (FC)**

All of these characteristics are significantly changed with an increase in tumor grade. **The number of FCs is considered a valuable diagnostic adjunct in defining malignancy of bladder tumors.**

Nucleolar organizer regions (NORs), related to regulation of protein synthesis and cell proliferation, are **more numerous in bladder cancer** samples. Their number appears to be significantly related to survival but **is only weakly correlated** with lymph node involvement and distant metastases, limiting its clinical usefulness.

3.5.3.1 CHROMOSOME ABERRATIONS, ONCOGENES, AND GROWTH FACTORS IN BLADDER CANCER (p. 321)

In TCC of the bladder, chromosome aberrations include

- a frequent **loss of 11p** sequences
- a trisomy for **chromosomes 1, 7,** and **11**
- a monosomy for **chromosome 9**
- a **loss of chromosome 17p**, which harbors the ***p53*** tumor suppressor gene
- other abnormalities involving **chromosomes 1, 5, 7, 8, 9, 11,** and **13**

The loss of chromosome 9 is the most frequent chromosomal aberration in low-grade TCC of the bladder, and appears to be one of the primary genetic events in TCC oncogenesis.

Underrepresentation of chromosome 9 in comparison to chromosomes 1, 7, and 11 correlates with **tumor progression**. ISH can be used in routine evaluation of cytogenetic changes occurring during the development of a malignant disease.

Point mutations in **codons 12, 13,** and **61** of the ***ras* gene** have been frequently observed in TCC of the bladder, but less common point mutations can be identified by **PCR** in **codons 59, 61,** and **63**.

Substitution of **valine for glycine (G → T mutation)** at codon 12 of the Ha-*ras* gene has been detected by PCR in some tumor samples of human urinary bladder carcinomas. Because this mutation was frequently found in the more aggressive urothelial tumors, the **performance of PCR amplification for the Ha-*ras* gene on voided urine specimens may serve as a helpful adjunct for identifying the aggressive variants of urothelial carcinomas**.

In a significant proportion of primary TCCs (16 out of 44), **c-*erb*B-2 overexpression** could be demonstrated using an immunohistochemical technique. Analysis of c-*erb*B-2 expression and a codon-282 mutation of the ***p53* gene** has been used to establish the origin of metastasized foci of a renal pelvis tumor.

The presence of a **high concentration of EGFR** is correlated with poor differentiation and invasiveness of human bladder carcinomas and, therefore, with poor prognosis.

Fingerprinting of malignant tumors emerges as a valuable tool for the evaluation of various events related to tumorigenesis, and/or the appearance of specific biological characteristics of a tumor that may have consequences to the host. Multiple highly polymorphic loci can be analyzed using **minisatellite probing**.

Changes in the intensity of the bands as well as **changes in the fine aspects of the fingerprint pattern** may reflect, as in the case of bladder cancer, various events related to chromosomal loss or rearrangements unique to a specific tumor and that are associated with changes in the behavior or aggressiveness of the tumor.

Chapter 3.6
Wilms' Tumor

3.6.1 INTRODUCTION (p. 323)

The group of **small round cell tumors** includes

- **neuroblastoma**
- **WT (nephroblastoma)**
- **Ewing's sarcoma**
- **rhabdomyosarcoma**
- **malignant lymphoma**

WT is a common renal tumor of childhood. The majority of cases are sporadic. A minority are hereditary, and are often bilateral and display an autosomal dominant trait with variable penetrance.

Characteristic immunohistochemical findings are

- positivity for **vimentin** (focally positive in the blastematous elements), **keratin**, and various **lectins for the epithelial elements**
- variable positivity of the **mesenchymal elements**, depending on the constituent components such as **desmin** and **myoglobin** in rhabdomyoblastomatous loci
- positive reactivity of the **neural elements** for **S-100 protein**, **NSE**, and **glial fibrillary acidic protein**

3.6.2 CHROMOSOME ABERRATIONS (p. 323)

Nonrandom structural abnormalities have been observed in **1p/1q, 11p, 7p/7q, 16p/16q, 12q,** and **17q**. The **11p13** region has been implicated in patients with the **WT-aniridia syndrome** as well as in those with **WT without aniridia**. Abnormalities of **chromosome 1** seem to be the most recently detected secondary structural change, with 80% of the breakpoints occurring in **1q** and 20% in **1p34** regions.

Trisomy 12 was the **most common abnormality**, being detected in 52% of all tumors, and particularly frequently in hyperdiploid tumors (81%).

The second most frequent chromosome aberration is **trisomy 18** (28%). **Secondary chromosome abnormalities**, especially numerical **(+12,+18,+6,+8)**, are also observed in other solid tumors and in leukemias.

A role for **genetic imprinting** in the pathogenesis of WT is demonstrated by the preferential LOH of the maternal alleles from 11p. Furthermore,

- the predominant alterations in sporadic WT are **point mutations or small deletions**
- a significant role is played by epigenetic changes such as **methylation-related imprinting**, which may explain the absence of detectable deletions in the *WT* gene itself

3.6.3 TUMOR SUPPRESSOR GENE(S) IN WT

Suppression of the tumorigenic phenotype of WT by **introduction of human chromosome 11** into the tumor cells strengthened the belief that TSG(s) may be located on chromosome 11. Other TSGs do not appear to be involved in WT.

3.6.3.1 *WT*-1 GENE

(p. 324)

Cytogenetic studies of WT patients as well as of those with Beckwith-Wiedemann syndrome pointed to, but did not limit, the involvement of 11p13 and 11p15.5. It appears that **at least three genes**, when mutated, predispose to or cause WT.

A **candidate gene** for the *WT* locus has been identified and sequences mapping in the *WT* area of band 11p13 have been isolated.

A 3-kb clone, designated **LK15**, has been isolated. The product of this clone appears to be a **DNA-binding protein** that could be a **translational regulator**. On the basis of the expression pattern of the LK15 transcript, its corresponding gene may play a part in normal kidney development.

The **WT-1 gene** has a restricted pattern of expression in certain **embryonal** tissues, but not in adult kidney. **Only one candidate gene for the 11p13 locus has been identified**.

In contrast to retinoblastoma (RB), in which the absence of a single gene appears to cause the malignancy, the study of individuals with WT pointed to an **autosomal microdeletion** that **removes more than one locus**, leading to a constellation of disorders that are otherwise known as single-gene, dominant disorders.

A further series of genomic and cDNA clones have been isolated that map within the constitutional and tumor deletions on chromosome **11p13 — the *WT* locus**. The identified product of the ***WT*-33 gene** was found to contain four **"zinc finger" domains** known to function as transcription regulators and showing a 51% similarity with the two human **early-growth response genes (*EGR*-1 and *EGR*-2)** involved in pathways controlling cell proliferation.

Similarities and distinctions with the *RB* gene have been noted:

- while the *RB* gene product appears to be expressed at significant levels in almost all types of normal dividing cells, the ***WT*-33 transcript** appears to be **limited predominantly to the kidney** and to a subset of hematopoietic cells
- inactivation of the *RB* gene occurs in a wide range of tumors, whereas this does not seem to be the case for the putative *WT* gene

- the difference extends to the function of the respective gene products: while **p105**^{RB} is a nuclear protein, it does not appear to bind to DNA specifically, and exerts control over growth indirectly through an interaction with other nuclear proteins
- the predicted WT-33 protein is a **likely direct transcription regulator**

Based on the location of the *WT*-33 gene, its tissue-specific expression, and the function predicted from its sequence, ***WT*-33 appears to represent the 11p13 *WT* gene.**

Inactivation of the 11p13 *WT* gene may be a **necessary, but not sufficient, event** for the development of malignancy.

Confirmation that the candidate *WT* gene is indeed the *WT* gene comes from the finding that this gene **(*WT*-2-1)** is expressed specifically in the condensed mesenchyme, renal vesicle, and glomerular epithelium of the developing kidney, in mesonephric glomeruli, and in cells resembling these structures in tumors.

The predominantly **blastemal** tumors, or those with epithelial differentiation, express high levels of the gene, whereas tumors with **stromal** predominance had low levels.

It is possible that the ***WT*-2-1 gene product acts as a transcriptional regulator** during normal kidney development to switch off genes involved in maintaining cell proliferation. Constitutional mutations in the *WT*-1 gene are seen in patients with WT and genital abnormalities, supporting the role of *WT*-1 as a recessive protooncogene that operates in mammalian development.

The cloned sequences of the *WT* gene may now find practical application for resolving differential diagnostic challenges and, potentially, be of prognostic value if correlations between the level of its expression and the clinical course of this malignancy are established.

3.6.3.2 N-*MYC* (p. 326)

Extensive overexpression of the N-*myc* gene without amplification, similar to that in neuroblastoma, is characteristic of WT. For example, in the case of an unusual small round cell tumor arising from a fallopian tube in a 15-year-old girl histological studies failed to distinguish WT from a neuroectodermal neoplasm. N-*myc*, however, proved to be overexpressed in the tumor, but not in normal tissues; this was not due to amplification of the gene, favoring the diagnosis of WT.

Likewise, the pattern of N-*myc* DNA probes allowed the diagnosis of **intrarenal neuroblastoma**; evaluation of N-*myc* gene amplification predicted a poor prognosis, subsequently confirmed by the clinical course.

Part 4
Gene Level Evaluation of Selected Endocrine Disorders

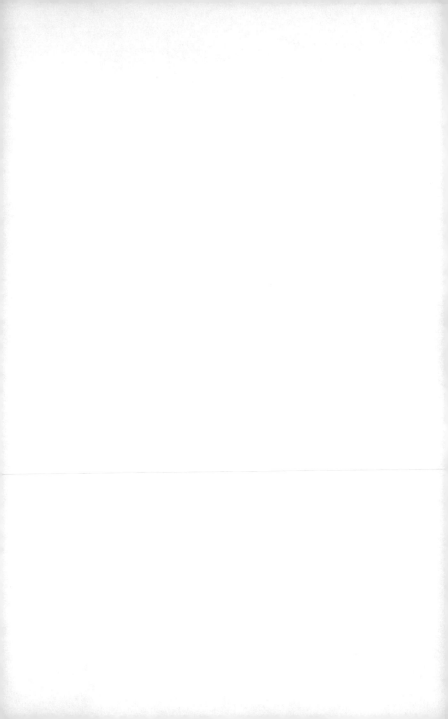

Chapter 4.1
Diabetes Mellitus

4.1.1 INTRODUCTION (p. 349)

Molecular genetic aspects of **non-insulin-dependent DM (NIDDM)** and **insulin-dependent DM (IDDM)** clearly point to an association of DM susceptibility with a certain pattern of arrangement of **HLA class II genes**. Conflicting observations abound and a number of hypotheses have been advanced. The complexity of interactions of various alleles is compounded by the so far limited understanding of the role of environmental factors (e.g., viral, chemical, and cultural).

Susceptibility to IDDM and NIDDM is discussed here only as an example of a variety of conditions in which the role of HLA is becoming progressively recognized. **DNA typing** will likely become a common approach to obtain additional and important information when working up common cases such as DM.

Important advances have already been made in unraveling the details of synthesis and function of **insulin, insulin receptor (IR)**, and **glucokinase (GK)** at the gene level. These offer a potential for evaluating clinical problems in diabetology at the gene level.

4.1.2 SYNOPSIS OF MOLECULAR PATHOLOGY OF DM

4.1.2.1 INSULIN-DEPENDENT DM (IDDM) (p. 349)

The key role of **autoimmune responses** resulting in the destruction of β cells has been defined and a working hypothesis of the etiology of IDDM proposed. It envisions **genetic predisposition (susceptibility) to IDDM** as determined by two genes on **chromosome 6**, one associated with **HLA-D/DR3**, the other with **HLA-D/DR4**. Susceptible individuals develop an abnormal response to the damage produced in β cells by a cytotoxic agent (a virus or a chemical). Immunologically determined destruction of the β cells or failure of the β cells to regenerate after the damage then follows.

The characteristic **mononuclear cell infiltrate** and **β-cell destruction (insulitis)** reflect an already advanced stage of the destructive process. The insulitis (13 to 75% of the islets destroyed) observed in recent-onset IDDM cases is more severe in younger patients, and is markedly reduced in long-standing IDDM.

Cell-mediated immunity involving **cytotoxic T cells** leads to the destruction of the β cells at the later stages, rather than during an initial assault. Early in the immune response **activated autoreactive T cells** recruiting natural killer (NK) cells and/or macrophages account for the specificity of recognition of β-cell **autoantigens**.

Humoral autoimmune reactions complement the cell-mediated immunity in the development of IDDM. None of these demonstrate absolute specificity for β-cells. The clinical significance of **islet cell antibodies, islet cell surface antibodies**, the **anti-64-kDa protein antibody**, and **insulin autoantibodies** as markers of IDDM remains to be elucidated.

Islet cell autoantibodies are thought to result from **polyclonal B cell activation** in response to the liberation of antigens from destroyed β cells.

The soluble products liberated from the infiltrating immune cells [**cytokines**, such as **interleukin 1 (IL-1)**] have been implicated in delivery of the lethal hit. The mechanism of IL-1 action is not entirely clear. The formation of damaging **free radicals** in response to overstimulation of β cell function by high local doses of IL-1 has been observed. It appears that **interleukin 6** (IL-6) is produced by the β cells of the islets and acts as a costimulator (along with IL-1) in T lymphocyte activation.

The molecular events leading to islet cell destruction seem to begin early on with the **hyperexpression of major histocompatibility (MHC) class I molecules** on islet cells and the **induction of MHC class II on β cells. Interferon α** (IFN-α) is implicated in mediating the MHC class I molecules on β cells; however, the molecular mechanism of induction of MHC class II molecules remains unknown. The hyperexpression of MHC molecules may constitute a nonimmune component of β cell destruction via a **transgene activation mechanism**.

Expression of the MHC class I and II molecules by β cells is **not sufficient for antigen presentation**, and the process of "**phenotype switching**" of the β cell is thought to induce many properties of antigen-presenting cells. The expression of MHC class I molecules is decreased in cells from IDDM patients, and T cell responses to self-antigens are defective, thus contributing to aberrant antigen presentation.

Associated with this process is lymphocyte adherence that is mediated on the lymphocyte surface by the **lymphocyte function-associated antigen 1 (LFA-1)** and by the **intercellular adhesion molecule 1 (ICAM-1)** on the target cell surface. The cytokine-induced ICAM-1 expression at sites of acute or chronic islet inflammation is thought to play a role in attracting lymphocytes to β cells.

Cytokines [**IL-1**, **interferon α (IFN-α)**, and **IFN-γ**], in addition to inducing hyperexpression of MHC, ICAM-1 and IL-6, directly affect β cell function.

4.1.2.2 NON-INSULIN-DEPENDENT DM (NIDDM) (p. 351)

NIDDM is usually defined as **maturity-onset diabetes**, characteristically displaying a mild hyperglycemia in response to a glucose challenge that is followed by late hyperinsulinemia. Both obesity and diabetes independently affect basal insulin levels. In addition to glucose, a number of other compounds may produce rapid insulin responses:

- **glucagon**
- **gastrointestinal inhibitory peptide (GIP)**
- **cholecystokinin (CCK)**
- **secretin**
- **amino acids**, in particular **arginine**

Insulin response to nonglucose stimulants is markedly reduced in NIDDM.

Insulin resistance, present in virtually all NIDDM patients, interacts with deficient islet function and leads to **compensatory hyperglycemia**, the degree of which is proportional to the extent of **β cell dysfunction**.

4.1.2.2.1 Islet Amyloid Polypeptide (IAPP) (p. 351)

Relative hyperproinsulinemia and the presence of **amyloid deposits** in the islets of NIDDM patients have been related to the inappropriate release of unprocessed **proinsulin**. The islet amyloid is composed of a peptide, called **islet amyloid polypeptide (IAPP)**, or **amylin**. This 37-amino acid neuropeptide-like molecule has 50% homology with the **neuropeptide calcitonin gene-related peptide (CGRP)**. Structural similarity with CGRP points to possible modulating effects of IAPP on **adenylate cyclase**.

IAPP is a normally secreted peptide produced along with insulin in β cells from a larger precursor, and its processing is accomplished by the same enzymes involved in converting proinsulin to insulin. IAPP does not appear to affect the secretion of insulin or other hormones. However, a restricting effect on the blood flow in acinar cells is a possibility.

The **level of IAPP markedly increases on glucose stimulation**, and is particularly elevated in obesity. There is no evidence, however, of an increase of IAPP in NIDDM. The genes encoding IAPP in NIDDM patients show no abnormality of either IAPP or its precursors.

IAPP is also deposited in normal individuals with aging. It is most likely that IAPP contributes to NIDDM by its accumulation as amyloid in β cells, which in turn leads to a reduced β cell mass.

For reasons not completely defined yet, in severely hyperglycemic states there is a **markedly greater responsiveness of IAPP secretion** relative to insulin.

4.1.2.2.2 Other Factors in NIDDM (p. 352)

Glucose transport function is affected in NIDDM, and is apparently related to the profound underexpression of **GLUT-2**, the **glucose transporter** on β cells.

At this time, the evidence strongly favors the view that **NIDDM is a heterogeneous disease syndrome characterized by both impaired islet function and insulin resistance.**

The etiology of NIDDM is probably related to hereditary risk factors modulated by changes in the environment. An intriguing finding has pointed to the possibility of persistent viral infection [e.g., cytomegalovirus (CMV)] selectively affecting β cells of the islets of Langerhans in persons with NIDDM.

One form of NIDDM, the so-called **maturity-onset diabetes of the young** (MODY), has an autosomal dominant mode of inheritance, and the gene responsible for MODY in one well-characterized pedigree is tightly linked to the marker **adenosine deaminase gene** (*ADA*) on chromosome 20q.

4.1.2.3 REGULATION OF INSULIN GENE EXPRESSION (p. 352)

The **insulin gene** is located on the short arm of **chromosome 11**. Transcription of the insulin gene occurs in the β cells of the Langerhans islets where the mRNA transcripts are processed. Translation of the mature insulin mRNA yields **preproinsulin**, which is proteolytically cleaved to produce **mature insulin** as well as **prepeptide** and **C-peptide** sequences.

Insulin and C peptide are secreted together in response to physiological stimuli. **Guanine nucleotide-binding proteins (G proteins)** play a significant role in both regulation of the insulin gene and modulation of insulin action in target tissues.

4.1.2.4 GLUCOKINASE (p. 352)

Stimulation of insulin gene expression by glucose appears to involve **glucokinase** as **glucose sensor** on the surface of β cells and hepatocytes. Glucokinase gene regulation is different in the liver and pancreas.

At least **three glucokinase isoforms** are generated by differential RNA processing of the glucokinase gene transcripts. Glucokinase gene expression may be involved in the determination of glucose homeostasis, including the dysregulation characteristic of diabetes.

Linkage analysis of the **glucokinase gene** and of its **tissue-specific promoter regions** in NIDDM pedigrees suggests that primary glucose refractoriness accounting for the symptomatology of NIDDM may involve **glucose sensor molecules** in β cells and hepatocytes (e.g., **glucokinase**, **phosphorylase a**, and **glucose-6-phosphatase**). In fact, a close linkage can be established between the glucokinase locus on chromosome 7p and DM in MODY patients.

4.1.2.5 GLUCOSE TRANSPORTER (p. 353)

Metabolic defects potentially involved in the pathogenesis of NIDDM include

- impaired **insulin binding** and **receptor kinase activity**
- **defective glucose transport**
- abnormal **glycogen synthesis**
- **insulin deficiency**

Glucose transporters are likely candidates for a highly specific genetic defect predisposing to insulin resistance. Genetic studies have so far, however, provided conflicting results:

- some favor an association between RFLP at the Hep G2 **glucose transporter locus** (*GLUT*) and NIDDM
- others demonstrate lack of such an association

Studies in mouse models and in humans with NIDDM, however, strongly support the view that deficiencies linked to the transporter locus may indeed exist.

4.1.2.6 INSULIN RECEPTOR (IR) (p. 353)

Interaction of insulin with the cell begins with IR specifically binding the hormone. Binding of insulin to IR leads to the appearance of the **insulin-dependent protein tyrosine kinase activity** of the IR that is essential, but not sufficient, for insulin signal transduction. This event initiates the physiological response of the cell, thereby **modulating the cell surface signal into a specific intracellular response**. Following ligand binding, IR is internalized and specific conformational changes of IR domains appear to be necessary for internalization.

A specific **insulin-degrading enzyme (IDE)** has been identified, cDNA produced, and the *IDE* gene mapped to human **chromosome 10**.

Specific structural sites on the IR molecule also determine other events such as recycling, retroendocytosis, receptor turnover rate, down regulation, and ligand degradation. Mutations of the IR can selectively disrupt the normal process of internalization, without affecting receptor recycling after internalization.

Interactions between different structural domains affecting IR function also in other IR molecules create the possibility of amplifying the functional consequences of the insulin-binding signal.

The amino acid sequence of IR has been established and the gene encoding IR has been identified, partially sequenced, and assigned to **chromosome 19**.

Several mutations in the *IR* gene associated with extreme **insulin resistance** are known. Linkage studies suggest that genetic variations in the *IR* gene are associated with an increased or decreased risk for NIDDM.

Characterization of the *IR* gene provides an opportunity to amplify individual exons of the gene by PCR and to **relate the specific sequence alterations to the symptomatology of the NIDDM patients** featuring variable degrees of insulin resistance.

So far, however, a linkage relationship between a specific RFLP and NIDDM failed to reveal any abnormality in the *IR* gene in a patient homozygous for this RFLP.

4.1.3 SUSCEPTIBILITY TO DM (p. 354)

The contribution of **genetic factors** to the development of DM has long been appreciated, and advances in molecular genetics have allowed direct testing for the postulated role alterations at the gene level.

DM associations have been studied predominantly by **linkage analysis** using RFLP markers for **candidate susceptibility genes**, which included

- genes in the **HLA locus**
- the **insulin gene**
- the ***IR* gene**
- the **glucose transporter gene**
- the **T lymphocyte receptor (TCR) β-chain gene**
- the **apolipoprotein genes**

The genetic factors accounting for susceptibility to IDDM and NIDDM have not yet been defined. Family and **monozygotic (MZ) twin** studies show the MZ concordance rate for NIDDM to be close to 100%, suggesting an **autosomal recessive** mode of inheritance. An NIDDM variant, a maturity-onset diabetes of the young (MODY), appears to be autosomal dominant. The **most powerful predictor of risk for diabetes** (either NIDDM or IDDM) **is being a monozygotic (MZ) twin of a diabetic person**.

In contrast to NIDDM, the MZ twin concordance rate for IDDM is much lower than 100%, in the range of 25 to 55%. This implies an important contribution by nongenetic, **environmental factors** such as "triggering" influences — chemical damage, viruses, as well as cultural traditions, diets, and so on. A number of hypotheses have been advanced to account for the nonlinear decrease in IDDM and NIDDM risk observed with progressive distance in relationship.

Most of the evidence on associations between diseases and the MHC genes has been obtained using **serological techniques** relying on complement-dependent lymphocytotoxicity assays.

On the other hand, molecular biological techniques, in particular **RFLP** analysis, **PCR** amplification, and **sequencing** of specific gene regions as well as **ASO** and **sequence-specific oligonucleotide (SSO) probes**, define in molecular genetic terms all of the specificities previously characterized serologically and by cellular methods.

One of the first **candidate genes** responsible for susceptibility to IDDM were the HLA genes or alleles, initially **HLA class I (HLA-B15** and **HLA-B8**) and later at the **class II HLA-DR locus** (**HLA-D/DR**) genes, in particular **HLA-DR3** and **DR4** (on the average twice as frequent among IDDM patients as in controls).

A particularly strong association has been observed between **insulin autoimmune syndrome** and HLA-DR4. Linkage association between **VNTRs in the 5′ region of the insulin gene (*INS*)** on chromosome 11p and increased risk of IDDM is seen in HLA-DR4-positive patients. **A gene(s) affecting HLA-DR4 IDDM susceptibility appears to exist in a region of INS-IGF-2.** Some alleles at the **HLA-DQ** locus, in particular **DQw8**, occur in over 95% of DR4-positive IDDM patients compared to 60 to 70% of healthy DR4-positive controls.

Comparison of **subtypes of HLA-DR4** emphasizes the primary MHC association with IDDM to reside in the DQ region in most DR4 patients. Testing with ASO probes for the MHC class II genes DQA1 and DQB1 in IDDM patients indicates

- The DQw1.18 allele of the DQB1 gene is associated with **inherited protection**
- On the other hand, the A3 allele of the DQA1 gene and the DQw2 allele of the DQB1 gene are associated with **susceptibility to IDDM**

Higher susceptibility to IDDM has also been observed in **DR3-associated haplotypes**. Among Caucasians 35 to 45% of IDDM patients have been shown to be DR3/4 heterozygotes, compared to 2 to 5% of healthy controls.

Restriction polymorphism defined by the *Bam*HI 3.7-kb fragment occurs more frequently in unaffected persons (37%) than in IDDM patients (0 to 2%).

Several studies have described **specific amino acid deletions** the HLA-DQβ chain (**aspartate** at position 57) to be **associated with protection against IDDM**. It appears that **specific combinations of DQβ and DRβ sequences play a major role in determining susceptibility to IDDM** .

In fact, the presence or absence of **Asp** at position 57 does not by itself determine the risk for IDDM. Therefore, **when testing for IDDM susceptibility, a combination of dot blot analysis using ASO probes not only for the HLA-DQ gene region, but also for the HLA-DR region, including also the α chains, is advocated**.

A particularly strong association has been found between **insulin autoimmune syndrome** and HLA-DR4. **VNTR probes** in the 5′ region of the **insulin gene** on **chromosome 11p** helped establish an association with this region and an increased risk of IDDM.

A genetic association between an RFLP of the insulin gene and **atherosclerosis** detected by *Bgl*II digestion has been reported in adult and pediatric populations.

One hypothesis emphasizes the **competition** of different class II molecules for a single peptide, for example, a **class II susceptibility gene product** such as **DQ 3.2**. Those individuals in whom DQβ 3.2 has the highest affinity for a diabetogenic peptide develop a predisposition to diabetes. If other genes (DR, or other DQ genes) generate products with higher affinity than DQ 3.2 for the same ligand peptide, then their competitive advantage in this example confers a relative resistance to IDDM. Furthermore, non-MHC genes may generate products competing with the diabetogenic peptides for binding to class II susceptibility genes.

Genetic markers have so far *failed to identify a single major gene* in NIDDM, and the proposed risk models visualize an interplay of a few genes with moderate effects.

Chapter 4.2

Thyroid and Parathyroid Disorders

4.2.1 THYROID GLAND

4.2.1.1 DNA PLOIDY (p. 359)

Diagnostic challenges can be encountered in combinations of two or more histological patterns, in poorly differentiated tumors, and in some **Hürthle tumors** as well as when distinctions are to be made between **benign** and **malignant follicular thyroid nodules**.

The practical value of **DNA content** evaluation by image cytometry (**ICM**) or flow cytometry (**FCM**) as an aid to differential diagnosis is not clearcut. Some workers suggest that **ploidy** of thyroid tumors can define their biological behavior, whereas others demonstrate that **aneuploidy occurs in normal glands, adenomas, and nodular goiter**, and fails to distinguish between benign and malignant **Hürthle cell tumors**.

In general, aneuploidy is correlated with poor survival in **papillary** and **follicular carcinomas**. However, even benign **Hürthle cell tumors** show aneuploid DNA patterns not associated with malignant behavior. Aneuploidy in **medullary thyroid carcinoma** (MTC), in contrast, is correlated with an aggressive growth pattern.

All **anaplastic carcinomas**, known for their poor prognosis, are aneuploid. On the other hand, in **multivariate analysis** of medullary carcinoma, when applied to an **individual patient**, none of the following parameters prove to be of significant predictive value with respect to 5-year survival: DNA content, patient age, hereditary background, calcitonin immunoreactivity, type of surgery, patient gender, clinical stage, histological subtype, and amyloid content of the tumor.

It appears that ploidy evaluation of thyroid lesions may be used **only as an adjunct** in diagnostic pathology, and is inferior to the prognostic value of age, tumor size, and extrathyroidal extension.

4.2.1.2 CHROMOSOME STUDIES (p. 360)

Some potentially useful associations have been noted between chromosome characteristics and specific histopathological types. In cytogenetically abnormal **papillary** adenocarcinomas, inversions, insertions, and balanced or unbalanced translocations were consistently observed in the **chromosome 10q** arm. Chromosome 10 appears also to be involved in **nonpapillary thyroid cancers** in association with other chromosome aberrations.

The gene for **multiple endocrine neoplasia type I2I (MEN I2IA)**, which involves medullary thyroid carcinoma, has been mapped to **10q11.2-10q21**. Furthermore, the ***PTC* oncogene** (for **p**apillary **t**hyroid **c**arcinoma) has also been mapped to 10q11-10q12.

The linkage information obtained using polymorphic **DNA probe RBP3** for the chromosome 10q1 pericentric region was adequate to provide genetic counseling for MEN IIA to 63% of families. When the power of RFLP analysis is limited by the small size of some families and the lack of informative matings, screening with **pentagastrin challenge** must be used.

Mapping for **MTC** without **pheochromocytoma** was found not to be precise enough to distinguish it conclusively from MEN I2IA.

Other chromosomal abnormalities of interest involve **chromosome 3**, where the **β thyroid hormone receptor (ErbA-2)** maps at **3p25-p21**. In an MTC cell line (MTC-SK), the predominant chromosome aberrations are terminal rearrangements affecting most frequently **chromosome 11p**. This is the region harboring the **Ha-*ras* protooncogene**, and the locus of the **calcitonin** and **calcitonin gene-related peptide genes**.

A characteristic instability of the centromeric region of **chromosome 16** and somatic pairing of the homologous chromosomes 16 have also been observed in MTC.

4.2.1.3 ANALYSIS OF SPECIFIC GENES IN THYROID DISEASE

4.2.1.3.1 *TSH*-β Gene (p. 361)

Studies at the gene level allowed the identification of a specific **thyroid-stimulating hormone β (*TSH*-β) gene** defect in members of five Japanese families with inherited **TSH deficiency**. Single-base mutations in **codon 29** of the *TSH*-β gene lead to the production of a defective β polypeptide that is unable to form a functional hormone through its association with the α subunit.

In another study of **hereditary hypothyroidism** in members of two Greek families, a **nonsense mutation in exon 2 of the *TSH*-β gene** has been identified. This led to the production of a truncated peptide incapable of functional association with the TSH α subunit.

The **TSH receptor (*TSHR*) gene** has been assigned by *in situ* hybridization (ISH) to **chromosome 14q31**. The association between this gene and that of immunoglobulin G heavy chain haplotypes, suggested by the autoimmune nature of Graves' disease, appears to be more complex than that emerging from chromosomal linkage studies.

4.2.1.3.2 Thyroid Peroxidase (*TPO*) Gene (p. 361)

Advances in the study of the pathogenesis of **thyroid autoimmune disorders** led to the identification of the thyroid microsomal-microvillar antigen as **thyroid peroxidase (TPO)**. One of the causes of **congenital hypothyroidism** is related to defects in the **organification of iodide**.

*Eco*RI VNTR DNA polymorphisms in the *TPO* gene probed with a 0.7-kb *TPO* cDNA have been used to analyze hereditary hypothyroidism cases in the Netherlands. Another probe, 3.0-kb *TPO* cDNA, revealed a different set of *Eco*RI RFLPs. Analysis of the *TPO* gene products by immunohistochemistry has been reported to be a good marker of malignancy in thyroid tumors as well.

4.2.1.3.3 HLA Associations (p. 361)

Molecular studies utilizing DNA probes to establish specific disease associations with a given pattern of RFLP provide essentially a new basis for the concept of genetic susceptibility in **Hashimoto's thyroiditis**. The strongest relative risk is associated with **DQw7**, found to occur in 56% of affected persons and only in 21% of controls.

Variations in DNA sequences in codons 45 and 57 of the ***DQB1* gene** are the critical point distinguishing *DQw7* from other *DQw* specificities.

Significant associations have also been established with the adjacent ***DQA1* genes**.

4.2.1.3.4 Insulin Growth Factors (p. 361)

The "**autocrine secretion**" observed in cancer cells allows them to establish a degree of autonomy independent of "external" growth factors. In an animal model that spontaneously develops MTC, the expression of **insulin growth factors I** and **II (IGF-I and IGF-II)** occurs in most differentiated MTC. Virtually **no transcription of these genes occurs in anaplastic medullary carcinoma**, suggesting a helpful differential pattern of expression of specific genes in two types of MTC.

4.2.1.3.5 *PTC, ras,* and *TRK* Genes (p. 362)

Activation of a specific oncogene has been observed in **papillary thyroid carcinoma** (PTC); the gene was designated ***PTC*** and assigned to **chromosome 10q**. This was the first oncogene specifically associated with thyroid cancer.

***TRK* oncogene** has also been activated in some papillary thyroid carcinomas. Because both *PTC* and *TRK* oncogene products display thyrosine kinase activity, the activation of this class of oncogenes may be specifically associated with *PTC*.

Activation of another oncogene, *ras*, has subsequently been shown to occur in high frequency in **microfollicular adenomas, follicular carcinomas**, and **undifferentiated carcinomas**, and with a low frequency in **papillary thyroid carcinomas**.

N-*ras* mutations are believed to occur only rarely in primary MTC.

The other *ras* oncogene, **Ha-*ras***, is activated in roughly the same proportions in benign and malignant thyroid neoplasms, suggesting that it may constitute an **early tumorigenic event**.

Analysis of point mutations of **Ki-*ras*** revealed that all three mutations detected in **papillary** cancers are different. In contrast, in **follicular** thyroid cancer a similarity of point mutations of the Ki-*ras* gene is noted. Therefore, **the pattern of point mutations of the Ki-*ras* gene appears to correlate significantly with the type of thyroid malignancy**. Strong correlations with the known difference in their biological behavior still remain to be established.

4.2.1.3.6 Receptors (p. 362)

Cytoplasmic and **nuclear receptors** are involved in the control of gene expression. Similarities of **thyroid hormone receptors** with **steroid hormone receptors** are seen not only in their mode of action but in their regulation of gene expression as well. At the structural level, the protein encoded by the human c-*erb*A gene is related to the family of steroid hormone receptors.

The expression of the c-*erb*A oncogene is markedly modulated by the degree of **methylation** and is directly controlled by the thyroid hormone at the transcription level. High-affinity thyroid hormone-binding proteins are found in many organs.

The introduction of the thyroid hormone receptor (c-*erb*A products, acting as thyroid hormone receptor) into cells known to be receptor deficient results in **specific gene expression of target genes** (e.g., the **growth hormone** and **prolactin genes**).

Likewise, **FAO hepatoma cells** characteristically do not have specific nuclear binding of T_3. On introduction of the chicken c-*erb*A gene, they display T_3-binding activity similar to that found in normal liver cells, and start expressing several hormone-dependent genes in response to T_3.

Fine analysis of the two c-*erb*A genes, c-*erb*A-2 and c-*erb*A-3, demonstrated that the so-called phenomenon of **generalized thyroid hormone resistance** is associated with the gene that is tightly linked to the **c-*erb*AB locus on chromosome 3**. Abnormalities of c-*erb*AB, which codes for a thyroid receptor in humans, account for various metabolic and growth abnormalities.

The role of thyroid hormone receptors in **benign thyroid** conditions or **neoplasia** still remains largely unknown. Some form of c-*erb*A derangement may have a role in the autocrine control of thyroid malignancy. We are not aware at this time, however, of specific findings to indicate this.

The roles of **TSH** and **LH/CG** (luteinizing hormone/choriogonadotropin) **receptors** in thyroid pathology include

- control of **thyroid cell metabolism**
- control of the growth and **development of the gland itself**
- **interaction with autoantibodies to TSHR**, which are responsible for the pathogenesis and hyperthyroidism of **Graves' disease**

Unexpectedly, the cloning thyroid library proved to contain several cDNAs encoding the human LH/CG receptor formerly thought to be present only in gonadal cells. Analysis of **the *LH/CG* genes showed similarity to**

the *TSH* gene. However, in the thyroid one finds an incompletely processed *LH/CG* mRNA, suggesting that **tissue-specific splicing** may be important. At any rate, the presence in the thyroid of the genes encoding GH/CG receptor may, in part, account for the observed elevation of T_3 and T_4 in early pregnancy, when TSH is not elevated.

4.2.1.3.7 Nuclear Oncogenes (p. 363)

No evidence could be provided of rearrangement or amplification of any member of the nuclear oncogenes (**c-*myc*, N-*myc*, L-*myc*, *fos*, *myb*,** and ***p53***) in **thyroid follicular cell neoplasia**. Changes in the expression of the oncogenes were thought to be the secondary consequences of the tumor phenotype.

The presense of mutations in the ***p53* gene** is correlated with the undifferentiated state of papillary adenocarcinomas of the thyroid: direct sequencing of the PCR-amplified products of exons 5 and 8 of the *p53* gene shows an **absence of mutations** in **differentiated tumors** and **specific mutations in codons 135, 141, 178, 213, 248, and 273 in undifferentiated neoplasms**.

In **C cell tumors**, on the other hand, **N-*myc*** expression assessed by ISH is viewed not only as a differentiation marker, but is interpreted as indicative of the role *N-myc* activation plays in neuroendocrine neoplasia.

c-*fos* gene expression is reduced in inverse correlation with the degree of differentiation of **follicular carcinoma**.

4.2.2 PARATHYROID GLAND (p. 364)

Differentiation between benign and malignant neoplasia of the parathyroid gland may present a diagnostic dilemma, and a combination of morphological and clinical features, including the presence or absence of metastases, is needed to recognize a malignancy.

Hyperplastic glands show aneuploidy in 22%, and up to 1.5% of cells are in the S phase. The S phase fraction (**SPF**) in **adenomas** constitutes up to 3.8%, whereas in **carcinomas** this value can be as high as 14.1%. All types of parathyroid neoplasia, but not normal parathyroid gland, demonstrate aneuploidy and tetraploidy, and therefore **adenomas cannot be distinguished from primary hyperplasia by DNA FCM**, although **an increase in the proliferative activity above 3.8 to 4% is highly suggestive of malignancy**.

According to other authors, **parathyroid carcinomas are more likely to be aneuploid than adenomas**.

It appears that in evaluation of parathyroid diseases FCM DNA should be **used only as an adjunct** to the histopathological diagnosis.

Chapter 4.3

Adrenal Gland and Multiple Endocrine Neoplasia 1 and 2

4.3.1 ADRENAL GLAND

4.3.1.1 DNA PLOIDY (p. 365)

One of the characteristic features of **adrenocortical carcinoma** is its usually large size, exceeding 100 g.

All **adenomas tend to be diploid**, whereas the majority (83.3%) of **carcinomas and of all the metastasizing tumors are aneuploid**. Aneuploid tumors are likely to be larger (over 50 g) and display several histological features of carcinoma.

DNA flow cytometry (FCM) can give prognostic information. However, the **inconsistency** of the FCM criteria is repeatedly stressed, and their greater prognostic power over that of size and histology is questioned.

Among the various histopathological parameters, such as nuclear grade, mitotic rate, architecture of the tumor, and invasive behavior, only the **mitotic rate** proves to be a reliable predictor of the outcome. It has been proposed that adrenocortical carcinomas with 20 or more mitoses be considered **high grade**, and those with fewer mitoses as **low-grade** neoplasms. **Because ploidy shows low sensitivity and specificity for prognosis and poor correlation with the outcome,** ***assessment of the biological behavior of adrenocortical tumors by FCM is not advised.***

Particular caution should be exercised in making a diagnosis of an individual patient solely on the results of DNA ploidy assay.

4.3.1.2 CONGENITAL ADRENAL HYPERPLASIA (CAH) (p. 366)

CAH is a manifestation of a constellation of metabolic derangements largely due to steroid **21-hydroxylase (21-OH) deficiency**. This enzyme of the **glucocorticoid pathway** converts 17-hydroxyprogesterone to 11-deoxycortisol, then eventually to cortisol. In the **mineralocorticoid pathway**, leading to the production of aldosterone, 21-OH converts progesterone to deoxycorticosterone. Deficient activity of 21-OH results in the **accumulation of precursors**, particularly **17-hydroxyprogesterone**, which stimulates the production of adrenal androgens clinically manifested as **virilization**. The other consequence of 21-OH deficiency is the defective production of aldosterone, leading to the syndrome designated as "**salt wasting**".

CAH is associated with aberrations in the two genes encoding 21-OH located on **chromosome 6** within the class III region of the HLA complex. They are flanked on both sides by class II (HLA-*DR*) and class I HLA genes (HLA-*A*, HLA-*B*, HLA-*C*).

Of the two genes for 21-OH, only one, the ***CYP21B* gene**, is functional and encodes cytochrome **P450c21**, which mediates 21-hydroxylation. The other gene, ***CYP21A***, located about 30 kb upstream from *CYP21B*, is a closely related **pseudogene** defective in gene expression due to several mutations.

The location of *CYP21A* and *CYP21B* in close proximity to the genes encoding **complement C4A** and **C4B**, respectively, is functionally significant. In addition, there is a high degree of sequence homology between the genes encoding 21-OH and C4.

The recombination events can lead to

- a gene **deletion haplotype — lacking the gene unit encoding 21-OH and C4**
- a gene **addition haplotype — having an extra gene unit encoding 21-OH and C4**

The **variable degree of clinical manifestations** of 21-OH deficiency depends on the specific mutation in a given case. For example,

- A single amino acid substitution **(Val-281 → Leu)** in the mild, "**nonclassic**" form leads to the enzyme having 20 to 50% of normal activity.
- Mutation **(Ile-172 → Asn)** results in a "**simple virilizing**" form with an enzyme having less than 2% of normal activity.
- Still **another cluster mutation** results in an enzyme without any detectable activity.

Pulsed-field gel electrophoresis indicates that the **21-OH deficiency is attributable to gene deletion in about one in three disease haplotypes**.

4.3.1.2.1 Prenatal Diagnosis of CAH (p. 367)

Prenatal diagnosis of CAH should begin with the measurement of steroid precursor levels. **Classic variants** of 21-OH deficiency can be diagnosed by the determination of 17-hydroxyprogesterone (17-OHP) in amniotic fluid. Elevation of 17-OHP can be diagnostic even in the first trimester of pregnancy.

Indirect identification of affected individuals relies on linkage analysis of polymorphisms at the gene loci mapping in the vicinity of the 21-OH locus, DNA-based HLA and C4 gene polymorphisms, and serologically defined HLA polymorphisms. For these methods to be informative **definitive diagnosis of parental disease and nondisease haplotypes is required**.

DNA extracted from **chorionic villi** as early as gestation weeks 8 to 10 can be analyzed for CAH in two ways:

- **indirectly**, by highly polymorphic HLA class I and class II probes

- **directly**, using the cDNA **probe pc21/3c** for the gene encoding 21-OH. This method can be used for all couples at risk, when DNA of the index case is available

Direct 21-OH gene probes identify mutations in the *CYP21B* genes that account for only about 40% of cases, and only about 15% of classic patients will have two haplotypes of this kind. This is **not sufficient for definitive prenatal diagnosis**. The direct 21-OH probe assay is not informative in some cases, so that this probe cannot be used routinely.

An **optimum strategy for prenatal diagnosis**, at this time, uses a combination of a 21-OH gene probe, a CYB probe, an HLA-*B* probe, and an HLA-*DR*β probe.

This combination identifies over 95% of families with a surviving affected child, who is informative for at least one of these probes.

4.3.1.2.2 Identification of CAH Carriers (p. 367)

Identification of carriers still relies on HLA typing involving the siblings or close relatives of a known 21-OH-deficient patient. DNA-based probes identify approximately 40% of the affected patients displaying either deletions in the *CYP21B* gene or large-scale *CYP21B* → *CYP21A* conversion haplotypes.

The **codon-281 mutation** has been found to be consistently associated with 21-OH deficiency of the nonclassic type displaying the HLA-*B14*, *DR1* haplotype. **This mutation has never been detected in normal controls.** The opposite, however, is not true: there are people with nonclassic 21-OH deficiency who do not have the codon-281 mutation.

The close physical proximity of the genes encoding 21-OH and C4 allows the prenatal diagnosis of 21-OH deficiency, using C4 DNA probes on chorionic villi. The **use of the C4 cDNA probe appears preferable to that of the HLA complex probes** because of the shorter distance (60 kb) between the C4B gene (where *Taq*I RFLPs are located), and the 21-OH B gene, compared to the distance between this gene and the HLA-*DR* and HLA-*B* genes (which are about 400 and 800 kb, respectively).

4.3.1.2.3 PCR in CAH Diagnosis (p. 368)

Direct characterization of the *CYP21B* and *CYP21A* genes can be accomplished by the PCR. Evaluation of the *CYP21B* gene by Southern analysis and PCR in the leukocyte DNA of CAH patients revealed the major abnormalities to be

- **deletion** of the *CYP21B* gene
- **gene conversion** of the entire *CYP21B* gene to *CYP21A* (its inactive counterpart)
- **frame-shift mutations** in exon 3
- an intron 2 mutation that causes **abnormal RNA splicing**
- a mutation leading to a **stop codon** in exon 8

Drastic simplification of laboratory diagnosis of CAH is due to the availability of specific primers for *CYP21B* and *CYP21A*, and automation of the PCR reaction and analysis of the PCR reaction products by gel electrophoresis.

4.3.1.2.4 Oncogenes in Adrenal Disorders (p. 368)

In laboratory animals, the overexpression of **c-*mos*** has been found in the lens and brain of animals developing pheochromocytomas, and in MTC simulating the human MEN 2 syndrome.

In malignant **Y-1 mouse adrenal cells**, the protooncogene **c-Ki-*ras*** is amplified, whereas **c-*fos*** does not appear to be amplified.

Southern analysis and specific DNA probes do not reveal amplification of **N-*myc*, L-*myc***, or ***erb*B-1** in **pheochromocytomas** of the adrenal medulla.

4.3.2 MULTIPLE ENDOCRINE NEOPLASIA 1 AND 2 (MEN 1 AND 2)

4.3.2.1 MEN 1 (P. 368)

MEN 1 syndrome is an autosomal dominant condition predisposing to **neuroendocrine neoplasms** of the parathyroid glands, the pancreatic islet cell tumors, and the anterior pituitary gland. The genetic defect has been assigned to **chromosome 11q13**. The development of MEN 1 conforms to **Knudson's two-hit hypothesis**, visualizing the recessive unmasking of an inherited mutation, when a somatic mutation or a deletion affects the other normal allele. In application to MEN 1 evidence supporting the gene deletion phenomenon has been provided.

Sporadic primary hyperparathyroidism appears to share the same mechanisms. Loss of heterozygosity (LOH) on **chromosome 11q13** in MEN 1 is quite high, reaching 82%.

4.3.2.2 MEN 2A (p. 368)

MEN 2A is an autosomal dominant syndrome that predisposes to the development of medullary thyroid carcinoma, pheochromocytoma, and hyperparathyroidism. The responsible gene has been assigned to the pericentric region of **chromosome 10p11.2-q11.2**.

LOH on **chromosome 1p** is always present in the tumors associated with MEN 2A, but does not appear to be a part of sporadic pheochromocytomas.

The centromeric probes **D10Z1** and **D10S94** have been described, and **presymptomatic diagnosis of MEN 2A syndrome is possible with pericentromeric DNA markers**. As demonstrated in families at risk for this syndrome, even in the absence of biochemical evidence of thyroid C cell hyperplasia **DNA probe analysis should be a part of routine screening for MEN 2A syndrome.**

Furthermore, the availability of informative markers for MEN 2A makes **prenatal diagnosis** a possibility. A marked sex difference noted in recombination frequencies in this pericentromeric region should be taken into account, however, when determining the genotypes at the D10Z1 locus.

The gene for **mannose-binding protein**, which maps to the pericentromeric region of chromosome 10, is an additional marker for MEN 2A.

4.3.2.3 MEN 2B (p. 369)

In addition to MTC and pheochromocytoma, the MEN 2B syndrome, also an autosomal dominant disorder, includes **neuromas** of the mucous membranes and **skeletal abnormalities**, including Marfanoid habitus with arachnodactyly.

Similar to MEN 2A, the gene for MEN 2B has been assigned by linkage studies to the centromere of **chromosome 10** by the D10Z1 probe. Both a **locus for familial MTC** and a **locus for MEN 2B** map to the same pericentromeric region of chromosome 10 as the MEN 2A locus.

4.3.3 SELECTED ANALYTICAL TECHNIQUES USEFUL IN GENE LEVEL DIAGNOSIS OF ENDOCRINE DISORDERS (p. 369)

The diagnostic application of ***in situ* hybridization (ISH)** in endocrinology has advantages over immunohistochemistry; in particular, ISH has the ability to

- **distinguish *de novo* synthesis** from uptake
- **identify specific genes** that are actively expressed, even if their final products are not detectable for a number of reasons (degradation, utilization, release and dissipation in tissue, etc.)
- **identify the status of a specific gene** without relying on identification of the final product, which can be misidentified by immunological techniques due to posttranslational modification

Further complication in using an immunological approach results from the fact that one mRNA species can be processed into various protein products. At least two well-known examples of this phenomenon are:

- **proopiomelanocortin mRNA** gives rise to messages translated into **adrenocorticotropin, β-endorphin, melanocyte-stimulating hormone, lipotropins** and other bioactive peptides
- **preproglucagon**, in addition to glucagon, is processed into a number of related peptides

Chromogranin A can be identified

- **immunologically**, by antibodies to its different portions (**pancreastatin** and **betagranin**)
- by **ISH**, detecting the intact chromogranin A mRNA coding for both pancreastatin and betagranin
- posttranslational modification of the parent molecules by proteolytic cleavage produces several **chromogranin A** and **B** peptides

While chromogranin A and B mRNAs are identifiable in a variety of tissues and neoplasms, evaluation of chromogranin expression by immunological techniques may be unreliable because **only small amounts of immunoreactive chromogranin A are usually expressed**. Chromogranin A and B have been detected by ISH in bronchial, rectal, and midgut **carcinoid tumors**.

Even if the gene product is abundant, as is the stored hormone secreted by **pituitary adenomas**, the identification of mRNA for **growth hormone** by ISH is a better diagnostic tool. For example, in clinically silent cases, the growth hormone mRNA is detectable in the tumors, whereas the product is either dysfunctional or not efficiently produced.

Given that observations of **thyroglobulin gene expression** in well-differentiated papillary and follicular neoplasms are confirmed, **thyroglobin mRNA levels** can be used for finer diagnosis and monitoring of thyroid tumors. Calcitonin mRNAs can be visualized by ISH using biotinylated probes.

Examples of the use of recombinant DNA techniques in identifying the molecular basis of endocrine disorders and in diagnostic applications include the following:

- Southern analysis to identify the presence or absence of the functional (*GH*-1) or variant (*GH*-2) **growth hormone gene**
- analysis of specific gene characteristics responsible for MEN 1 and MEN 2A complements prenatal sex determination, in cases where genetic anomaly is predominantly expressed in either sex
- identification of point mutations by PCR in **vitamin D receptor gene analysis**
- use of **oligonucleotide probes** complementary to neurotensin mRNA in unstimulated cells for the diagnosis of pheochromocytoma
- identification of gene expression in various endocrine tissues, where cells may appear secretion depleted (thus failing to reveal the tested gene products) because of their **rapid turnover**; examples of such false-negative assays are
  - the failure to detect ectopic production of ACTH by SCLC
  - the failure to detect glucagon synthesis by malignant islet cell tumor

Some of the **advantages of using oligonucleotide probes over larger fragments of DNA or RNA** are

- the high **degree of consistency of tissue assessment** over time
- the easily **controlled concentration**
- the better **control over the specific activity**
- the ease of customized synthesis of common or unique sequences
- the **absence of a need for subcloning**
- **no requirement for denaturation step** (as needed with double-stranded cDNA)
- the **reduced needs for pretreatment of target tissue** (Triton X-100, proteinase K, and acetic anhydride treatments can be omitted)
- the **long shelf life** at 4°C

Among the **disadvantages of oligonucleotides** for ISH are

- the relatively **low specific activity**
- the **lower binding of** DNA-RNA hybrids than RNA-RNA hybrids, and consequently
- the **lower sensitivity** than that achieved with cRNA probes

Reverse transcription (RT)-PCR sequencing is widely employed for a variety of **research objectives**, such as

- **cloning of** genes
- **detection of gene transcripts**
- **detection of chromosomal translocations** (e.g., in leukemia)

RT-PCR is more sensitive than northern blotting and ISH for detecting specific mRNAs in endocrine pathology. Besides, this approach obviates the need to isolate mRNAs by oligo-dT chromatography.

Although **single-stranded cDNAs** seem to give a high background signal due to entrapment of nonhomologous bacterial DNA sequences, they do not self-hybridize, unlike the double-stranded probes. The latter must be unwound prior to and during the hybridization process. As a result, two processes — the self-annealing and hybridization to the target sequences — are in competition.

Cloned RNA probes are appealing because

- they are usually **longer than cDNA**
- they have **higher thermal stability**, allowing higher stringency conditions to be used
- the **high-affinity of RNA-RNA hybrids**
- **probe sizes are constant**
- the probes **do not contain vector sequences**
- a **sense probe** can be prepared and used as a control. A sense probe has a sequence identical to the target mRNA and, therefore, is unable to hybridize with the target

An ingenious extension of *in situ* evaluation of gene expression has been the development of ***in situ* transcription (IST)**, which allows

- a specific cDNA to be synthesized *in situ* by the enzyme **reverse transcriptase**, following the addition of appropriate primers
- the cDNA thus produced to be isolated from the tissue section and analyzed by a variety of techniques
- higher sensitivity compared to conventional ISH
- the relative ease of **quantitation of mRNAs**
- the **preservation of the morphology** of the tissue section when isolating nucleic acid sequences, as is required for conventional PCR

It remains to be explored to what extent IST can be used for diagnostic purposes in a clinical setting, particularly using nonradioactive primers.

Part 5
Gene Level Evaluation of Selected Neurologic and Connective Tissue Disorders

Chapter 5.1

Muscle, Bone, and Skin Disorders

5.1.1 DUCHENNE AND BECKER MUSCULAR DYSTROPHIES: OVERVIEW OF MOLECULAR BIOLOGICAL FINDINGS (p. 385)

The triumphant discoveries of the gene and its product, **dystrophin**, responsible for **Duchenne muscular dystrophy (DMD)**, and a related milder, allelic form, **Becker muscular dystrophy (BMD)**, allowed the development of molecular tools for unequivocal differentiation of these muscular disorders from a number of clinically similar diseases.

A **brief summary** of the major findings related to the molecular biology of DMD/BMD is given here only to emphasize the relative merits of specific diagnostic approaches of current and potential clinical value.

Essentially three major groups of muscular dystrophies are recognized:

- X-linked — DMD, BMD, and **Emery-Dreifuss dystrophy**
- autosomal recessive — the **autosomal recessive childhood muscular dystrophy, adult limb girdle dystrophy,** and **congenital muscular dystrophy**
- autosomal dominant — **myotonic dystrophy, fascioscapulohumeral dystrophy,** and **oculopharyngeal muscular dystrophy**

Revision of the formerly perceived differences between these conditions may follow the insights gained with molecular diagnostic tools.

In DMD/BMD the identification of the gene led to the discovery of its principal product, although the fine aspects of pathobiology of the disease at the level of muscle still remain to be elucidated. DMD and BMD share the same genetic locus, and share a comparable molecular pathology. Clinical differences in clearcut cases justify distinction between these two diseases, which require different management and genetic counseling approaches.

DMD characteristics are:

- manifestation **predominantly in boys**
- incidence of **1 in 3000**
- **early onset**, usually before the age of 3 years
- **progressive muscle weakness** reaching its maximum before age 12 years
- frequent **mental retardation**
- **cardiac involvement**

- increase in **creatine kinase** (CK) up to 20,000–30,000 IU/ml
- **death** usually following **pulmonary complications** at the age of 15 to 25 years

BMD is a milder form of a related disorder:

- The incidence is **1 in 30,000**.
- It has a variable time of onset, often in **early adult life**.
- **Loss of ambulation** occurs in the third and fourth decades.
- Cardiac involvement and mental retardation are **infrequent**.
- **CK elevations are more pronounced in childhood**.
- **Calf hypertrophy** is often striking, more so than in DMD.
- Death is rare before 30 years of age.

Up to one third of DMD/BMD cases are represented by new mutations. Very rarely, carriers can manifest DMD/BMD. Such patients present a challenge of differentiating the condition from autosomal recessive limb-girdle muscular dystrophy.

Application of restriction fragment length polymorphism (RFLP) analysis to flow cytometry (FCM)-sorted X chromosomes from unusual patients led to the finding of deletions and translocations of the middle of the short arm of **the X chromosome**. Cytogenetic studies narrowed the location of the *DMD* gene to **Xp21**, at least for cases of "simple" DMD. Genetic and structural studies demonstrated that **both the *DMD* and *BMD* loci map at Xp21**, establishing that these two diseases represent allelic conditions.

Testing clones derived from the Xp21 region of the DNA from a particularly informative patient by a **phenol emulsion reassociation technique** allowed Kunkel and coworkers eventually to identify potential coding sequences for DMD, which were subsequently confirmed by other researchers.

The complete cDNA for *DMD* gives rise to a 14-kb transcript, encoded by over 70 small exons spread over nearly 2500 kb of genomic DNA. This proved to be the **largest known gene**, its size accounting for the high mutation rate. About 65% of DMD and BMD patients have deletions or duplications of one or more exons of this gene. Estimates reveal that a **duplication** of part of the gene occurs in 6% of all DMD cases.

Identification of deletions in DMD/BMD cases offers a definitive diagnosis in 65% of cases, which can be performed prenatally in pregnancies at risk.

- Analysis of **deletions** at the DMD/BMD locus revealed **a correlation between the phenotype and the type of deletion mutations**. A common pattern of deleted exons (0.5, 1.5, and 10 kb) has been related to a mild phenotype.
- However, **the pattern of deletions and their extent cannot be used as a prognosticator of the severity and clinical course of the disease**.
- Many "**in-frame" deletions** of the dystrophin gene may not be detected because they fail to produce typical DMD/BMD phenotypes.
- **Altered processing of the message** can also contribute to the phenotype: analysis of the PCR-amplified dystrophin transcripts that give rise

to truncated dystrophin molecules from muscle specimens of DMD and BMD patients demonstrate that **alternative splicing patterns correlate with the observed clinical phenotypes.**

The **DMD gene product, dystrophin,** located at the inner surface of the myofiber cell membrane, is related to other **cytoskeletal proteins,** such as **α-actinin** and **spectrin.** In skeletal muscle dystrophin constitutes a major component of the sarcolemmal cytoskeleton.

The identification of dystrophin made the definitive diagnosis in proband families possible. Most significantly, evaluation of the dystrophin presence and molecular weight in muscle biopsies shows

- **patients with DMD consistently demonstrate the absence of detectable dystrophin**
- **patients with BMD invariably have dystrophin, which is altered either in size or quantity**

Importantly, the **replicative life span of DMD myoblasts is severely reduced compared to that of normal controls.**

The **tissue-specific expression** of the dystrophin gene in muscle tissue and brain is explained by the fact that the **brain promoter** is active only in neurons, whereas the **muscle promoter** operates in skeletal, cardiac, and smooth muscle as well as in glial and neuronal cells.

Moreover, selective reduction in dystrophin expression in the brain or muscle may be due to a **deletion of either brain-specific or muscle-specific promoter.** In CNS, dystrophin is localized at postsynaptic membrane sites, and alterations in its level at those sites may be the basis for the cognitive impairment observed in DMD. It appears that **specific loss of the brain promoter may account for some cases of X-linked mental retardation.** In the dystrophin complex there are four **glycoproteins (156K, 50K, 43K,** and **35K), integral components** of the dystrophin oligomeric complex (~18S). **The reduction of the 156K glycoprotein is thought to be the first step in the molecular pathogenesis of DMD.**

Important insights have been gained in the structure and function of the **43K** and **156K dystrophin-associated glycoproteins (DAGs),** termed **dystroglycans.** Absence of dystrophin in DMD leads to the loss of dystroglycans that bind muscle fibers to the extracellular matrix (**laminin**), this disruption accounting for the eventual muscle cell necrosis.

5.1.1.1 STRATEGIES FOR DMD/BMD TESTING (p. 388)

The basic premise of DMD/BMD testing is the observation that the vast majority of patients with muscle disorders other than DMD and BMD like the normal controls have unaltered levels of normally appearing dystrophin.

Essentially two groups of diagnostic approaches are used for DMD/BMD:

- **assessment of the gene product** (dystrophin) by western blot and immunofluorescence
- **evaluation of the gene** itself by Southern analysis with cDNA probes, PCR, or RFLP-based linkage studies

5.1.1.1.1 Immunoblotting (p. 389)

Dystrophin immunoblotting requires only 10 to 15 mg of tissue obtained by either needle or open biopsy. The tissue is immediately transferred to a minus 70°C environment. Following SDS electrophoresis and transfer to the nitrocellulose membrane, the protein is detected by specific antibodies and visualized by the immunoperoxidase technique.

- The **complete absence of dystrophin is predictive of DMD with over 99% accuracy**.
- The **reduction of dystrophin** compared to normal levels, and/or **the detection of larger or smaller sized molecules** of the protein by SDS electrophoresis, **establishes a diagnosis of BMD with over 95% accuracy**.
- An **intermediate phenotype** has been recognized with dystrophin levels under 20% of normal levels. In some cases, testing with multiple antisera is needed to establish the diagnosis of BMD.

Dystrophin is a rodlike protein consisting of a long **central domain** that separates an **N-terminal actin-binding domain** from a **cysteine-rich domain** and **a C-terminal domain**. It appears that deletions or alterations involving the cysteine-rich and C-terminal domains have **the most severe pathological consequences**. On the contrary, mutations extending up to 46% of the molecule in areas not affecting the interactions of dystrophin with other muscle proteins, such as actin, produce only **very mild muscular dystrophy**.

Identification of the **integrity of the cysteine-rich and C-terminal domains** is important, necessitating the use of specific **monoclonal antibodies (MAbs)** in immunoassays for dystrophin.

Two classes of MAbs have been raised using recombinant DNA technology:

- **one stains dystrophin in normal muscle extracts**
- **others are dystrophin specific and stain nothing in the dystrophic muscle**

The immunoblotting assay for dystrophin is sensitive enough to be diagnostic for DMD and BMD in boys, but it is not capable of detecting the smaller reductions of dystrophin levels in most DMD carrier females.

Alternatively, **culturing of myoblasts** from suspected carriers can be performed to reveal the dystrophin expression. Although this method is lengthy and expensive, it is thought to represent **the most sensitive tool for detecting DMD and BMD carriers**. The life span of myoblasts from DMD patients is severely reduced. At present, several laboratories offer dystrophin evaluation. One of these is Genica Pharmaceuticals Corporation, the Neuromuscular Disease Company (Worcester, MA). A large number of laboratories offer tests for dystrophin gene mutations [*e.g., see Supplement 2*].

5.1.1.1.2 Immunofluorescence (p. 389)

Dystrophin visualization in biopsy material from frozen sections reveals the sarcolemmal protein as a **homogeneous ring around the periphery of all muscle fibers in normal individuals**.

The following staining patterns can be seen:

- muscle fibers from **DMD patients** show **no mmunofluorescent staining**
- in specimens from **BMD patients** staining varies from **normal to patchy** and is significantly **lighter**
- in **DMD carriers** patches of **negative fibers are scattered among positive fibers**
- muscle biopsies from **symptomatic DMD carriers** can be identified by a distinct **mosaic staining of surface membranes** of fibers when 2 to 8% of fibers show partial immunofluorescence
- **false-negative results** are produced when all the fibers stain positively, either due to **sampling errors** or to nonrandom X inactivation with **selective loss of dystrophin-negative fibers**
- no specific immunofluorescence pattern can identify **BMD carriers**
- dystrophin staining can be analyzed even in **freeze-dried tissue**

5.1.1.1.3 Detection of Dystrophin Gene Mutations (p. 390)

At present, two methods are used to identify alterations of the dystrophin gene:

- **PCR** (DNA can be amplified by **multiplex PCR from Guthrie spots**)
- **Southern** (DNA isolated from peripheral blood is probed with cDNA on Southern blots)

For the 65% of DMD/BMD patients who have detectable mutations (duplications or deletions), **Southern analysis is definitive**. It is also capable of

- **carrier detection**
- **prenatal diagnosis**, on either chorionic villi or amniocentesis-derived cells
- an **estimate of the severity** of a patient's disease in about 92% of cases
- if the mutation introduces a frame-shift in the protein translation, the disease phenotype is that of **DMD** with **little or no functional dystrophin produced**

In **BMD**, mutations usually occur "in-frame," producing an internally deleted or duplicated **partially functioning protein**.

Southern analysis of the dystrophin gene using cDNA **requires up to seven different subcloned probes** to examine at least two different restriction digests of genomic DNA. Because most of the deletions cluster in two hot spots, a subset of approximately 70 exons is sufficient to identify the majority of deletions.

The preferential cluster of dystrophin gene deletions at the "hot spots" allows **rapid screening** of the deletions by **multiplex PCR analysis with primers for only a subset of exons**.

The PCR multiplex system developed by Chamberlain and coworkers amplified up to nine exons predominantly deleted in the majority of DMD/BMD patients. Another nine exons can be amplified with other **primers to cover all of the known deletions in BMD, and over 97% of deletions in DMD patients**.

PCR-based detection of carriers, prenatal diagnosis, and mutation detection can be performed substantially faster than the conventional Southern blot analysis. Furthermore,

- PCR works even on highly degraded DNA unsuitable for Southern blotting
- the entire procedure can be performed in 1 or 2 days with a sensitivity of up to 98%, thus virtually eliminating the need for Southern blotting in the 65% of patients with deletions
- the assay can be run on fixed, or even embedded, tissue specimens such as those from deceased patients, relatives, and so on

The PCR does not eliminate the need for linkage studies on the remaining 35% of DMD/BMD patients, where polymorphic markers must be used.

For diagnostic purposes, however, the more reliable Southern analysis is recommended at this time.

Determination of carrier status in DMD/BMD by Southern analysis relies on the assessment of the **relative abundance/decrease in specific bands** detected by an array of cDNA probes. Efforts have been made to quantitate this technique through **densitometry of the autoradiographic bands**, and the **use of multiple probes**.

Given the **poor reproducibility** of high-quality Southern blots as well as the **intrinsic limitations** of the technique, **quantitative PCR using allele-specific oligonucleotides (ASO) primers** offers several advantages when new deletions occur in families for which probes may not yet be available.

Generally speaking, to meet the challenge of identification of the full spectrum of DNA sequence alterations possible in diseases with high new mutation rates, **a combination of three PCR-based techniques is advocated:**

- The first technique, **multiplex amplification**, allows the rapid detection of alterations in a large sized gene such as the *DMD* gene;
- The second method, **chemical mismatch cleavage**, is used for screening DNA fragments of up to 1.5 kb in size;
- The full spectrum of mutations is detected by a third approach — **direct automated sequencing** — which is applicable to DNA pieces of up to 350 bp.

A combination of all three techniques is recommended for routine, rapid diagnosis of any genetic conditions for which the defective gene had been cloned along **an algorithm for diagnostic strategies in new mutation disorders** (Figure 5.1). Extending this general approach, a **multiplex PCR-based linkage analysis of DMD families** has also been developed.

- A **nonradioactive method** for the detection of deletions or duplications in the dystrophin gene has been developed based on **peripheral blood lymphocyte mRNA**. The entire coding region of the dystrophin mRNA can be amplified with a set of 10 reverse transcription (RT) and overlapping nested PCR reactions (**RT-PCR**). The amplified products are directly visualized on acrylamide minigels by ethidium staining, and

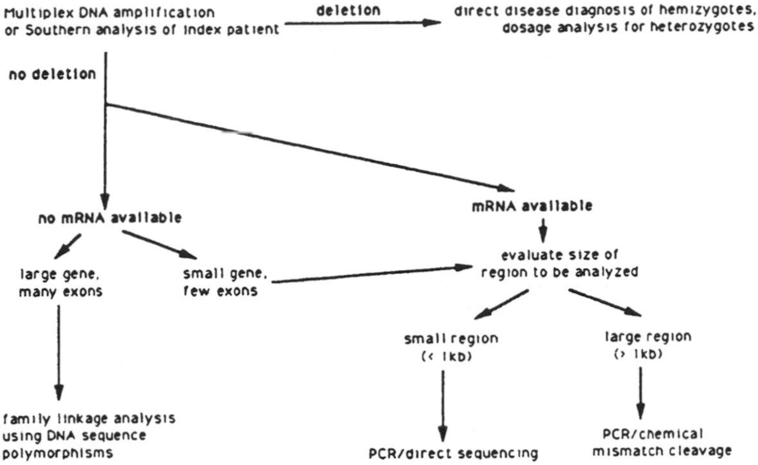

FIGURE 5.1

major mutations are identified by the appearance of a band of different size than that of the wild type.

- Screening for DMD can be accomplished by combining PCR amplification of exon 45 of the dystrophin gene with **single-strand conformational polymorphism (SSCP) analysis** of the amplified product. The absence of the amplified product denotes the presence of a characteristic deletion, whereas the SSCP data help to resolve the carrier status of the relatives.
- Yet another approach to performing linkage analysis in DMD families is the detection of the **CA repeat polymorphisms** by PCR at the extreme 5′ terminus of the dystrophin gene which display a high degree of polymorphism.
- The detection of dystrophin gene deletions can also be accomplished by **RT-PCR** amplification of dystrophin mRNA even on **lyophilized biopsy tissue**, affording a higher degree of resolution than is usually attained with Southern blot or multiplex PCR analysis of DNA.

5.1.1.1.4 Linkage Analysis of DMD/BMD (p. 393)

As noted above, in 65% of patients with DMD or BMD alteration (deletion or duplication) of the dystrophin gene affects the exons. In other cases, point mutations or other alterations occur in noncoding sequences, and may require indirect identification methods. It requires, however, an appropriate number of individuals from a family with a positive history of the disease, and extensive testing of multiple DNA samples by restriction analysis. **Amino- and carboxy-terminal antisera specific for dystrophin**

have been found inadequate for differential diagnosis of disease severity based solely on dystrophin quantitation.

At present at least **19 intragenic RFLPs** within the cDNA have been identified and some 45 molecular deletions within the dystrophin gene were recognized.

In **prenatal testing for DMD/BMD carrier status**, DNA RFLP analysis was performed and proved informative in the vast majority of cases. A number of important diagnostic pitfalls, including nonpaternity, karyotypic anomalies, and gonadal mosaicism, have been found to complicate carrier detection and prenatal diagnosis for DMD and BMD with DNA analysis.

The use of **intragenic RFLP analysis** to estimate the carrier risk of possible female carriers enabled 78% of possible carriers to be assigned high or low risks. A combination of genomic and cDNA probes for the dystrophin gene is advocated.

It is recommended that the **clinical evaluation of DMD/BMD families** include

- a genetic register for X-linked muscular dystrophies
- banking of DNA samples
- thorough carrier testing by creatine kinase analysis
- deletion testing
- linkage prediction
- counseling

PCR-based assays are being developed to allow analysis of polymorphisms informative in constructing linkage relationships.

An algorithm combining linkage studies and direct detection of gene alterations by PCR technique currently employed by Beggs and Kunkel in their evaluation of clinical cells is shown in Figure 5.2.

Dosage analysis has proved to be a suitable alternative, and in some cases even more efficient than linkage, to determine carrier status of female relatives of DMD patients known to have a deletion within the *DMD* gene.

5.1.2 MYOTONIC DYSTROPHY (DM) (p. 394)

DM is an autosomal dominant disorder and is the most common form of adult muscular dystrophy, with a variable age of onset and clinical severity.

Linkage studies in this disorder point to the proximal long arm of **chromosome 19** near the apolipoprotein C-II gene (*apo* C-II) and the gene encoding enzyme peptidase D. A 10-cM region of chromosome 19 was established that contains the DM locus.

Using **probes for *CKMM*, *apo* C-II**, and the repair genes ***ERCC*-1** and ***ERCC*-2**, and separation of large-sized DNA restriction fragments by field inversion gel electrophoresis, a long-range restriction map of the 19q13 region has been constructed.

Genetic linkage studies in DM as in other diseases can establish the carrier status, but may not predict the severity and progression of the disease, especially because DM is known for its variable expressivity and age-dependent penetrance. A **chromosome 19** marker, *ERCC*-1, that shows

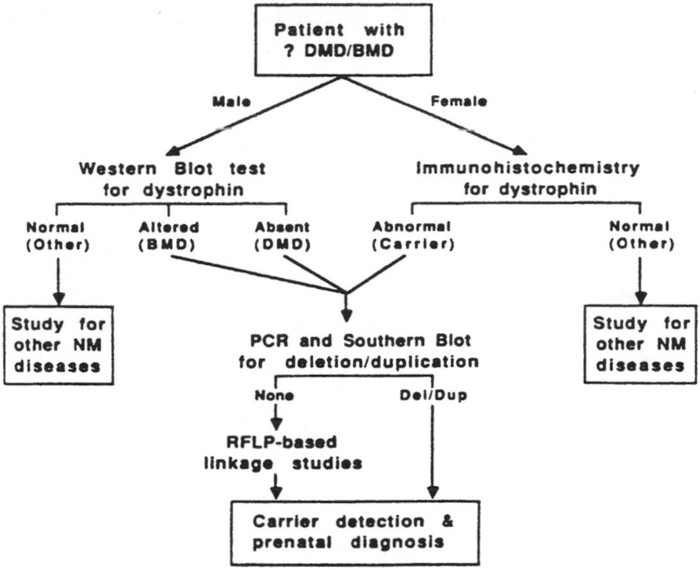

FIGURE 5.2

close linkage to DM has been characterized. **PCR-based protocols** have been developed for the direct identification of *CKMM* haplotypes.

The variable nature of the severity and age of onset of MD is related to the presence and size of a restriction fragment that is larger in affected persons than in unaffected individuals. The unstable fragment containing the characteristic **CTG repeat motif** is located at the 3′-untranslated region of an mRNA that encodes a polypeptide of the protein kinase family expressed in the tissues of MD persons. In severely affected individuals the triplet is amplified up to several kilobase pairs compared to 5 to 27 copies in normal individuals. **The larger the unstable fragment the more severe the disease symptomatology.**

5.1.3 CHRONIC SPINAL MUSCULAR ATROPHY AND X-LINKED SPINAL AND BULBAR ATROPHY (p. 395)

Spinal muscular atrophy (SMA) is a group of heritable degenerative diseases that selectively affects the **α-motor neurons.** SMAs with onset in childhood are second to cystic fibrosis among autosomal recessive disorders and are the leading cause of heritable infant mortality.

All three forms of SMA, formerly distinguished on the basis of onset

and phenotypic severity, are due to different mutations at a single locus on chromosome 5.

The biochemical defect underlying this disorder has not been defined at the time of this writing.

Kennedy's disease (**X-linked spinal and bulbar atrophy**) is an adult-onset form of motoneuron disease, related to **androgen insensitivity**, in which the gene encoding the **androgen receptor** (**AR**) is the candidate gene. By analogy with fragile X syndrome and DM, this gene harbors mutations consisting of an increase in size of a polymorphic **tandem CAG repeat** in the coding region in affected individuals. **Unaffected persons** have fragments 405 to 450 bp in length, whereas the fragments of **affected individuals** are about 500 bp.

5.1.4 FASCIOSCAPULOHUMERAL MUSCULAR DYSTROPHY (LANDOUZY-DEJERINE DISEASE) (p. 396)

Fascioscapulohumeral muscular dystrophy, an autosomal dominant disorder with a prevalence of 1 in 20,000, manifests at the end of the first or in the second decade with **facial or shoulder girdle weakness**. The biochemical defect resides in a deficiency of complex III of the mitochondrial respiratory chain. Using the **microsatellite marker Mfd22 (D4S171)** allowed the assignment of the causative gene to **chromosome 4**. Linkage studies argue against genetic heterogeneity as being responsible for the phenotypic heterogeneity of this disease.

Using the DNA marker D4S171 and probe **D4S139** on the distal region of 4q, **presymptomatic detection** and **prenatal diagnosis** of fascioscapulohumeral disease have been shown to be clinically possible.

5.1.5 EMERY-DREIFUSS MUSCULAR DYSTROPHY (p. 396)

Emery-Dreifuss muscular dystrophy is an X-linked disorder characterized by relatively **benign muscle wasting** of the humeral and peroneal muscle groups as well as limb girdle muscle, and involves the heart at an early stage. The candidate gene has been firmly assigned to the distal arm of the X chromosome, and appears to be located in Xq28, distal to **DXS305**.

5.1.6 MARFAN SYNDROME (p. 397)

This serious connective-tissue disorder is inherited as an autosomal dominant trait, with the diagnosis usually made clinically during childhood and sometimes in adulthood.

The **heterogeneous manifestations** of the syndrome, affecting a variety of organ systems, may have more than one cause and one mutant gene locus.

The **microfibrillar fibers** accumulate in Marfan skin and are produced by cultured fibroblasts from Marfan syndrome patients. **Fibrillin**, a major structural protein of microfibrils, has been studied in tissues from affected individuals by indirect immunofluorescence. A consistently deficient accumulation, and/or production and assembly, of immunoreactive fibrous components has been detected. This approach can be used for diagnostic purposes.

Topical alterations of the microfibrillar fibers manifested in a marked reduction in microfibrillar antigen in tissues taken from the affected side account for the **unilateral manifestations** of the Marfan syndrome.

Immunofluorescent abnormalities in the microfibrillar fibers cosegregate with the Marfan phenotype, while **all nonaffected family members did not show fibrillin abnormalities.**

Fibrillin cannot account for all mutations in the pleiotropic phenotype of Marfan syndrome, as demonstrated by the immunofluorescent studies.

Also, it appears that **fibrillin immunofluorescence is not absolutely specific for Marfan syndrome**, because false-positive results have been observed with other connective tissue disorders — **cutis laxa** and **homocystinuria.**

Advances in Marfan syndrome research include

- assignment of the defective gene to **chromosome 15** at **D15S45.** Chromosome 15 also contains genes coding for **the type I collagen receptor, chondroitin sulfate proteoglycan I core protein,** and **cardiac muscle X-actin**, which could be viewed as candidate genes for the mutation in Marfan syndrome
- **a partial sequence of a candidate gene for the Marfan syndrome has been defined**
- *de novo* **missense mutations** have been detected in the fibrillin gene in some Marfan syndrome cases
- **DNA polymorphism** in the fibrillin gene has been closely linked to sporadic cases of Marfan syndrome

Specific markers for the fibrillin gene — **$(TAAAA)_n$** for the gene on **chromosome 15** and **$(GT)_n$** for the gene on **chromosome 5** — have been used in PCR amplification to establish that **Marfan syndrome is produced by mutations in a single fibrillin gene on chromosome 15.** Consequently, the diagnosis of Marfan syndrome can now be established by direct molecular evaluation of the responsible gene.

5.1.7 OSTEOSARCOMA (OS)

OS is the most common malignant tumor of bone, excluding plasma cell myelomas; it arises most frequently in the second decade of life and affects the knee area, particularly the distal femoral metaphysis.

5.1.7.1 CHROMOSOME ABERRATIONS (p. 398)

Alterations of the *RB* gene at 13q14 occur in a high percentage of cases. *RB* involvement in OS is thought to conform to the **two-hit Knudson's model**:

- the **primary event** is the alteration of one of the alleles at 13q14
- **subsequent chromosomal rearrangements** or **mutations** affect the other allele, leading to tumorigenesis

Loss of heterozygosity (LOH) at various other loci on chromosome 13 is also present in OS.

5.1.7.2 THE *RB* GENE (p. 398)

A total or partial deletion of the *RB* gene has been detected in 43% of the cases. The LOH on chromosome 13 by either homozygosity or hemizygosity was found in 64% of all informative cases, including those without detectable structural alteration of the *RB* gene. The highest frequency (77%) of LOH on other chromosomes was detected particularly for **chromosome 17**.

Deletions in *RB* correlate with the grade of OS: they are found predominantly in high-grade tumors but not in low-grade ones.

5.1.7.3 THE *p53* GENE (p. 398)

The involvement of the ***p53* gene** on the short arm of chromosome 17 may prove to be coordinated with *RB* gene activity in the pathogenesis of OSs.

- In a murine model of OS and in human OS cell lines, inactivation of the *p53* gene plays a greater role in the development of OS compared to that of the *RB* gene.
- The **wild-type** form of *p53* appears to be **lost** from the tumor cell lines.
- Experimental data points to the causative role of the ***p53* gene as a recessive cancer susceptibility gene** in the development of OS.

5.1.7.4 PROTOONCOGENES (p. 399)

Among other genes reported to be characteristically associated with OS, some at least in cell cultures, are **c-*sis***, ***met***, **PDGF**, **c-*myc***, and **v-Ki-*ras***. In contrast to the heterogeneous expression of **N-*myc*** in human neuroblastoma, **c-*myc* is uniformly expressed in OS** and is apparently involved in the control of osteogenic differentiation of transformed cells.

The elevated c-*myc* mRNA level in some tumors is accompanied by a low expression of **osteopontin (bone sialoprotein)** and of **bone Gla protein**, another marker of highly differentiated osteogenic cells.

In radiation-induced murine OS, N-*myc*, c-*sis*, **c-*mos***, and **c-*fos*** appear normal structurally and are not amplified.

Three groups or types of OS tumors are distinguished with respect to c-*myc* expression:

- The first type shows **no apparent alteration** of either the gene structure or its expression, suggesting that c-*myc* is probably not involved in tumorigenesis;
- The second includes malignancies with **enhanced c-*myc* mRNA expression**, apparently due to amplification or rearrangements in the c-

myc gene; c-*myc* transcripts contribute to the enhanced growth of the tumor cells;

- The third group consists of tumors in which the c-*myc* mRNA transcripts are increased, but **osteogenic markers osteopontin and bone Gla protein are at low levels.** This lowered expression of osteogenic differentiation markers may be correlated with the constitutive c-*myc* expression, resulting in a block of cell differentiation.

Studies of c-*myc* involvement in OS have so far not addressed the possible interaction of c-*myc* with the *p53* and *RB* genes. In view of the established role of the latter in OS such an interaction may be highly probable. If the above discussed findings are extensively confirmed in human OS, **the degree of c-*myc* gene expression or alteration may be used in evaluating the level of dedifferentiation of osteogenic tumors in clinical situations.**

Interactions between several protooncogenes are a frequent occurrence. For example, in soft-tissue sarcomas (**leiomyosarcomas, rhabdomyosarcomas,** and **malignant fibrous histiocytomas**) concurrent abnormalities of the *RB*-1 and *p53* TSGs can be detected.

5.1.7.5 DIFFERENTIAL DIAGNOSIS (p. 400)

Positive staining for **vimentin intracellularly** and **type I collagen extracellularly**, as well as focal positivity for **S-100 protein** and **type II collagen**, are characteristic immunohistochemical features.

Osteopontin and **bone Gla protein** appear to be reduced in inverse relationship to the degree of differentiation of OS. It appears that analysis of ***RB*, *p53*, and c-*myc* genes and their products, combined with that of the osteogenic markers osteopontin, osteonectin, and bone Gla protein, may offer a practical approach to resolving some of the diagnostic and prognostic challenges presented by OS.**

5.1.8 SKIN

5.1.8.1 MALIGNANT MELANOMA (MM)

5.1.8.1.1 DNA Ploidy (p. 400)

The increasing incidence of **cutaneous MM** is related to **solar ultraviolet radiation.** The major precursor of MM is considered to be the **dysplastic nevi. MM** is often recognized at early stages of tumorigenesis, and that accounts for the significant success of curative surgery. However, approximately 20% of all cases relapse within 5 years of treatment.

Analysis of **nuclear DNA** content revealed that 57% of MM cases are diploid. The **S phase fraction (SPF)** of over 15% has bad prognostic significance in MM. **Aneuploidy** and the occurrence of **hypertetraploidy** and **multiclonality** are other significant prognostic factors of a poor outcome. **DNA ploidy ranked fifth** after growth pattern, ulceration, thickness, and pathological stage in its prognostic power in stage I MM.

In stage II MM with **metastatic spread** to lymph modes, aneuploidy reaches 86%, the rest of the cells being tetraploid. The development of

multiclonality is associated with tumor progression. Quantitation of these characteristics by digital imaging systems is thought to provide useful prognostic information.

5.1.8.1.2 Chromosome Studies (p. 401)

The most significant findings related to events at the gene level in skin diseases have been made in MM. The progressive nature of the disease has been linked to the **accumulation** of **karyotypic abnormalities**. Characteristically, frequent random karyotypic changes and abnormalities of **chromosomes 1, 6,** and **7** are observed, particularly in metastatic MM lesions.

The majority of segregations (66.7%) were due to loss of the same allelic fragments on **chromosome 9** and the **X chromosome**. All cells derived from the metastases were aneuploid. **Monoclonality** of the metastic lesions has been confirmed by RFLPs in all cell lines. Frequent and random LOH in metastatic MM cells contrast with LOH at specific loci, presumed to be closely linked to recessive tumor genes in other human primary tumors [RB, Wilms' tumor (WT), etc.].

A **high-resolution chromosome banding** study correlated structural chromosomal abnormalities occurring 10 or more times with patient **survival**. The most frequent sites of abnormalities were found on

- **chromosome 1** (31 times)
- **chromosome 6** (15 times)
- **chromosome 2** (14 times)
- **chromosomes 7** and **9** (11 times each)
- **chromosomes 11, 21**, and **3** (10 times each)

No correlation is found between specific chromosomal changes and histopathological parameters of the diagnostic biopsy samples; however, **patients with structural abnormalities of chromosome 7 or 11 had a significantly shorter survival.**

Nonrandom alterations of **chromosome 6** in patients with MM, in particular, appear to have relevance to malignancy. **Introduction of this chromosome into MM cells using a microcell hybridization technique** leads to drastic changes in their malignant phenotype features and the cells fail to induce tumors in nude mice. It appears that **one or more genes on chromosome 6 may be involved in the expression of the malignant characteristics of MM.**

In **uveal MM** with poor prognosis, the characteristic aberrations are

- monosomy of **chromosome 3**
- multiplication of **chromosome 8q**

The **gene for susceptibility to melanoma-dysplastic nevus** was assigned to a location on chromosome 1 between an anonymous DNA marker (**D1S47**) and the gene locus for **pronatrodilatin (*PND*)**, a commonly used reference gene, in chromosome band **1p36**.

So far, all the cytogenetic and chromosomal studies have succeeded in providing only tentative evidence on a rather wide range of correlations between chromosomal aberrations and the biology of cutaneous MM. **No specific, diagnostically helpful changes at the chromosome level can be used at present in differential diagnosis or patient management in MM.**

5.1.8.1.3 *RAS* Protooncogene in MM (p. 402)

No correlation could be found between the frequency of particular **Ha-*ras*** protooncogene alleles with MM occurrence using ***Bam*HI** or ***Msp*I/*Hpa*II** digestion. Further analysis of the same DNA for a different polymorphion using ***Taq*I** sites identified a group of allele variants in the VNTR region of H-*RAS*-1, named **Tp**, which were significantly more frequent in patients with MM than in normal controls.

No mutations in **codons 12, 13,** and **61** of the *ras* gene in **sporadic** human MM could be identified except for one specimen.

On the contrary, a correlation between *in vivo* exposure to solar radiation and **N-*ras*** mutation at codons 12 and 13 has been found in some patients with MM. **Primary tumors with activated N-*ras* were localized exclusively in continuously sun-exposed body areas**. The affected sequences were found in the codons containing **pyrimidine dimers (TT or CC)**, known to be produced by UV irradiation.

Subsequent PCR analysis confirmed these findings, and established a correlation with specific phenotypic characteristics of tumors and precursor lesions. **N-*ras* occurred 10 times as frequently as Ha-*ras* and always contained a mutation at codon 61**. No *ras* mutations could be found in normal nevi or dysplastic nevi, considered to be precursors of melanomas.

Evaluation of **p21 expression** in tissue sections of tumors revealed **no correlation with tumor progression** or any growth characteristics of melanomas with or without *ras* mutations. *ras* **mutations seem to be the result, rather than the cause, of transformation**.

These findings contrast sharply with the **immunohistochemical demonstration** by other workers of a strong p21 protein presence in cutaneous MM, which correlates with the degree of malignancy. The **anti-$p21^{ras}$** reactivity is particularly high in **nodular melanoma, epithelial-type melanoma**, and deeply **invading tumors**. No reactivity is found in **melanocytic nevi with junctional activity** or in nevus cells of **compound nevi**.

5.1.8.1.4 Other Genes in MM (p. 403)

A number of observations have been made, some of which may have the potential of establishing clinically relevant diagnostic tools. Among these are:

- DNA sequences with partial homology to the **HPV 9** and **EBNA 1** genomes
- an alteration in the ***myb*** protooncogene that correlates with a 6p22 chromosomal abnormality
- a tumor-specific deletion in the gene encoding **α-type protein kinase C (PKC)**
- polymorphic alleles for the gene encoding **EGFR**

- the expression of the **cAMP-dependant protein kinase** gene
- **consistent overexpression** (9- to 14-fold) of the **c-*myc*** gene
- **no detectable level of expression** in **c-*fms***, **c-*abl***, **v-*src***, **c-*erb*A-1**, **c-*erb*B**, **v-*mos***, transforming growth factor β (***TGF*-β**) gene, and **c-*myb***
- a level of expression of **c-Ha-*ras***, **N-*ras***, **c-*fos***, and **c-*sis*** equal to that in normal human melanocytes and metastatic MM
- significant **intercorrelations** between **c-*myc***, ***p53***, and **c-*src*-1** levels, and between *p53* and c-*erb*B-2
- in at least some cell lines, the detection of expression of **c-*mel***, **c-*erb*B-2**, **c-*myc***, **c-*src*-1**, ***p53***, the gene encoding **PDGF A** chain, ***gro***, **TGF-α**, ***EGFR***, and the gene encoding **tissue plasminogen activator**
- a preliminary observation on the possible role of **c-*jun*** and **c-*fos*** in the transformation of cultured human primary melanocytes to MM melanoma
- ***Taq*I RFLP** analysis of the human ***TGF*-α locus**, revealing a significantly higher frequency of a 2.7-kb allele in MM
- **increased expression** of closely related growth factors TGF-α, TGF-β, and EGFR mRNA
- release of a **low molecular weight human TGF in the urine** of patients with MM. **Urinary TGF** levels appear to parallel **tumor burden**. In fact, the release of TGF-α in the urine also appears to serve as an **early marker** for melanoma
- the observation that expression of the **EGFR** in various **hyperproliferative skin lesions** accompanying MM markedly declined following removal of melanoma
- some **53 new human genes** found in human MM cells by the technique of molecular subtraction in an effort to identify specific MM tumor markers. Their characterization has not yet been completed at the time of this writing

MM detection can be accomplished by assaying **peripheral blood** for the presence of MM cells by identifying the gene for **tyrosinase** specific to melanocytes. A single melanoma cell can be detected in 2 ml of blood by PCR.

Another gene specific for melanoma cells that can be targeted for detection encodes the **antigen MZ2-E**, which is recognized by cytotoxic T lymphocytes.

5.1.8.2 SKIN DISEASES OTHER THAN MM (p. 404)

A coherent picture of key molecular events in skin diseases involving epidermis is not available at present. The diversity of skin pathologies and lack of concentrated effort lead only to fragmentary observations on the biology of keratinocytes affected by some skin disorders.

- In chemically induced skin transformation by **papillomas** and **SCC**, a definite role for activated **Ha-*ras*** oncogene has been shown.

- Activated Ha-*ras* genes appear, however, to be a common aberration **not restricted to a specific type of cancer**.
- While Ha-*ras* is implicated in the **initiation** of skin carcinogenesis, the ***fos*** oncogene appears to be involved in malignant **progression** of premalignant skin cell lines.
- Risk of SCC is significantly associated with **HLA-B** mismatching and **HLA-DR** homozygosity.
- Introduction of c-Ha-*ras* alone into **cultured human keratinocytes** markedly affects the growth potential of the cells, but no direct correlation between high levels of expression of its gene product, p21, and malignant growth can be observed.
- The **v-*jun*** oncogene appears to operate as a dominant negative regulator in mouse epidermal tumor cells.
- The role of **EGF** and **EGFR** has been demonstrated by Southern analysis in **epidermoid carcinoma** AY31 cells.
- In skin fibroblasts from members of a cancer-prone family with the **Li-Fraumeni syndrome** the level of **c-*myc*** expression was three- to eightfold that of controls accompanied by an apparent activation of **c-*raf*-1** gene.
- **Defective DNA repair** operates in **xeroderma pigmentosum** (see DNA Repair).
- The majority of **basal cell carcinoma (BCC)** tumors have a high level of **nonclonal and clonal structural chromosomal aberrations.**
- **Erythema multiforme**, in particular its recurrences, has been associated with **herpes simplex virus (HSV)** infection, as demonstrated by serological, immunofluorescent, electron microscopic, and recently also by PCR techniques.
- The candidate gene for the **nevoid BCC** (the **Gorlin syndrome**) has been assigned to 9q22.3-q31, allowing surveillance of presymptomatic patients for the development of skin cancer.

In **psoriasis**:

- linkage analysis has assigned four RFLPs informative in a large **psoriasis** kindred to a region on **chromosome 6p** with a probable translocation defect in that area
- **no increase** in the expression of **c-*myc*, c-Ha-*ras*, c-*erb*B (EGFR), c-*jun*,** or **TGF-β** transcripts has been detected in psoriatic lesions; the expression of **c-*fos*** was lower, while that of **TGF-α** was markedly increased
- in psoriatic lesions as well as those of **pemphigus** and **bullous pemphigoid** there is an elevated *in situ* expression of **tissue-type plasminogen activator (tPA)**

5.1.8.3 CHEMICAL CARCINOGENESIS OF THE SKIN (p. 406)

Activation of the c-Ha-*ras* gene, particularly in codon 61, is thought to constitute one of the early events in malignization of **chemically induced** skin **papillomas**. In chemically induced tumors, progression from papilloma to carcinoma may involve the amplification of mutated **Ha-*ras*** alleles. The level of expression of viral Ha-*ras* in **virally induced carcinomas** is particularly high.

The other protooncogene amplified and expressed in a chemically-induced **epidermoid carcinoma** cell line is **c-*erb*B.**

Chapter 5.2

Neurological and Mental Disorders

5.2.1 HUNTINGTON'S DISEASE (HD) (p. 409)

Huntington's disease is an autosomal dominant, incurable disorder which affects the central nervous system. Onset is usually in the fourth or fifth decade of life. HD is characterized by choreic movements, cognitive impairment, behavioral changes, and, in the advanced stages, by physical debilitation, resulting from premature nerve cell death throughout the brain, being most prominent in the striatum.

The **biochemical defect** responsible for the neurological manifestations of HD still awaits identification. It appears that the premature death of selective medium-sized spiny, GABA-containing neurons may be due to **defective repair**.

Linkage associations established for an anonymous DNA locus **D4S10**, identified by the **DNA probe G8** with the disease assigning it to **chromosome 4**. The linkage data obtained in large informative pedigrees established that this locus (D4S10) displays about 4% recombination with HD. Family studies suggested that **over 95% of HD cases are caused by a mutation on chromosome 4**, the **4p16** band, and later pointed to the **flanking markers**. The location of the *HD* gene was then placed farther toward the telomere from the most distal tested marker **D5 (D4S90)**. It appears that the *HD* mutation may be distal even to this locus.

The **phenotypic severity** of motor, cognitive, and psychiatric symptoms in HD **is not related to the gene dosage**, as could be expected in dominant disorders. **Patterns of inheritance** in relation to the age of onset of HD symptomatology suggest an effect of **paternal imprinting**.

5.2.2 TESTING FOR HD (p. 410)

In spite of the limitations of the presently available DNA probes for HD, the **presymptomatic and prenatal identification of probable gene carriers can now be established in informative families with a high degree of probability (up to 98%).**

At this stage, the test is based on linkage studies, which identify the allele, or a DNA marker (D4S10 or others), that is genetically linked to the *HD* gene. To this end, the segregation of the DNA marker is determined among affected and unaffected persons in a pedigree where many people must be involved for the result to be informative.

A detailed protocol incorporating linkage data and age at onset, adjusted for censored observations, sex of affected parent, and familial correlation, has been developed into a **computer program MLINK** for calculating the risk of having HD. Pre- and posttest counseling, psychological testing, and paternity testing are part of a formal presymptomatic evaluation protocol.

Interestingly, **neurological, psychiatric, psychological,** and **social variables** failed to predict the genetically affected individuals reliably. A multitude of problems, ranging from inappropriate use of the test prior to or postevaluation to loss of samples in transit to the laboratory to social, psychological, insurance, and ethical issues, are to be dealt with when testing for HD.

Testing of minor children is deemed inappropriate for a number of reasons:

- it is not possible to obtain an informed decision from the person to be tested
- because no treatment is currently available to improve the condition of the affected person with manifestations of HD, the effects of such testing on the child's social interactions are not predictable

Prenatal testing can be performed when a parent is affected or is at increased risk for HD, and when the fetus has a 50% change of having inherited the *HD* gene.

With the addition of newer markers (**D4S43** and **D4S95**) to D4S10, virtually all families became fully informative in the **Edinburgh** predictive testing program. The DNA markers D4S10 and D4S43/S127 firmly established the identity of the Japanese *HD* gene with the Western gene, in spite of the lower prevalence of HD in Japan. RFLP analysis of HD in **Finland,** where the prevalence of the disease is very low (~5 in 1 million), demonstrated the genetic homogeneity of HD.

The **most efficient strategy** in presymptomatic, prenatal, and exclusion testing for HD incorporates the use of seven closely linked DNA markers: **D4S10, D4S81, D4S95, D4S43, D4S111, D4S125** and **D4S115**.

A **combination of exclusion prenatal testing** and **definitive testing of a fetus**, when a parent is determined to be at increased risk of carrying the *HD* gene, or when he/she is already affected with HD, has been under study in British Columbia, Canada.

5.2.3 NEUROFIBROMATOSIS (p. 412)

Neurofibromatosis is a degenerative disease of neural crest origin, characterized by predominant hyperplasia of Schwann cells. This single-gene disorder occurs in two forms:

- **peripheral neurofibromatosis (NF-1)**, also known as **von**

Recklinghausen disease (VRNF)
- **central** or **bilateral acoustic neurofibromatosis (NF-2)**

5.2.3.1 NF-1

(p. 412)

NF-1 is one of the most common autosomal dominant disorders (incidence, 1 in 3000), manifested in multiple **benign neurofibromas** and **café-au-lait spots**. The clinical diagnosis is established when two or more of the criteria proposed by the **NIH Health Consensus Development Conference on Neurofibromatosis** (1987) are met. Some of these are
- six or more **café-au-lait macules**
- **optic glioma**
- two or more **Lisch nodules (iris hamartomas)**
- two or more **neurofibromas** of any type or one **plexiform neurofibroma**
- a **first-degree relative** with NF-1, and so on

NF-1 patients also have an increased risk of developing other malignancies, including leukemias. All these features suggest the similarity of NF-1 to other hereditary cancer syndromes, such as **retinoblastoma** and **Wilms' tumor**.

The following has been determined about the ***NF*-1 gene**:
- *NF*-1 has been mapped to **chromosome 17 (17q11.2)**
- *NF*-1 encodes an **11- to 13-kb transcript ubiquitously expressed** and denoted **NF-1 LT**. It is frequently disrupted in NF-1 patients due to mutant alleles, a premature termination, or an insertion
- the **rate of mutations** at the *NF*-1 locus is **high** (around 10^{-4}), apparently resulting from the large size of the locus
- in contrast to retinoblastoma (RB), **all NF-1 cases result from the inheritance of a mutant allele**
- by analogy with RB, ***NF*-1 is expressed in all tissues studied so far,** although a high frequency of neoplasia is expressed only in a limited type of tissues

Members of the *NF*-1 family products have been previously shown to act as **transcription regulators** and **initiation factors**. The disruption of normal NF-1 function could lead, among other effects, to
- growth stimulation
- interference with a differentiation pathway

A similarity of the *NF*-1 gene to a **tumor suppressor gene** is strongly suggested by
- the frequent disruption of the *NF*-1 gene in VRNF patients
- the autosomal dominant mode of inheritance

The **negative regulatory role** of the NF-1 family of proteins has also been demonstrated in their interaction with the **CRE1 element of the human proenkephalin inducible promoter**.

Sequence analysis of the encoded protein points to a homology with the catalytic domain of the human *ras* **GTPase activating protein (GAP)**, and an even closer homology to the **yeast *IRA* gene** product. It appears that NF-1 may act as a significant regulator of $p21^{ras}$ activity.

In **benign neurofibromas** and **malignant neurofibrosarcomas** the role of the *NF*-1 gene is not understood. A similarity has been suggested with the FAP second allele inactivation in benign colorectal adenomas that eventually leads to malignization.

A number of probe-restriction enzyme combinations allow an informative **linkage analysis** for **carrier and prenatal diagnosis**. Now **direct testing for the *NF*-1 gene is possible**.

PCR-based analysis of an **intragenic *Alu*-repeat polymorphism** allows the conduction of more efficient prenatal evaluation in a family with only one affected person.

5.2.3.2 BILATERAL ACOUSTIC NEUROFIBROMATOSIS (NF-2) (p. 413)

Similar to the clinical presentations of **RB** and **Wilms' tumor**,

- **Unilateral lesions** occur in **sporadic NF-2** cases
- **Bilateral acoustic neuromas** invariably develop in **familial NF-2**

In familial NF-2 cases, the dominant pattern of inheritance is due to the **presence of one already-defective allele at the *NF*-2 locus**. By analogy with RB, one more hit appears to be enough to destroy the remaining functional allele and for tumor formation in NF-2.

In fact, LOH on **chromosome 22** has been found in acoustic neuromas of NF-2 patients and in leukocytes of patients with sporadic unilateral acoustic neuromas, and in a **meningioma** of a patient with central neurofibromatosis.

Early diagnosis of NF-2 is complicated, to a greater degree than in NF-1, by the lack of diagnosable manifestations. Timely surgical intervention can alleviate adverse consequences of the disease. The **flanking markers can be used for accurate presymptomatic and prenatal diagnosis** before the defective gene(s) is isolated and direct probes are constructed.

5.2.3.3 RETINOBLASTOMA (RB)

5.2.3.3.1 General Features (p. 413)

Retinoblastoma is a malignant tumor affecting the retina of predominantly very young children. Two forms are known:

- **hereditary** — both eyes are involved
- **sporadic** — only one eye is affected

RB is an autosomal dominant condition with approximately 90% penetrance: there is a 50% chance that an offspring of affected parents will also develop the tumor.

For the malignancy to appear, **two events must occur**:

- the presence of **a mutated, tumor-predisposing allele**
- either a somatic mutation, leading to **loss of the normal *RB* allele**, or **some other chromosomal alteration**

Eighty-five percent of patients with only one affected eye have not inherited the *RB* gene. An inherited mutant *RB* allele predisposes to **osteosarcoma**, **soft-tissue sarcoma**, **melanoma**, and **bladder cancer**. Relatives who do not carry the mutation have **no excess risk** of developing cancer. The role of **environmental factors** (e.g., X-rays) in the development of the nonfamilial, sporadic form of RB is not clear.

5.2.3.3.2 Chromosome Aberrations (p. 414)

The predisposition to the development of tumors in survivors of hereditary RB is 1000-fold higher than in the general population. The evaluation of **chromosomal characteristics** in the tumor may be of great prognostic significance.

The presence of a **homozygous deletion** has been established in region 13q14 by **high-resolution banding** of RB tumor cells.

Genomic imprinting may play a role in this disease, as shown by the **paternal origin** of the affected chromosome 13, carrying a reduction of homozygosity in both bilateral and unilateral tumors.

5.2.3.3.3 The *RB* Gene (p. 414)

The human **RB susceptibility locus** located within **chromosome 13q14** spans approximately 200 kb. The wild type contains 27 exons and 26 introns.

A number of **CpG islands**, frequent sites of **methylation**, have been identified in that region. CpG islands are found at the start of many genes, therefore aberrations in the vicinity of *RB*-1 may be relevant to the development of **many malignant tumors** if they harbor deletions and/or translocations. In fact, **genetic changes at 13q14 have been identified not only in RB, but also in tumors of various tissues**. For example, leukocytes of RB carriers show LOH of **chromosome 13** markers. This LOH is also frequently observed in **osteosarcomas** from patients with and without a history of retinoblastoma.

The **mutant gene has variable expression** regulated at the level of transcription and manifested in the absence of a functional *RB*-1 gene product. The *RB* gene 4.7R on chromosome 13q14 contains, in some cases, homozygous internal deletions yielding corresponding **truncated transcripts**. The **loss of a splice junction** is noted in the majority (60%) of tumors with no identifiable structural changes in the gene.

Somatic mutations at the *RB*-1 locus may be involved in the formation and development of tumors in RB patients. Sometimes a deletion at the *RB*-1 locus can be detected in **metastatic carcinoma cells**, but not in normal unaffected tissues.

Inactivation of both alleles of the *RB*-1 gene is produced by

- small **deletions**
- **duplications**

- **point mutations**

These lead to splice alterations and, consequently, loss of an exon from the mature mRNA.

A specific association between **p105**^{RB} and simian virus 40 **(SV40) T antigen** has been demonstrated in an **animal model of heritable retinoblastoma**.

5.2.3.3.4 The RB Protein (p. 415)

The *RB* gene encodes a protein of 928 amino acids, called **p105** or **p110**, which is localized to the nucleus and displays DNA-binding activity. This is a ubiquitous protein identified in all vertebrate species examined, but **conspicuously absent from many tumor cell lines**.

Interaction of the RB protein with oncogene products causes its inactivation and transformation of cultured calls. In addition to the *RB* gene product, two ***RB*-associated proteins** have been identified that could be important either in *RB* gene regulation or in interaction with it in a dependent fashion.

The extent of the **RB protein phosphorylation** is important in the control of its transforming potential. The phosphorylated RB protein lacks the growth-supporting function, and it is the **dephosphorylated product that apparently prevents tumors from developing**.

Mutant RB protein is unable to become hyperphosphorylated and fails to bind to **nuclear structures** and to the **E1A oncoprotein**. The state of phosphorylation of the RB protein varies over the cell cycle:

- the RB protein is phosphorylated from the S to M phases
- it is dephosphorylated in the G_1 phase
- the protein kinase **cdc2**, essential for the cell cycle transition, specifically phosphorylates the RB protein
- RB phosphorylation also changes markedly during **differentiation**: induction of the differentiation state is associated with the predominance of the dephosphorylated form of the RB protein

There are also **cell-specific**, or **lineage-specific**, differences in both the regulation of RB phosphorylation and *RB*-1 gene expression at the mRNA level. The **RB activity** is regulated primarily **posttranscriptionally**.

A cellular protein (**RbAP46**) has been identified that specifically associates with the RB protein. Competitive studies with oncoproteins (e.g., SV40 T antigen) suggest this protein may be crucial in regulating the RB protein function in the cell.

Analysis of the **effect of aging** on the RB protein shows that **a change in the regulation of *RB* transcription is not part of senescence**.

Specific antibodies against normal and truncated *RB* gene products and associated proteins are available. These may provide clinically useful information on tumor samples, especially if differences are found in **RB-positive** and **RB-negative tumors** in their biological behavior, aggressiveness, or resistance to therapeutic modalities.

5.2.3.3.5 The *RB* Gene in Other Tumors (p. 416)

The *RB* gene represents the **prototype of a recessive gene**, in which the absence of both alleles, or a loss of function affecting both alleles, results in tumorigenesis.

The *RB* gene is inactivated in some human **breast cancer cell lines**. The *RB* gene product, $p105^{RB}$, cannot be detected with specific antibodies in these cell lines.

A highly sensitive **immunological assay** for the *RB*-encoded p105 protein confirms that **inactivation of the RB product is universal** in RB cells and is nearly as frequent in small cell lung carcinoma (**SCLC**). On the other hand, the **majority of other pulmonary neoplasms have readily detectable levels of intact *RB* mRNA and $p105^{RB}$.**

One third of the **bladder carcinoma cell lines** tested carried altered or absent $p105^{RB}$.

Only infrequent *RB* inactivation is detected immunologically in human **colon carcinoma, breast carcinoma,** and **melanoma.**

Structural alterations of the *RB*-1 gene have been observed in 33% of **soft-tissue sarcomas**, and particularly in **leiomyosarcomas**. The pattern of expression of the RB protein appears to correlate with survival:

- homogeneous — survival is markedly increased
- heterogeneous or no expression at all — survival is markedly decreased

RB* gene inactivation, if present, may act in concert with other mechanisms, such as oncogene amplification** (e.g., N-*myc* is amplified in retinoblastoma and osteosarcoma cells in RB patients; ***p53 is altered in an osteosarcoma cell line).

Introduction of the *RB* gene drastically affects

- cell morphology
- growth rate
- soft agar colony formation
- tumorigenicity of cultured cells in nude mice

Replacement of suppressor genes in tumor cells opens a practical approach to gene therapy of clinical malignancy, at least in selected cases.

5.2.3.3.6 Detection of *RB* Gene Abnormalities (p. 417)

Diagnosis of the involvement of the *RB* gene in RB or other malignancies has relied on RFLP analysis using probes from the cloned gene.

An **intragenic *Bam*HI RFLP** within the 5′ end of the *RB*-1 gene appears to be superior to other polymorphic probes within *RB*-1 for genetic counseling of familial and nonfamilial RB patients and their relatives. Intragenic probes can also be used for **prenatal screening** for the inheritance of mutant alleles. Densitometric analysis of Southern blots may be used for qualitative and quantitative estimates of *RB*-1 gene deletions in patient samples.

A different **screening strategy** has been developed using **PCR amplification** of the *RB* gene regions, followed by direct DNA sequencing.

An ***Xba*I RFLP** within the human RB susceptibility gene can be used for analysis by the PCR. Another pair of primers has been reported for the analysis of a polymorphic ***Bam*HI** site in intron 1 of the human *RB* gene. Qunatitative estimates of *RB*-1 gene deletions can be made in patient samples by densitometric analysis of Southern blots.

Combining PCR with direct sequencing of well-characterized loci offers a viable diagnostic approach for detecting useful DNA polymorphisms that complement RFLP analysis, **denaturing gradient gel electrophoresis**, and **RNase mismatch** methods. Moreover, it is procedurally convenient, rapid, and capable of detecting essentially all polymorphic sites.

Combining **RT-PCR with SSCP**, the identification of abnormal *RB* transcripts can be used for detecting possible nucleotide substitutions, deletions, and so on.

5.2.4 NEUROBLASTOMA (NB)

Neuroblastoma

- produces **highly malignant** neoplasms
- arises in the **adrenal medulla** and **sympathetic ganglia** of the retroperitoneum and posterior mediastinum
- occurs **predominantly in infancy** and early childhood (under 3 years)
- is composed of **small**, **round to fusiform cells** usually with large hyperchromatic nuclei

5.2.4.1 DIFFERENTIAL DIAGNOSIS (p. 418)

Other small round cell tumors are to be considered:

- **Ewing's sarcoma**
- **Wilms' tumor**
- **malignant lymphoma**
- **rhabdomyosarcoma**

Immunohistochemical and/or ultrastructural studies are helpful in poorly differentiated neoplasms. NB shows positivity with

- **neuron-specific enolase (NSE)**
- **synaptophysin**
- **chromogranin A**

Additional markers can be helpful: **PAS**, **cytokeratin**, **leukocyte common antigen** and **lymphocyte markers**, **myoglobin**, **S-100 protein**, **adhesion molecules L1**, and the **highly sialylated neural cell adhesion molecule isoforms**.

Gene level markers, particularly **amplification of N-*myc* vs. c-*myc***, can help in differentiating NB from tumors with similar morphology.

N-*myc* may not be as helpful for differential diagnosis of other types of tumors, however, N-*myc* amplification and/or high expression have been reported in **retinoblastomas**, **SCLC**, **Wilms' tumor**, **rhabdomyosarcoma**, **astrocytoma**, and **thyroid C cell tumors**.

Multiparameter analysis of cellular and molecular characteristics is particularly emphasized in the classification of the heterogeneous group of small round cell tumors.

5.2.4.2 PROGNOSIS (p. 419)

Although the determination of **total catecholamines** is marginally the best single test to diagnose NB, comparison of a number of parameters demonstrated that **clinical staging** and **DNA content** could be used as independent prognostic indicators, particularly in stage III and IV patients.

The **age** of patients under 1.5 years appears to be the best prognostic indicator, followed by the degree of tumor differentiation. The immunohistochemical pattern is of no additional help.

Aneuploidy is associated with a better prognosis, possibly due to a better response to treatment.

5.2.4.3 N-*myc*

5.2.4.3.1 N-*myc* and Catecholamine Metabolism (p. 419)

An **inverse relationship** exists between N-*myc* **amplification** and **catecholamine metabolism** in patients with advanced NB. **Mass screening based on urinary catecholamine metabolites, particularly if performed at only 6 months of age may be inadequate for the detection of high-risk NBs**. Based on chromosome and N-*myc* amplification studies, a mass screening program is recommended to test at 6 and 12 months of age for the early detection of high-risk NBs.

5.2.4.3.2 N-*myc* and DNA Index (p. 419, 421)

N-*myc* evaluation alone is insufficient to predict the course of disease in all cases.

The determination of **DNA content** or **chromosomal analysis** may provide more useful information.

On the other hand, **no correlation** was found between the **DNA index** (DI) and N-*myc* gene amplification or expression. Also, the **DI of tumors as determined by flow cytometry (FCM) does not correlate with disease behavior**.

A **combination of DI** and N-*myc* amplification level is advocated for routine management of NB patients, because **only these two parameters in combination proved to be significantly associated with a high risk of relapse** independent of patient age.

The NB tumors with N-*myc* amplification represent a subset of nonaneuploid tumors. The **degree of amplification** (20 to 1500 copies) **is not related to ploidy**, and independently **predicts poor prognosis** regardless of the histological signs of differentiation.

5.2.4.3.3 Mechanism of N-*myc* Amplification (p. 419)

One of the **initial steps** leading to **N-*myc* amplification** is excision of the

N-*myc* locus from the vast majority on **chromosome 2**. Other chromosomal abnormalities found in NB include the frequent deletion of the short arm of **chromosome 1**.

As new facts emerge, **the role of N-*myc* amplification in NB becomes progressively complicated.** For example, the morphology of cultured NB cells and alteration in N-*myc* gene expression appear to be related.

Because N-*myc* amplification appears to play such a prominent role in determining the course of NB, the characterization of the N-*myc* **amplification units** is important in understanding the role of this oncogene in the disease. Relatively few rearrangements within the N-*myc* amplification unit have been found. Moreover, there is a significant homogeneity among the amplification units within one tumor, and a great similarity between the amplicons in different tumors.

Gene probes for N-*myc* can be especially useful when measuring low levels of amplification in bone marrow samples during remission.

N-*myc* amplification is an attractive tool for diagnosing, monitoring, and predicting the course of disease in a clinical setting because it is

- **stable**
- **uniform** within a tumor
- **similar** in its characteristics in different patients
- **consistent** with respect to the N-*myc* gene copy numbers within a given tumor not only at **different sites**, but also at **different times** *in vivo*
- also supported by the **increased stability of the N-*myc* mRNA**

5.2.4.3.4 N-*myc* and Prognosis (p. 420)

As noted above, amplification of N-*myc* in NBs has been associated with an **advanced stage** of the disease and **poor prognosis**. The elevated expression of N-*myc* contributes to the acquisition of an **enhanced malignant phenotype**.

N-myc expression contributes to malignant progression possibly by stimulating the expression of **autocrine growth factor(s)**. Neutralization of this autocrine growth factor activity may offer an avenue for the treatment of human NBs.

The N-*myc* **gene load** correlates with a **prognosis** of NB:

- **a significantly better, progression-free survival is noted in patients with single-copy tumors compared to those having multicopy tumors**

Caution: A highly tumorigenic NB cell line has been described that had undetectable N-*myc* expression.

5.2.4.3.5 N-*myc* and Drug Resistance of NB (p. 421)

Tumors with amplified N-*myc* are initially sensitive to chemotherapy; later, however, they acquire resistance to therapy and tend to recur.

The use of N-*myc* amplification for patient management necessitates that **its evaluation be performed prior to administration of any treatment modalities**, because not only the histology of tumors may be modified by treatment, but the assessment of N-*myc* amplification may also become inaccurate.

Paradoxically, an **inverse correlation** between the expression of N-*myc* and ***mdr*-1** has been found in the same tumors: the *mdr*-1 mRNA was expressed at a **lower level** in the advanced stage tumors and in the histologically undifferentiated NBs. Thus, **the acquisition of drug resistance by these neoplasms is not associated with an increase in *mdr*-1 expression.**

This is in contrast with the reported increase in the level of *mdr*-1 expression in human NBs **in response to chemotherapy**.

The reason for these conflicting observations is not clear at this time. Other mechanisms may be involved in drug resistance of NB:

- expression of the **epidermal growth factor**
- increased activity of **topoisomerase II**
- increased activity of **glutathione *S*-transferase**

Evaluation of the presence of **P-glycoprotein** prior to therapy allows prediction of the success or failure of therapy for nonlocalized NB.

Specific **antitumor agents** directed against N-*myc* (N-*myc* **antisense oligomers**) demonstrate effective growth inhibition and a change of the phenotype characteristics of cultured neuroepithelioma cells.

5.2.4.3.6 Other Oncogenes Related to Prognosis of NB (p. 422)

In addition to N-*myc*, a marked **decrease in the expression** of **c-*myb*** and **c-Ha-*ras*** is observed in NB cells induced to differentiate.

The opposite is true of **c-*ets*-1** and **c-*fos***, whereas the expression of c-*fos* parallels development of an extensive network of neuritic processes.

In contrast to N-*myc*, the expression of the Ha-*ras* gene is **positively correlated with good prognosis**. The amount of the Ha-*ras* p21 protein detected by a simple immunohistochemical procedure can be of clinical importance.

Expression of the protooncogene **c-*src***, in particular the neuronal product of the gene, **pp60**$^{c\text{-}srcN}$, is associated with **good prognosis**, especially in the evaluation of infants. The expression of **pp60**$^{c\text{-}srcN}$ is limited to NBs and RBs, whereas **pp60**$^{c\text{-}src}$ is found in most tumors studied.

The expression of the **neuropeptide Y mRNA** in NB tumors may be associated with a **better prognosis** among patients under 1 year of age at the time of diagnosis.

An **oncoprotein**, designated **Op18**, has been described in acute lymphocytic leukemia and some solid tumors, including NB. This protein appears to be similar, if not identical, to the phosphoprotein **prosolin**, and another phosphoprotein called **stathmin** — all these proteins being related to the state of differentiation.

In NB the **state of phosphorylation of Op18** is inversely related to the level of N-*myc* amplification. The level of Op18 phosphorylation increases with differentiation of NB.

p25A^{smg} (a p21^{ras}-like GTP-binding protein) gene is expressed in neural crest-derived tumor cell lines and NB tissues. Its level of expression is closely related to the state of neuronal differentiation in tumors.

Yet another marker of differentiated NB cells is **β_2-microglobulin**, which is not associated with N-*myc* expression.

No association has been observed between N-*myc* expression and diminished **levels of class I major histocompatibility (MHC) antigen expression** in NB.

The gene encoding **nerve growth factor receptor (NGFR)** is apparently normal in NB cell lines, but defects in the NGF/NGFR pathway are thought to play an important role in the initiation or maintenance of the undifferentiated state of NB.

The known **tumor suppressor genes**, such as ***RB*-1** and ***p53***, have so far **failed to correlate with tumorigenic expression** in hybrid NB cell lines.

The ***ret*** protooncogene has been specifically associated with NB cell lines, although its role remains to be defined.

5.2.5 OTHER SELECTED TUMORS OF THE NERVOUS SYSTEM

In relatively frequent tumors of the nervous system, such as **gliomas**, the accumulated molecular genetic evidence so far has failed to implicate specific etiological factors. Certain regularities in the **spatial and temporal expression of protooncogenes** in normal neural tissues and tumors of neural origin have been observed.

5.2.5.1 CHROMOSOMAL ABNORMALITIES (p. 423)

Structural alterations of **chromosomes 7, 10,** and **22,** and of the **sex chromosomes**, are the most common in **malignant gliomas**.

Other chromosomal abnormalities in glial tumors involve the short arm of **chromosomes 9, 3,** and **11** as well as **chromosomes 1, 6,** and **19**; **double minutes (DMs)** are present in about 50% of gliomas.

LOH has been repeatedly demonstrated in **meningiomas** in the CNS, and in peripheral neural tumors, **schwannomas**, and **neurofibromas**, pointing to the possible involvement of TSGs.

Glioblastoma multiforme (GBM) displays a characteristic overrepresentation of **chromosomes 7** and **22** (harboring genes for the **A** and **B chains of PDGF**).

A lower-grade malignant tumor, **anaplastic astrocytoma**, shows focal areas of **overexpression of the gene encoding the PDGF A chain** that are substantially more pronounced in GMB. This marker could potentially be used for differentiating the histologically equivocal cases.

RFLP analysis of malignant astrocytomas revealed LOH on all autosomes except chromosome 21, and the number of chromosomes with LOH correlated with tumor histopathology. A particularly frequent event, especially in GBM, was loss of broad regions of **chromosome 10**. Also frequent was LOH on **chromosome 17p**.

Frequent aberrations of **chromosome 22** are noted in meningioma patients. A **meningioma locus** has been tentatively mapped, based on the deletion data, within **22q12.3-qter**. Consistent with the **hormonal effects in the pathogenesis of meningioma**, the male patients showed a higher percentage of tumors with no detectable aberrations on chromosome 22.

5.2.5.2 ONCOGENES AND GROWTH FACTORS (p. 424)

Oncogene amplification is a **relatively uncommon** mechanism of oncogene activation **in the pediatric group.**

Even within one group of tumors, the **medulloblastomas**, mechanisms of oncogene activation may vary.

In **astrocytomas**, a positive correlation has been observed between the level of **epidermal growth factor receptor (EGFR) expression**, the **phosphotyrosine kinase activity** associated with the EGFR, and the **degree of malignancy**.

An enhanced expression and a **positive correlation** between N-*myc* expression and poor prognosis has been demonstrated in **medulloblastoma tumors**, which can be used in patient management.

Other genes noted in various brain tumors include:

- the ***ROS*-1** gene rearrangement observed in **glioblastoma**
- **Ki-*ras*** gene expression in **meningiomas**
- rare **Ha-*ras*** and **c-*mos*** alleles in **intracranial tumors**
- **N-*ras*** and ***EGRF*** genes in **glioblastomas**
- **c-*myc*** amplification in **medulloblastoma** cell lines and transplants, and **glioblastoma**
- v-*sis*, **v-*myc***, and **v-*fos*** in primary brain **tumors of neuroectodermal origin**
- rearrangement of the ***abl*** gene in **glioblastoma**
- coexpression of **PDGF β chain** and **PDGR-R** in **glioblastoma**
- coexpression of ***TGF*-α** in **malignant glioma**
- the presence of ***p53*** mutations appears to be associated with the clonal expansion of cells in the progression of brain tumors (**low-grade astrocytomas** and **high-grade glioblastomas**)

Table 5.1 summarizes the **protooncogenes** and **growth factors** identified so far in tumors of the central and peripheral nervous systems.

5.2.6 SELECTED MENTAL DISORDERS

5.2.6.1 INTRODUCTION (p. 424)

Practical application of gene level testing for the predisposition to or definitive diagnosis of the majority of psychiatric disorders is **still a matter of the distant future.**

Specific genetic linkages are few and reflect the complexity both at the clinical and genetic levels. Although the characterization of **brain-specific gene expression** has been attempted, **no disease-specific associations have been firmly established so far**. The progress in molecular characterization of the fragile X syndrome gives definite hope for success.

TABLE 5.1 (p. 425)
Oncogenes and Growth Factors in Tumors of the Nervous System

Tumor	Probe	Source-DNA/RNA	Result[a]	Ref.
Astrocytoma	c-*myc*	Tumor-DNA	A, RA, EE	Sauceda et al. (1988)
	N-*myc*	Tumor-DNA	A, EE	Garson et al. (1985)
	N-*ras*	Tumor-DNA/RNA	A	
	Ha-*ras*	Tumor-DNA	Rare alleles	Diedrich et al. (1988)
	c-*mos*	Tumor-DNA	Rare alleles	
	v-*sis*	Tumor-RNA	EE	Fujimoto et al. (1988)
	v-*fos*	Tumor-RNA	EE	
	v-*myc*	Tumor-RNA	Minimal expression	
	c-*ros*	Tumor-RNA	A	Wu and Chikaraishi (1990)
Glioblastoma	c-*myc*	Tumor-DNA	Minimal Expression	Sauceda et al. (1988)
	abl	Cell line-DNA	RA	Heisterkamp et al. (1990)
		Cell line-DNA/RNA	A, RA, EE	Trent et al. (1986)
	v-*myc*	Tumor-RNA	EE	Fujimoto et al. (1988)
	N-*ras*	Tumor-DNA/RNA	A, EE	Sauceda et al. (1988); Gerosa et al. (1989)
	Ha-*ras*	Tumor-DNA	Rare alleles	Diedrich et al. (1988)
	c-*mos*	Tumor-DNA	Rare alleles	
	PDGF	Cell lines-RNA	EE	Nister and Westermark (1986); Nister et al. (1988)
		Tumor-RNA	EE (regional)	Hermansson et al. (1988)

TABLE 5.1 (continued)
Oncogenes and Growth Factors in Tumors of the Nervous System

Tumor	Probe	Source-DNA/RNA	Result[a]	Ref.
	EGFR	Tumor-DNA/RNA	A, EE	Gerosa et al. (1989); Libermann et al. (1985)
		Cell lines-DNA/RNA	A, RA, EE	Nister and Westermark (1986); Nister et al. (1988; Wonu et al. (1987)
	TGF-α	Cell lines-RNA	EE	Nister et al, (1988); Rutka et al. (1984)
	c-*ros*	Cell lines-DNA/RNA	A	Birchmeier et al. (1987); Wu and Chikeraishi (1990)
Meningioma	c-*myc*	Tumor-DNA	A	Sauceda et al. (1988)
	N-*myc*	Tumor-DNA/RNA	A, RA	
	Ha-*ras*	Tumor-DNA	Rare alleles	Diedrich et al. (1988)
	c-*mos*	Tumor-DNA	Rare alleles	
	Ki-*ras*	Tumor-RNA	A	Carstens et al. (1988)
Medulloblastoma	v-*myc*	Tumor-RNA	EE	Fujimoto et al. (1988)
	N-*myc*	Tumor-protein	EE	Garson et al. (1989)
	N-*ras*	Tumor-DNA/RNA	A	Sauceda et al. (1988)
	*erb*Bl	Cell line-DNA	A	Wasson et al. (1990)
	N-*myc*	Cell line-DNA	A	Wasson et al. (1990)
	N-*myc*	Tumor-DNA	A	Nisen et al. (1986); Rouah et al. (1989)

	c-*myc*	Cell line-DNA	A	
		Tumor-DNA	EE	Bigner et al. (1989); Bigner and Vogelstein (1990); Friedman et al. (1988)
Neuroblastoma	N-*myc*	Cell lines-RNA	A, EE	Cohen et al. (1988); Seeger et al. (1984); Thiele et al. (1987a)
	c-*ets*	Cell lines-RNA	Minimal expression	Thiele et al. (1987a)
	c-*myb*	Cell lines-RNA	Minimal expression	Thiele et al. (1987a)
Adenoma	c-*myc*	Tumor-DNA/RNA	A, RA	Sauceda et al. (1988)
	N-*myc*	Tumor-DNA/RNA	A, EE	
	N-*ras*	Tumor-DNA/RNA	A, EE	
	Ha-*ras*	Tumor-DNA	Rare alleles	Diedrich et al. (1988)
	c-*mos*	Tumor-DNA	Rare alleles	
Oligodendroglioma	c-*mos*	Tumor-DNA	Rare alleles	Diedrich et al. (1988)
	Ha-*ras*	Tumor-DNA	Rare alleles	
Ependymoma	c-*mos*	Tumor-DNA	Rare alleles	Diedrich et al. (1988)
	Ha-*ras*	Tumor-DNA	Rare alleles	
Neuroepithelioma	c-*sis*	Cell line-DNA	RA	Thiele et al. (1987b)

Abbreviations: A, amplification; EE, enhanced expression; RA, rearrangement.

Adapted and expanded from Rutka, J. T. et al. (1990). Cancer Invest. *8*:425–438. With permission.

Among the relatively reliable associations of a subset of bipolar affective disorders are those pointing to **X-linked transmission**. Other chromosomes implicated in mental disorders are **chromosomes 11, 6,** and **16**. The once suspected association of the susceptibility to schizophrenia with the **chromosome 5q11-q13** region has been reevaluated.

New linkage data continue to appear on various genetic diseases, bringing the possibility of developing clinically useful tests for diagnosis and monitoring of neurological diseases closer every day.

Examples include the following:

- A linkage has been demonstrated between the gene causing **familial amyotrophic lateral sclerosis (Lou Gehrig's disease)** and four DNA markers on the long arm of **chromosome 21;**
- Assignment of the gene for **tryptophan oxygenase** to **4q31** allowed further exploration of the role of this gene in **alcoholism** and certain **depression** conditions;
- Cloning of the D_1 and D_3 **dopamine receptors** has helped unravel the molecular aspects of **Parkinson's disease, schizophrenia**, as well as **drug addiction** and **alcoholism;**
- The amplification of **cholinesterase-encoding genes** noted in hematopoietic disorders, ovarian carcinomas, and malignant gliomas points to the **role of neurotransmitters in malignancy;**
- Positive DNA diagnosis has been possible in **familial amyloid polyneuropathy** (FAP) pedigrees. The variant **prealbumin (transthyretin) gene** was found to be closely linked to clinical manifestations of the disease in both sporadic cases and familial nonsymptomatic gene carriers.

Only **selected neurological conditions** will be briefly summarized to illustrate the progress made toward clinically useful gene level evaluation.

5.2.6.2 FRAGILE X SYNDROME (p. 427)

The **fragile X syndrome**, also known as the **Martin-Bell syndrome**, is the most common form of mental retardation after Down syndrome, afflicting 1 in 1250 males and 1 in 2000 females.

Affected males

- are **mentally retarded**
- have **long faces, large ears**, and **testes**

A number of unusual features complicating the pattern of inheritance define **variable penetrance** of this X-linked disease.

In at least 20% of males that carry a fragile X chromosome the **syndrome is not expressed**. The trait is passed to their daughters, who are also asymptomatic.

Daughters of normal transmitting males (NTMs) are obligate carriers of the fragile X chromosome, and **mental retardation** is first expressed among the grandchildren of transmitting males.

The fragile X chromosome in affected males or females **is always inherited from their mothers**. It has been proposed that two independent events are required for the syndrome to be expressed:

- first, the **fragile X mutation occurs** (known to have a high mutation rate — 4 to 5 × 10^{-4})
- the X chromosome is **inactivated** in pre-oogonial cells

The syndrome is not expressed as long as the X chromosome remains inactivated. Normal **reactivation** of the X chromosome prior to oogenesis is blocked by the mutation. It is this **block to reactivation** that leads to mental retardation in affected progeny.

The ***XIST* gene**, located in the **X-inactivation center** on the human X chromosome, is implicated at least in some aspects of X inactivation.

The **fragile site** in this syndrome **(*FRAXA*)** has been located on the **long arm of the X chromosome** at band position **q27.3**. Cytogenetic analysis has a limited value in identifying fragile X:

- it detects only 55% of carrier females
- it detects even fewer (30%) of the mentally normal individuals
- it usually fails to identify phenotypically normal males

Some of the better linkage markers **(DXS304)** lie within 5 centimorgans (cM) of *FRAXA*.

Methylation of the fragile site in affected individuals apparently leads to a reduction in transcriptional activity in methylated areas. Others, however, fail to observe any differences in DNA methylation in the vicinity of *FRAXA* on active and inactive X chromosomes and in fragile X males.

These observations, although not refuting **Laird's imprinting hypothesis**, suggested that **mechanisms other than methylation may operate to allow for an imprint to persist through generations**.

In the meantime, probes close to the fragile X locus revealed that **rare-cutter restriction sites** (the so-called **CpG islands**) were resistant to digestion in fragile X males due to **hypermethylation**. A single CpG island containing a cluster of rare-cutter sites between markers **DXS463** and **DXS465** has been identified. While in affected males the CG-containing restriction sites surrounding this CpG island are predominantly methylated or even hypermethylated, these remain largely **unmethylated in normal males** and **unaffected transmitting males**.

A gene, designated ***FMR*-1** (for fragile X mental retardation), has been identified from the cDNA library of the fragile X region. The *FMR*-1 gene is highly conserved in evolution, and shows a strong similarity to a family of late-onset neurodegeneration genes, ***deg*-1** and ***mec*-4**, which encode proteins termed "**degenerins**" in the nematode *Caenorhabditis elegans*.

Related human genes are thought to be involved in the process of neurodegeneration and cell death in various human neurodegenerative conditions (e.g., **Huntington's disease** and **amyotrophic lateral sclerosis**).

A DNA region that spans ***FRAXA*** appears to be critically involved in the expression of the clinical phenotype. Moreover, immediately distal to the CpG island a 7-kb ***Eag*I-*Bgl*II** fragment has been identified and designated **St B12**.

Analysis with ***Pst*I** reveals a **region of instability** localized to a trinucleotide repeat, **p(CCG)$_n$**, whereas ***Eco*RI**, ***Hind*III** or ***Bgl*II** defines the **(CGG)$_n$** repeat:

- **unaffected** individuals have 2 to 50 copies
- **normal transmitting males** have 55 to 125 copies
- **affected males** have markedly increased numbers of the repeats — up to 6000 copies.

The probe/enzyme combinations **St X21E** or **St A22/*Ban*I** reveal a 1.15-kb fragment in normal males, an additional 1.7-kb fragment in normal females, and abnormal patterns in all carrier females. The fragment is much larger (over 2.2. kb) in females with X expression. **Abnormal fragments appear only if the restriction site is methylated.**

The NTMs pass the **DNA insertion** on to their daughters. **This insertion** is the same size, or only slightly larger, in the daughter but **becomes significantly increased (1.5 to 3.5 kb) in their children expressing the phenotype**.

With these discoveries direct diagnostic possibilities became available. **A combination of *Eco*RI and *Eag*I restriction unambiguously distinguishes in a single test between normal genotype, the premutation, and the full mutation.**

Other diagnostic probe/enzyme combinations are:

- **pfxa 3/*Pst*I** or **pfxa 3/*Eco*RI** and ***Sac*II**
- **VK 21A/*Taq*I**
- **St B12.3/*Eco*RI** and ***Eag*I**

PCR-based amplification of the CGG repeat can be combined with probing by the **5′-(CGG)$_5$-3′** oligonucleotide.

5.2.6.3 DOWN SYNDROME (DS) (p. 429)

DS is the most common chromosomal disorder in humans, occurring in 1 in 600 to 1000 live births. DS is also the most common genetically defined cause of **mental retardation**, and a major cause of **congenital heart defects**.

It is the trisomy of specific genes only, not of any sets of genes, on chromosome 21 that accounts for the Down phenotype. Parts of bands 21q22.1, q22.2, and **q22.3** are responsible for certain phenotypic characteristics of DS.

Specific chromosomal assignments of various features constituting the DS phenotype define the minimal region on chromosome 21 that, when duplicated, accounts for the majority of phenotypic characteristics. So far, **no single chromosomal defect accounting for the DS phenotype has been identified**.

Diagnosis of trisomy 21 can be made by the following methods:

- Conventional **cytogenetic banding** of metaphase chromosomes may be used;
- ISH using specific DNA probes may be used to recognize some cases of DS, for example, those caused by a **translocation at 21q22-21qter — trisomy of *this subregion only* is sufficient to produce a Down phenotype;**
- Using a single **cosmid clone probe (c519),** the **quantitation of trisomy 21** in interphase cells is possible;

- Estimation of the **gene dosage** is possible with a **slot blot method** — intensities of the signals generated on blots probed by the chromosome 21 DNA sequences (**D21S11** and **D21S17**) are quantified by **densitometric scanning** calibrated to reflect the **copy number of unique chromosome 21 sequences**. The accuracy of the method is 100%;
- An unequivocal diagnosis can be established on a relatively small sample of **fetal tissue**. A diagnostic approach carrying an even lower risk for the fetus uses specific probes in combination with **PCR** on **fetal reticulocytes in maternal circulation**. A **prenatal diagnosis of DS** can be established by PCR of unique sequences on chromosome 21 in a **recombination-based assay (RBA);**
- The **parental origin** of the supernumerary 21 chromosome can be established using **DNA polymorphisms**. The extra chromosome 21 is **maternal in origin** in about 95% of the cases, and **paternal** in only about 5% — much less than conventionally estimated by cytogenetic methods.

5.2.6.4 ALZHEIMER'S DISEASE (AD) (p. 430)

AD is the most common form of dementia in older people. **Degenerative neurological changes** are associated with progressive **impairment of memory**, **reasoning**, **orientation**, and **judgment**.

Generally AD occurs as a sporadic disorder of late onset. However, some cases, designated **familial Alzheimer's disease (FAD),** are transmitted in an autosomal dominant fashion suggesting a genetic defect.

The **characteristic histopathological feature** of AD — numerous **neuritic plaques** and **tangles** — appear to correlate in their frequency of occurrence with the degree of intellectual impairment of the patients.

The neuritic plaques represent aggregates of altered axons and dendrites that surround extracellular deposits of **amyloid filaments**. Amyloid filaments are recognized in

- **normal aging** human brain
- the neuritic plaques and cerebrovascular amyloid deposits in **Down syndrome** patients
- **aged mammals** of some other species

β-Protein or **β-amyloid**, a 4.2-kDa polypeptide, constitutes the principal protein of the amyloid filaments. The locus encoding the β-amyloid maps to **chromosome 21** in the vicinity of the genetic defect causing FAD.

The **β/A4 protein**, the principal component of the amyloid fibril, is derived from a large membrane protein, **Alzheimer amyloid protein precursor (APP)**. **Defective processing** of APP leads to deposition of amyloid. **Normal cells** *in vivo* and *in vitro* have been shown to produce and release soluble β-amyloid peptide.

It appears that **a mutation within the *APP* gene** causing **Val → Ile substitution** can account for at least some cases of early-onset AD. There may be another gene on chromosome 21 predisposing to FAD.

The result of the isoleucine substitution can be

- altered anchoring of APP in the membrane
- deregulation of the control of translation of APP, leading to overproduction of the protein

The **clinical implication** of this mutation is open to speculation at this time. Mutations at codon 717 (Val → Gly and Val → Phe) of the β-amyloid precursor protein gene appear to seggregate with FAD. Another line of molecular studies concentrates on **β APP processing** by **lysosomal proteases.** A possibility exists that aberrations in this process may produce amyloidogenic peptides and play a critical role in the pathogenesis of AD.

Neuropathological similarities between AD and DS raised the possibility that AD could be caused by overexpression or duplication of one or more genes on chromosome 21. No evidence of increased chromosome 21 DNA dosage could be observed in AD patients.

Several probes map to within 1.5 Mb of one another on the proximal arm of chromosome 21 in most centromeric loci, providing the most tightly linked probes to the **candidate genes for FAD.**

The **inheritance mode** of five polymorphic DNA markers — **D21S16, D21S13, D21S52, D21S1**, and **D21S11** — in pedigrees with FAD suggested that AD is not a single entity. It is more likely that it is a consequence of multiple genetic defects on chromosome 21 and elsewhere as well as of nongenetic factors.

It is significant, however, that linkage analysis of FAD using chromosome 21 markers in kindreds considered homozygous for FAD fails to support the view that a chromosome 21 gene is resonsible for late-onset and even some early-onset FAD.

Similarity in the changes found typically in AD and in normal aged brains suggested that AD may be a **variant of aging**, or that **aberration of the normal aging process** may account for the AD phenomenology. Along these lines, attention has been drawn to the **cellular membranes**. In fact, a selective **alteration of band 3 in red blood cells (RBCs)** suggests that the normal aging process of these cells from AD patients is disturbed.

The discovery of production of β-amyloid by normal cells opens new possibilities for **monitoring** its levels, studying its involvement in disease and for the development of appropriate **pharmacological strategies.**

5.2.6.5 MITOCHONDRIAL ENCEPHALOPATHIES (p. 432)

A large group of neurological diseases, the so-called **mitochondrial encephalopathies**, can be diagnosed now at the gene level.

Mutations in **mitochondrial DNA (mtDNA)** have been described, for example, in

- **chronic progressive external ophthalmoplegia (CPEO)**
- **Kearns-Sayre syndrome**
- **myoclonus epilepsy with ragged red fibers (MERRF)**
- mitochondrial encephalopathy with active acidosis and strokelike episodes (**MELAS**)

- another mitochondrial disease that, in addition to dementia, features ataxia, neurogenic muscle weakness, and retinitis pigmentosa, and is a maternally inherited condition

One of the most frequently detected mutations is an A → G transition at position 3234 in the **mitochondrial tRNA^{Leu} gene**.

It appears that **mtRNA dysfunction** is a key feature of mitochondrial encephalopathies. Importantly, this mutation can be **detected noninvasively** in blood samples and even urine. Another adult-onset myopathy and cardiomyopathy without involvement of the nervous system that is maternally inherited is genetically determined through a mutation (**A → G3260**) of the tRNA$^{Leu(UUR)}$ gene.

PCR amplification of the region harboring the position-3234 mutation as well as the position-8344 MERRF mutation can be efficiently applied for the molecular genetic diagnosis of mitochondrial encephalopathies from **blood samples**, even when abnormal histological findings in muscle biopsy specimens are absent.

This screening test is advocated for accurate diagnosis and genetic counseling in patients with

- myoclonic epilepsy
- ataxia
- other undiagnosed encephalopathies
- in early or atypical strokes

Part 6
Gene Level Evaluation of Infectious Diseases

Chapter 6.1

Introduction

The gene level analysis of infections may offer the fastest, most accurate, and sensitive method of diagnosis and epidemiological characterization of known or yet undisclosed etiological agents.

Adaptations of analytical and preparatory techniques of molecular biology for diagnostic purposes, the ever-expanding arsenal of specific **nucleic acid probes**, and the accumulation of **sequence data** allow a vast range of infectious agents to be reliably identified in the **environment**, in **vectors** and **patients**.

The number of commercially available gene level diagnostics is rather limited at present, and they will most likely complement the **conventional immunological, culture,** and **microscopic techniques**. The on-going technology transfer has led already to

- the development of progressively **user-friendly procedures**
- the introduction of **nonradioactive labels**
- refinements in the specificity and sensitivity of the available **kits**
- adaptation of **amplification methodologies**, such as **PCR**, **3SR**, **LCR**, and so on (see Part 1)

New dilemmas are to be resolved as to what degree of sophistication the diagnostic information is clinically relevant, and under what circumstances (e.g., as in the identification of HPV infections).

The **genotyping** of microorganisms allows heretofore unavailable precision in the **epidemiological characterization** of isolates, sometimes redefining the concept of patient management.

6.1.1 IDENTIFICATION OF PATHOGENS IN CLINICAL SPECIMENS (p. 455)

Identification of a pathogen in clinical material can be achieved in a number of ways:

- the probes can be **applied directly** to a specimen
- alternatively, the tissue can be **made more accessible** to the probe (e.g., by protease digestion)
- the pathogen can be identified with or without **amplification** of the nucleic acids of the pathogen

Amplification methods are particularly attractive for the detection of low concentrations of infectious agents, as in latent conditions, or for the fastest identification of the pathogen without laborious isolation and/or purification of target nucleic acids.

Adaptation of molecular biological techniques to the clinical laboratory results in hybridization protocols that can be run in a format similar to the more familiar immunological techniques (e.g., ELISA) — for long a mainstay of diagnostic clinical laboratories.

Examples are:

- a simple **dot hybridization** and **fluorimetric hybridization** can be performed in **microdilution plates** for rapid and routine genotyping of **viridans group streptococci**
- an almost 100% sensitive and specific **digoxigenin-labeled oligonucleotide probe** assay for the identification of *Streptococcus pyogenes*, and for typing **group A streptococcal isolates**, is certainly preferable to the serological approach
- among clinical pneumococcal isolates, a large number of strains **cannot be characterized** by serotyping and antibiotic susceptibility testing. A specific DNA probe containing a 0.65-kb fragment coding for a position of the **major pneumococcal autolysin (amidase)** has been constructed for **specific** identification of *Streptococcus pneumoniae*
- DNA probes for culture confirmation of *S. agalactiae*, *Hemophilus influenzae*, and *Enterococcus* species are commercially available (e.g., Accuprobe, Gen-Probe), eliminating the need for **subculturing**
- **restriction endonuclease digestion patterns** are probed with **rRNA gene-specific probes (ribotyping)**, may be used in the typing of **coagulase-negative staphylococci**; although the results are easier to interpret, they show less discriminatory power than **total DNA fingerprinting**
- a specific probe for the detection of methicillin-resistant *Staphylococcus aureus* has been constructed
- **restriction endonuclease analysis (REA)** has been successfully used in the elucidation of the genetic relatedness of *A. pleuropneumonia* to selected members of the family Pasteurellaceae
- direct application of **REA** of *Pseudomonas aeruginosa* to patient management helped establish that treatment failure was due to the development of drug resistance rather than to reinfection

6.1.1.1 *IN SITU* HYBRIDIZATION (ISH) (p. 457)

ISH has long been used for viral diagnosis in tissues, particularly for the **demonstration of low levels of viral burden**. ISH allows the **semiquantitation** of the viral load in individual cell populations.

By combining ISH with electron microscopy (EM), the analysis of **ultrastructural features of viral replication** has been possible, for example, in

- **CMV**-infected fibroblasts
- identification of **HPV** in tissue sections from cervical condylomata, revealing a progressive increase in viral burden with epithelial differentiation
- identification of CNS involvement with **slow viruses** and **HIV**

6.1.1.2 rRNA GENE-SPECIFIC PROBES (p. 457)

To improve the **sensitivity** of nucleic acid probes the highly **conserved RNA sequences** can be targeted.

- A combination of **rRNA-DNA solution hybridization** with **immuno-detection of the hybrids** has been adapted to a latex bead format. When alkaline phosphatase serves as the reporter molecule, **quantitative estimates** become possible. A broad range of bacterial species can be detected with a sensitivity of 500 cells/assay.
- A **quantitative slot blot** is capable of detecting 3 pg rRNA in its radioactively labeled version, and 50 pg rRNA when a digoxigenin-labeling system is used. Digoxigenin-labeled probes were developed for rapid biotyping of *Campylobacter* (*Helicobacter*) species in clinical isolates.
- rRNA can also be used **to probe RFLP patterns** (e.g., in a study of a **nosocomial infection** with *Corynebacterium jeikeium*; it should be noted that patterns of antibiotic susceptibility correlated with RFLP profiles of *C. jeikeium* strains).
- **Fluorescent-oligonucleotide probes** constructed to identify 16S rRNA sequences can be **directly applied to nonfractionated material** from the intestine for bacterial strain identification using fluorescent microscopy.

6.1.1.3 PCR (p. 458)

Amplification of the nucleic acid material of pathogens has become a **method of choice** for a large number of organisms. Some of the **advantages** are

- **speed**
- **specificity**
- **low cost**
- the possibility of **identifying a pathogen without isolation** and purification
- the opportunity to **avoid culturing organisms** — many cannot be cultured with ease at the present time
- the possibility to conduct **retrospective epidemiological studies** on fixed, or archival, material
- the possibility to **monitor the therapeutic efficiency of pharmacological interventions** when the pathogenic organisms may not be viable, but still present
- the availability of **automated procedures**, which substantially reduces the "hands on" time
- the **simultaneous amplification** of several target sequences of different nature (e.g., viral and human) in a **multiplex PCR** format, which significantly enhances the efficiency and economy of diagnostic laboratories

Some of the **shortcomings of PCR** for infectious disease evaluation include

- the danger of **contamination** and "**carryover**"
- the **lack of easy quantitation**, particularly pronounced when applied to clinical material without purification of target DNA

Reappraisal of old notions is stimulated by PCR with respect to

- the **prevalence** of various pathogens
- what constitutes a **latent infection**
- what should be considered a "**gold standard**"

The list of infectious agents that have been successfully detected and/or characterized by PCR amplification is growing exponentially. A sampling of reported studies at the time of this writing is given in Table 6.1.

6.1.1.4 EPIDEMIOLOGICAL ANALYSIS (p. 458)

In the evaluation of infections the concept of **clonality of pathogens** incorporates characteristics defined by phenotyping and genotyping methods.

Phenotyping methods rely on

- the pattern of **pathogen response** to a variety of substrates
- **interaction with antisera**
- **interaction with chemical challenges** (e.g., drugs)
- **multilocus enzyme electrophoresis**

Genotypic specificity of an organism can be defined by

- **plasmid restriction patterns** (that suffer from intrinsic variability)
- **chromosome restriction patterns**

Chromosome fingerprinting is

- a substantially **more permanent** characteristic, and the specific differentiation of a bacterial species **does not require the expression of a phenotype**
- **too complex** (about 1000 bands in an average fingerprint), making multiple comparisons cumbersome and poorly reproducible; therefore, **restriction patterns of selected genetic regions** can be used. These are, however, largely **limited to species-specific or strain-specific identifications**

A substantially broader spectrum of hybridization specificities is afforded by probing chromosomal DNA with **ribosomal RNA (rRNA)**. This **ribosomal DNA (rDNA) typing** has been successful in

- **subtyping strains** isolated in sporadic outbreaks
- **monitoring the incidence** of various isolates in both sporadic and epidemic cases

Examples of REA of genomic DNA with or without PCR amplification include the molecular analysis of

- ***Streptococcus viridans*** strains in a case of myocarditis
- ***P. aeruginosa*** strains in sputum samples of cystic fibrosis patients and in immunocompromised children
- ***Listeria monocytogenes*** in epidemic outbreaks
- ***Vibrio cholerae*** strains
- ***Shigella*** strains
- ***Salmonella*** strains

TABLE 6.1 (p. 459)
Selected Assays for Infectious Agents

Pathogens	Ref.
Viral pathogens	
Cytomegalovirus	Demmler et al. (1988)
Epstein-Barr virus	T. Saito et al. (1989)
Herpes simplex virus	Brice et al. (1989)
Human herpes virus 6	Qavi et al. (1989)
Human papillomaviruses	Shibata et al. (1988); Wagatsuma et al. (1990); Griffin et al. (1990); McNichol and Dodd (1990)
Parvoviruses	Salimans et al. (1989a,b)
Rotaviruses	Xu et al. (1990); Wilde et al. (1990, 1991); Gouvea et al. (1990)
HIV-1	Greenberg et al. (1989); Shibata et al. (1989); Loche and Mach (1988); Hart et al. (1988); Laure et al. (1988); Hufert et al. (1989); Abbott et al. (1988); Hewlett et al. (1989, 1990); Carman and Kidd (1989); Lai-Goldman et al. (1988); Albert and Fenyo (1990); Holodniy et al. (1991); Ou et al. (1988); Dahlen et al. (1991); Kellogg et al. (1990); Clarke et al. (1990); Cassol et al. (1991); Dickover et al. (1990); Conway et al. (1990); Wages et al. (1991); Horsburgh et al. (1990); Lifson et al. (1990)
HIV-2	Rayfield et al. (1988)
HTLV-I	Abbott et al. (1988)
HTLV-II	Ehrlich et al. (1989)
Enteroviruses	Rotbart (1990a,b)
Rhinoviruses	Gama et al. (1988a,b)
Hepatitis A virus	Jansen et al. (1990)
Hepatitis B virus	Larzul et al. (1988, 1989); Sumazaki et al. (1989); S. Kaneko et al. (1989a,b, 1990a); Matsuda et al. (1989)
Hepatitis C virus	Weiner et al. (1990a,b); S. Kaneko et al. (1990b)
Influenza viruses	Bressoud et al. (1990)
Mumps virus	Forsey et al. (1990)
Adenoviruses	Allard et al. (1990)
Bacterial pathogens	
Chlamydia spp.	Dean et al. (1990); Pollard et al. (1989)
Mycoplasma pneumoniae	Bernet et al. (1989); Jensen et al. (1989)
Rickettsia rickettsii	Tzianabos et al. (1989)
Mycobacterium leprae	Hartskeerl et al. (1989)
Mycobacterium tuberculosis	Brisson-Noel et al. (1989); Shankar et al. (1989)
Borrelia burgdorferi	Rosa and Schwan (1989)
Treponema pallidum	Noordhoek et al. (1990, 1991); Burstain et al. (1991)
Enterotoxigenic *Escherichia coli*	Olive (1989); Pollard et al. (1990)

TABLE 6.1 (continued)
Selected Assays for Infectious Agents

Pathogens	Ref.
Shigella	Frankel et al. (1989)
Bordetella pertussis	Houard et al. (1989)
Clostridium difficile	Wren et al. (1990); Kato et al. (1991)
Leptospira	van Eys et al. (1989)
Helicobacter pylori	Valentine et al. (1991)
Listeria monocytogenes	Deneer and Boychuk (1991)
Yersinia enterocolitica	Wren and Tabaqchali (1990)
Coliform bacteria in water supply	Bej et al. (1990)
Naegleria fowleri	McLaughlin et al. (1991)
Legionella pneumophila	Bej et al. (1991); Mahbubani et al. (1990)
Fungi	
Cryptococcus spp.	Vilgalys and Hester (1990)
Protozoal pathogens	
Toxoplasma gondii	Savva et al. (1990)
Drug-resistant *Plasmodium falciparum*	Zolg et al. (1990)
Parasites	
Echinococcus multilocularis	Gottstein and Mowatt (1991)
Pneumocystis carinii	Wakefield et al. (1990)
Trypanosoma cruzi	Moser et al. (1989)

Adapted and expanded from Stoker, N. G (1990). Trans. Roy. Soc. Trop. Med. Hyg. *84*:755–756. With permission.

- ***Acanthamoeba*** strains
- ***Leptospira*** strains
- ***Candida*** species
- penicillin resistance of ***Neisseria meningitidis*** strains
- **human adenovirus type 3** isolated from ocular lesions

to name just a few published examples.

The **specificity of hybrid formation** with the diagnostic probe underlies all taxonomic, epidemiological, and diagnostic assessments. Full understanding of the role of various parameters in hybrid formation and their potential in producing **experimental error in DNA-DNA pairing** is mandatory.

6.1.1.5 PULSED-FIELD GEL ELECTROPHORESIS (PFGE) (p. 460)

Bacterial **strain identification** can be based on PFGE of genomic restriction digests. For example, PFGE restriction profiles using *Sma*I-digested genomic DNA allow a reliable differentiation between *Campylobacter jejuni* from *Campylobacter coli*. This offers a better alternative for epidemiological studies than a combination of conventional DNA fingerprinting with RNA hybridization procedures.

Chapter 6.2

Enteric Pathogens

6.2.1 ENTEROVIRUSES (p. 461)

Human enteroviruses include **polio viruses, coxsackie viruses, echoviruses,** and **hepatitis A virus,** which affect the nervous, respiratory, cardiovascular, and gastrointestinal systems, respectively.

The lack of adequate speed, sensitivity, and specificity in traditional diagnostic methods for enterovirus identification prompted the development of nucleic acid-based techniques. The earlier cDNA, when directly applied to clinical specimens of cerebrospinal fluid (CSF) or stool, had low sensitivity (33%). Subsequent developments resulted in clinically useful assays incorporating

- improvements in **target retention**
- the use of **single-stranded RNA probes**
- **target amplification** by the PCR

6.2.1.1 ENTERIC ADENOVIRUSES (EA) (p. 461)

The identification of **EA types 40** and **41**, which cause gastroenteritis in infants, by gene probes directly in clinical specimens can be performed in a **direct spot test** format using radioactive and nonradioactive probes. Although slightly lower in sensitivity than the **EM methods** following **ultracentrifugation** of specimens, this approach represents a useful diagnostic test for patient management.

A higher diagnostic sensitivity on diluted stool samples is offered by a **PCR** assay that compared favorably with **cell culturing** and with **REA** of viral DNA.

6.2.1.2 ROTAVIRUSES (RVs) (p. 461)

Rapid diagnosis has traditionally been made by **immunological methods**, visualization of the virus by **EM**, and viral RNA detection by **polyacrylamide gel electrophoresis** and silver staining.

Several ELISA systems (e.g., **EIA**, International Diagnostic Laboratories; **Pathfinder**, Kallestad Laboratories; **Rotaclone**, Cambridge BioScience) and a nonradioactive oligonucleotide **assay kit** (**SNAP**, Molecular BioSystems) are available for RV diagnosis.

Some fastidious nongroup A RV **cannot be cultivated** in tissue culture and are not detectable by conventional immunological techniques. Earlier

dot blot assays used radioactively labeled probes and, more recently, **biotinylated ssRNA** probes have been used in dot blot and **northern analysis** of RV double-stranded RNA (dsRNA) genomes.

Alkaline phosphatase-conjugated oligonucleotide probes have also been used to characterize extracted viral RNA from stool samples. It compared favorably with **polyacrylamide gel electrophoretic** and immunological [enzyme-linked immunosorbent assay (**ELISA**)] analysis of RV.

Drawbacks of oligonucleotide probes include

- an **overnight incubation** for maximum sensitivity
- **laborious isolation** of viral nucleic acids

Characterization of **RV serotypes** can be achieved by **sequencing** the genes encoding serotype-specific VP7 glycosylated products of gene *7*, *8*, or *9*, or by **RNA segmental oligonucleotide mapping**.

Hybridization with **oligonucleotide probes** and **solution RNA-RNA hybridization** have been used for molecular **characterization of RV isolates** in analyzing **epidemic outbreaks**.

A new amplification typing approach (**PCR typing**) complements grouping of RV by **electrophenotyping**.

PCR is helpful in determining **RV shedding** by children, substantially improving RV detection compared to ELISA (58 vs. 36% positivity, respectively).

The **inhibitory substances** present in unpurified fecal samples may prevent identification of RV by PCR in clinical specimens. **Chromatographic cellulose fiber powder** (CF11 powder) appears to eliminate these interferences.

6.2.2 HEPATITIS B VIRUS (HBV)

Molecular probes have enabled scientists and clinicians to make important observations on the **biology of HBV** and **pathogenesis of HBV infection** that are helpful in patient management.

6.2.2.1 HBV IN SERUM (p. 462)

One of the traditional **biochemical markers** of liver involvement in HBV infection is the **level of transaminases**, in particular **alanine aminotransferase (ALT)**. Hybridization studies placed the peak of serum **HBV DNA** shortly before or simultaneously with the maximum elevation of ALT in over 90% of the acute **exacerbations** of chronic HBV hepatitis.

Demonstration of the **HBeAg** and the rising of the homologous antibody (**anti-HBe**) is taken as serological evidence of the replicative phase of HBV infection. However, as shown by the HBV DNA presence, HBV replication can occur in the presence of anti-HBe in carriers suffering from **major liver damage**. The atypical **concomitant presence of HBV DNA** and **anti-HBe** in serum correlates with histologically demonstrated **severe periportal hepatitis** leading to cirrhosis. Therefore,

- HBV DNA is present in the plasma of 85.7% of people positive for both HBsAg and HBeAg, and in 11.8% of those positive for HBsAg and negative for HBeAg. In a longitudinal study, HBV DNA was found to be a better **index of impending chronicity** than HBeAg

- a positive HBV DNA test in patients with positive HBsAg and anti-HBe serology is **significant in predicting an unfavorable outcome**
- the presence of HBV DNA also indicates **chronic progression** of hepatitis. In addition, in 83% of cases, the liver tissues obtained during clinical exacerbations show **free replicative forms** of HBV DNA as demonstrated by Southern blot hybridization.

Some workers report a high percentage (69%) of **blood donors** negative for all serological HBV markers tested positive for HBV DNA in serum by dot blot hybridization.

Based on the HBV DNA hybridization studies, the **infectivity** in acute hepatitis B is characteristic of the **presymptomatic** and **early symptomatic** period.

PCR analysis in HBsAg carriers demonstrates that 100% of the HBeAg-positive individuals have HBV DNA in their blood. Among the HBeAg-negative carriers, those with CAH were 90% positive for HBV DNA, the healthy carriers were 80% positive, and patients with cirrhosis had the lowest rate of HBV DNA presence — 69.2%.

It appears that viral replication is most active in carriers with CAH **despite seroconversion** to anti-HBe status. It is anticipated that simplified PCR assays using nonradioactive probes will prove indispensable in mass screenings for HBV DNA.

Evaluation of HBV DNA in serum has also been used in

- **monitoring steroid treatment** of patients with hepatitis
- establishing **maternal-infant HBV transmission**
- **screening** medical workers
- the identification of **epidemiological relationships** between cases of HBV infection **by REA** and **sequencing** of HBV DNA from serum
- presenting evidence of **intrafamilial transmission** using probes for specific regions of the viral genome

The diagnostic potential of molecular hybridization and PCR on serum specimens is superior to serology and biochemical markers in establishing infectivity and predicting the development of HBV infection.

6.2.2.2 HBV DNA IN WHITE BLOOD CELLS (p. 464)

HBV DNA has been detected in

- **bone marrow** cells
- **mononuclear blood cells** (PBMCs) of homosexual men, hemophiliacs, and intravenous drug users
- **lymphocytes** from peripheral blood, bone marrow, semen, and lymph nodes from AIDS patients
- **leukocytes** from AIDS patients

Both integrated and free, but not replicative, HBV DNA forms were identified in these cells.

Judging the presence of **1.0- to 3.2-kbp fragments** generated by *Eco*RI digestion to be indicative of the replicative forms, **HBV DNA could be detected in serum of only those HBsAg carriers whose PBMCs contained replicative forms.**

HBV DNA replicates in PBMCs. Replication and integration of HBV DNA occur in both **T lymphocytes** and **non-T lymphocytes.** HBV **replication** takes place in hepatocytes and PBMCs. Although it is more successful in hepatocytes, **integration** occurs more readily in PBMC. A difference in the molecular forms of HBV DNA is found between PBMCs and the liver, and **the HBV DNA forms in PBMCs do not correlate with serum HBV markers.**

RT-PCR analysis of **polyadenylated RNA** in HBV-infected **PBMCs** definitively establishes the occurrence of transcription of viral DNA in peripheral blood cells.

Confirmation of HBV DNA **transcription** in PBMCs was also provided by **ISH** demonstrating RNA production in HBsAg-positive and -negative patients with liver diseases. It appears that **persistence of the virus in peripheral blood constitutes a part of the infectious process at an early stage.**

6.2.2.3 HBV DNA IN THE LIVER (p. 465)

The presence of HBV DNA in **formalin-fixed** liver sections detected with biotinylated probes positively correlates with the HBV serological markers and histopathological features. **HBV DNA has always been detectable in the liver of chronic hepatitis patients, regardless of the serological status.** On the other hand, no HBV DNA can be identified in sections showing a normal histological pattern, despite positive anti-HBc and anti-HBs.

The relationship between **HBV gene expression**, **liver disease**, and HBV **RNA** shows a more consistent correlation. The **integrated DNA** has a 21S RNA transcript and HBsAg as its sole expression, whereas **free DNA** gives rise to a 29S RNA transcript and expression of HBcAg in the liver and of HBeAg in the serum.

Extrachromosomal monomeric viral DNA seems to persist in patients serologically immune to HBV infection.

The presence of **free replicative forms** of HBV DNA in the liver of patients with CAH has been repeatedly correlated with **greater tissue damage**. In contrast, **integration of HBV DNA**, which occurs early in CAH, was found predominantly in livers showing **no necrosis** or **inflammatory activity** in patients positive for anti-HBe.

Evaluation of the state of integration and expression of HBV DNA in the hepatocyte and the serum has been used to develop **optimal therapeutic protocols**:

- Antiviral therapy can be **advised for HBsAg-positive CAH patients with episomal HBV DNA**, irrespective of the presence of integrated sequences;

- The selection and evaluation of patients for antiviral therapy should consider the HBV DNA state (**episomal vs. integrated**) found in liver biopsy, because the **existence of HBV DNA in the episomal state in the liver is not always reflected by the presence of HBV DNA in the serum**;
- **Parallel disappearance of HBV DNA** from the liver, PBMCs, and serum has been observed in children and adults receiving interferon therapy.

6.2.2.4 EXTRAHEPATIC LOCALIZATION OF HBV (p. 466)

Aside from the liver and PBMCs, the extrahepatic distribution of HBV has been described in the **brain, saliva, pancreas, kidney, skin,** and **vessels** of animals and humans.

In **patients who died from fulminant** and **resolving hepatitis**, HBV nucleic acids were identified in **lymph nodes, spleen, gonads, thyroid grand, kidneys, pancreas,** and **adrenal glands**. The strongest signal was detected in the lymph nodes. Importantly, **little or no HBV nucleic acid was detected in the serum or liver in these patients**.

6.2.2.5 HEPATOCELLULAR CARCINOMA (HCC) (p. 466)

The **number of HBV DNA-positive cells varies from tumor to tumor**.

HBV DNA **nuclear staining** did not correlate either with the gene expression markers (HBsAg and HBcAg) or with the type or degree of differentiation of the tumor.

Southern analysis of total tumor genomic DNA shows **multiple integration events**. ISH showed HBV DNA was present in hepatocytes **adjacent** to HCC foci, suggesting that **integration of viral DNA precedes tumor development**.

In a large number of HCC cases, integration of HBV DNA **could not be** convincingly shown, and no free viral DNA could be seen in neoplastic areas in earlier studies.

Fine analysis of **HBV DNA integration** in **HCC in children** demonstrated **multiple integration sites** possibly related to the expression of **oncogenes**. In contrast, others believe there is a **single integration site** accounting for clonal proliferation of such infected cells.

The overwhelming evidence, however, indicates that various rearrangements (**deletions**, **duplication**, and **insertions**) are seen in integrated HBV sequences and in the cellular **flanking sequences**.

Although the viral **integration occurs at random** in the host genome, it appears that HBV DNA preferentially integrates in areas with **abundant repetitive sequences**.

Amplification of integrated HBV and of its **flanking cellular sequences** occurs in HCC. Amplification of oncogenes triggered by the viral integration cannot be excluded as contributing to carcinogenesis. No final word has been pronounced on that so far.

In **noncancerous** regions of the liver, **heterogeneous sites** of HBV DNA integration into hepatocyte DNA appears to **precede hepatocarcinogenesis** and leads to **multifocal clonal cell proliferations** of the hepatocytes harboring integrated HBV DNA.

The role of HBV integration in HCC does not seem to be as direct as that of viruses with oncogenes or that of viruses acting by a "promoter-insertion" mechanism. **Hepatocyte injury** and **regeneration** are important in hepatocarcinogenesis related to HBV.

Methylation of the HBV genome plays a role in the **expression of the HBV genome** in HCC and chronic hepatitis. Methylation per se does not seem to be an absolute determinant of the HBV genome expression in human HCC, and additional factors must contribute to HBV expression in the malignancy.

It is apparent that a combination of morphological evaluation of liver biopsy with a sensitive HBV DNA status analysis may offer a substantially more precise assessment of HBV infection and its sequelae.

6.2.2.6 HYBRIDIZATION METHODS (p. 467)

Dot blot hybridization of serum samples has been performed either following **phenol-chloroform extraction**, or using a **direct application** of radioactively labeled DNA probes. Relatively small samples of serum (less than 500 µl) were sufficient to generate visible spots by autoradiography, when **cloned HBV DNA** was used as a probe.

Evaluation of the positive results in this approach can be **semiquantitative** at best. **Reproducibility** of the method even in the same laboratory has never been thoroughly studied. In my experience, dot blot hybridization with some of the commercially developed kits suffered from poor reproducibility, beginning with the collection of serum samples on membranes: parallel samples took different times to filter and wash even within the same run, the autoradiographic spots showed sometimes uneven distribution of radioactivity, and quantitation of the signals was rather subjective.

A serial dilution of a standard HBV DNA-positive sample can be used, against which the test spots are compared, giving an **indirect approximation** of the number of viral particles. The sensitivity of the dot blot method is 1×10^5 DNA-containing particles per milliliter of serum.

A **spot hybridization** test cannot identify the **physical nature** or **size** of DNA being tested. It can be used only for the fast identification of patients with relatively high levels of HBV in their blood.

The **superior sensitivity** of HBV DNA hybridization over that of conventional serological methods had already become evident in these earlier studies:

- it is **more sensitive than the DNA polymerase and HBeAg assays** in detecting complete viral particles
- it allows the determination of the **level of infectivity** in patients. A thorough, quantitative study of the **HBV genome titers** in sera of infected individuals distinguished **virus carriers** with high ($>5 \times 10^7$ HBV genomes), moderate (10^5 to 5×10^7), and low ($<10^5$) infectivity
- it helps in monitoring the **response to therapy**

A potential drawback is that

- **false-positive** results have been observed in HBV DNA dot hybridization assays due to the **residual portions of the plasmid** used to construct the probe; it hybridized to the contaminating plasmid-containing bacteria in the samples

Improvements in HBV DNA characterization in patient specimens come from

- development of better **oligonucleotide** probes
- **solution hybridization**
- use of **nonradioactive** labels
- the **PCR** technique

A combination of **biotinylated** DNA probe with the detection system using **avidin-β-galactosidase complex** detected as little as a few picograms of target DNA, being comparable to the conventional ^{32}P-labeled probe method. The biotin-labeled probes can be successfully stored at –20°C for over 1 year without loss of sensitivity or specificity.

A **streptavidin-alkaline phosphatase conjugate** has been used as a detector system with biotinylated DNA probe in a dot blot hybridization format.

Another nonradioactive HBV DNA probe used **2-acetylaminofluorene (AAF)** as a reporter molecule. The sensitivity limit of the AAF modified probe, which used an **anti-AAF monoclonal antibody** detection system, was found to be 5×10^6 viral particles per milliliter serum compared to 5×10^5 particles per milliliter for the radioactive probe.

The respective sensitivities for HBV DNA probes labeled with **biotin, photobiotin**, and **digoxygenin** tested on serial dilutions of plasmid DNA processed as serum samples were

- 1 to 5 pg with ^{32}P and the digoxigenin-labeled probe
- 5 to 10 pg with the biotin-labeled probe
- 50 pg with the photobiotin-labeled DNA

Unfortunately, a high rate of 23% **false-positive** results was noted with the biotin-labeled probes. However, phenol extraction followed by ethanol precipitation proved to eliminate this nonspecific hybridization completely.

Digoxigenin-labeled riboprobes show the detection limit of 0.5 to 1.0 pg HBV DNA and a high specificity.

Chemoluminescent substrate for **alkaline phosphatase** linked to a DNA probe for HBV allows the detection of 1.18×10^6 copies of HBV_c plasmid DNA in 30 min. This favorably compares with a similar assay (sensitivity, 9.8×10^7 copies of DNA) employing a **tetrazolium-based** detection system.

A **commercial** HBV DNA **solution hybridization** system has been introduced for research use in the United States by Abbott Laboratories (**Abbott Genostics™ hepatitis B viral DNA**). The Abbott Genostics system with the **^{125}I-labeled DNA probe** displays a slightly higher sensitivity than the spot test. The Abbott kit is well standardized, providing a quantitative evaluation of HBV DNA level in serum. The cut-off level of sensitivity for a positive result is equal to or greater than 1.5 pg HBV DNA per milliliter. It favorably compares with dot blot hybridization and, although somewhat less sensitive, still allows monitoring of antiviral treatment.

The **ISH** approach offers advantages unavailable in serum assay systems for establishing correlations between **morphological alterations** and the extent of tissue involvement with the virus.

The cytoplasmic HBV DNA visualized by ISH in hepatocytes positive for HB_cAg correlated with the level of cytoplasmic HB_cAg, but not with the presence of nuclear HB_cAg.

ISH shows that HBV DNA is **diffusely distributed** in the liver during the early stage of chronic liver disease, whereas the **site of HBV replication becomes localized in the advanced stage of the disease.**

6.2.2.7 PCR IN HBV DETECTION (p. 470)

The PCR allows detection of HBV early in an infection, when it is present at a level of only 4×10^4 particles/ml serum.

PCR assay identifies as little as 0.4 fg viral DNA, corresponding to about 130 genome equivalents.

An even finer PCR system for HBV DNA has been developed using an **AAF**-labeled DNA probe, with a detection limit of 3 to 30 particles.

The extent of HBV infection can now be estimated, when a definitive serological diagnosis (anti-HBc positivity) is impossible, and when even direct hybridization assays fail to detect HBV DNA.

Southern analysis of the amplified products allowed the resolution of a single HBV DNA molecule. Southern blot hybridization with the PCR-amplified product allowed a sensitivity of 3.6×10^{-15} g, and better to be achieved. The **need for high-stringency** conditions to minimize false-positive results is emphasized.

Sample preparation is critical:

- **thermal treatment** of the sample for 5 min at 115°C to disrupt purified particles enhances the detection so that a single viral particle can be identified
- when applied to **crude serum** the detection limit was approximately 2×10^5 viral particles after 40 amplification cycles with the *Taq* polymerase
- either **entrapment** of viral particles in the clot formed during heating, or the presence of **inhibitors** of the PCR in serum, accounts for a decrease in the assay sensitivity in crude specimens
- **phenol-chloroform extraction** of samples as well as removal of heparin by heparinase and removal of proteins and interferences on glass beads can markedly enhance the efficiency of PCR-based detection

Because **rigorous purification of viral DNA is not required for PCR amplification**, a release of HBV DNA from viral particles appears to suffice and is achieved by

- incubation of serum with 0.1 *M* **NaOH** for 1 h at 37°C

The HBV DNA sequences amplified by the PCR for 30 cycles are then detected by **2% agarose gel electrophoresis** and **ethidium bromide** staining. The detection limit of that method is 10^{-5} pg HBV DNA; it can be completed in 1 day and does not require the use of radiolabeled reagents.

A marked increase in sensitivity of the assay was attained by a **second round of amplification** on the products of the first PCR cycle. **Combining PCR with Southern analysis** of the amplified material using radiolabeled probes allows as few as three HBV DNA molecules to be identified.

Various PCR protocols have been described for **typing HBV genomes** with and without restriction analysis of the amplified products. Effective PCR amplification has been demonstrated on DNA isolated from **formalin-fixed, paraffin-embedded** liver tissues.

An **automated PCR** procedure for HBV using only **two thermal steps** and employing three primer pairs, was performed efficiently under highly stringent conditions and detected a **single HBV DNA molecule** with **X region-specific primers**. The X antigen is a protein responsible for **transcriptional trans-activation**; it is expressed during HBV infection.

HBV DNA has been detected by the PCR using primers for the ***S*** and ***C*** regions of the HBV genome, in sera from patients negative for HBV DNA by conventional hybridization technique. This PCR assay could estimate the extent of assembly of viral particles.

A **microtiter sandwich hybridization assay** has been developed for the simplified detection of the PCR-amplified product. Using an **ELISA-like format** with **biotin-labeled probes** a sensitivity of five copies of HBV DNA, equal to that of Southern hybridization, was obtained with sandwich hybridization. The sandwich assay can have the additional advantage of yielding **quantitative** estimates of the HBV DNA-amplified product.

6.2.3 HEPATITIS C VIRUS (HCV)

6.2.3.1 IMMUNOLOGICAL TESTS (p. 471)

The etiological agent of **non-A, non-B (NANB) hepatitis**, first proposed as a cause of posttransfusion hepatitis in 1975, has been identified and referred as **hepatitis C virus (HCV)**. The Chiron Corporation and Center for Disease Control (CDC) scientists succeeded in the identification of one **clone (5-1-1)** reactive with chimpanzee serum known to be positive for NANB hepatitis antibody.

HCV characteristics are:

- the nucleic acid is a **single-stranded RNA** molecule up to 10,000 nucleotides long and **enclosed in a lipid envelope**
- HCV is similar to **togaviruses** and **flaviviruses**; it is an unusual virus that is **most related to the pestiviruses**
- significant **genetic diversity** exists, particularly within the putative 5′ structural gene region of different HCV isolates; already five types are recognized that show different prevalence in different parts of the world

A testing antibody followed the production of a **clone, C100**, comprising the four initially isolated **clones 5-1-1**, **32**, **36**, and **81**.

The **polypeptide, C-100-3**, containing 363 viral amino acids, was used first in a **radioimmunoassay**, and then in a commercially developed **ELISA** offered by Ortho Diagnostic Systems.

This assay demonstrated that

- about 80% of patients worldwide clinically diagnosed to have NANB hepatitis are positive for the antibodies to HCV
- the **overall prevalence of HCV antibodies** in blood donors is **low** — 0.1% in Germany, 0.3 to 0.7% in the U.K., 0.7% in France, 0.9% in Italy, and 1.2 to 1.4% in the U.S. and Japan
- intravenous drug abusers (from 48% in Germany to 81% in the U.K.), hemodialysis patients, and hemophiliacs have a **high frequency of seropositivity**. Hemophiliacs who have received untreated blood products showed anti-HCV positivity — 64% in Spain and 85% in the U.K. In contrast, **hemophiliacs given only dry-heated factor VIII apparently are not infected**
- prevalence estimates may now be subject to revision due to emerging limitations of serological methods employed to assess HCV infection in different countries and populations

6.2.3.2 LIMITATIONS OF SEROLOGICAL HCV TESTING (p. 472)

- There is a **"window" of nonreactivity** in the range of 4 to 32 weeks after onset of hepatitis or even up to 1 year. It is prudent to refrain, based on that assay, from excluding HCV infection in antibody-negative patients up to 6 months after the onset of symptoms.
- The **HCV antibody assay is apparently not sensitive enough**, because **PCR** amplification confirms the presence of the virus in anti-HCV-negative and anti-HCV-positive patients.
- The **first generation** of anti-HCV test measured an **IgG antibody**.
- The initially introduced tests yielded **false-positive results** in patients with active autoimmune chronic liver disease or the so-called type 2 autoimmune hepatitis.
- A substantial **increase in the specificity** of the original ELISA procedure can be achieved by an **8 *M* urea wash**.

The **next generation** of immunological anti-HCV test was based on the **recombinant immunoblot assay (RIBA)** developed by the Chiron Corporation and Ortho Diagnostic Systems. The RIBA assay was designed against the two antigens (5-1-1 and C100), and when it is used to assess infectivity the reactivity even to one of the antigens correlates with infectivity.

The newer generations of RIBA assays are **RIBA-2** (using two antigens) and **4-RIBA** (using four antigens). When serological testing is indeterminate PCR is recommended.

Another assay, the **peptide test**, developed on synthetic peptides derived from immunodominant regions of both capsid and nonstructural proteins, detects HCV antibody earlier, by 4 to 10 weeks.

6.2.3.3 IMMUNOLOGICAL TESTING OF BLOOD PRODUCTS FOR HCV (p. 473)

The position of the **European community** and that proposed by **FDA** in this country differ toward screening of sources of plasma for HCV.

The **French** workers argue that antibody-positive donors should be excluded from contributing to the plasma pools used in the preparation of blood products such as albumin, despite the use of apparently effective procedures for inactivation of HCV.

The **FDA** proposed, however, to conduct a scientific experiment, arguing that exclusion of anti-HCV positive donors may substantially lower the amount of antibody.

Support for the latter position is based on the overwhelming evidence that the current techniques of preparing immunoglobulins ensure adequate safety of the blood-derived products.

It is apparent, however, that the current inactivation procedures may be inadequate for total elimination of HCV as shown by transmission of the virus in some factor VIII concentrates purified by monoclonal antibody adsorption followed by exposure to 60°C for 30 h. Also, concern has been voiced about the validity of FDA experimental findings in chimpanzees. **Organic solvent-detergent treatment** of blood preparations is now widely used to reduce or eliminate HIV, HBV, and HCV from these products.

Up to 52% of **intravenous immunoglobulin** (IVIG) preparation lots have been found to be positive for anti-HCV as tested by the Ortho ELISA. Depending on the brand and the country of origin of the donor plasma the prevalence of anti-HCV antibody varied from 0 to 100% of the lots.

The development of chronic hepatitis has been documented in several patients who had received IVIG preparations with high titers of anti-HCV antibodies. The seroconversion rate reached 88% in recipients given anti-HCV positive blood when monitored for up to 12 months following blood transfusion.

The reported prevalence of anti-HCV positivity of plasma pools was

- 0.4% from the U.K. sources
- 1.5% in Spain
- up to 100% from the U.S. sources, apparently due to the use of paid donors

Using PCR assays, HCV RNA could be detected in all batches of **commercially** available factor VIII specimens. **FVIII preparations produced from plasma donations by volunteers was uniformly negative for HCV RNA by the PCR.**

A 5-year follow-up of hemophiliacs receiving preparations dry heat treated or exposed to wet heat or solvent-detergent treatments revealed that virucidal treatments of concentrates are effective in preventing HCV transmission.

6.2.3.4 IMMUNOLOGICAL HCV TESTING OF LOW-PREVALENCE GROUPS (p. 474)

Relatives of anti-HCV-positive patients enrolled in an interferon trial study tested positive for anti-HCV, whereas none of the relatives of anti-HCV-negative patients showed the presence of the antibody.

Occupational transmission of NANB hepatitis in health-care workers accounts for its increased prevalence (around 5%) among U.S. patients with NANB hepatitis. Documented cases of needlestick HCV seroconversion have been reported in a surgeon and a nurse. In dental practice HCV RNA has been detected in saliva, although at lower levels than in the blood of HCV carriers.

When screening a **population with a presumably low prevalence** of HCV, a discrepancy has been observed between the assay results obtained by **test kits of three different manufacturers** (Ortho, Abbott, and Chiron). The apparent false positivity of findings in some studies has not been satisfactorily explained, and the use of caution was advocated in interpreting the results of the Ortho ELISA testing.

The prevalence of anti-HCV in recipients of blood products before transfusion is comparable to that found in the blood products (less than 1%) (1/383 vs. 6/5150).

6.2.3.5 IMMUNOLOGICAL HCV TESTING IN LIVER DISEASE (p. 474)

Anti-HCV testing in autoimmune chronic active hepatitis by the Ortho ELISA test showed a **high degree of false positivity** (65%). None of the patients with primary biliary cirrhosis tested positive.

Histological changes characteristic of posttransfusion hepatitis in the livers of hemophiliac patients are associated invariably with anti-HCV positivity of the patients' sera.

The reverse is not true, however; the absence of histological features consistent with hepatitis does not exclude anti-HCV positivity, as almost 50% of such patients had anti-HCV antibodies.

The strong association of HCV infection with the development of HCC in patients negative for HBV infection suggests a significant role of HCV as a causative factor. Moreover, modes of **transmission of HCV other than blood transfusion** are suggested, because only about 30% of the HCV antibody-positive HCC patients had a history of blood transfusions. The prevalence of HCV infection in sexually promiscuous groups (female prostitutes, 8.97%; clients of prostitutes, 16.36%; homosexual men, 5.48%) is markedly higher than in voluntary blood donors (0.48%), suggesting that **HCV can be transmitted by sexual intercourse.**

A high prevalence of HCV antibody positivity was found in patients whose HCC had been ascribed to **alcohol abuse.**

6.2.3.6 HCV RNA ASSAYS (p. 475)

Reverse transcription of HCV RNA followed by **PCR amplification (RT-PCR)** of the derived cDNA has been used to compare HCV isolates from Japanese and American patients. HCV-J has a 22.6% difference in nucleotide sequence, and a 15.1% difference in amino acid sequence, compared to the American isolate.

PCR amplification offers an approach to **direct HCV RNA identifica-**

tion. Amplification of the viral sequences is superior to viral nucleic acid identification by hybridization, because the **hybridization signal is not sensitive enough** to detect HCV RNA. Besides, HCV RNA nucleotide sequences show **significant variability** among different isolates, a known property of single-stranded RNA viruses, and thus some HCV genomes fail to be detected.

Improvements in the **selection of primer sequences** have been continuously made by different researchers. The detection rate of a newer, **"nested" PCR** was significantly increased compared to that of the original assay described by the Chiron Corporation.

The possible **primer self-annealing** and **extension** may account for discrepant results. It is proposed that in addition to the use of **nested PCR** to increase sensitivity in the detection of HCV RNA, a **sequence-specific hybridization** appears to be essential to ensure the bands of amplified material visualized by **ethidium bromide** are true positives, and not **primer artifacts**.

Other RT-PCR protocols with different primer sequences for the detection of HCV RNA have been described as first using sets of "outer" and then "inner" primers. When this assay was applied to the detection of HCV in **clotting factor concentrates** all the unheated preparations except one were found to contain HCV RNA. No HCV sequences were identified in any of the "superheated" (80°C, 72 h) batches, whereas "less heated" concentrates (60°C, 32 h) did have the viral sequences.

Marked diagnostic power has been achieved with a **new set of primers** designed and synthesized based on the **highly conserved** (>99%) 5′-noncoding region of the HCV genome. The original set of primers detected HCV DNA in 35% of serum samples from hemophiliacs treated repeatedly with unheated commercial factor VIII concentrate. A PCR assay with the new set detected the viral sequences in 85%.

A PCR assay demonstrated the existence of **symptom-free HCV carriers** who had no detectable anti-HCV. HCV RNA has also been detected by PCR in the **saliva** of patients with posttransfusion HCV infection.

It is thought that cDNA/PCR assays for HCV RNA are **likely to underestimate** the true incidence of the virus:

- they may not be as sensitive as PCR assays for DNA genomes
- the methods for specimen retrieval, storage, and preparation for the assay may affect the number of viral genomes available for cDNA analysis

Nevertheless, the PCR assay for HCV RNA **by far exceeds** the detection capabilities of immunological techniques and shows that the majority of anti-HCV-positive patients are carriers of HCV.

6.2.4 HEPATITIS DELTA VIRUS (HDV)

6.2.4.1 THE BIOLOGY OF HDV (p. 477)

The **delta hepatitis antigen** was first described in 1977 in HBV carriers.

- The **route of transmission** is similar to that of HBV.

- HDV infection is endemic in the **Mediterranean** area.

Epidemiology of the HDV infection shows a **marked variation its prevalence** in different parts of the world:

- It is **rare in the Far East**.
- It is extremely **frequent in the Middle East** countries, **Brazil**, parts of **Africa**, and in the **South Pacific islands**.
- In **Europe** and the U.S. the infection is widespread, predominantly among **high-risk groups** such as intravenous drug abusers and hemophiliacs.
- HDV is a **defective RNA virus** displaying hepatotropism and depends exclusively on HBV.
- **HDV RNA** is circular, single-stranded, and composed of 1678 nucleotides. It requires the coat proteins of HBV or other related hepadnaviruses.
- In contrast to HBV, **HDV is invariably pathogenic**.
- HDV causes an **acute self-limited infection or chronic liver disease.**
- The highest degree of **tissue damage** is found before the elimination of the virus. **Histological changes** in the livers of patients infected with HDV seem to correlate with **the HDV load**.
- HDV produces the primary or complicating forms of severe HBsAg-positive acute and chronic liver disorders.
- The virus produces an antigen, **HDAg**, and antibody to this antigen (anti-HD) can be detected in infected persons.
- Infection produced by HDV appears to occur only **when the HBV load becomes infectious**.
- **Superinfection** by HDV is more prevalent in chronic HBsAg carriers. It results in a fulminant form of hepatitis.
- HDV superinfection may induce the **clearance of HBV-associated antigens**.
- There is **no mutual inhibition** between HDV and HBV replication.

Histological changes in the livers of patients infected with HDV seem to correlate with the **HDV load**.

- The **immune response** to HDV replication appears to have a major role in determining the outcome of the infection.
- There is **no sequence homology** between HDV RNA and HBV DNA and HCV RNA.
- Many **open reading frames (ORFs)** are found in the HDV genome.
- **HDV proteins** in the serum appear to be the same size as those in the liver and they can be recognized by a human monoclonal antibody.
- The **immune response of the individual** plays an important part in the pathogenesis of HDV hepatitis.

6.2.4.2 IMMUNOLOGICAL METHODS (p. 478)

Initially, **total delta antibody evaluation** was performed by a **solid-phase blocking radioimmunoassay (RIA)**. Subsequently, **commercially available kits** for a blocking RIA assay of anti-HDV became available [e.g.,

Abbott Laboratories (North Chicago, IL); Deltassay (Noctoch, Dublin, Ireland)].

6.2.4.3 HDV RNA DETERMINATION (p. 478)

Hybridization protocols have been developed to evaluate HDV RNA in serum in clinical situations following infection, avoiding the necessity to perform liver biopsy for this purpose. A **spot hybridization** technique and a **"slot blot" assay** can be used to detect the HDV RNA in serum using a cDNA as a probe.

By using **northern blot hybridization** with a **riboprobe** derived from a cDNA fragment, HDV RNA can be demonstrated in the serum in 92% of those who had HDAg in their liver biopsy specimens.

A **riboprobe** derived from a cDNA fragment of a 650-bp HDV RNA sequence has been developed for **spot hybridization**. Its advantages are:

- it can be used on large batches of specimens more easily than the **northern blot** hybridization assay using a cDNA of HDV RNA
- results of riboprobe spot hybridization correlate well with those of northern blotting
- the riboprobe assay offers a substantial **enhancement in sensitivity** (83 vs. 63%)
- it offers significant savings in time compared to the northern blot assay using a cDNA probe
- when detecting **HDV viremia**, the riboprobe method proved to be the most sensitive of the hybridization-based methods described so far

A novel **ISH** method applicable to the detection of HDV RNA in **formalin-fixed, paraffin-embedded** liver biopsies complements the earlier ISH assays adapted for **frozen sections**.

An **RT-PCR**-based method amplifies a region of the HDV genome that includes the **entire HDAg ORF**. It allows detection of HDV RNA in 10 pl samples of serum or in less than 0.1 pg of liver RNA.

PCR assays proved to be 10,000 times more sensitive than slot blot hybridization on serial dilutions of known concentrations of HDV RNA.

6.2.5 HEPATITIS A (HAV) (p. 479)

ssRNA probes with radioactive (^{32}P) and nonradioactive (biotin) labels have been constructed for **dot blot** hybridization to detect hepatitis A virus.

Using **differential hybridization** with probes for positive- and negative-sense strands (compared with HAV genomic RNA), a positive:negative ratio was used to develop a **semiquantitative evaluation** of HAV presence.

The same ssRNA probes can be used for **ISH** evaluation of infected cells.

Genomic fingerprinting of HAV isolates has been reported and revealed only a minor (9%) variability of the viral genome.

6.2.6 OTHER GASTROINTESTINAL PATHOGENS

6.2.6.1 *ESCHERICHIA COLI* (p. 479)

Escherichia coli strains causing gastrointestinal disease have been classified into four major groups depending on the predominant pathogenetic route by which these bacteria cause diarrhea:

- **enterotoxigenic (ETEC)**
- **enteroinvasive (EIEC)**
- **enteropathogenic (EPEC)**
- **enterohemorrhagic (EHEC)**

Nucleic acid probes identify these organisms in **pure culture** and **directly in clinical specimens**.

A variety of probes have been constructed for the detection of **plasmid genes** encoding enterotoxin [heat labile **(*LT*)** and heat stable **(*ST*)**], genes encoding invasiveness factors and adherence, and chromosomally encoded genes for Shiga-like enterotoxin **(*SLT*).**

Although the earlier probes for the **pInv** (**plasmid encoding invasion factors**) were more effective than serotyping techniques they were more effective in colony hybridization assays than in direct stool blots.

DNA hybridization using **synthetic oligonucleotides** is the method of choice, particularly for analysis of ST-ETEC strains, in spite of their somewhat lower sensitivity (78.8%) compared to **ELISA** and the **infant mouse assay**.

A **single hybrid RNA probe** containing sequences complementary to the *ST* and *LT* alleles, labeled either radioactively or with biotin, showed a 100% correlation with the biological assay.

Strains of *E. coli* producing SLT or **verotoxin (VTEC)** often include representatives of both EPEC and EHEC. The earlier probes for ***SLT*-related genes** have been supplemented with PCR-based assays. Oligonucleotide probes for the **EPEC adherence factor** proved to be more sensitive and specific, and produce clearer results on hybridization than the corresponding DNA sequences.

Nonradioactively labeled probes for various genes of *E. coli* included such reporter molecules as

- **biotin**
- **alkaline phosphatase**
- **horseradish peroxidase**
- **digoxigenin**

Of these, digoxigenin-labeled DNA fragments proved to be more sensitive and produced less **background noise** than biotinylated probes for ETEC genes. Pretreatment of colonies with **sodium dodecyl sulfate** and **proteinase K** significantly reduces nonspecific hybridization of horseradish peroxidase-labeled trivalent polynucleotide probe directed against *LT*-h, *ST*-Ia and *ST*-Ib genes of ETEC cells. This probe appears to be suitable for routine use in clinical laboratories for the detection of ETEC.

A **PCR** assay for LT-producing *E. coli* amplifies target genes with assays performed directly on *E. coli* colonies in a relatively pure culture. Diagnosis in clinical isolates was sensitive and specific enough not to require posthybridization with radioactively or alkaline phosphatase-labeled oligonucleotide probes.

6.2.6.2 *CAMPYLOBACTER* AND *HELICOBACTER* SPECIES (p. 480)

- **Ribosomal gene-specific probes** have been constructed for *Campylobacter* species analysis.
- A commercial detection system for the identification of three *Campylobacter* species (***C. jejuni, C. coli***, and ***C. laridis***) is offered by Gen-Probe (San Diego, CA) (see Supplement 1). Positive results exclude just those three species, and conventional biochemical and microscopic evaluations should be performed on organisms negative with the probe, but suggestive of campylobacters by **morphology**, **motility**, and **oxidase positivity**.
- Another commercially available system is specific for *C. jejuni* and *C. coli*, but not for other *Campylobacter* and ***Helicobacter pylori.*** It uses **alkaline phosphatase** as label (**SNAP**; Molecular BioSystems, San Diego, CA).
- The **digoxigenin-labeled probes** with an alkaline phosphatase detection system have been constructed to differentiate between *C. jejuni*, causing the majority of human gastroenteritis cases, and *C. coli*, implicated in only 2 to 5% of cases. These probes, when applied to fecal specimens, had a sensitivity of 89 and 93% and a specificity of at least 84.3%.
- On the basis of **16S rRNA** sequence data oligonucleotide probes have been constructed for and are capable of distinguishing *Campylobacter fetus* and *Campylobacter hyointestinalis*.
- Other **ribosomal genes probes** have been developed for the identification and subtyping of aerotolerant *Campylobacter* species by RFLP analysis.
- *Helicobacter pylori* **DNA** and **RNA probe assays** complement the culturing, histological, serological, and biochemical tests.
- **A PCR assay** has been described for the identification of *H. pylori* directly in gastric biopsy and aspirate specimens. It was positive in 93% of positive tissues and in 85% of aspirate specimens positive by microbiological culture and histological examination.

6.2.6.3 *CLOSTRIDIUM* SPECIES (p. 481)

Clostridium difficile is the etiological agent of **pseudomembranous colitis**, and may be in part responsible for antibiotic-associated diarrhea.

Whole-cell DNA fingerprinting and **species-specific oligonucleotide probes** for rRNA can distinguish between various isolates and related species.

The diagnosis of *C. difficile* is confirmed by

- **isolation** of toxigenic *C. difficile* from stool
- **detection** of toxins A (**enterotoxin**) and/or B (**cytotoxin**)
- **cell culture assay**
- **latex agglutination**

- **ELISA**
- **counterimmunoelectrophoresis**

A **PCR** amplification assay for a fragment of the gene encoding toxin A in *C. difficile* has been described. Although highly sensitive, that assay showed some cross-reactivity with ***C. sordellii.***

To avoid cross-reactivity, primers to both repetitive and unique sequences of the *C. difficile* toxin A gene have been constructed and a PCR assay with these primers proved to be highly specific for *C. difficile*.

Only minimal processing of stool samples may be needed for specific amplification of the *C. difficile* toxin A gene sequences.

Identification of the toxin CPE-producing strains of ***C. perfringens*** is difficult, and four oligonucleotide probes for **the *CPE* gene** sequences have been developed. Because *C. perfringens* appears to produce only one enterotoxin, and *C. perfringens* strains normally colonize intestine without causing a disease, positive hybridization with these probes testifies to the presence of toxigenic *C. perfringens* strains.

C. botulinum is identified by its ability to produce **neurotoxin** and studies are in progress to clone the gene encoding toxin B and other *C. botulinum* genes for the construction of appropriate probes.

6.2.6.4 *YERSINIA* SPECIES (p. 482)

Yersinia enterocolitica is an enteric pathogen frequently implicated in a variety of diseases from diarrhea to hepatosplenic abscesses and septicemia. **DNA and oligonucleotide probes** have long been developed for the identification of *Y. enterocolitica*. A **PCR**-based assay has been developed for *Y. enterocolitica*.

6.2.6.5 *SALMONELLA* SPECIES (p. 482)

Clinicians need to have positive identification of *Salmonella* within hours of infection; however, the presently available methods produce results no sooner than within a few days. Epidemiological evaluation of *Salmonella* infections requires a simple and fast definitive identification of the organism.

Blood culture is the standard method for diagnosis of typhoid fever, detecting 40 to 70% of ***S. typhi***-infected patients. The low bacteremia in typhoid patients (15 *S. typhi* organisms per milliliter of blood) may be below the sensitivity of the probe (500 bacteria). Following **concentration of the bacteria** from blood samples by lysis-centrifugation of blood and subsequent **growth amplification**, the detection level of the DNA probe becomes roughly comparable to that of the culture method.

A **DNA probe** constructed for the **via B antigen locus** of ***Citrobacter freundii*** identifies bacteria that synthesize **Vi antigen**, the capsular antigen of *S. typhi*, by ***in situ* colony hybridization**. This **8.6-kb Vi DNA probe** proved to be highly specific and sensitive on freshly isolated colonies from clinical specimens.

A different DNA probe for *Salmonella* species, **SAL6**, capable of identifying all 70 *Salmonella* serotypes, has been used in a **rapid DNA hybridization** procedure.

6.2.6.6 *LISTERIA MONOCYTOGENES* (p. 483)

DNA probes targeted to **rRNA** sequences, the **β-hemolysin gene**, and the **DTH (delayed-type hypersensitivity) factor** have been developed.

The DTH probe identifies only ***Listeria monocytogenes***, and nonpathogenic *Listeria* species do not interfere.

DNA fingerprinting proved valuable in epidemiological studies of **neonatal listeriosis.**

Newer probes tagged with horseradish peroxidase are suitable for the detection of *L. monocytogenes* by **direct colony hybridization.**

Using primers to the **listeriolysin *O* gene** the specific amplification of *L. monocytogenes* has been accomplished by **PCR** detecting fewer than 50 organisms.

6.2.6.7 *LACTOCOCCUS* AND *ENTEROCOCCUS* SPECIES (p. 483)

There is an increasing need to differentiate *Enterococcus* species contaminating the starter *Lactococcus* species cultures in the dairy processes.

A set of oligomers to the **23 rRNA** sequences has been constructed. The probes allowed specific identification of the cocci and their quantification.

6.2.6.8 COLIFORM BACTERIA IN THE ENVIRONMENT (p. 483)

These are used for monitoring the bacteriological purity of **water supplies** as indicators of potential human fecal contamination and thus the danger of the presence of enteric pathogens.

A highly sensitive **PCR** amplification assay is capable of detecting as little as 1 to 10 fg of genomic *E. coli* DNA and as few as one to five viable *E. coli* cells in 100 ml of water.

6.2.6.9 *GIARDIA LAMBLIA* (p. 483)

At least two types of **DNA probes** for ***Giardia lamblia*** have been reported:

- a **cloned fragment**
- a **total genomic DNA probe** showing similar sensitivity and specificity to the cloned probe

The sensitivity of the total genomic probe was found to be 10 ng of *G. lamblia* DNA, 10^5 trophosoites, and 10^4 guanine thiocyanate-treated cysts.

The detection of *G. lamblia* cysts in water can be accomplished using **dot blot** hybridization with a short cDNA probe from the small subunit of rRNA; it is capable of detecting as few as one to five cysts per milliliter of water sample concentrates.

6.2.6.10 *ENTAMOEBA HISTOLYTICA* (p. 484)

REA of extrachromosomal circular DNA molecules containing **rRNA** genes as well as clusters of **tandemly reiterated sequences** identified two types of sequences and displayed specificity to pathogenic and nonpathogenic ***Entamoeba histolytica.***

PCR studies established that repetitive sequences characteristic of pathogenic *E. histolytica* are also present in low copy numbers in a nonpathogenic strain.

A **cDNA library** has been constructed on the basis of RNA isolated from an axenically grown strain of *E. histolytica* expressing a **pathogenic isoenzyme pattern (zymodeme). Clone C2** is capable of differentiating between pathogenic and nonpathogenic *E. histolytica* zymodemes in a dot blot format assay and detects as few as 100 amoebic trophosoites.

Yet another approach is the amplification of a 482-bp segment that contained identical 5′ and 3′ ends, but different internal cleavage sites for restriction endonucleases in pathogenic and nonpathogenic strains. The strains are distinguished by **REA** of the amplified product. The **genotypic patterns** correlate closely with the clinical picture produced by the organisms in patients.

Other enteropathogenic organisms identifiable with gene probes (*Vibrio* species, *Staphylococcus aureus*, etc.) have been extensively discussed elsewhere.

Chapter 6.3

Respiratory Pathogens

6.3.1 RESPIRATORY VIRUSES

6.3.1.1 INFLUENZA VIRUS, ADENOVIRUS, AND RESPIRATORY SYNCYTIAL VIRUS (p. 485)

Laboratory identification of **influenza A virus** is hampered by the need for cumbersome **cell culture** methods.

RT-PCR of a 441-bp region encoding viral hemagglutinin allowed the detection of influenza virus H1 subtype in nasopharyngeal lavages. Cultivation of the virus becomes unnecessary, amplification of the relatively genetically stable region of the **hemagglutinin gene** markedly reduces the rate of false negativity.

The detection of **adenoviruses** in cells by **dot blot** hybridization utilizes either selected DNA clones or total adenovirus type 2 and 16 (Ad2 and Ad16) genomic DNA. Total Ad2 DNA and cloned radioactive probes show similar sensitivity of 10 pg of adenovirus DNA.

A commercial kit for **Ad2** (Pathogen II; Enzo Biochemical) uses biotinylated total Ad2 DNA as probe.

To apply DNA probes for respiratory adenoviruses directly to clinical material without preliminary amplification in cell culture, cross-reactivities with other subgroups must be eliminated, and the sensitivity should be increased.

Respiratory syncytial virus (RSV) causes lower respiratory tract infections in infants and young children worldwide. A relatively simple hybridization assay using **cDNA** probes specific for subgroups A and B of RSV detects the viral RNA in fixed, virally infected cells.

An **RT-PCR** with nested PCR detects RSV RNA by targeting the **F glycoprotein** of the virus in effusions in otitis media patients.

6.3.1.2 RHINOVIRUSES (p. 485)

Human rhinoviruses (HRVs) are the major cause of the common cold. The virus can be detected by **culture**, although it is difficult to grow, or by an **ELISA** for viral antigens. The ELISA is type specific, and because there are over 100 distinct serotypes there is a need for a broad diagnostic assay.

A cDNA to the 5′-noncoding region of the HRV RNA genome has up to 90% homology between different HRV serotypes and allows only homologous serotypes to be identified. **Synthetic oligonucleotide probes** constructed for the totally conserved region among different serotypes are capable of detecting HRVs in nasal wash samples.

cDNA fragments have also been used for grouping HRV according to their genetic relationships, and for **ISH** of nasal epithelial cells. cDNA probes, however, do not offer adequate sensitivity for detecting HRV in clinical specimens without prior preamplification of the virus in cell culture.

PCR-based methods for HRV identification may replace cell culture amplification of HRVs.

6.3.1.3 HERPESVIRUSES

6.3.1.3.1 *Herpes Simplex* Virus (HSV) (P. 486)

Cloned viral DNAs probes are able to detect at least 5 pg of homologous DNA. An even higher sensitivity (3 pg of DNA) can be achieved in a dot-blot hybridization assay using *in vitro* synthesized, radioactively labeled single-stranded RNA (ssRNA) transcripts as probes.

Single-stranded, short (22 bp) **oligonucleotide probes**, homologous to unique regions of HSV types 1 and 2, detected HSV in vesicle material and genital swabs. At a sensitivity of 10^4 to 10^5 HSV infectious units they showed 99% agreement between hybridization and monoclonal antibody typing.

Radioiodinated DNA probes have been developed for detection of acyclovir-resistant HSV isolates.

Biotinylated HSV probes for detecting the virus by **ISH** in genital lesions, eyes, and in cutaneous SCC have been constructed.

dsDNA probes labeled with **photobiotin** or **alkaline phosphatase-labeled synthetic oligonucleotide probes** can be used for ISH on paraffin-embedded tissue.

Commercially available HSV DNA probes may improve the routine diagnosis on clinical material (e.g., bronchoalveolar lavage and in paraffin-embedded tissue), particularly when cytological diagnosis is equivocal.

In one estimate, the **Enzo kit** has an overall sensitivity of 92% in comparison to tissue cultures. However, other workers reported the **poor performance** of the biotinylated Enzo HSV DNA probe (sensitivity, 24.5%; specificity, 88.3%) in comparison to a **shell vial culture assay** enhanced by a **direct fluorescent (HSV-monoclonal) antibody** stain, and conventional **tube cultures** argued against the its routine use for HSV identification.

The **Ortho DNA probe HSV confirmation kit** (Ortho Diagnostic Systems, Inc, Raritan, NJ) for detection of HSV in clinical specimens from various sources also was **not considered to be a suitable substitute** for conventional tube cell culture for detecting HSV in clinical specimens showing a sensitivity of 82% and a specificity of 100%.

Nevertheless, HSV DNA probes can be of help. Among the **complicating morphological features** resolvable by HSV DNA probes are

- the **similarity of HSV inclusions** to those of cytomegalovirus (CMV) and adenovirus
- **dual infections** with HSV and CMV

- **multinucleation** produced by viruses other than respiratory tract pathogens

HSV can be identified in **placenta** by **ISH** in the absence of characteristic morphological findings.

PCR has proved to be of great clinical value for the diagnosis and typing of HSV DNA in

- **chorioretinal inflammatory diseases**
- the early diagnosis of **HSV encephalitis**
- **occult dermatological lesions**

It also detects the presence of HSV type 1 in the **brains** of normal people as well as in those of Alzheimer's disease patients.

6.3.1.3.2 Varicella-Zoster Virus (VZV) (p. 487)

Traditional methods of diagnosing VZV rely on

- the **cytopathic phenomenology** in cell cultures, which may take 3 to 14 days
- **immunofluorescence microscopy (IF)**
- **electron microscopy (EM)**

IF and EM either require highly specific sera or lack viral specificity (EM).

^{32}**P**-, **tritium**- or **biotin**-labeled **DNA probes** derived from the VZV genome detect VZV in infected human cells in a **dot blot** hybridization assay.

The sensitivity of the radioactive VZV DNA probe ranges from 5 to 10 pg of viral DNA. The biotinylated probe is comparable in sensitivity (10 to 15 pg of viral DNA), and offers the results much faster than does the radioactively labeled probe.

PCR amplification demonstrated **latent VZV infection** in trigeminal and thoracic ganglia in a postmortem study, and identified the virus in vesicle samples, crusts, and throat swabs from patients with clinical **varicella**. It can be detected in PBMCs of adults; it is absent, however, from umbilical cord blood or from PBMCs of young children.

6.3.1.4 *CYTOMEGALOVIRUS* (CMV) (p. 487)

Human **CMV**, a typical member of the herpesvirus family,

- causes debilitating conditions in **newborns**, and is accompanied by growth retardation, microcephaly, and psychomotor disorders following primary infection in mothers during pregnancy
- may be present in multiorgan infections in **immunocompromised patients** and in **AIDS patients**
- has been detected in **the vascular tree** of otherwise healthy individuals
- may play a role in vascular damage and the formation of **atherosclerotic lesions**
- has been implicated in the pathogenesis of **type 1** and **type 2 diabetes**

The conventional diagnosis of CMV relies on

- isolation of the virus in **tissue culture**
- detection of virus-specific antibodies by **immunofluorescence**
- **ELISA**
- **fluoroimmunoassay**
- **latex agglutination**
- **indirect immunofluorescent assay**
- **enzyme-labeled antigen assay (ELA)**

6.3.1.4.1 Immunological Assays (p. 487)

Interference from **anti-CMV IgG antibodies** significantly undermines the demonstration of specific **IgM** in the patient's serum. These can be removed by

- **column chromatography**
- treatment of the specimens to be tested with goat **anti-human IgG** antibodies prior to anti-CMV IgM testing

The detection of **anti-CMV IgM** appears to be **of little value** for rapid diagnosis of CMV infection. Other, faster immunological methods such as the **rapid shell viral assay** should be preferred.

Antibody preparations from a number of manufactures vary in their quality. One comparison, for example, revealed that only the **Du Pont monoclonal antibody** showed 100% sensitivity and specificity and an absence of nonspecific reactions.

The detection of CMV viremia by its specific **cytopathic effects** takes on average over 36 days. In a large number of cases with laboratory-proven CMV infections, no cytopathic effect can be observed.

Early antigen immunofluorescence (IF) reduces the length of time required for virus identification from a typical 10 weeks to 4 to 6 days.

Monoclonal antibodies have been used for the **detection of early antigen fluorescent foci (DEAFF)** in cell cultures. Characteristic **nuclear fluorescence** is detected microscopically. In comparison with conventional cell culture assays taking several weeks this test can be performed within 24 h.

The **disadvantages of this test** are

- false-negative results due to a low titer of the virus
- false-negative results due to lack of specificity of the monoclonals
- failure of fibroblasts to grow
- variable susceptibility of the cells to the virus
- inability to test patients on antiviral therapy
- contamination

On balance, the DEAFF on urine, saliva, and blood samples fails to produce results in 6.5% of cases, whereas the cell culture method fails in 10.5% of cases.

Comparison of various immunological methods revealed that screening for CMV **early antigen antibody**, but not for CMV-specific IgM, can identify infected donors.

ELISA formats are considered convenient for large-scale screenings although the sensitivity of the CMV one-step assay should be increased. The latex agglutination test has the advantage of being extremely rapid.

An **indirect immunofluorescence test** has been used to resolve discrepancies between results of various immunological tests. The enzyme immunoassay and the latex agglutination tests had the highest sensitivity. For **large screenings** the latex agglutination was considered preferable.

The detection of antibodies to CMV can be achieved by **FCM** with a **biotin-streptavidin amplification** procedure using **phycoerythrin** as the fluorescent label. **Microsphere-associated fluoresence** can be quantitated by FCM and the sensitivity is at least sixfold that of conventional assays. However, it is difficult to anticipate wide acceptance of this approach in routine clinical laboratories.

Direct identification of CMV by monoclonal antibody has been adapted to a **dot immunoperoxidase assay** performed on nitrocellulose paper dotted with urine cell-free pellets.

It is apparent that the traditional immunological methods for CMV detection are relatively **lengthy**, **labor intensive**, and **unreliable**, when the virus is either not abundant or inactivated in transport to the testing facility or by the therapeutic measures, or when the patient fails to mount an immune response of adequate strength.

6.3.1.4.2 Blot Hybridization (p. 489)

Using different combinations of **urine sample preparation** involving extraction of the virus from **low speed**- and **ultracentrifuged** and untreated specimens, CMV DNA in urine was detected as a **free virus particle** as well as in a **cell-associated form**.

Nonradioactive versions of the assay included:

- **horseradish peroxidase** as label had a sensitivity of **80 pg** of homologous DNA
- a biotinylated *Bam*HI AB DNA probe of CMV Towne strain coupled with **horseradish peroxidase-streptavidin conjugate** detection was capable of identifying **60 pg** of CMV DNA
- a **biotinylated probe** derived from *Bam*HI restriction **fragment B** could detect **30 pg** of CMV DNA
- the biotinylated *Hin*dIII L fragment of CMV AD169 strain and **[^{125}I] streptavidin conjugate** detected **30 pg** of CMV DNA
- **alkaline phosphatase**-labeled probe detecting **10 pg** of isolated CMV DNA, when tested directly on urine samples
- spot hybridization with ^{32}P-labeled 32 kb probe representing the *Xba*I **fragment C** of the Towne strain. The probe was specific, did not cross-hybridize with other herpesviruses or human DNA, and could detect at least **10 pg** of CMV
- using a 11.7-kb *Hin*dIII CMV DNA **fragment L** as a probe derived from the laboratory strain **AD169 CMV**, as well as *in vitro*-synthesized RNA transcripts, a **dot blot** hybridization detected up to **3.2 to 10 pg** of CMV DNA
- other probes used two fragments representing the short and the long

unique segments transcribed early and late, respectively, during CMV infection. The sensitivity of the probe was **1 pg** of DNA — higher than that of the standard tissue-culture techniques

- ^{32}P-Labeled fragments of CMV DNA: *Hin*dIII L fragment (11.7 kb) of the coding sequence for the late viral polypeptides and *Eco*RI J fragment (10.6 kb) containing the coding sequence for the major immediate early RNA and other transcripts of the immediate-early region have been used. The purified probes were able to detect **500 to 750 fg** of cloned CMV DNA after 4 to 12 h of autoradiography

The **problems of probe assays** include

- a high rate of **false-negative** reactions
- **nonspecific interferences** producing **false-positive** reactions
- the length of the **extraction procedure** — 36 h
- **cross-reactivity** with HSV types 1 and 2
- **lack of correlation** between the amount of CPE in cell culture and the intensity of the color reaction of the probes
- **interference** of substances found in urine samples noted when using biotinylated probes
- **high cost**
- **length of performance** of assays

The **advantages of probe assays** include

- **Despite all the deficiencies the probe assay was superior to serology and culture in detecting CMV DNA.**
- Positivity in a dot blot hybridization assay proved to be an **earlier marker** than early antigen (EA) detection.
- A dot blot hybridization assay proved to be less tedious and to offer an excellent alternative to the **plaque reduction assay** in the evaluation of **antiviral activity** of various drugs against CMV.
- Dot blot hybridization has been adapted to produce **quantitative assessment** of CMV DNA in blood leukocytes of viremic patients by employing a **video densitometer-based technique** to monitor the effects of **antiviral therapy**. Up to 10 pg of CMV DNA could be detected even in the presence of microgram quantities of host cell DNA.
- **Antiviral drugs** do not interfere with the DNA probes that, in addition, allow the detection of CMV in sections of Kaposi's sarcoma by **ISH**, when culture methods fail.

The **mode of radiolabeling** CMV DNA probes affects their performance:

- **nick-translated** probes detect **1 to 10 pg** of homologous DNA, or the equivalent of 10 to 50 CMV-infected cells
- **random hexanucleotide-primed DNA probes** are more sensitive, with the detection limits of **0.1 to 0.5 pg**, or one to five CMV-infected fibroblasts

The diagnostic efficiency of **five different methods** for CMV infection of renal transplant and dialysis cases is as follows:

- DNA hybridization with radiolabeled probes — in 31.9% of patients
- immunofluorescence with monoclonal antibodies on centrifugation vial cultures — in 25.2%
- complement fixation test — in 23.4%
- virus isolation in conventional tube cell cultures — in 19.2%
- electron microscopy (EM) — in 2.1%

6.3.1.4.3 ISH (p. 492)

ISH is more sensitive in detecting CMV infection than **routine light microscopy** in Kaposi's sarcoma cases.

Sequential detection of CMV DNA [by ISH using a biotinylated commercial CMV DNA probe (Enzo Biochem, NY) and CMV antigens (immunological staining and a two-color technique for the identification of the virus in formalin-fixed, paraffin-embedded tissues)] allowed the definition of the **intracellular localization of** viral DNA. The detection of CMV infection by the probe proved to be a **more sensitive** technique than immunocytochemistry.

ISH assays for CMV in **PBMCs** yielded positive results 1 to 2 weeks **before the conventional assays** (virus isolation, CMV-specific IgM and IgG serology, and complement fixation).

The Enzo Biochem CMV DNA probe has been used to detect the virus in **bronchoalveolar lavage** (BAL) specimens by ISH. When compared to the results of tissue culture, a sensitivity of 90% with a specificity of 63% was obtained. Conventional **cytological examination** was markedly less efficient, revealing CMV inclusions in only 23%.

In contrast, other workers find that in **BAL** samples, the Enzo PathoGene CMV kit with biotinylated label proved to have a high specificity but poor sensitivity (62%).

DNA ISH is considered to be **less useful** for diagnosis than immunological assays because it was usually positive in cells displaying **cytopathic changes**, whereas EA immunostaining was helpful in the evaluation of early or focal cases of CMV colitis.

The differences between the results obtained with the **monoclonal detection system** and **ISH** are ascribed in large measure to the natural course of a CMV infection.

When two versions of nonradioactive CMV DNA probe labeling — **biotinylation** and **direct linkage of enzymatically active horseradish peroxidase (HRP)** — are compared to **indirect fluorescent monoclonal antibody** (IFA) and **direct FA staining**, only the IFA and ISH with direct horseradish peroxidase give consistent and reliable CMV detection as early as 45 h postinoculation.

Biotin-label ISH (PathoGene II; Enzo Biochem) performed **much worse** than did the HRP-DNA probe test (Vista-Probe, Digene Diagnostics). [*See Supplement 1*.]

ISH using biotinylated probes with two different detection systems (**alkaline phosphatase** and **peroxidase**) (Enzo Biochem) proved to be more difficult to perform, and in some instances more difficult to interpret, than immunohistochemistry for detection of CMV (HSV) in routinely processed tissues.

A **combination** of **digoxigenin** label with an **alkaline phosphatase**-labeled detection system using **anti-digoxigenin Fab fragments** has been found to be sensitive, specific, and to provide good resolving power on both **frozen** and **paraffin-embedded** tissue sections for detection of CMV infection in AIDS patients.

Radioactively labeled, but not biotin-labeled, probe is capable of identifying **reactivation** of CMV infection in liver biopsy where serological, histological, and culture methods fail.

ISH for CMV in **liver biopsies** using the biotinylated label in the PathoGene Kit (Enzo Biochem) proved to be **less sensitive**, although highly specific, when compared to immunostaining with a monoclonal antibody against an early CMV antigen.

In a different study, the same Enzo Biochem biotinylated CMV kit for ISH proved to be **as sensitive and fast as immunohistochemical evaluation** of immediate early antigen. Both techniques were comparable or even better than the serological detection of CMV.

The **consensus** appears to be that, given

- a simplification of the sample preparation for probe assay is achieved
- the label is made nonradioactive

the molecular biological approach will assume the leading role in clinical laboratory for the diagnosis and monitoring of CMV infection in tissues.

6.3.1.4.4 PCR (p. 494)

A PCR protocol for the identification of CMV in formalin-treated **peripheral blood PMNs** of patients with AIDS detects CMV DNA in 20 μl of blood. Amplification of a portion of the CMV genome before amplification of the detection probe allowed the detection of the PCR products from as few as 1 plaque forming unit or 10 CMV plasmid copies.

- The PCR has been used to determine HIV and CMV viruses on **fixed autopsy tissues** from AIDS patients following extraction of DNA.
- With primers specific for the **major immediate-early** and **late antigen genes** of the CMV genome, the PCR was able to detect the virus in 1 μl of urine supernatants from newborns without any pretreatment of the samples as well as in tissue culture of urine specimens.
- **Direct gel analysis** identifying 400- and 435-bp bands was possible on a 5 μl aliquot of a 10^{-3} dilution of the tissue culture mixture.
- **Dot blot hybridization** with ^{32}P-labeled oligonucleotide probes was positive at a 10^{-4} dilution of the same culture.
- **No cross-amplification** could be detected with other members of the herpes family of viruses (HSV, VZV, EBV) by direct gel analysis or by dot-blot hybridization of the PCR-amplified products.

Importantly,

- urine samples contain an **inhibitor** substance(s), which interferes with the PCR if not sufficiently diluted. Five and even 2 μl aliquots of urine yielded negative PCR results. A smaller volume (1 μl) reduced the inhibitory effect of the urine, and allowed the amplification products to be detected by direct gel analysis
- the overall **sensitivity** was 93% and the **specificity** was 100% relative to tissue culture results
- **dot blot assay of the amplified product increased the sensitivity** to 100%, raising the predictive value of a negative result from 90 to 100%

Advantages of PCR over tissue culture are as follow:

- the test can be completed within 24–48 h vs. 2–28 days, using a minute amount of urine
- the virus does not have to be viable
- the viral DNA does not have to be intact
- no special treatment of the sample is required

CMV PCR using primers produced by Collaborative Research (Waltham, MA) for the conserved region of the CMV major immediate-early gene detects six gene copies in 5 μl of **preheated urine** used directly in the reaction without further treatment before amplification. The assay could be completed within 5 h and showed no interference by other herpesviruses. Detection of the amplified products was made by direct gel observation. This PCR consistently detected CMV **directly in urine** specimens of renal transplant patients, when ELISA, nonradioactive DNA hybridization, and virus isolation results were variable.

Nested PCR assays significantly **reduce the background** that is usually produced by false positivity due to amplification of fragmented DNA. Discrepancies in reports from different laboratories on the efficiency of CMV PCR assays can in part be explained by the interference in the reaction. One ingenious method dramatically increases the efficiency of PCR by absorbing protein and other interfering substances on **glass beads** prior to recovery of DNA for PCR.

A PCR assay for CMV DNA can now detect as few as **three copies of the viral genome**, representing a 1000-fold increase in sensitivity over conventional DNA hybridization assays. By using two primer sets, including those for a CMV-specific **pp150 structural protein** not found in other herpesviruses, that particular PCR protocol was found to be highly reliable.

PCR is by far superior to culture assays in **monitoring ganciclovir therapy**.

Differences among strains of human CMV can be established by amplifying the **a-seq** region of the CMV genome with specific sets of primers. Isolates from different patients proved to be different. At this time no clinically significant correlations have been provided to CMV strain differences.

An **RT-PCR** detects **CMV mRNA** as early as 6 h after infection.

6.3.1.5 EPSTEIN-BARR VIRUS (EBV)

Interpretation of **serological findings** of EBV infection is particularly complicated in the presence of immunosuppression. Detection of the viral genome may be an invaluable tool in establishing the pathogenetic role of the virus.

6.3.1.5.1 Hybridization Methods (p. 496)

Southern blot and **ISH** definitively established the causative role of EBV in producing **liver cell necrosis** in a pediatric case of fulminant **infectious mononucleosis**.

A **slot blot** hybridization assay in **Hodgkin's disease (HD)** used the *Bam*HI W probe containing EBV DNA from a **variably repeated region** present in the genome of different isolates. Another probe (*Eco*RI B), which detected a **single-copy DNA** sequence in EBV DNA, was employed to **estimate the number of EBV genomes** present in positive cases. EBV DNA was detected both in cases **with** and **without clonal gene rearrangements** as well as in both the **nodular sclerosing** and **mixed cellularity**-type lymphomas.

Southern analysis detected EBV DNA in the lymph nodes of 29% of **HD patients**. Both nodular sclerosis and mixed cellularity histological types were positive for the virus. Significantly fewer (only 8%) of the **diffuse large cell** lymphomas harbored EBV.

An **ISH** study that used **tritium-labeled EBER-I antisense RNA** on archival formalin-fixed, paraffin-embedded HD specimens demonstrated that the majority of mixed cellularity and nodular sclerosis tumors displayed an intense signal in virtually all of the tumor cells. This finding suggests the virus has a role in growth regulation of Reed-Sternberg cells.

Indeed studies on **EBV-encoded latent gene products** suggest that EBV may be associated with more cases of HD than previously suspected.

Using two types of **biotinylated EBV DNA probes**:

- one cloned from the *Bam*HI-V (W) **large internal repeat region**
- the other cloned from the *Not*I/*Pst*I synthetic oligonucleotide probe from **tandem repeat regions**

a **stronger signal** confined to the nuclei of fewer lymphocytes was observed with the **synthetic** oligonucleotide probe pointing to **heterogeneous distribution** of gene amplification in EBV-infected cells.

DNA from **T cell lymphomas** was subjected to restriction endonuclease analysis (**REA**) to reveal T cell receptor and immunoglobulin gene rearrangements, and ISH was performed with a biotinylated *Bam*HI-V (W) fragment of EBV genome as probe. Detection of EBV DNA was combined with the identification of cell surface phenotypes. The study suggested that examination of T cell lymphomas for EBV DNA is indicated on a wider scale.

Dot-blot hybridization with the 28-kb *Eco*RI A fragment of EBV was used to estimate in oropharyngeal specimens the viral load in patients with **AIDS-related complex (ARC)** and **AIDS**. Limited sensitivity of the assay precluded consistent detection of EBV DNA in specimens with fewer than 10^5 to 10^6 genome equivalents per milliliter, as is the case in ARC patients.

Actually, the hybridization assay was inferior to the **lymphocyte transformation assay** for use in general screening for the presence of EBV, but adequate for **gross approximation** of EBV load in specimens with more than 10^6 EBV genome equivalents per milliliter.

Absence of the EBV DNA sequences from lymph nodes in HIV-positive patients and those with **persistent generalized lymphadenopathy (PGL)** suggests the virus apparently is not directly involved in the pathogenesis of PGL at the cellular level.

ISH for EBV DNA on swabbed specimens from **oral hairy leukoplakia** proved to be almost equal sensitivity to that of ISH on biopsy specimens. This offers an early alternative to identification of EBV in suspicious lesions usually appearing before overt manifestations of AIDS, thus avoiding undesirable invasive biopsy procedures.

The etiological role of EBV in **nasopharyngeal carcinoma** has been studied by an ***in situ* cytohybridization assay** with nonradioactive **digoxigenin-labeled** *Bam*HI fragments of EBV DNA as a probe. The results are available within 96 h compared to 5 weeks for ^{3}H-labeled ISH probes used in the **autoradiographic** approach. EBV DNA is found in malignant epithelial cells of nasopharyngeal carcinoma as well as in the lymphocytes infiltrating the tumor and in some morphologically unaltered epithelial cells in the tumor biopsies.

6.3.1.5.2 PCR (p. 497)

PCR amplification of the **long internal direct repeat region** of the EBV genome has proved its applicability to formalin-fixed, paraffin-embedded tissues. The use of PCR on **paraffin-embedded** tissue for EBV identification **reduces the sensitivity** of detection. Optimally, DNA extracted from **frozen** tissue should be used.

Primers constructed for the conserved regions of the EBV genome encoding **capsid protein gp220** and **EBNA-1** afford a sensitivity of 10 to 50 copies of the EBV genome.

Further sophistication of a PCR assay for EBV DNA comes from the construction of **three pairs of primers** specific for genomic sequences coding for the two forms of **EBNA**, 2A and 2B, and for a DNA sequence from the *BamZ*/*Bam*R region that ensures a reliable identification of type A and B viruses.

PCR identified EBV DNA sequences in 41 undifferentiated **nasopharyngeal carcinoma** cells, in 2 of 4 moderately differentiated tumors, and in 3 of 5 keratinized specimens.

Interestingly, up to 23% of **normal healthy adults** have been found by PCR to harbor EBV DNA in samples of **oropharyngeal cells**.

An undifferentiated **lymphoepithelial gastric carcinoma** also resembles that of the nasopharynx in that it harbors EBV DNA detectable by PCR, in distinction to **adenocarcinomas with lymphoid stroma**.

Tissue specimens retrieved from **infectious mononucleosis** patients, **renal** and **heart transplant** patients, and a **nasopharyngeal carcinoma** patient all showed the presence of EBV. Histologically normal spleen and lymph node tissue used as controls were negative for EBV sequences.

PCR identifies EBV DNA in blood and tissue biopsies from the **Sjögren's syndrome**, **infectious mononucleosis**, **HD**, **immunocompromised**, and **transplant** patients, and in patients with various **lymphoproliferative disorders** (**lymphomatoid granulomatosis** and **angioimmunoblastic lymphadenopathy**).

The detection of EBV DNA obtained by ISH, Southern analysis, and PCR has been compared in **lymphoid tumors**, **benign lymphadenopathy**, and normal **thymuses**:

- ISH was equal to Southern blot analysis
- both ISH and Southern analysis were inferior to PCR in their efficiency to detect EBV DNA
- **Southern analysis** may have a **greater discriminatory power** in assigning the presence of EBV DNA sequences to a specific cell type (e.g., T cells vs. B cells).

6.3.2 *MYCOPLASMA* SPECIES (p. 499)

Deficiencies of **cell culture** methods and mycoplasma DNA **staining** for identification of these pathogens include

- **poor reproducibility**
- **ambiguity** (ca. 11%)
- **false positivity** (ca. 2%)
- **false negativity** (0.7 to 2.4%)

The detection of mycoplasmas by **DNA probes** directed to the **total DNA** or **rRNA gene** sequences has long been described. A **commercially available** DNA probe assay (Gen-Probe, San Diego, CA) for ***M. pneumoniae*** favorably compares with culture methods.

Although little, if any, cross-hybridization could be noted with specific DNA probes for *M. pneumoniae* and ***M. genitalium*** using radioactive labels. The same probes labeled by **biotinylation** or **sulfonation** have an order of magnitude lower sensitivity.

PCR amplification assays for *M. pneumoniae* offer a highly species-specific identification of the organism in **bronchoalveolar lavage (BAL)** with a sensitivity of 10^2 to 10^3 organisms, or even better — 35 genome copies.

6.3.3 *LEGIONELLA* SPECIES (p. 499)

Cultivation of *Legionella* is laborious and identification requires the use of **direct fluorescent antibody tests** or **gas chromatography**.

DNA probes selected by subtractive hybridization have been tested in **colony** and **dot hybridization**. They are also potentially useful for direct identification of ***Legionella pneumophila*** in clinical and water samples.

A rapid and simple dot blot hybridization assay using photobiotin-labeled DNA probes can be performed in a clinical laboratory. Among 500 isolates, 20 strains were identified as *L. pneumophila* where serological methods failed.

The Rapid Diagnostic System for *Legionella* offered by Gen-Probe utilizes radioiodinated cDNA as a probe for rRNA sequences specific for members of the genus *Legionella*. The sensitivity and specificity of the Gen-Probe assay were found to be 93.9 and 97.4%, respectively, in an experimental system.

The DNA probe was considered a practical alternative to a direct fluorescent antibody assay for rapid diagnosis of *Legionella* infections.

The occurrence of false-negative results emphasized the importance of testing multiple specimens and underscored the role of culture in *Legionella* diagnosis.

A **microplate technique** using microdilution plates with immobilized unlabeled reference DNAs has been developed for the identification of *Legionella* at the species level.

DNA hybridization allowed the identification of a new isolate from a cooling water tower that proved to be a new *Legionella* species given the name ***L. adelaidensis.***

Restriction analysis has been useful in tracking down the source of various isolates linked to nosocomial *Legionella* infections and proved to be easier to perform than **alloenzyme typing**.

Pulsed-field gel electrophoresis (PFGE) using *Not*I and *Sfi*I digests of *L. pneumophila* DNA has been used for the detection of nosocomial *L. pneumophila* in water supplies on the basis of restriction patterns.

By selecting **PCR** primers to the **macrophage infectivity potentiator (*mip*) gene**, a good correlation was established with culture methods in the differentiation between viable and dead *L. pneumophila*. **PCR** amplification for detecting *L. pneumophila* in water is possible even following biocide or elevated temperature treatment. **Nonviable** cells did not show PCR amplification, whereas culturable and **viable** noncultivable cells could be detected by the PCR.

6.3.4 *BORDETELLA* SPECIES (p. 500)

The identification of ***Bordetella pertussis*** and ***Bordetella parapertussis*** by a DNA hybridization assay without preculturing the organisms was only marginally successful (38% positives compared to serological assays) and compared to the 50% sensitivity for culture methods.

A higher sensitivity of the assay reaching 69% could be achieved at the expense of specificity, however, following 24 to 72 h **preculture** of nasopharyngeal aspirates from patients with suspected pertussis.

6.3.5 *MORAXELLA (BRANHAMELLA) CATARRHALIS* (p. 500)

Moraxella (Branhamella) catarrhalis is a respiratory pathogen identified in patients with underlying **pulmonary disease**, in **immuno-compromised** patients, as well as in children with **otitis media** and **sinusitis**.

A **synthetic oligonucleotide DNA probe** derived from a random fragment of **chromosomal DNA** 100% specific to the *M. (B) catarrhalis* species has been developed. This gene probe can make *M. catarrhalis* testing definitive on pure cultures.

6.3.6 FUNGI

6.3.6.1 *HISTOPLASMA CAPSULATUM* (p. 501)

Using **differential hybridization**, several yeast phase-specific genes of ***Histoplasma capsulatum*** have been cloned including Yps-3, the one apparently determining the pathogenicity of the organism. The specificity of this 1.85-kb nuclear DNA probe is helpful not only in classification studies, but may be used for the identification of *H. capsulatum* as well.

6.3.6.2 *CRYPTOCOCCUS* SPECIES (p. 501)

A detailed study of **restriction patterns** of **rDNA** sequences of several *Cryptococcus* species amplified by **PCR** proves the utility of this approach for

- **taxonomic** and **phylogenetic evaluation**,
- **species**- and **strain-specific** identification within *Cryptococcus* species

REA of the three separate regions of the **rDNA repeat** (**A**, **B**, and **C**) revealed that ***C. neoformans*** had no variation among the 19 strains tested.

While *C. neoformans* REA patterns show only minor variations, considerable variability exists among **PCR fingerprints** in the other *Cryptococcus* and *Candida* strains examined.

6.3.6.3 *NOCARDIA ASTEROIDES* (p. 501)

Because the nocardiae may produce infections indistinguishable from those caused by the **mycobacteria** in both **immunocompetent** and **immunocompromised** hosts, a gene level evaluation of this slow-growing organism may be helpful.

The detection of diagnostic **metabolites** of ***N. asteroides*** frequently may be outside the capabilities of routine clinical laboratories: **HPLC** can be helpful, as are **serological** methods.

Two **recombinant clones** have been constructred that identified 31% of the *N. asteroides* strains in a reference collection. Additional clones were then selected to provide pooled DNA probes capable of identifying all the strains tested.

Efforts are being directed toward the development of an amplification procedure for *N. asteroides*, using some of the probes as primers.

6.3.7 *MYCOBACTERIUM TUBERCULOSIS* (p. 501)

The resurgence of **tuberculosis (TB)** infection adds to the challenge of the fastest identification of specific mycobacterial infection.

The **Ziehl-Neelsen staining** is insensitive and nonspecific, requiring over 10^4/ml organisms to be present in the specimen. Only half of patients with pulmonary TB and a quarter of those with extrapulmonary disease are identified by the Ziehl-Neelsen and **rhodamine-auramine stains.**

Over half of the estimated 8 million TB cases per year worldwide are smear-negative pulmonary and extrapulmonary diseases.

Detection by **latex agglutination, radioimmunoassay**, **enzyme-linked immunoasorbent assay**, and **serological methods** also suffers from insensitivity and low specificity.

Improvement of **immunological methods** enhance sensitivity and specificity and offer procedural simplicity of **serological** approach appealing to routine clinical laboratories.

One of the latest improvements of existing **ELISA**-based techniques directed to **immumodominant specific antigens** of *M. tuberculosis* is the **sandwich ELISA** adaptation of the **solid-phase antibody competition test** (SACT-SE).

The **superior diagnostic ability of the RFLP** technique over serotyping was demonstrated by failure of the latter approach to establish the genetic identity of mycobacteria.

6.3.7.1 DNA PROBES (p. 502)

Whole-chromosomal DNA probes for rapid identification of ***Mycobacterium tuberculosis*** and ***Mycobacterium avium*** complex have been used in **dot blot** hybridization. Assays are completed within 48 h and the results can be significantly improved by lysing the cultures prior to application of the probes.

A **biotin**-based detection system recognized not only *M. tuberculosis*, but also **mycobacteria other than tuberculosis (MOTT)**. One of these probes recognizes the closely related *M. tuberculosis*, *M. bovis*, and *M. microti*, but no other mycobacterial species. The second probe was specific for *M. malmoense*, and yet another probe, in addition to *M. tuberculosis*, hybridized to *M. intracellulare, M. malmoense, M. scrofulaceum, M. simiae, M. xenopi, M. avium, M. szulgai, M. kansasii*, and *M. hemophilum*.

Differentiation of *M. tuberculosis* from *M. bovis* can be performed by REA using ***Pst*I** and ***Hae*III RFLPs.**

Certain DNA fragments can be used either as *M. tuberculosis* **complex-specific probes** or as **panmycobacterial probes**, depending on the stringency condition of hybridization and washing conditions:

- at **high-stringency** conditions (0.1 × SSC, 65°C), a random fragment of *M. bovis* DNA generated by sonication and cloned in the M13 vector hybridizes to 27 isolates of *M. bovis* and *M. tuberculosis*, but not to DNA from *M. avium, M. fortuitum*, or *M. phlei*
- at **low stringency**, the probe also hybridizes to nonmycobacterial species

Adaptation of **quantitative microdilution plate hybridization** to immobilized reference DNAs for species identification of mycobacteria uses colorimetric detection of hybrids.

A combination of **sonication of mycobacteria in chloroform** followed by probing for the **rRNA genes** in **dot blots** of the lysate gave the highest sensitivity in detecting sequences homologous to the DNA probes. The probes generated by restriction digestion of the complete rRNA operon from *M. smegmatis* afford the detection of 200 organisms, which is comparable to some PCR based methods.

A **kit** available from Gen-Probe uses ^{125}I-labeled *M. avium* and *M. intracellulare* DNA probes for hybridization with rRNA from a test organism. The kits are fast, highly sensitive, and specific in identifying *M. avium-M. intracellulare* complex and *M. tuberculosis* grown in culture.

By combining the Gen-Probe method with the **BACTEC** radiometric technique, most isolates of *M. tuberculosis, M. avium*, and *M. intracellulare* can be identified within 4 weeks in contrast to the 9 to 11 weeks usually required with conventional media and biochemical identification.

A combination of BACTEC and Gen-Probe assays allows over 83% of specimens containing clinically relevant mycobacteria to be identified within 7 days for *M. avium-M. intracellulare* isolates and within 18 days for *M. tuberculosis* isolates.

Further improvement is offered by **lysing** and **concentrating** mycobacteria from blood specimens grown in BACTEC 13A bottles prior to testing with the Gen-Probe diagnostic system. **Parallel subculturing** of positive BACTEC bottles is suggested for review of mixed infections, such as are encountered in AIDS patients, positive for both *M. tuberculosis* and *M. avium* complex.

Although this technology still requires cultivation of the organism from the patient isolate, the lack of nonspecific reactions and savings in time and labor make the **DNA probe approach a valuable alternative.**

6.3.7.2 PCR DIAGNOSIS OF MYCOBACTERIA (p. 504)

Because a reliable identification by the Gen-Probe method requires the presence of at least 10^6 organisms in the test sample, the practical sensitivity is still too low to make this or other such methods appealing in a clinical setting. They can be successfully used on cultured specimens for the fastest final identification, but not directly on a patient's sample. For example, the **time factor** becomes particularly important in the outcome of **tuberculous meningitis**.

Specific primers and reaction conditions for identification of the *M. tuberculosis* complex bacteria by the PCR have been developed for various targets, making it the method of choice for the fastest, most specific, and most sensitive detection of the organism.

The first primers for PCR assays were based on the evolutionarily conserved **65-kDa heat shock gene**. Selective **oligonucleotide probes** specific for *M. tuberculosis-M. bovis, M. avium-M. paratuberculosis*, or *M. fortuitus* strains were used for specific identification of the amplified PCR product.

The **insert IS 6110** of the **repetitive sequences** in the genome of members of the *M. tuberculosis* complex *(M. tuberculosis*, 10 to 12 copies; *M. bovis*, 1 to 3 copies) has been the target of **PCR** amplification for diagnostic purposes. Oligonucleotides derived from this sequence were used as primers for the PCR assay.

Significant conservation of the insert IS 6110 across strain and species lines allows **RFLP analysis** of this sequence to be used in fingerprinting various strains of the *M. tuberculosis* complex.

An *M. tuberculosis* **DNA insert in the λ gt11 gene** library has been identified that specifically hybridized with DNA from mycobacterial species belonging to the *M. tuberculosis* complex. Parts of this insert (**158 bp**), which was found to occur in many copies in the mycobacterial chromosome, were used as primers for the PCR.

These PCR assays, performed on sputum samples and lung tumor biopsy, had a detection limit, assessed on *M. tuberculosis* **chromosomal DNA**, of 100 fg, corresponding to approximately 20 bacteria. When whole bacteria are used for reference material, the sensitivity fell to 200 bacteria for cells suspended in buffer and to 1000 bacteria for cells suspended in sputum.

Even the level of sensitivity of 1000 bacteria represents an **improvement of three orders of magnitude** over the detection methods that use DNA or oligonucleotide probes only for hybridization assays.

This amplification assay remained positive 16 weeks after specific **isoniazid-rifampin-pyrazinamide** treatment, when no bacteria could be cultured or detected by Ziehl-Neelsen staining.

Another PCR method targets the nucleotide sequences of the gene for the 38-kDa protein antigen **b (Pab)** of *M. tuberculosis*. No cross-reactivity was noted with 27 strains of other mycobacteria, and the sensitivity of the assay was less than 10 mycobacterial cells on serial dilutions of a culture containing approximately 4×10^6 *M. bovis* BCG cells per milliliter.

A promising approach to **direct** and **specific identification** of mycobacteria (although still applied to cultured bacteria) is based on the **amplification of rRNA**. rRNA sequences are characteristic for a given organism and can be used for fine taxonomic differentiation as well as for the identification of microorganisms, especially considering the abundance of rRNA in the cell (10^3 to 10^4 molecules per cell).

Based on the **16S RNA** sequences, a highly sensitive and specific **RT-PCR assay** has been developed for distinguishing closely related taxa within the genus *Mycobacterium*. The subsequent identification of the amplified DNA fragment was performed by **hybridization with oligonucleotides**, the specificity of which was defined at a genus, group, or species level. Fewer than 100 mycobacteria could be detected by this assay.

No evaluation of this assay has been carried out on clinical specimens at the time of this writing but the general idea of targeting rRNA sequences for enhancing the sensitivity of an assay has been used by Gen-Probe testing systems. In fact, the **Gen Probe rapid diagnostic system** has been used to reassign various serovar strains of the *M. avium* complex.

PCR amplification has targeted **repeated nucleotide sequences** not only in rRNA, but also in mycobacterial DNA. Using a segment of one of the cloned repeats of *M. tuberculosis*, a **PCR assay based on the repetitive 123-bp fragment** has been developed. The detection limit of the assay was 1 fg of purified target DNA, which is equivalent to about one copy of the *M. tuberculosis* chromosome (~3000 kb). The assay, however, was greatly affected if applied directly to the patient's sample, rather than to purified DNA.

Two **genomic DNA probes** for PCR identification of mycobacteria have been isolated:

- one cloned fragment (**pMAv 17**) was recognized by the three mycobacterial species (*M. tuberculosis, M. avium*, and *M. intracellulare*), and was selected as a **genus-specific** probe
- the other two clones (**pMAv 22** and **pMAv 29**) hybridized predominantly to *M. avium*, and are used as species-specific probes

These probes offered the following **advantages**:

- the banding pattern in gel electrophoresis produced by the amplified product of the *M. tuberculosis* complex was distinct from that of other mycobacteria
- PCR amplification reduced the time of required X-ray film exposure from 24 h to 30 min
- the sensitivity was 1 fg *M. avium* DNA by PCR, which corresponds to **less than a single genome** of mycobacterial DNA. In contrast the direct assay, without PCR amplification and using only specific DNA probes, shows a sensitivity of about 1 mg of DNA. This corresponds to 2.5×10^6 mycobacteria — the number of organisms that can be detected by light microscopy

Some PCR procedures allow the fastest identification of a *Mycobacterium* strain on the basis of **agarose gel electrophoresis patterns** generated by the amplified products without the need for confirmation by hybridization.

Some other reported PCR assays are:

- a detection protocol unique for *M. tuberculosis* based on the PCR product generated at **low primer-annealing temperatures**
- another assay targets a 240-bp fragment that encodes part of the **MPB 64 protein** and a 17-mer oligonucleotide complementary to the central portion of the amplified product used as a probe
- amplification of a **336-bp repetitive fragment** in the *M. tuberculosis* chromosome formed the basis of an assay capable of detecting the amount of DNA corresponding to **less than 10 organisms**. This assay proved to be **at least as sensitive as conventional culture** detection methods on the same specimens
- the application of PCR to **uncultured clinical specimens** using as primers two 24-bp synthetic oligonucleotides derived from a sequence coding for the 65 kDa antigen of *M. tuberculosis*. The assay distinguishes *M. tuberculosis* and *M. bovis* BCG from all other species of

mycobacteria. As few as 40 *M. tuberculosis* bacteria (0.1 pg of DNA) can be identified even in the presence of DNA equivalent to 10^6 human cells

- a **nested PCR** using two pairs of primers to the insertion sequence IS 6110 for the detection of *M. tuberculosis* in a case of **tuberculous pericarditis**. DNA had to be extracted from pericardial fluid, and the diagnostic band could be identified in the patient's specimen within 2 days
- in the case of **tuberculous meningitis (TBM)**, the PCR assay compares favorably with conventional bacteriology and ELISA for cerebrospinal fluid (CSF) antibodies. In fact, it proved to be the most sensitive technique, exceeding by some 20% the efficiency of detection by culture (75 vs. 55%). The 20 to 30% false-positive rate of both antigen and antibody detection assays in high-prevalence areas is definitely inferior to almost 100% specificity of the PCR
- **yet another PCR assay** for TBM uses primers for the MTB 64 protein gene sequences
- compared to the standard amplification protocol, the **nested PCR** assay for the 65-kDa antigen was much more sensitive, yielding 13 positive results among 35 samples from tuberculosis patients that tested negative by culture. The standard PCR procedure was positive in only one of six culture-negative specimens
- a PCR procedure using genus-specific or *M. avium*-specific probes has been adapted to identify **whole mycobacteria** grown under different culture conditions without prior lysis or isolation of nucleic acids from the organisms
- significant **reduction of false-negative** results and improved efficiency of amplification in specimens containing **inhibitors** of PCR can be achieved by pretreatment of clinical samples with **guanidium thiocyanate**

6.3.8 *MYCOBACTERIUM PARATUBERCULOSIS* (p. 506)

The role of ***M. paratuberculosis***, the etiological agent of **Johne's disease** and a chronic enteritis of ruminants, in human inflammatory bowel disease has caused some controversy.

The difficulty of separating *M. paratuberculosis* from ***M. avium*** complex organisms is due to the close homology of their DNAs; however, a clone, **pMB22** containing an **insertion sequence IS900** that is a highly specific marker for *M. paratuberculosis*, has been defined. It is present in multiple copies in *M. paratuberculosis* strains, but is absent from *M. avium* DNA.

A PCR assay for the identification of *M. paratuberculosis* targeted this insert in the first fast (a few hours) and accurate test for the assessment of *M. paratuberculosis* in animal and human disease.

Analysis of archival material of **tissues from Crohn's disease by PCR failed, however, to identify any mycobacteria, including *M. paratuberculosis*.**

6.3.9 *MYCOBACTERIUM LEPRAE* (p. 507)

In the case of *Mycobacterium leprae* neither **monoclonal antibodies** nor **DNA probes** for hybridization offer the required sensitivity.

A **nested PCR assay**, using DNA isolated from cultivated organisms, detects as few as 40 bacilli of *M. leprae*.

Another **two-step PCR** assay for *M. leprae* uses two nonradioactive outside and two inside **nested primers**. As little as 3 fg of *M. leprae* genomic DNA could be detected, which corresponds to a single bacillus. In **crude lysates** of *M. leprae* bacilli, the assay was capable of identifying as few as 20 organisms.

Amplification of a 530-bp fragment of the gene encoding the proline-rich antigen of *M. leprae* can be performed on DNA from *M. leprae* in clinical **fixed** or **frozen biopsy** samples from leprosy patients.

6.3.10 PARASITES

6.3.10.1 *PNEUMOCYSTIS CARINII* (p. 507)

Pneumocystis carinii has acquired particular notoriety as the major cause of fatal pneumonia in immunocompromised patients due to either **immunosuppressive therapy**, **cancer**, or **AIDS**.

So far, conventional identification of the organism relies essentially on

- **direct microscopy** of silver- or Giemsa-stained preparations of bronchoscopic lavage or sputum
- **monoclonal antibody reagents**
- **immunofluorescence**

A **PCR** assay for *P. carinii* includes oligoblotting for enhancing the specificity. **Oligoblotting** increased the sensitivity of this PCR assay 100-fold compared to that obtained with **ethidium bromide** band staining.

An **ISH** protocol, using biotinylated oligonucleotide probes for **rRNA**, has demonstrated its utility in complementing conventional staining and immunohistochemical methods for the diagnosis of *P. carinii*.

6.3.10.2 *ECHINOCOCCUS MULTILOCULARIS* (p. 507)

Using differential antibody screening of cDNA clones of *Echinococcus multilocularis*, two potentially **immunodiagnostic antigen gene clones** designated EM2 and EM4 have been identified.

While the EM2 clone proved to be nonspecific, the EM4 native antigens are coded for by a single-copy gene showing no homology with other parasites.

A highly specific **PCR** assay for *E. multilocularis* that uses two primer sets has been developed; it is capable of detecting 50 pg of parasite DNA in a variety of sources. Oligonucleotides used in the assay can also be helpful in assessing strain variation, RFLPs and other manifestations of genetic variation in *E. multilocularis*.

Chapter 6.4

Sexually Transmitted Diseases

6.4.1 *TRICHOMONAS VAGINALIS* (p. 509)

A number of diagnostic modalities exist for the diagnosis of *Trichomonas vaginalis*:

- **microscopy** of wet mounts or stained preparations
- *in vitro* **culture**
- **immunological methods**

All these methods have difficulty identifying the organism when a low number of *T. vaginalis* are present.

A **2.3-kb *T. vaginalis* DNA fragment** has been cloned and used for the diagnosis of trichomoniasis by **dot blot** hybridization on vaginal discharge specimens. This radioactively labeled probe was sensitive enough to detect 200 *T. vaginalis* isolates.

6.4.2 *HEMOPHILUS DUCREYI* (p. 509)

The identification of *Hemophilus ducreyi* in chancroid lesions is hindered in part by the **autoagglutination** of the organism in clumps when suspended in liquid. This precludes the use of **serological agglutination** tests and interferes with the interpretation of **fluorescent antibody** tests.

The epidemiology of chancroid lesions can be studied by **ribotyping** *H. ducreyi* using genomic fingerprints probed by rRNA from *Escherichia coli*.

A **DNA probe assay** for this organism using radiolabeled DNA fragments from a virulent strain of *H. ducreyi* could identify the organism with a sensitivity of 103 colony-forming units (cfu) in **colony hybridization assays** in pure or mixed cultures.

An even higher sensitivity must be attained, however, to detect the bacilli directly in a clinical specimen with an **amplification assay**.

6.4.3 *CHLAMYDIA* SPECIES (p. 509)

A variety of laboratory techniques are available for the isolation and detection of chlamydiae, which frequently present significant diagnostic challenges.

Automated fluorescence image cytometry has been described for DNA quantitation and detection of chlamydiae.

Total **sulfonated** *Chlamydia trachomatis* **DNA** has been used as a probe for the ***in situ* hybridization (ISH)** detection of the parasite on microscope slide-mounted, experimentally infected cells as well as in clinical specimens.

Visualization of the probe was done by **enzymatic sandwich immunodetection** with alkaline phosphatase as the reporter molecule. **Intracellular inclusions** could be detected within 8 h of inoculation of cultured cells by using a high probe concentration (10 μg/ml). The ISH results on clinical specimens correlated with the identification of *C. trachomatis* by culture.

Synthetic oligonucleotide probes have been constructed for the 7.4-kb plasmid of *C. trachomatis*.

A multistep assay using a **dioxetane detection system** offers substantial advantages over colorimetric detection with other substrates, including **horseradish peroxidase:**

- following **solution hybridization** with unlabeled probes, the probes-target duplex can be captured onto a microtiter dish surface and, using a **novel signal amplification technique** and an **alkaline phosphatase-labeled probe**, the organism can be quantitated
- eventually the **chemiluminescent reporter dioxetane substrate** is triggered by the enzyme and the signal is registered by a luminometer or by exposure to instant film. This 5-h assay is specific for all serovars of *Chlamydia*, detecting as few as 10 to 20,000 chlamydial elementary bodies

A **commercial DNA probe** for *C. trachomatis* (Gen-Probe **PACE**, Gen-Probe, San Diego, CA) uses single-stranded DNA (**ssDNA**) labeled with **acridinium ester**, complementary to *C. trachomatis* **rRNA**.

However, based on the number of probe-positive results that could not be confirmed by culture, the Gen-Probe PACE was not recommended for use in screening for *C. trachomatis* in women with a low to moderate risk of infection. A comparison of the Gen-Probe method with culture in **direct testing** of clinical specimens revealed a particularly diminished sensitivity (68%) and specificity (75%) in the male population when compared to culture. The other disadvantage of the probe assay is that the quality of the specimen cannot be assessed.

PCR amplification adapted for *Chlamydia* species is capable of detecting one chlamydial DNA molecule in 105 cells. No cross-reactivity was observed. This extremely fast (less than 1 h) and sensitive (10^{-18} g of a sequence of *C. trachomatis* plasmid DNA) assay allowed the resolution of one bacterium-equivalent, with the bands clearly visible on ethidium bromide-stained gels.

PCR amplification of DNA sequences from the chlamydial genome using **degenerate oligonucleotide primers** has been used to study regulation of *C. trachomatis* RNA polymerase production.

A **PCR-EIA protocol** combines detection of PCR-amplified DNA by an enzyme immunoassay, and proved to be highly sensitive and specific even when applied **directly to cervical specimens**.

The **advantages** of PCR-EIA are:

- solution hybridization
- capture of the amplified product by an immunological method
- absence of radioactive labels
- quantifiable format

These features make the PCR EIA approach in general an attractive system for the evaluation of infectious processes in a clinical setting, and it has been used to diagnose and monitor **trachoma.**

PCR assays for *C. trachomatis* infection show almost complete agreement with the results of culture methods in evaluating the effect of **doxycycline** treatment.

PCR amplification combined with **restriction endonuclease analysis (REA)** of the amplified *C. trachomatis* DNA can be used for **genotyping**.

Chlamydia psittaci has been specifically detected by a two-step PCR that did not cross-react with *C. trachomatis* and ***Chlamydia pneumoniae.***

6.4.4 *NEISSERIA* SPECIES

6.4.4.1 *NEISSERIA GONORRHOEA* (p. 511)

Five **rapid immunological diagnostic systems** have been evaluated:

- **quadFERM** (API Analytab Products, Plainview, New York)
- **rapid NH system** (Innovative Diagnostic Systems, Atlanta, Georgia)
- **Gonochek II** (E.I. Du Pont de Nemours and Co., Wilmington, Delaware)
- **RIM-N kit** (Austin Biological Laboratories, Austin, Texas)
- **Phadebact Monoclonal GC OMNI test** (Pharmacia Diagnostics AB, Uppsala, Sweden)

The best overall performance was found with the quadFERM test; however, two other kits, RIM-N and Rapid NH, also were 100% sensitive and specific on clinical isolates.

A **dot blot** DNA-DNA hybridization format and **rRNA-derived DNA probes** specific for the less conserved regions of rRNA were combined in an assay system that could either exclusively react with *Neisseria gonorrhoea* DNA or could detect other bacterial species.

rRNA-derived DNA probes differentiated *Neisseria* isolates at the species, and even **subspecies, level** without the use of Southern analysis.

Several groups have conducted comparative studies on the Gen-Probe **PACE (Probe Assay-Chemiluminescence Enhanced)** system for *N. gonorrhoea*. It demonstrated sensitivity, specificity, and positive and negative predictive values of 93, 99, 97, and 99%, respectively.

The **PACE2** Gen-Probe assay for *N. gonorrhoea* exhibited increased sensitivity and performed virtually identically to the culture method in both symptomatic and asymptomatic populations. It is considered, therefore, suitable for the screening and diagnosis of gonorrheal genital infections in women.

A different **DNA digoxigenin probe assay** is designed to evaluate a **cryptic plasmid-encoded segment of gene *cppB*** sequences, which are also found in the chromosome. The level of sensitivity of that assay is 25 pg of

DNA or 500 cfu of *N. gonorrhoea*. All *N. gonorrhoea* strains, even those free of the plasmid, tested positive with this probe. The sensitivity of the assay was 95%, with a specificity of 98%, in an STD clinic patients. The entire assay takes less than 5 h to perform.

Other probe systems for the detection of *N. gonorrhoea* using genomic DNA sequences coding for the **gonococcal pilin** showed cross-reactivity with *Neisseria meningitidis*.

6.4.4.2 *NEISSERIA MENINGITIDIS* (p. 512)

A **DNA probe, pUS210,** containing a 2-kb insert from *N. meningitidis*, has been used for **epidemiological studies** of outbreaks of meningococcal disease.

A different DNA probe, **reacting with all meningococci tested**, has also been described.

The results obtained with these two different types of probes were **discordant**, emphasizing the need for careful analysis of RFLP data in epidemiological studies. A **PCR assay** targeting flanking sequences of the **dihydropteroate synthase gene** allows the earliest diagnosis of **meningococcal meningitis** when CSF cultures are negative.

6.4.5 *TREPONEMA PALLIDUM* (p. 512)

A **PCR** amplification assay capable of detecting 0.01 pg of purified *T. pallidum* DNA and even a single treponeme in a suspension has been applied to serum, CSF, and amniotic fluid from syphilis patients as well as to paraffin-embedded tissue.

The test is somewhat complicated by the need for **confirmatory DNA-DNA hybridization** with the **tpp 47-specific probe** for the enhancement of its sensitivity and specificity; the test does not rely solely on **ethidium bromide** identification of the amplified product on the gel.

Compared with the rabbit infectivity test, the PCR test for *T. pallidum* is 100% specific. The combined sensitivity for all clinical specimens is 78%.

Another PCR assay directed to the DNA sequence of **the 39-kDa *bmp* gene** is capable of detecting *T. pallidum* DNA in CSF for up to 3 years following intravenous administration of penicillin.

6.4.6 HUMAN IMMUNODEFICIENCY VIRUS 1 (HIV-1)

6.4.6.1 IMMUNOLOGICAL ASSAYS (p. 512)

The initial approach to the identification of HIV infection was to use **electroimmunoassay (EIA), enzyme-linked immunosorbent assay (ELISA), western blot,** and **immunofluorescence** — all of which determine the presence of antibodies produced in the patient against various structural components of the virus.

However, instances of misidentification of HIV-2 proteins by western blot analysis have been reported. More recently, **recombinant** or **synthetic peptides** have been used, instead of native proteins, in the development of serological tests.

Specific antigen epitopes are targeted in the most informative assays designed to evaluate the status of infection and, possibly, predict its future progression. Several **immunodominant proteins** have been the subject of close scrutiny in this regard: **p24 (gag), p66/p51 (pol), gp41 (transmembrane),** and **gp120 (envelope)** proteins and, more recently, **p17 (gag** or **shell** protein). Peptides representing **HIV-1 p17 epitopes** have been used to develop novel assays to measure HIV-1 p17-related antibodies and antigens in patient serum. The definitive test for an active HIV-1 infection is **the recovery of HIV or provirus DNA from the infected person**.

The **coculture** methods are the most sensitive in isolating HIV-1 from patients. The genetic variability of HIV-1 accounts for the variability in proviral *gag* DNA epitope sequences in the cytotoxic T cells. These may not be recognized by autologous T lymphocytes, leading to **immune escape.**

6.4.6.2 HIV HYBRIDIZATION ASSAYS

6.4.6.2.1 Solution Hybridization (p. 513)

Extraordinarily low concentrations of HIV-1, on the order of 1 infected cell in 100,000 PBMCs, occur in asymptomatic infected individuals.

One approach to enhance the hybridization signal uses a probe, which has a **replicatable reporter** — the part of the probe that generates the signal when a specific hybrid is formed. This assay uses the high specificity of **QB replicase** for its target on the RNA attached to the HIV probe (see Part 1). This assay is directed to a conserved region of the **HIV-1 *pol* gene.**

Quantitation of the viral molecules in the sample by **QB replicase** assay is based on the kinetics of amplification of the replicable RNA in the probe. The **limit of detection** is about **10,000 target molecules**.

Analogous to **sandwich hybridization**, except for reversibility of the link between hybrid complexes and the affinity support, "**reversible target capture**" aims at improving the detection limit of hybridization assays and **quantitating** the target sequences. **Subpicogram** quantities of HIV-1 *pol* gene RNA could be measured with a **signal:noise ratio of over 10**.

Quantitation of HIV-1 in microculture of infected PBMCs has been described using **RNA-RNA hybridization**.

Detection of HIV-1 and HIV-2 nucleic acids provides not only definitive evidence of the viral infection, but is also used to identify the viruses in blood products.

6.4.6.2.2 *In Situ* Hybridization (ISH) (p. 514)

Early **ISH** detected HIV in frozen brain tissue from AIDS patients, infected lymphocytes in lymph nodes, and peripheral blood.

Detection of HIV-1 by a **nonisotopic technique**, using a **streptavidin** and **alkaline phosphatase reporter system**, makes it better suited for clinical laboratories.

A **fluorescent modification** of the HIV-1 ISH technique detects target sequences as small as 5.3 kb with a sensitivity allowing the identification of HIV proviral DNA, when only 5% of the cells in a population are infected.

HIV-1 ISH on **bone marrow** of patients with AIDS and AIDS-related complex (ARC) has been adapted, by using radiolabeled DNA and RNA probes, to **formalin-fixed, paraffin-embedded tissues, cell blocks**, and **smears** of cultured cells.

Commercially available HIV-1 ISH kits are produced by several companies (e.g., Enzo Biochemical, Inc.; Applied Biotechnology, Inc.).

6.4.6.3 HIV PCR: ADVANTAGES (p. 514)

By using **PCR** primers to the **highly conserved sequences** flanking the 5′ end of the gp41 encoding region of the ***env* gene**, this virus can be diagnosed in all patients. In contrast, using primers to the **variable sequences** leads to false negatives in some patients.

Exposure to HIV-1 can lead to a **latent form** of infection persisting up to several years in the absence of active virus replication. Less than 1% of both T lymphocytes and monocytes in blood has been found by PCR to carry a latent infection at all stages of HIV-1 infection.

Immunological methods may be inadequate for the diagnosis and monitoring of latent infection because antibody response to HIV-1 may vary greatly. Quantitation of the viral load is also not always possible by immunological methods.

Culture methods can be used but their efficiency varies with different isolates, sample handling variations, and the inherent variability of duration and magnitude of viremia at different stages of infection.

The sensitivity of **antigen capture methods** may often be below that required for the detection of the low levels of virus present in most clinical specimens.

Conventional nucleic acid hybridization assays using **^{32}P-labeled probes** detect 10^4 to 10^6 copies of HIV-1, and **nonradioactive labels** lower the assay sensitivity even further.

The antigen detection and unamplified nucleic acid assays are suitable for virus detection only during the **active stage** of infection, and may fail to detect HIV-1, particularly in **asymptomatic infections**.

The above-described ISH and hybridization assays using a **reversible target capture** approach are more sensitive and can detect as few as 10 to 100 infected cells per 10^6 uninfected cells. However, the presence of **integrated viral sequences** in the host genome in only one or a few copies may not be detected by either ISH or other hybridization assays.

6.4.6.3.1 Primers and Amplification Conditions (p. 515)

PCR has been shown to possess the required sensitivity and specificity for the detection of HIV-1-related sequences in clinical blood specimens.

The **standard conditions** for performing PCR amplifications using the original set of primers have been published and widely used in clinical studies. Significant effort has been given to the selection of optimal

- **amplification buffers**
- **annealing temperatures** and **time**
- **chain elongation temperatures** and **time**
- **cycle number**
- **amounts of *Taq*I polymerase** needed
- **concentrations of primers**
- **electrophoretic conditions** for analysis of amplified products

Deoxynucleotide and primer concentrations and chain elongation time proved to be not rate limiting so long as they were kept above a critical level.

Under a certain set of conditions, the cycling of the reaction through **three temperatures can be avoided**, because both annealing and chain elongation can be done at 60°C.

In the **oligomer restriction technique**, in addition to the initial pair of primers, a third, radioactively end-labeled primer (such as **K7** or **SK03**) is added to the amplified product. This **primer contains a restriction enzyme site**, which is then recognized by a corresponding enzyme that generates a restriction fragment of a specific size detectable on the gel.

Coamplification using a number of primer pairs to sequences of interest is a useful approach, particularly when only a limited amount of clinical material is available, or when unrelated sequences are added as an internal control of the reaction.

As few as 10 copies of the HIV-1 RNA and 10 to 20 copies of integrated proviral DNA can be detected using **simultaneous addition** of primer pairs to the *gag, env, tat,* and *nef* regions of the HIV-1 genome in the same reaction mixture. Analysis of PCR products is done by liquid hybridization with end-labeled oligonucleotides and gel electrophoresis. This approach can be used for the simultaneous detection of HIV-1 and HTLV-I, or of other unrelated sequences.

Addition of **several pairs of primers** also increases the **detection rate** compared to that observed with two primer pairs (86 vs. 75%).

Detection and quantitation of HIV-1 RNA in the **cell-free portion** of the blood by the PCR is described as a useful assay for evaluating disease progression or in monitoring antiviral therapy.

Refinements of reaction conditions and **primers** are continually being made. For example, lysis of PBMCs is accomplished directly in a buffer containing SDS, Triton X-100, and proteinase K; this **crude lysate** is then directly amplified in a **two-step nested PCR** using two pairs of primers. The amplified product is **directly visualized by ethidium bromide** staining of agarose gels. Again, the reaction used multiple sets of primers added simultaneously to amplify HIV-1 *gag*, env^{gp120}, env^{gp41}, and *pol* sequences.

6.4.6.3.2 Amplified Product Analysis and Quantitation of HIV-1 (p. 516)

Retention of the amplified products can be achieved by

- **affinity-based hybrid collection**
- **DNA sandwich hybridization**

The **affinity-based hybrid collection** technique uses 5′-**biotinylated primers**; the amplified DNA fragments are detected by liquid hybridization using radiolabeled probes. For quantitation, the hybrids are collected on polystyrene microparticles or onto microtiter wells using the **biotin-avidin interaction**.

In the **DNA sandwich hybridization** approach, the capture of the amplified product is achieved via hybridization to the **immobilized portion** of the viral genome attached to a microtiter well.

Both methods are highly sensitive, detecting as few as 10 to 30 molecules of DNA.

Detection of the PCR-amplified product can be achieved by

- **oligomer restriction**
- **liquid hybridization**
- **filter hybridization**
- **direct nucleotide incorporation**

A combination of PCR amplification with the subsequent resolution of the amplified products by **capillary DNA chromatography** has been developed and is aimed at the future automation of screening of blood for HIV-1.

Quantitative estimates of target sequences can be obtained by

- **scanning densitometry** of slot blot hybridization signals
- calibrated **solution hybridization**

Titration of HIV-1 and **quantitative analysis** of virus expression can be achieved with either liquid **RNA-RNA** hybridization, or capture of the PCR-amplified products first in solution hybridization using a **europium label**, and then onto streptavidin-coated microtitration wells. The europium label is then quantitated in a **time-resolved fluorometer**.

Quantitation of the PCR-amplified products by **direct incorporation of radioactive labels** yielded ambiguous results.

Solution hybridization may provide false-negative results due to the sequence diversity of retroviruses. Therefore the recommended approach was to use multiple primer pairs with a comparative analysis of blot and liquid hybridization and, ideally, all positive samples should be confirmed from an independent DNA sample.

Quantitation of the PCR-amplified products is used to extrapolate the amount of target present in the assayed sample. The **electrochemiluminescence** protocol has been used to quantify the HIV-1 *gag* gene. **Standardization** of the assay can be achieved by introducing either internal or external standards.

Internal standards use **parallel amplification**, in which a eukaryotic gene sequence is amplified in the same reaction as the target (e.g., a region of the HLA locus, *DQ*α) in a coamplification assay with a portion of HIV-1 (e.g., *gag*) to achieve **simultaneous amplification** and **quantitation** of both target sequences.

A less rigorous standardization uses β-globin sequences amplified **in a**

separate reaction alongside amplification of the HIV-1 sequences. The amplified products can be analyzed by oligomer hybridization, dot blot, or sandwich hybridization using appropriate probes for the internal standard and the target.

In the **external standard approach**, known amounts of the target are **amplified in different tubes**.

One of the **drawbacks** of internal standardization is the different efficiencies of amplification of the target and standard sequences, as they may be present in different amounts.

6.4.6.3.3 Quality Control (p. 517)

Contamination of the PCR may produce a signal comparable to that expected in a positive sample, but which is amplified from **spuriously introduced target** sequences. This **carryover** is the main problem that plagues PCR assays in general (see Part 1).

The **measures to avoid carryover** include

- **physical separation of facilities** for handling the test sample from those where the amplification reaction is carried out
- **dedicated laboratory tools,** segregated by separate stages of the PCR assay work flow
- the use of **reagents** dispensed in small aliquots
- avoidance of movements or procedures that are capable of **aerosolizing** solutions or specimens

Other factors important in the quality control of PCR include

- **standardization** procedures
- **multiple primers** to ensure quantitation and required sensitivity and specificity of the assay
- a set of **negative controls**
- **parallel testing** of related sequences to ensure maximum specificity
- maintenance of appropriate stringency of hybridization conditions during identification of the amplified products
- in screening, selection of primers for the **conserved regions of the viral genome** (because of the **sequence variability** of the retroviral RNA). One recommended pair of primers (**SK38/SK39**) is designed for the amplification of a 115-bp region of HIV-1 *gag*. The other pair (**SK145** or **SK150/SK101**) amplifies a 130-bp region of *gag* that is conserved among the HIV-1 isolates and HIV-2 (isolate ROD). The confirmatory probes for the SK38-39 and SK145-101 products are **SK19** and **SK102**, respectively
- the finding that **single internal mismatches** have **no significant effect** on the PCR product yield. Mismatches that occurred at the 3′-terminal base had varied effects: A-G, G-A, and C-C **mismatches reduced overall PCR product yield** about 100-fold, and A-A mismatches about 20-fold. In spite of that, even the presence of two mismatches allowed significant amplification to proceed

The detection efficiency of HIV-1 RNA by **RT-PCR** is markedly (68 vs. 26%) enhanced following treatment of samples with **heparinase**.

6.4.6.3.4 Comparison of HIV-1 Detection by PCR and Serological Methods (p. 517)

Virus isolation by PCR from anti-HIV antibody-positive hemophiliacs was more sensitive than **serological detection of p24 antigen** or the **decline of p24 antibody**. **PCR** detected the (probably latent) virus in some patients from whom virus could not be isolated.

The detection of HIV-1 by the PCR in **seropositive** individuals raises little controversy. It is definitely more accurate than tissue culture isolation methods, particularly early in the infection.

PCR assays in **seronegative** individuals, however, raise some questions:

- HIV-1 sequences have been reported in peripheral blood lymphocytes from patients seronegative for anti-HIV antibodies, and even in seronegative sexual partners of seropositive individuals
- **low viral load** has been cited as a cause of some false-negative PCR results
- PCR testing of HIV-1-**seronegative homosexual men**, in which three sets of primers to the *gag*, *pol*, and **LTR regions** were used, did not detect the virus
- likewise, HIV-**seronegative hemophilia patients** tested by the PCR with the same three sets of primers failed to show the presence of HIV DNA
- furthermore, **sexual partners of blood donors** with isolated and persistent HIV-1 core antibodies did not have HIV-1 DNA according to a PCR assay using one primer for the *gag* region and two primers for the *pol* region. The latter findings were interpreted to mean that persons with **persistent core antibodies** (anti-p24 or p17) are not infected with HIV-1. Similar results have been reported by others
- persons at high risk for HIV infection were tested by a PCR with the *gag* region primer pair SK38/39 and the *env* region primer pairs SK68/69 and CO71/72; only 3.4% had HIV DNA detectable by PCR among HIV-seronegative persons

The controversy is not yet resolved, because some of the previously HIV-seronegative persons on retesting were HIV seropositive, whereas others remained seronegative. Thus, the possibility of **false-positive PCR** results (3%) cannot be excluded.

The other reported possibility, that of the presence of **persistent infection without a demonstrable antibody** response, could not be confirmed by later studies.

- In another study, 98% of the HIV-1-seronegative patients were also PCR negative. In two cases, **PCR positivity could not be reproduced on repeat testing**. A generally good agreement has been observed between serological tests and the PCR.

It is believed that the **low concentration of the virus** in latent infection and the **effects of freezing and thawing** (the testing being performed largely on frozen samples) may have contributed to the observed discrepancies.

- A similar **concordance of serological and PCR results** has been observed using three pairs of primers in testing HIV-exposed seronegative hemophilia patients.
- However, in a different study, a high proportion (11 out of 27) of sex partners of seropositive hemophiliacs were seronegative, but tested positive by the PCR, suggesting these persons had a latent infection.
- Subsequent seroconversion of the PCR-positive, antibody-negative persons suggests that **amplification by PCR using multiple pairs of primers is capable of detecting latent infection.**

It appears that **neither PCR nor antibody testing is completely specific for detecting the presence of HIV infection**. Testing of an **independent sample** and **a clinical follow-up** are advised when discordant results are obtained.

Detection of HIV-1 RNA or DNA by PCR amplification is of particular importance in **cases when immunological response cannot be expected**, as in the case of **newborn babies**. Comparison of HIV detection by lymphocyte cultures with PCR amplification of viral RNA and DNA clearly established that HIV DNA detection by the PCR is far more sensitive for the early diagnosis of HIV infection in offsprings of seropositive mothers. The PCR assay provided evidence of infection of infants born to seropositive, and in some cases, seronegative mothers well before the development of any AIDS-related symptoms.

It is advocated that **prenatal screening** should include not only women at risk for having HIV infection, but be extended to all pregnant women. Analysis of regions of the *env* gene in mother-infant pairs suggests that only a minor subset of maternal virus infects the infant.

6.4.6.3.5 Detection of HIV by PCR in Various Tissues (p. 519)

Proviral HIV DNA has been predominantly detected in **CD4** lymphocytes, rather than in **CD8** cells, so that every hundredth CD4 cell may be infected with HIV-1. DNA amplification for the detection of HIV-1 proviral sequences is possible in **crude cell lysates** of PBMCs by **bracketed** and **nested PCR** protocols using colorimetric and chemiluminescent substances for the identification of the amplified products in dot blot hybridization or agarose electrophoresis and Southern blotting.

PCR can be applied to **formalin-fixed, paraffin-embedded** tissues.

The presence of HIV-1 has been demonstrated in

- **lymph nodes, lung, periodontal lesions, dried blood spot specimens, aorta, CSF, brain**, and **urine**
- the cool **aerosols** generated by some surgical instruments, as effectively demonstrated by immunological and cultural methods
- **saliva** and **semen** in the **absence of infectious virus**

The complex **topological arrangements of DNA strands** found in unintegrated HIV-1 DNA as well as the **extreme sequence variability** of the virus even in the conserved regions may affect the identification and quantitation of the virus by PCR.

In summary:

- for **screening purposes** the immunological tools may be adequate, whereas the appropriateness of the PCR for screening is not clear
- the evaluation of **latent forms** of HIV-1 infection as well as the resolution of some controversial findings in **low-level infections** can benefit from PCR-based methodology
- further improvements of amplification techniques will undoubtedly make this modality a **method of choice** for the diagnosis and monitoring of the disease, as well as for the identification of HIV-1 in blood products

6.4.7 HUMAN PAPILLOMAVIRUSES (HPVs)

6.4.7.1 INTRODUCTION (p. 520)

The putative role of HPV as a causal agent of premalignant and malignant lesions seemed to gain support from **epidemiological, molecular biological**, and **clinical** observations.

Molecular biological evidence links a **predominant HPV type** with a characteristic **anatomical site** or **lesion**. Based on **high-stringency hybridization** over 60 HPV types and subtypes have been identified so far.

The **epidemiology** of HPV infection has been extensively studied in various countries. No significant changes in the prevalence of HPV DNA or HPV types have occurred over the period from the 1920s to the 1980s.

Significant **indirect evidence** points toward association of cervical cancer with predominantly "**high-risk**" HPV types 16, 18, and, to a lesser extent, types 31, 33, 35, 41–45, and 51–56. However, no firm cause-effect relationship has been unequivocally established so far.

Dissenting arguments have been repeatedly voiced **advocating abandonment of mass screening for HPV**.

The preferred epidemiological approach appears to be a prospective longitudinal study with accumulated cytological preparations examined for HPV DNA by various techniques ranging from ISH to PCR.

In the meantime, wider HPV typing correlated with tissue morphology and quantitative evaluation of at least some tentatively corroborating factors such as hormonal status, oncogene activity, concomitant infections, and tobacco metabolites will enhance our understanding of the role of HPV infections in human disease.

6.4.7.2 HPV INFECTION IN MALES (p. 522)

Understanding the importance of HPV infection in males in the etiology of either **penile cancer** or cervical lesions of their sexual partners is limited. The molecular biological approach appears to be particularly appropriate because neither location, gross appearance, nor cytological details of penile condylomata are indicative of the presence of HPV 16 and HPV 18.

HPV infection is **sexually transmissible**. Sixty-five percent of male sexual partners of HPV DNA-positive women also harbored the virus. Males either with signs of infection, and/or those whose **sexual partners** are known to have lesions or HPV, should be routinely evaluated.

HPV 6 was detected in a **primary carcinoma of the urethra** in men.

In men with **genitoanal warts**, the high-risk type of HPV (mostly HPV 16) was found in 11%, along with some degree of **dysplasia**. HPV has been detected in the **semen** of males with severe chronic wart disease (HPV 5 or HPV 2), suggesting at least a possibility that genitoanal warts may be involved in transmission of HPV infection.

Penile intraepithelial neoplasia (PIN) harboring HPV 16 and 33 was seen in 32.8% of partners of women with **cervical intraepithelial neoplasia (CIN)**.

A high percentage (93.5%) of the male sexual partners of women with genital tract abnormalities had visible genital lesions, and 6.1% of those biopsied showed histological evidence of PIN.

In the male partners of women with warty atypia or condylomata, HPV 6/11 was most common in penile lesions. HPV types 16 and 18 were most commonly seen in the partners of women with CIN. **PCR** reveals the presence of HPV 6 and 11 in **penile lesions lacking halos and nuclear atypia**, but located in areas of relative hyperkeratosis and in the thickened granular layer in epithelial crevices.

Although obvious and subclinical HPV-related lesions in men can be **successfully treated**, controlling the infection in men does not seem to influence the rate of failure in the treatment of cervical dysplasia in their female sexual partners.

6.4.7.3 TISSUE TROPISM OF HPVs (p. 524)

Different HPV types are associated with particular kinds of **cell transformation** and **body sites**. HPV types 16 and 18 are found in the higher-grade lesions.

HPV 18 is thought to play a role in the enhancement of **progression** of cervical cancer and tends to induce **adenodifferentiation**, whereas **HPV 16** leads to **squamous maturation**.

The **multicentric nature** of HPV infections of the anogenital tract is an established fact: up to 81% of women with **cervical** disease also have **vulvar** disease.

Nineteen percent of **anal cancers** harbor HPV 6, 11, 16, 18, and 31, as detected by **ISH** with biotinylated DNA probes for HPV DNA.

Most of the lesions containing HPV 16 display only subtle **koilocytotic atypia**. Fluorescent ISH (FISH) and ISH identified HPV 6 and 11, in equal proportion, in 55% of squamous cell carcinomas (SCCs) arising in the condyloma acuminatum of the vulva.

Condylomata arising on the **oral mucosa** have been found to contain HPV 6, 11, or related genomes in 85% of the lesions. A significant proportion of oral SCCs (11.8%) and dysplastic lesions (28.6%) contain HPV DNA. The most frequently detected HPV types were 6, 11, 16, and 18.

In juvenile **laryngeal papillomata** all cases studied contained HPV 6 and/or 11. HPV 35 has been also detected in laryngeal SSC.

Immortalization of **cultured nasal** and **nasopharyngeal cells** has been demonstrated for HPV 16.

PCR assays for **the *L1* gene** of HPV (see Section 6.4.7.4.2) showed 71% of patients with **esophageal carcinoma** to have HPV DNA either in the tumor itself or in adjacent tissue. Interestingly, 15% of patients without esophageal malignancy also demonstrated HPV DNA in the biopsy material. Half of these were cases of **clear cell acanthosis**.

Anatomical predisposition to certain HPV types, although not being exclusive of other types or location, remains an **enigma** at present.

Among the factors thought to be responsible for tissue tropism is the **local hormonal microenvironment**. It is speculated that an **elevated level of progesterone**, as found in oral contraceptive users and in pregnant women, accounts for the high prevalence of HPV in these groups of patients.

The presence of HPV in **morphologically unaffected** cervices and progression of selected groups of patients to **malignant transformation** implicated additional factors as contributing to the process, such as

- **hormonal** influences
- **smoking**
- **herpesviruses**
- **hygiene**

6.4.7.4 MOLECULAR GENETIC ASPECTS OF HPV INFECTIONS (p. 525)

The precise mechanism of HPV-related malignization still remains to be established.

Papillomaviruses (PVs) belong to the **Papovaviridae** family of small **DNA viruses**, which have distinct **species, tissue type**, and **differentiation stage tropism**.

The nucleic acid is a **single, circular, double-stranded DNA** molecule, on the average 7800 bp long. Three genomic regions are recognized in PVs:

- the **early coding (E)** region, composed of eight genes **(*E*1-*E*8)**
- the **late coding regions (*L1*** and ***L2*)**
- a **noncoding segment** — the **long control region (LCR)** — which is 0.4 to 1.0 kb long and separates the E and L regions

The LCR harbors a number of **regulatory elements**:

- several RNA polymerase II promoters
- constitutive and inducible transcriptional enhancers
- binding sites for cellular transcription factors
- several copies of the palindrome $ACCN_6GGT$

6.4.7.4.1 Role of HPV Integration (p. 525)

HPV DNA can exist in the **extrachromosomal** form, as an **episome**, or **integrated** in the cellular genome.

Some HPV types exist exclusively as episomes (e.g., HPV 6 and HPV 11), whereas others (e.g., HPV 16 and HPV 18) are found as free DNA molecules and as integrated DNA sequences in invasive cervical cancers.

Integration of HPV 16 DNA sequences apparently occurs as an **early event**.

In contrast to **premalignant** lesions containing the HPV DNA in episomal form, **most carcinomas appear to harbor viral DNA integrated into the host cell chromosomes**.

Integrated HPV DNA occurs either as **single copies** (such as HPV 16 and HPV 18 in the cell lines SiHa and C4-I, respectively) or **amplified** up to 10- to 50-fold, as is HPV 18 DNA in the cell lines HeLa and SW756. Integrated viral sequences replicate along with those of the cellular genome.

Integration of HPV DNA is **not a random event** and **specific integration sites** are recognized (e.g., **chromosome 12** at band **q13** for HPV 18 DNA). The **opening sites** on circular viral DNA consistently occur within the **3′ end** of the ***E1* open reading frame (ORF)**, or the **5′ end** of the adjacent ***E2* ORF**.

Little is known concerning the **specificity of the site of disruption** for different HPV DNA types and correlation of these with the development of carcinomas other than in the anogenital region.

Sometimes integration occurs near cellular **oncogenes** such as c-*myc*, c-*src*-1, c-*raf*-1, *abl*, and *sis*. In cell culture models cooperation with oncogenes has been demonstrated for EJ-*ras*, v-*fos*, and c-*myc*.

Certain HPV genes such as ***E7*** are able to interact with the products of **tumor suppressor genes**, such as ***RB*-1** and **p53**, as well as with **SV40 large T**, **polyoma large T**, and **adenovirus E1A** in addition to the product of activated *ras* oncogene. Binding of the *E7* gene product to the product of the *RB* gene accounts, at least in part, for its transforming potential.

The **flanking sequences** may have an effect on the expression of the viral genome.

Both episomal and integrated forms of HPV 16 are present in invasive cervical cancer:

- up to 70% of invasive lesions showed only the episomal form of HPV 16 DNA, without integration, as shown by the analysis of *Pst*I restriction patterns. Some specimens (30%) showed **integrated multimeric forms** of viral DNA either without the episomal form or with the concomitant episomal form

No correlation has been found between the forms of viral DNA and the **clinical stages** of tumor.

Integrated DNA shows the same pattern in primary tumors and the nodes, whereas episomal DNA was reduced in the metastases. Attempts are being made to use analysis of HPV DNA integration as a diagnostic marker for the detection of **early node dissemination.**

6.4.7.4.2 Transcriptional Regulation of HPV (p. 526)

The **nonrandom incorporation of HPV near or in the *E1/E2/E4/E5* regions** disrupts or deletes these ORFs and is important for regulation of the **transcription pattern of HPVs**.

Only ***E6*** and ***E7*** transcripts are found in **invasive** tumors.

In **CIN**, in addition to these transcripts, there are also transcripts from the ***E1, E2/E4/E5*** coding regions. Late-region transcripts cannot be detected in either case.

E7 induces a **stronger transformation effect** than *E6*, although in some cases both genes are required for the transformation to occur. It appears that both the *E6/E7* and the *E2/E4/E5* regions are required for the production of the complete transformed phenotype.

The ***E7* gene transcripts are more abundant in cancers than those of *E6***, and both are present in cervical carcinomas, whereas in lesions produced by the benign HPV types 1a, 6, and 11 no *E7* message could be identified. This is reflected in the increased prevalence of HPV 18 **anti-E7 antibody** in cervical cancer patients. The *E7* gene product is a phosphoprotein resembling E1A proteins and the T antigen of other papovaviruses. *In vitro*, the *E6* and *E7* genes of HPV6 display only a weak immortalizing activity.

Common to all PVs, the ***E2* ORF** plays a regulatory role in controlling the expression of **transformation** and **replication functions**. A portion of the *E2* ORF encodes a protein termed **E2-TR**, which, although identified only in BPV1, specifically **represses trans-activation**. The *E2* ORF of HPV 16 and 11 presents similar activities.

The **transcriptional regulation by *E2*** is mediated through direct binding to a viral *cis* element, specifically to the **short palindromic repeated sequences** $ACCN_6GGT$.

Binding sites for the E2 protein are found in the noncoding LCR region of the viral genome. These **binding sites are absent in oncogenic HPVs** type 16, 18, and 33, suggesting a different type of transcriptional regulation in these viruses.

The predominant **function of the *E2* gene product** appears to be transcriptional repression, which correlates with what is presently understood as the tentative mechanism of malignization by HPV. In a simplified model **the loss of the repressor function of the *E2* ORF leads to unimpeded transcription of ORFs *E6* and *E7***.

In studies on primary human keratinocytes, the differential immortalization activities of HPV 16 and 18 have been assigned to the **regulatory region** located **upstream of the *E6*** and ***E7* genes**.

The two late genes ***L1*** and ***L2*** are **expressed only in terminally differentiated keratinocytes**.

L1 appears to be the **most highly conserved PV gene** and codes for the major capsid protein carrying the major antigenic determinants for **group-specific cross-reactivity** among different groups of PVs.

Diagnostic use of *L1* is visualized

- in **antibodies** raised against the *L1* gene product for the detection of specific HPVs in clinical lesions
- in **PCR** amplification of *L1* gene sequences, intended for the **most broad** identification of HPVs

6.4.7.5 HPV AND HORMONES (p. 528)

LCRs contain the **constitutive enhancer elements** active only in some epithelial cells, and the **glucocorticoid response element (GRE)**, which confers strong inducibility by dexamethasone. The GRE can also function as the **progesterone response element (PRE)**.

Therefore, a relationship between HPV infection and hormone use may be important. Relevant findings include the following:

- only a slightly elevated risk of invasive cancer is observed among women who had been using **oral contraceptives** for a long time
- the use of contraceptives and the length of use is correlated with an elevated risk of **high-grade** squamous intraepithelial lesions
- **no synergism** is observed **between the presence of HPV 16 and HPV 18 and oral contraceptive**
- although HPV 16 and 18 seemed to be more readily detected in **oral contraceptive users**, the etiological role of either factor could not be established
- the response of *E6-E7*, particularly to **hormonal stimulation**, varies in different cells
- viral elements outside the LCR [or the **upstream regulatory region (URR)**], viral-cellular junction fragments, or the flanking cellular sequences at the integration site of the viral DNA are able to override the response of the HPV promoter elements to glucocorticoid hormones
- **estrogen receptor (ER) positivity** is reduced in cervical cells containing the carcinogenic HPV 16 and HPV 18, suggesting that the virus genes may suppress ER expression

6.4.7.6 HPV-HOST CELL INTERACTION (p. 529)

The predilection of HPV types for **specific anatomical locations** so far has not been explained in molecular terms.

As epidemiological studies show, **the sheer presence of HPVs is not sufficient for anogenital cancers to develop.**

Production of **mature HPV particles** is associated with the process of **host cell differentiation** and proceeds only in **highly differentiated keratinocytes**, although HPV DNA can be detected in **basal cells** as well.

In stratified cultured keratinocytes, the amplification of HPV 31b was found to parallel epithelial differentiation. Likewise, the concentration of HPV 33 mRNA increases in condyloma acuminata with the degree of differentiation of the keratinocytes.

The role of **host cell factors** in controlling HPV transforming activity also comes from the finding of a specific translocation event involving **chromosome 11** in HPV-transformed cultured cells.

The concept of an **intracellular surveillance system** controlling the expression of HPV in host cells, developed by zur Hausen in 1987, may account for the observed latency of the HPV infection and the high prevalence of the infection in a clinically normal population. The existence of a putative **cellular interfering factor (CIF)** is invoked, which controls HPV transcription in growing cells *in vivo*.

***E6-E7* transcripts** are seen as mediating proliferative changes, and the **cellular regulatory mechanisms** controlling early HPV transcription are incorporated in a coherent **model of a stepwise transforming process**.

Cell-specific factors apparently contribute to the relative abundance of *E7* over *E6* in cancers.

The role of **cell-mediated immune responses** in HPV infection is suggested by the association of HLA-*DQw3* and HLA-*DR5* and the biology of SCC of the cervix:

- **HLA-*DR5* is associated with increased risk.**

Although the specifics of such associations may be contested, **HLA polymorphisms do appear to be associated with HPV-related disease processes**, as shown in animal models and in humans.

6.4.7.7 RISK FACTORS (p. 530)

Although no general consensus has been reached on a number of issues concerning risk factors a few **general correlations** seem to emerge:

- **no direct "cause-effect" relationships** have been demonstrated
- a **constellation of factors** is certainly at play in modulating the potentially oncogenic effects of HPV
- the **variability** of genital HPV infections contributes to the markedly divergent prevalence estimates reported by various authors
- **sampling differences** reflecting the known **heterogeneous distribution** of HPV DNA in epithelia confound the incidence data
- **genomic variability** of HPVs, although recognized, has not been adequately used in epidemiological studies

Among the **risk factors** implicated are:

- an early **age at first intercourse**
- **number of sexual partners**
- **low socioeconomic level**
- **number of pregnancies**
- **smoking**
- certain **venereal diseases**
- patterns of **male sexual behavior**
- **herpesviruses**
- **HIV-related immune deficiency**, which appears to allow **reactivation** of HPV, eventually leading to epithelial abnormalities

The weight of epidemiological data **fails to implicate HPVs as the only cause** of cervical neoplasia.

Analysis of the role of HPV in anogenital cancer is continually being refined, taking into account the **multiple risk factors**.

6.4.7.8 LABORATORY METHODS FOR THE EVALUATION OF HPV INFECTION (p. 531)

Clinically manifest HPV infection can be diagnosed by

- **colposcopy**
- **Pap smear**
- **punch biopsy**

At this level of detection, the prevalence is about 2 to 3% in the general population (age range, 25 to 60 years).

It is in the **subclinical** and/or **latent** infections that limitations of various techniques become more pronounced. The presence of HPV DNA may not be expressed in morphological alterations of the tissue and, therefore, will remain undetectable by colposcopy or Pap smears.

The introduction of newer techniques, such as the PCR, which are capable of detecting a single viral particle, dramatically **changes our ideas about the prevalence of HPV** in clinically unaffected individuals.

In fact, its exquisite sensitivity challenges the conventional notions of the epidemiological relevance of HPV presence and requires a reappraisal of the accumulated evidence.

The **absence of HPV infection** by definition implies the absence of virus even when evaluated by the most sensitive technique — **PCR** amplification.

Electrochemiluminescent labels allow quantitation of PCR-amplified products as shown also for HPV 16 and 18.

The database for establishing a relationship between HPV presence and eventual development of malignancy must also be obtained at this sensitive level.

On the other hand, practical considerations of **mass screenings** of unselected populations favor the use of a significantly less sensitive method, for example, the Pap smear. Even with a follow-up with more sensitive techniques in selected cases, the bias introduced can affect meaningful extrapolations to general populations.

One of the **suggested compromises** is the use of ISH on the tissue material collected at mass screenings.

Biotin-labeled probes are known for their markedly lower sensitivity (100 to 800 copies per genome) compared to that of ^{3}H-labeled probes (1 to 10 copies per genome) in the DNA-DNA hybridization format.

Significant enhancement of sensitivity can be obtained by a **three-step amplification technique** (including a rabbit anti-biotin antibody, a biotinylated goat anti-rabbit antibody, and a complex of streptavidin-alkaline phosphatase, streptavidin-gold, or streptavidin-fluorescein).

A **combination** of biotinylated capture probes and ^{32}P-labeled detector probes can be used in a **solution hybridization** method suitable for diagnosis of infections with a low copy number of HPV.

ISH is more suited for diagnosis of infections with a high copy number of HPV, even if present in only a few infected cells.

Nonradioactive labeling of DNA by chemical insertion of a **sulfone group** on cytidine residues can be used in a dot blot hybridization format in a clinical laboratory setting.

Fingerprinting HPV DNA can help trace subsequent malignancies to the earlier lesions.

A more detailed **summary of comparative features** of a selection of HPV diagnostic studies using hybridization methods is given in Table 6.3.

In spite of the abundance of studies examining HPV infection in various tissues, only a few reports are available comparing different molecular diagnostic methods (see Table 6.4).

6.4.7.8.1 Commercial Kits for HPV Assays (p. 540)

Commercial kits are available for the identification and typing of HPV by conventional hybridization procedures, for example, Enzo Diagnostics, Genemed Biotechnologies (San Francisco, California), Digene Diagnostics (Gaithersburg, Maryland), as well as testing offered by major reference laboratories (e.g., Roche Biomedical, Specialty Labs, SKBS). [*See Supplements 1 and 2.*]

A commercial kit for HPV typing of smears and scrapes proved to be a convenient, reproducible and dependable tool, especially if combined with appropriate controls (e.g., for the DNA content of the material tested).

The **mode of specimen collection** has a significant influence on the results of testing. Sampling with a **Cytobrush** introduces blood into the specimen, but the cellularity has always been adequate. Sampling of cells with a spatula, on the other hand, is frequently inadequate.

The results of HPV typing by dot blot hybridization using **scrapes and biopsy** specimens show a poor correlation between the two methods of specimen collection. The scrapes detected more of the HPV 18 and HPV 31 DNA than did the biopsies. These differences possibly reflected **heterogeneity of the distribution** of different HPV types within the cervix.

- **Rigorous standardization of specimen collection for HPV DNA typing is an obvious necessity.**

Cervicovaginal lavage is advocated as a more sensitive method of cell collection than the scrape-swab technique for HPV DNA typing.

6.4.7.8.2 ISH in HPV Diagnosis (p. 540)

ISH is best suited for the demonstration of target nucleic acids and their **relationship to morphology** of cells and tissues.

Although **less sensitive** than filter hybridization assays, specific probes for ISH allow the use of relatively uncomplicated procedures on **conventional tissue sections** routinely processed in pathology laboratories.

6.4.7.8.3 Sensitivity of ISH (p. 540)

The **^{35}S-labeled RNA** probe in ISH of HPV 16 RNA in **precancerous lesions** displayed almost an order of magnitude higher sensitivity than **biotin**-labeled probe.

TABLE 6.3 (p. 533)
Comparison of Different Hybridization Techniques for Human Papillomavirus Diagnosis

Technique	Sensitivity[a] (copies detected per cell)	DNA needed	Label and testing time	Advantages	Disadvantages
Southern blot	0.1	10 μg	^{32}P, 4–5 d Biotin, 2 d	1. Detects new types 2. Integrated/episomal 3. High/low stringency	1. Very laborious 2. Not suitable for screening
Dot blot	1.0	500 ng	^{32}P, 4–5 d Biotin, 2 d	1. Rapid 2. Suitable for screening	1. False positive 2. High stringency only
Reverse blot	1.0	500 ng	^{32}P, 4–5 d Biotin, 2 d	1. High/low stringency 2. Several types can be identified at the same same	1. Requires individual labeling reactions
Tissue *in situ*	20–50	Few cells	^{35}S, 6–8 d Biotin, 1–2 d	1. Cellular localization 2. Correlation of HPV type with morphology 3. Use of routinely fixed tissue	1. Relatedness only

TABLE 6.3 (continued)
Comparison of Different Hybridization Techniques for Human Papillomavirus Diagnosis

Technique	Sensitivity[a] (copies detected per cell)	DNA needed	Label and testing time	Advantages	Disadvantages
Filter *in situ*[b]	$1–5 \times 10^{4–5}$ HPV DNA molecules	Few cells	^{32}P, 1–2 d	1. No DNA extraction 2. Rapid	1. Background problems 2. Relatively high copy numbers per cell detected
Sandwich hybridization	$1–5 \times 10^{5}$ HPV DNA molecules	$10^5–10^6$ cells	^{32}P, 1 d Biotin, 6 h	1. Rapid 2. Detection with liquid scintillation	1. Only one type/sample can be analyzed
PCR	0.00001	10 pg	5 h	1. Highly sensitive 2. Semiautomated	1. Contamination 2. Relatedness only 3. Requires sequence information

[a] Sensitivities given are for radioactive labels.
[b] Only for exfoliated cells.

From Syrjanen, S. M. (1990). APMIS *98*:95–110. With permission.

TABLE 6.4
Comparative Features of Representative Molecular Human Papillomavirus Diagnostic Studies

Ref.	Tissue	Lable; HPV type; time	Sensitivity and specificity	Operational features	Findings	Comments
Henderson et al. (1987)	141 patients: cervical and anal scrapes	^{32}P; HPV6, 11, 16, 18		Radiolabeled *Alu* probes were used for relative quantitation of the signals	HPV DNA present in 16% of women without past or present history of lesions, in 24% of those without visible genital dysplasia, 30% with dysplasia of CINI	HPV group-specific antigen could be detected only in 60% of HPV DNA-positive samples
Webb et al. (1987)	66 patients: scrappings and smear biopsy	^{32}P; HPV6, 11, 16, 18; and 31	1 HPV genome per cell for HPV6 and HPV16	Reverse blotting	96% of 54 patients with condylomas or dysplasia had HPV DNA, mostly type 16	Suitable for detecting multiple defined sequences in small quantities of DNA
de Villiers et al. (1987)	9295 smears	HPV6, 11, 16, 18		FISH. Cytologically normal smears had HPV DNA in 10%, CINI–III had HPV DNA in 35–40%, mostly in women under 30 years	FISH underestimates the total rate of HPV infection by a factor of 2 to 3	
Bergeron et al. (1988)	28 endometrial samples	^{32}P; HPV6/11, 16, 18; 31		Southern blot	No HPV DNA was identified in endometrial samples, either normal, hyperplastic, or neoplastic	
Battista et al. (1988)	88 biopsies: cervix, uterus, vulva, larynx	^{32}P; HPV6, 11, 16, 18; 48 h to 2 weeks		Southern blot	Types 6/11 are present as episomes in benign lesions, types 16/18 are found as episomes and/or integrated in benign, premalignant, and malignant lesions	Analysis of the state of HPV DNA in lesions is suggested as an adjunct tool in monitoring HPV infection of anogenital areas

TABLE 6.4 (continued)
Comparative Features of Representative Molecular Human Papillomavirus Diagnostic Studies

Ref.	Tissue	Lable; HPV type; time	Sensitivity and specificity	Operational features	Findings	Comments
Demeter et al. (1988)	19 patients: scrapings and biopsy	^{32}P; HPV6, 11, 16, 18; 1–3 days	24.8 ± 11.5 µg of cell DNA; 1 pg HPV DNA	High correlation between scrappings and biopsy. 96-well manifold. Amount of DNA related to *Alu* DNA signals. Pepsin digestion is replaced by alkali detergent lysis of tissue sections. DNA neutralized for nitrocellulose binding. Alkaline DNA solution	No HPV DNA in histologically normal tissues. Formalin-fixed, paraffin-embedded cervical biopsies were suitable for HPV detection and typing by FISH. Alkaline DNA solution binds to nylon. Alkaline lysis improved filterability of samples	Can be used for retrospective studies also correlated with immunohistochemistry. Can be rehybridized with different probes. An alternative to ISH
Nuovo et al. (1988)	39 vaginal condylomata biopsies	^{32}P; HPV6, 11, 16, 18		Southern or dot-blot hybridization	63% of lesions with koilocytotic atypia contain HPV DNA; without koilocytotic changes, 11.7%	Lack of HPV DNA detection by these methods was interpreted as evidence of HPV-unrelated etiology of the lesions without koilocytotic atypia (see, however, PCR data below)
Caussy et al. (1988)	23 condyloma patients, 23 patients with cervical cancer, 33 patients after hysterectomy for nonneoplastic diseases	^{32}P; HPV6, 11, 16,18	Percentage detection: 82% — Southern, 62% — FISH, 72% — ISH; specificity of ISH: 72% for condylomas, 30% for invasive cancer. Overall relative sensitivities, 61–66%; relative specificities, 86%; agreement 77–80%. FISH, 0.1 pg HPV DNA	FISH modified to digest cellular proteins and mucus. ISH most vulnerable to sampling variation. 10 copies of HPV16 genome, detected in SiHa cells, not detectable by ISH	Prevalence of HPV DNA in cancer: 89%, FISH; 70% ISH. Majority of cancers in some lesions have less than 500 copies of the HPV genomes per call	Southern requires unfixed tissue, tedious. FISH is fast, but many nonspecific reactions. ISH works on paraffin-embedded tissue, low sensitivity in a large number of invasive cases. Suitable for retrospective studies on multiple sections

Cornelissen et al. (1988)	SiHa, Caski cell lines, cervical smears, biopsy	^{32}P; HPV16	Dot blot: 10^3 Caski cells, 10^5 SiHa cells; FISH, 50 pg HPV DNA; dot blot, 1–5 pg DNA	Dot blot is 10–50 times more sensitive than FISH. Binding capacity of filters saturated at 10^{-6} cells	5 pg of homologous DNA detectable in presence of 500 pg of heterologous DNA. Dot-blot finding coincided with Southern blot analysis for HPV16 DNA. Findings in smears are confirmed by biopsy. No correlation between HPV prevalence and severity of lesions	False positives on smears if vector probes are used. Prevalence by dot blot and Southern is twice that determined by FISH
Gerber-Huber et al. (1988)	Scrappings	^{32}P; HPV6, 11, 16, 18; quantitation by *Alu* hybridization. 6–7 days	10^4 HPV molecules	Slot blot is 100 times more sensitive than FISH or Southern blot. Southern analysis used to define HPV type when questionable	90% HPV positive in dysplasia outpatient clinic patients. 83% positive for HPV16/18 alone (42%) combined with other types (41%). 17% positive for HPV6/11	Possible cross-reaction with pBR vector. In slot blot the purified cellular DNA is loaded onto a small area concentrating it and leading to increased sensitivity
Duggan et al. (1989)	Colposcopy scrappings	Biotin (Enzo Biochemical) HPV6, 11, 16, 18, 33	1–10 pg viral DNA = 10^5–10^6 HPV DNA molecules. Similar to Southern blot	Dot blot. Limited cocktail of probes. Fails to identify unrecognized HPV types. Cervical scrapes — low cell yield for good hybrid signal. Dilution of infected cells by uninfected cells leads to false negative	Low-risk types infrequent (7%), high risk (41%), condyloma/CIN I (75% of cases). Typable HPV DNA in 60% cases of condyloma. Nondysplastic noncondylomatous changes suggest latent virus	Unexpectedly low frequency of HPV6/11 and high frequency of HPV16/18/33 in first time colposcopy
Nuovo and Richart (1989)	205 patients: biopsy, smears	ViraPap; Vira Type 6/11, 16/18, 3/33, 35			FISH is superior to ISH in equivocal cases or when other HPV types are present. ISH helpful in differentiating HPV6/11 from oncogenic types. Slot blot and Southern are highly concordant	ISH fails to identify HPV presence in equivocal histology

TABLE 6.4 (continued)
Comparative Features of Representative Molecular Human Papillomavirus Diagnostic Studies

Ref.	Tissue	Lable; HPV type; time	Sensitivity and specificity	Operational features	Findings	Comments
Melchers et al. (1989a,b)	1963 scraping samples	^{32}P; HPV6, 11, 16, 18, 31, 33	1 pg HPV DNA	FISH, alkaline denaturation, and neutralization prior to application onto the membrane increases sensitivity (5-fold for episomal HPV DNA, and 16-fold for integrated HPV DNA)	FISH and Southern analysis show a 100% correlation. 92.2% of all specimens tested by FISH showed no HPV6/11 or 16 confirmed by Southern analysis. 3.6% were equivocal. 3.2% had both HPV6/11 and 16	Reducing surface of sample application on the membrane 28 mm^2 vs. conventional 314 mm^2; 54 samples/membrane. Improved filterability of sample
Colgan et al. (1989)	30 patients: biopsies of cervical lesions and unaffected area	^{32}P; HPV 16; 10 days	1 pg HPV DNA = 0.1 HPV gene copy per cell; occasional cross-reaction with HPV 31	Southern analysis	28% of patients with CIN or condyloma had HPV16 DNA in lesions and adjacent epithelium. 57% were HPV DNA positive if invasive carcinoma was present	
Auvinen et al. (1989)	467 biopsy specimens	^{32}P; HPV6, 11, 16, 18	Single-stranded RNA probes are more specific than DNA in typing clinical specimens	Spot blot and Southern. PEG 6000-containing hybridization solution suitable for typing specimens, comparable to formamide. Increasing formamide decreases sensitivity, but specificity increases 10%. Dextran increases specificity without loss of sensitivity	Total percentage positive 33.5%; HPV16/18, 14.7%; HPV6/11, 18.2%. High SDS concentration is necessary to diminish background. 10–15% PEG 6000 + 7% SDS is optimal. In clinical samples superiority of PEG over dextran sulfate is not evident	

Duggan et al. (1990a,b)	119 patients; scrappings	Biotin; HPV6, 11, 16, 18, 33; 1–10 pg HPV DNA or 10^5–10^6 HPV DNA molecules	HPV DNA positivity rate almost equal for wood (30%) and plastic (32%) spatulas	Dot blot	Wood spatulas collect more cells (over 10^5 cells) than plastic, optimizing the sensitivity of the detection	Nonpurple dots in scrapes by wood spatulas do not affect detection rate
Duggan et al. (1990a,b)	401 patients: cervical scrapes	Biotin; HPV6, 11, 16, 18, 33		Dot blot	41.3% of CINI condylomas had HPV16/18/33; 65.4% in CINII and III	Possible underdiagnosis of HPV6/11 due to lower sensitivity of dot blot
Neumann et al. (1990)	18 patients: scrapings, biopsy	^{32}P; HPV16, 18		FISH less sensitive than ISH	ISH correlates with cytology and histology	
Ikenberg et al. (1990)	18 patients: vagina, lymph node, metastases	^{32}P		Southern analysis	56% with invasive carcinoma of vagina had HPV DNA, mostly type 16, in 0.5 to 50 copies per cell, some lymph nodes were positive for HPV DNA. 56% of HPV-positive patients were alive without recurrence, 100% of HPV-negative patients died from disease within 13 months	No correlation established between the grade of differentiation, clinical stage and HPV positivity, HPV type, or copy number
Nuovo et al. (1990)	100 patients: 130 biopsies	^{32}P; HPV6, 11, 16, 18, 31, 35, 51		Southern analysis	Absence of CIN has 34% positive for HPV. Presence of HPV DNA in cervical lesions lacking CIN characteristic histology predicts current or future CIN in such patients	ISH deemed insensitive for this type of study. Dot-blot technique may replace Southern analysis for this purpose
Ranki et al. (1990)	178 patients: smears, swabs	AffiProbe kit, ^{35}S; 1 day ViraPap/	AffiProbe and ViraPap have equal sensitivity cut-off: 1.5 times the mean of the	Solution hybridization. 10^9 unrelated HPV DNA molecules are needed for	Internal positive standards define the cutoff values. Comparable to	No purfication is needed for AffiProbe. RBCs and mucus do not interfere, no

TABLE 6.4 (continued)
Comparative Features of Representative Molecular Human Papillomavirus Diagnostic Studies

Ref.	Tissue	Lable; HPV type; time	Sensitivity and specificity	Operational features	Findings	Comments
		ViraType 6/11, 16, 18; 4 days to 2 weeks	background. 1.5×10^7 HPV DNA molecules	cross-reactivity	ViraPap/ViraType but faster, suitable for screenings	cross-reactions with unrelated bacteria or viruses
David et al. (1990)	Biopsy	^{32}P; HPV6, 11, 16; 18 days	1 ng to 0.01 pg; HPV6 and 11 cross-react. Slot blot is more sensitive than dot blot and Southern. FISH, 5×10^4 HPV DNA molecules; Southern, 10^7 HPV DNA molecules, and needs µg of DNA material, labor intensive	Slot blot is faster than Southern, increased stringency allows discrimination of HPV6 from HPV11	Detection rate overall, 35.2%. FISH: not quantitative, subjective evaluation. Southern allows highest specificity	
Bartholoma et al. (1991)	50 swabs, 11 biopsies	^{32}P-Oncor; HPV6, 11, 16; 18, 31, 33, 35; up to 2 weeks. Variability in exposure time needed for different specimens.Dot blot could be held at room temperature for up to 2 weeks, and transported without refrigeration	Semiquantitive	Southern blot (Oncor), dot blot (LTI) with the Probe Tech I instrument up to two membranes simultaneously (18 specimens) for Southern. 21 specimens per run of LTI dot blot with triplicate assays	Overall agreement 78.8%, 8 of 13 specimens identified by dot blot, but not by Southern. Blood-tinged specimens give false negatives. Overall prevalence, 50.8%. 78.6% HPV DNA positive in histologically adnormal specimens	Dot blot preferred for clinical specimens. Southern failed to identify in some swab specimens. Southern required freezing throughout. Unexplainable smearing of pattern in Oncor Southern analysis; difficult to interpret

However, other reports indicate that both biotin and ^{35}S-labeled probes can detect one or two copies of HPV16 per SiHa cell using ISH; therefore, the only advantage of radiolabeled probes is the possibility of quantitation of the autoradiographic signals.

A comparison of the Enzo ISH kit, which uses biotinylated HPV 6/11, and 16 DNA probes to Southern blot hybridization, demonstrated the sensitivity of ISH to be 88 to 89%, with a specificity of 99%.

A markedly lower sensitivity of biotinylated (Enzo) DNA probes for ISH compared to that of tritium-labeled probes for HPV 6, 11, and 16 has been demonstrated in vulvular, vaginal, and cervical carcinomas.

ISH with **digoxigenin**-labeled probes for HPV 11, 16, and 18 shows the **same sensitivity** as that of radiolabeled probes on paraffin-embedded sections. The **topographical localization** of hybrids is in fact even better with digoxigenin probes than with ^{35}S-labeled probes.

Fixation in **Bouin** solution markedly reduces the rate of detection of HPV DNA by ISH compared to that in tissue fixed with **buffered formalin**. Fixation and processing of tissues may impede the penetration of probes to the viral DNA.

ISH with **biotinylated** DNA probes is strongly influenced by **proteinase pretreatment** of the specimen:

- an increase in proteinase concentration to 1 to 5 μg/ml from 0.1 μg/ml increased the detection limit for HPV 16 DNA from 30 to 40 copies per carcinoma cell to at least 20, at the expense of poor preservation of morphology

ISH fails to detect HPV DNA in equivocal vulvar lesions, and **Southern analysis** can be the method of choice in such cases.

6.4.7.8.4 ISH in HPV Research (p. 541)

Single-stranded sense and **antisense RNA** radiolabeled probes give higher signals in ISH, which correlate with the degree of cellular differentiation, the koilocytotic cells showing the strongest signal.

HPV mRNA can be detected even in relatively undifferentiated dysplastic cells and in invasive carcinomas.

ISH with single-stranded RNA (ssRNA) probes for **early** and **late regions of HPV genomes** reveals that early regions are present even in mature epithelia, whereas late regions are detected in the oldest and most differentiated keratinocytes.

Using subgenomic **riboprobes** for HPV 16 mRNA, a **strong antisense RNA signal** could be detected in some genital SCC.

An ISH study with biotinylated DNA probes showed that the **intensity of signals** for HPV 16 and 18 in the malignant lesions was **less** than in the dysplastic or benign lesions, consistent with suppression of replication of HPV DNA in transformed cells.

The relationship of a malignancy to **precursor lesions** revealed by ISH with biotinylated probes suggests the origin of cervical carcinoma from **areas of dysplasia**.

ISH with biotinylated probes helps to differentiate between the possible **origins** of malignancy: **cervical adenocarcinomas** (positive for HPV 16/18) can be differentiated from **endometrial adenocarcinomas** (negative for HPV 16/18).

6.4.7.8.5 ISH in Clinical HPV Studies (p. 542)

ISH alone or in combination with other techniques helps to establish the **source** of infection. For example:

- in a **longitudinal analysis** of anogenital condylomata with an unusual bladder tumor, ISH analysis identified HPV 6 and 11 in both types of lesions, suggesting a **common source** and **spread**
- **conjunctival papillomas** of children and young adults, but not adult **conjunctival dysplasias**, were found to contain HPV 6 or HPV 11 apparently acquired during passage through an infected birth canal
- **condylomata acuminata** in children are associated with the same types of HPV found in anogenital lesions in adults, including types 16 and 18

In analyzing the relationship of **metastatic lesions** to the **primary tumor**, ISH offers certain advantages over Southern analysis. Specifically:

- although Southern blot hybridization can establish the presence of HPV in both cases, it requires fresh frozen tissue and cannot assign HPV to individual cell types. ISH helps to locate HPV precisely within tumor cells and in its lymph node metastases in formalin-fixed, paraffin-embedded tissues

A **combination of Pap smears with ISH** can be used for **screening** and **follow-up** of patients with cervical lesions.

6.4.7.8.6 HPV PCR (p. 542)

PCR has been applied to identify HPV DNA where ISH has failed. PCR offers unparalleled sensitivity with the **specificity essentially determined by the oligonucleotide primers to target sequences**. Therefore, detection of unknown types of HPV DNA would not be possible if the primer does not share common sequences with the target.

To bypass this limitation, primers corresponding to **highly conserved HPV sequences** capable of detecting both known and unknown HPV DNA types can be used. These **universal primers** can be used to detect HPV in **metastasis** in women treated for cervical cancer.

It is not clear at this stage, however, to what extent the identification of HPV DNA in metastatic lesions may have clinical significance that would contribute to improved treatment of patients.

PCR detects HPV DNA in 70% of cervical scrapes from women with positive cytology, compared to 46% detected by the modified filter ISH and Southern analysis.

PCR detected HPV DNA in 5% of the control group women with normal smears and no history of cervical lesions.

The overall **prevalence of HPV in cytologically normal scrapes** was estimated to be 6%, and in those showing **cytological dysplasia** the HPV prevalence went up to 60%. In biopsies of **SCC of the cervix**, the HPV prevalence was 90%; all cells contained only HPV 16 and 18. Analysis of **paraffin-embedded tissues**, although 10 to 40 times **less sensitive** than analysis of cytological preparations, revealed a significant level of carriage of high- and low-risk HPV types in both normal and cancerous tissues.

A study of reaction conditions in which **genotype-specific primers** for HPV 6b, 16, and 18 were used emphasized the importance of a particular set of parameters, other than selected primers, which affect the **sensitivity** and **specificity** of amplification. For example, in a **multiplex PCR** different annealing temperatures were found to be optimal for amplification of HPV 16 (50°C, but not 60°C) and HPV 18 (60°C), whereas HPV 6b primers could be annealed at either 50 or 60°C.

A **multiplex PCR assay** designed for the simultaneous amplification of HPV sequences specific for types 6b, 11, 16, and 18 in a single tube has been also described. That amplification system works on cervical scrapes, as well as on paraffin-embedded material.

The HPV PCR amplification systems using **type-specific primer sets** identify respective viral DNA in the target by the product of a particular size.

The presence of **variant or novel viral types** can be identified using **consensus primers**. Using consensus primers designed for the late HPV region, ***L1***, amplification of HPV DNA from a wide range of types, including 6, 11, 16, 18, 31, 33, 35, 39, 40, 42, 45, 51 to 59, as well as 25 other secondary or novel HPV types, can be performed.

Another consensus primer set designed for a 240-bp region of the **early *E6* gene** adds the detection of a gene consistently associated with HPV-related tumors.

The **positive-strand primers** selected for the *E6* region are complementary to sequences just upstream of the *E6* ORF, and contain sequences considered to be the target of the regulatory interaction with the ***E2* ORF product**.

The **negative-strand primers** are designed for the ***E6* coding region**. Other primers have been constructed to target **transcriptional activators**.

For various primer sets some of the amplification reaction parameters had to be **established empirically**, such as

- the optimal primer ratios
- $MgCl_2$ concentrations
- the cycling parameters

Simultaneous amplification of the β-globin fragment can serve to assess DNA **integrity** and the relative **quantity of DNA** in the sample.

Although bands of expected sizes could be visualized in the gels in the majority of tumors, in some cases a **high background** necessitates that Southern hybridization be performed to identify amplification products. **Disintegrated target DNA** and **nonspecific priming** were thought to be responsible for the background signals.

An additional **advantage of the second set of consensus primers** for the

E6 region is that they help detect degraded HPV DNA sequences, as found in archival material and paraffin-embedded tissues.

To expand the HPV types detectable by PCR the **general primer-PCR (GP-PCR)** has been developed to complement the **type-specific (TS)-PCR**. It increased HPV detection from 50 to 80% in mild dysplasia, and from 67 to 88% in severe dysplasia. In cytologically normal scrapes, the prevalence of unsequenced HPV types was about 10%.

The possibility of using GP/TS-PCR amplifications on **crude cell suspensions** makes this system applicable for large HPV-screening programs.

Pitfalls in PCR application for clinical diagnosis are numerous, among which contamination of the testing system by the previously generated PCR products is one source of potential misdiagnosis. A **more rigorous assay** in one discrepant situation indicated that the high estimates of prevalence had resulted from an accidental contamination (or carryover), to which PCR is vulnerable.

6.4.7.8.7 Comparison of PCR with Other Modalities in HPV Diagnosis (p. 545)

Discrepant observations include the following:

- some 91% of dot blot-negative specimens proved to be positive for HPV DNA by PCR, the majority containing HPV16 DNA sequences
- the PCR assay identified HPV DNA in 34.8% of biopsy specimens found negative by ISH
- in another study, dot blot hybridization could identify only 44% of HPV-infected persons. Some 19% of specimens were dot blot negative but PCR positive
- a significant number of histologically **abnormal** biopsies **failed** to reveal HPV DNA by PCR (15%) or dot blot (21%)

The **discrepancies between detection efficiencies** observed can be attributed to

- the difference between the nature of the specimens tested: **cervicovaginal lavage** vs. **scrapings**
- **primer failure**
- the presence of **inhibitors** of PCR amplification
- **false-positive dot blot estimates**
- frequently **subjective estimates** of positivity
- **cross-hybridizations**

Observations have been reported on the as yet unexplained absence of HPV DNA in a subset of tumors carrying a poor prognosis. Dissenting opinions maintain, however, that HPV DNA-negative tumors do not carry a worse prognosis and the presence or absence of HPV DNA in cervical carcinomas has little or no influence on clinical outcome. However, in a retrospective study, two groups of patients could be discerned — a younger, HPV RNA-positive group with a better prognosis, and an older, HPV RNA-negative group with a poorer prognosis.

A large comparative study of randomly selected women tested for HPV types 6, 11, 16, 18, 31, 33, 35, 42 to 45, 51, 52, and 56 was made by commercial hybridization kit [*see Supplement 1*] and PCR. The **higher sensitivity of the PCR diagnostic assay** was in part limited by the narrow range of oligomer reporter probes used in hybridization rather than by the consensus amplimer pair used. Identification of amplified products by gel **electrophoresis** or the **restriction digest** protocol **broadened the range of detectable HPVs**, although both ViraPap and PCR yielded about a **10% rate of uninterpretable results.** The overall agreement between the two methods was 77.6%, with the PCR identifying 1.7 times as many positives as ViraPap.

An important medical observation made in this study was that a **substantial proportion of the population harbors the less common HPV types.**

Methodologically, the identification of amplified products **solely by their sizes** in electrophoretic gels proved to be **unreliable**. REA of the amplified product helped to resolve the occurrence of multiple HPV infections difficult to distinguish only by hybridization methods.

The recommendation for using PCR for screening of HPV infections should await further developments in the amplification method, making it more inclusive and verifiable. A step in this direction is clearly seen in the introduction of **simultaneous amplifications** of the ***L1*** and ***E6*** **ORF consensus sequences.**

S. Syrjanen advocates a reexamination of the concept of the existence of HPV DNA in "healthy women," because

- **latent** infections of the anogenital tract are **extremely common**
- the only **adequately sensitive** method to diagnose such infections is PCR
- a **single Pap smear** is **grossly inadequate** in excluding the presence of clinical, subclinical, or latent HPV infection (or even CIN)
- colposcopic examination with the **acetowhite technique cannot distinguish HPV** infections from changes unrelated to HPV
- currently used **light microscopic criteria fail** to adequately predict subclinical and latent HPV infections

Over 90% of lesions with **koilocytotic atypia** found in condylomata of the **lower female genital tract** contain HPV DNA. Even those lacking koilocytotic atypia harbor HPV DNA in 7% of vulvular and 3% of cervical lesions when ISH is used. When Southern analysis is performed these values are 19 and 36%, respectively. PCR assays give estimates somewhat **lower** than those by Southern hybridization: 27 and 17% for vulvular and cervical lesions, respectively. The discrepancy is apparently due to the presence of a relatively high proportion of **"novel" HPV types.**

The higher estimates by PCR than ISH are attributed to the low copy number of viral DNA, apparently below the threshold of the ISH analysis used.

The relatively high number of viral particles required for ISH to be informative can be used to distinguish **presumably infectious tissues**, lacking koilocytotic atypia, from tissues without "active" viral proliferation when HPV ISH is negative.

PCR detects the presence of HPV DNA in up to 20% of the **penile condylomata** lacking koilocytotic atypia negative for HPV by ISH.

The majority of penile lesions identifiable only by the **acetowhite method** were found to lack HPV DNA by ISH and PCR assays.

While **ISH** identifies 2.4% of cervical and vulvular lesions as containing **two** or **more different HPV types**, a higher rate of occurrence (18%) of two different HPV types has been detected by a **type-specific PCR**.

PCR detected HPV 18 DNA in **anal adenocarcinomas**. In contrast, no HPV DNA could be shown in any adenocarcinomas of the **rectum** and **colon**, or in the **adenomatous polyps** of the colon. ISH assays of the same lesions were positive in only one case of anal SCC and in one case of adenocarcinomas of the cervix.

The high rate (84.6%) of HPV DNA presence in anal invasive SCC has been demonstrated by PCR, whereas ISH was positive only in 50% of cases.

While PCR was able to detect **herpes simplex virus (HSV)** in **advanced** cancers (38.5%) and in high-grade anal intraepithelial lesions (75%), ISH failed to detect HSV DNA at all.

Skin tests using the **HPV16 ORF *E4*** and ***L1* proteins** identify women with CIN as clearly positive for the C-terminal part of L1 ORF.

The informative value of testing for HPV infection by **major capsid protein** expression still remains to be established.

Chapter 6.5

Pathogens Causing Multisystem Infections

6.5.1 *AEROMONAS HYDROPHILA* (p. 549)

Aeromonas hydrophila can cause both diarrheal and extraintestinal infections with hemolytic and soft-tissue necrosis, and wound infections. The major toxin is **aerolysin**.

A **PCR** assay targeting a 209-bp fragment of the ***aer*** gene sequence has been developed, and shows a 100% concordance with biological assays for hemolytic, cytotoxic, and enterotoxin activity of *A. hydrophila*.

Cross-reactivity with streptococcal fragments could be differentiated by ***Nci*I digestion**.

Furthermore, consistent amplification of the streptococcal sequences required 60 cycles with **1 μg** of nucleic acids, whereas only 30 cycles of amplification using **10 ng** of nucleic acids was adequate in the PCR for aerolysin-positive *A. hydrophila*.

6.5.2 *CANDIDA ALBICANS* (p. 549)

Routine laboratory diagnosis of nosocomial candidemia and disseminated candidiasis is relatively slow.

To target the *C. albicans* gene responsible for the conversion of lanosterol to ergosterol, which is inhibited by commonly used antifungal drugs (e.g., miconazole, ketonazole), a **PCR** amplification assay has been developed.

6.5.3 RUBELLA VIRUS (RV) (p. 549)

RV is a single-stranded RNA virus that usually produces an acute, mild systemic illness with fever and exanthema (German measles), but in some cases may result in acute and chronic rheumatological, neurological, and autoimmune disorders.

Isolation of RV in chronic cases and identification by characteristic **cytopathic effects** or by an **indirect enterovirus interference assay** are slow, unreliable, and labor-intensive procedures.

The **ISH** detection of rubella RNA and the diagnosis of fetal rubella by **nucleic acid hybridization** have been described.

Two **PCR**-based assays of reverse-transcribed RV RNA are capable of fetal diagnosis of rubella infection:

- a **nested PCR**-based assay for the detection of both acute and persistent RV infections detects a segment of the RV gene that encodes the E1 membrane glycoprotein of RV
- following **reverse transcription-based PCR (RT-PCR)** the DNA product can be **sequenced** to confirm the specificity of the RV PCR, which can also be established by REA and Southern analysis

The nested RV PCR proved to reduce background amplification and improve the efficiency of amplification. The level of **sensitivity** of the nested RV PCR was 5 fg of total cytoplasmic RNA from RV-infected cells, allowing the **detection of one molecule of viral RNA**.

6.5.4 B19 PARVOVIRUS (p. 550)

B19 parvovirus causes a number of diverse conditions such as the common childhood exanthem called fifth disease, acute joint symptoms, absolute reticulocytopenia, and anemia. Infection during pregnancy may result in **hydrops fetalis**.

A **dot blot** hybridization assay for B19 parvovirus has been used in screening of blood donors.

In a case of chronic bone marrow failure due to persistent B19 parvovirus infection, the presence of B19 virus in the erythroid progenitor cells has been demonstrated by **ISH** and quantitated by dot blot hybridization, while the replicating forms were identified by **REA**.

A dot blot hybridization assay using **digoxigenin-labeled probes** can be used for routine screening of B19 parvovirus DNA in clinical specimens.

A dot blot hybridization assay with radiolabeled RNA probes is capable of detecting as little as **0.3 pg** of viral DNA. The sensitivity of the nonradiolabeled DNA probe is much lower — **3 pg** of viral DNA.

A **PCR** assay is capable of detecting **100 fg** of viral DNA without hybridization of the amplified product. After hybridization with a radioactively labeled probe up to 10 fg of B19 DNA can be detected.

Eighty-three percent **of placental tissues** from women diagnosed to have B19 infections during pregnancy were positive for B19 DNA by PCR. The amplification reaction was 107 times more sensitive than dot blot hybridization with an internal radiolabeled probe.

Not all sets of probes detect all the isolates of B19 virus.

Because PCR detects the virus only during the period of viremia, a more comprehensive evaluation should also **include anti-B19 IgM**.

The PCR assay for B19 DNA is of particular benefit for the evaluation of **chronic infection** with the parvovirus in immunocompromised patients, who fail to mount IgG or IgM responses.

A **PCR** assay on amniotic fluid, fetal serum, and maternal serum for B19 DNA in conjunction with B19 IgM and IgG assays allows identification of acute and chronic infections for the evaluation of high-risk pregnancies.

6.5.5 HUMAN HERPESVIRUS 6 (HHV-6) (p. 551)

HHV-6, also known as **human B-lymphotrophic virus (HBLV)**, was first isolated from patients suffering from AIDS and other lymphoproliferative disorders. HHV-6 leads to **interstitial pneumonitis**, the formerly unexplained lung disease in immunocompromised persons.

HHV-6 differs from known human herpesviruses biologically, immunologically, and by molecular analysis.

Elevated antibody titers have been observed in patients with certain malignancies, Sjögren's syndrome, sarcoidosis, chronic fatigue syndrome, and some B cell lymphomas, and is frequently associated with HIV-induced immunodeficiency. **ISH** of tumors of B cell origin revealed the presence of HHV-6 in some of these.

A **PCR**-based assay for HHV-6 demonstrated the prevalence of this virus among AIDS patients to be 83%.

HHV-6 is also detectable by PCR in the saliva samples of healthy people, and persists the majority of peripheral blood samples following primary infection.

The salivary gland appears to be the **site of replication of HHV-6** and a potential **site for HHV-6 persistence**.

The prevalence of anti-HHV-6 antibody was found to be 54% among normal blood donors, 63% in patients with non-Hodgkin's lymphoma, and 83% in 25 Hodgkin's cases.

HHV-6 is also frequently detected by PCR assays in the **PBMC** from kidney transplant patients.

Short-term cultures offer a reliable **test for reactivation** of HHV-6.

6.5.6 PATHOGENS AFFECTING THE HEART (p. 552)

The presence of **coxsackie B virus**-specific RNA has been demonstrated in myocardial biopsies from patients with chronic dilated cardiomyopathy and myocarditis.

An **ISH** procedure using radiolabeled coxsackie **B3 virus cDNA** demonstrated this approach can be used for an unequivocal diagnosis of enteroviral heart disease.

ISH using a cDNA probe derived from coxsackie **B4** virus-infected cell RNA can reveal different enteroviruses, including **coxsackie A** and **B** viruses, **echoviruses**, and **poliovirus**.

The **coxsackie virus RNA** in cardiac biopsies from patients diagnosed to have the infection is localized in areas **other than histologically recognized sites of affected tissue**.

ISH using **strand-specific probes** for enteroviral RNA of broad specificity may allow a more informative evaluation of the state of viral infection. Total RNA of **acutely infected** cells contains about 100-fold more viral genomic **plus-strand RNA** than complementary **minus-strand RNA**. After the acute phase of infection is over the amounts of plus-strand and minus-strand RNA become similar. Strand-specific hybridization may help in elucidating the virus role in this pathology.

A **PCR** assay detected the presence of enteroviruses in 5 of 48 patients with clinically suspected myocarditis or dilated cardiomyopathy.

6.5.7 ORAL PATHOGENS (p. 552)

Although the major effort has been placed on the study of **streptococci** in oral diseases, a number of other pathogens have also been analyzed at the gene level.

The most prevalent species **in periodontitis** was found to be *Prevotella intermedia*. **REA** demonstrated different digestion patterns among all isolates of *Actinobacillus actinomycetemcomitans*, *Prevotella gingivalis*, and *P. intermedia*, suggesting that no cross-infection occurred among the subjects studied.

As for the **oral spirochetes**, four cultivable *Treponema* species, *T. denticola*, *T. socranskii*, *T. vincentii*, and *T. pectinovorum*, have been identified by **chromosomal DNA probes** isolated from representative strains within each species. The probes were also capable of identifying *T. denticola* in uncultured plaque samples.

Oligonucleotide probes from the **16S rRNA** of oral bacteria identify hypervariable regions. **Species-specific probes** are useful for the rapid detection and identification of *Bacteroides gingivalis* in highly mixed samples from subgingival sulci.

The efficiency of DNA probe assays was found to be an **order of magnitude higher** than that of indirect **immunofluorescence**. It was particularly better for *B. gingivalis* and *B. intermedius*. These probes, however, did not identify any samples as positive for *A. actinomycetemcomitans*.

The importance of identifying oral pathogens, and the extraordinary sensitivity and specificity of gene probes, has led to the establishment of a diagnostic service for oral pathogens (formerly Biotechnica Diagnostics, now OmniGene, Inc., Cambridge, MA).

6.5.8 SOME OF THE PATHOGENS AFFECTING THE NERVOUS SYSTEM

6.5.8.1 HERPES SIMPLEX VIRUS (HSV) (p. 553)

HSV can cause severe **focal encephalitis** that, if not treated early, can be fatal. Because **viral cultures** of CSF are only infrequently productive and **serological tests** are rarely positive in early stages of infection, a **PCR** assay developed for **HSV in CSF** samples can be life saving.

Using primers derived from a **highly conserved region** of the **DNA polymerase gene** of HSV viral DNA isolated from CSF by proteinase K/chloroform-phenol extraction can be amplified.

The detection of viral DNA in the CFS supernatant of ultracentrifuged samples indicated the presence of free viral DNA rather than intact viral particles.

A set of primers to a different region of the HSV genome flanks a conserved region within the **glycoprotein D gene**.

PCR and isoelectric focusing with affinity immunoblotting are suggested as routine diagnostic modalities for adequate and timely recognition of **HSV encephalitis**.

6.5.8.2 RABIES VIRUS (p. 553)

Infection with rabies virus produces eosinophilic cytoplasmic inclusions, called **Negri bodies**, which are not always detectable.

Immunofluorescent techniques and **immunoperoxidase staining** as well as **cell culture** techniques and **mouse inoculation** tests have traditionally been used for the diagnosis of rabies infection.

ISH can also be applied to paraffin-embedded tissues. This technique, using **minus-strand radiolabeled RNA probes**, was developed in an experimental animal system, but it has the potential for clinical application as a diagnostic test for rabies.

6.5.8.3 MEASLES VIRUS (p. 554)

Subacute sclerosing panencephalitis (SSP) is a rare, chronic inflammatory consequence of the involvement of the brain in persistent infection with the measles virus.

ISH using **cloned measles virus DNA** demonstrated the measles viral RNA not only in peripheral blood mononuclear cells, but also in nerve cells and cells from the perivascular infiltrates in brain biopsies from a patient with SSP.

6.5.8.4 EPSTEIN-BARR VIRUS (EBV) (p. 554)

EBV has so far been identified in the lymphoproliferative conditions such as Hodgkin's disease, lymphoepithelioma, infectious mononucleosis, posttransplant B cell lymphoproliferations, hairy cell leukoplakia, angioimmunoplastic lymphadenopathy, lymphomatoid granulomatosis, lymphoepithelial carcinoma of the stomach, in the blood and oropharynx of healthy adults and HIV-positive individuals, and in nasopharyngeal carcinoma.

EBV is detected by **ISH** with biotinylated probes on **formalin-fixed, paraffin-embedded primary CNS lymphomas** from immunocompromised, but not immunocompetent, patients. The extent of virus-positive cells in the tumors showed no correlation with survival.

In a different study, EBV sequences have been found in CNS lymphomas in immunocompromised patients secondary to renal transplantation, HIV infection, leukemia, and Wiskott-Aldrich syndrome. That study used **radiolabeled EBV probes** in ISH, and it is possible that the higher sensitivity of radiolabeled probes accounted for the detection of EBV sequences in lymphomas from immunocompetent patients (see also Section 6.3.1.5).

6.5.8.5 *NAEGLERIA FOWLERI* (p. 554)

Naegleria fowleri is a ubiquitous free-living ameba isolated from heated aquatic environments worldwide. Amebic meningoencephalitis produced by *N. fowleri* is often fatal and because only this species is known to be pathogenic in humans, accurate and early diagnosis is essential.

A **PCR** assay for the specific detection of this pathogenic ameba targets **mitochondrial DNA** sequences from *N. fowleri*.

6.5.9 OTHER VIRUSES (p. 554)

A high rate of infection, particularly viral infections, is observed in **heart-lung transplant recipients** undergoing immunosuppressive therapy. **ISH** with biotinylated probes on lung biopsies and one heart biopsy revealed **CMV DNA**.

Lassa virus causes an often fatal disease transmitted to humans by contact with persistently infected rodents in West and Central Africa. The laboratory diagnosis is usually accomplished by **cell culture**. A combination of **reverse transcription** of the Lassa virus **RNA sequences** followed by **PCR** works on blood and urine specimens, offering a rapid, alternative method of diagnosis for this dangerous systemic infection.

Chapter 6.6

Pathogens Transmitted by Arthropods

6.6.1 VIRUSES

6.6.1.1 ARBOVIRUSES, DENGUE VIRUS (DV) (p. 557)

DV causes the most important mosquito-borne viral infection of humans, affecting millions of people annually in tropical countries. *Aedes* mosquitos are the common vectors of DV.

The laboratory diagnosis of DV relies on **serology** and **virus isolation**, procedures that are either lengthy or nonspecific, although **nucleic acid probes** for *DV2* have been described.

In **spot hybridization assays** with RNA extracted from cells infected with one of 14 different flaviviruses or Semliki Forest virus, DV was detectable by DV-specific **photobiotin-labeled probes**.

Radioactively labeled RNA probes have been constructed for **mixed-phase** and **solution hybridization assays** for *DV2* RNA in mosquito vectors. Solution hybridization was faster (2–3 h vs. 16–20 h) and had a broader probe specificity than mixed-phase hybridization.

To detect the commonly found low-level viremia, a combination of **PCR** amplification of the *DV* cDNA with type-specific probing of the amplified product can be used. The sensitivity of the PCR assay is 20 times higher than that of cell culture.

6.6.1.2 COLORADO TICK FEVER (CTF) VIRUS (p. 557)

CTF virus is the only orbivirus that causes disease in humans, mainly in the western United States and western Canada; its primary vector is the soft-bodied tick *Dermatocenter andersoni*.

Molecular hybridization studies have long been conducted for the genetic characterization of this virus, and an **RNA-RNA dot blot** format has been used to analyze relatedness of various strains.

6.6.1.3 DUGBE (DUG) VIRUS (p. 557)

DUG virus is transmitted by *Amblyomma variegatum* ticks that infrequently infect humans, causing Crimean-Congo hemorrhagic fever, a mild, febrile disease.

A **PCR**-based assay following **reverse transcription** of the viral RNA and combined with **dot blot** analysis of the amplified product with a **DUG-specific cDNA probe** is still inferior in sensitivity to the **intracerebral inoculation** of mice.

The PCR assay, however, yields results within 48 h compared to at least 8 days needed for the inoculation test.

6.6.2 RICKETTSIAE (p. 558)

PCR amplification assays have been described for two members of the Rickettsiae:

- *Rickettsia typhi*, the etiological agent of murine typhus
- *Rickettsia rickettsii* and *Rickettsia conorii*, the etiological agents of Rocky Mountain spotted fever and boutonneuse fever, respectively

Both assays already display sufficient sensitivity and specificity to be regarded as supplementary to, if not **replacing, candidates for routine early diagnosis** of respective pathogens in a clinical setting.

The sensitivity of the assay is around 50 rickettsiae. Very low levels of the organisms at early stages of infection and as yet undefined interferences may partly account for the occasional PCR failures.

Compared to **ELISA** for *R. typhi* the PCR assay is more sensitive when evaluated on rickettsiae isolated from fleas.

6.6.3 SPIROCHETES

6.6.3.1 *BORRELIA BURGDORFERI* (BB) (p. 558)

BB is the etiological agent of the tick-borne **Lyme borreliosis**, affecting most commonly the skin, joints, heart, and nervous system.

The laboratory diagnosis of Lyme borreliosis can be accomplished by **IFA** and **antigen detection**. **Serological evaluation** of Lyme disease patients includes

- **cultivation** of BB from CSF
- **western blot analysis**
- **ELISA**

REA and **DNA hybridization** characterized BB from different isolates in North America and Europe and demonstrated genotypic heterogeneity within this genus and species. It appears that REA is an accurate and reliable method for the identification of Lyme disease agent from different isolates.

PCR assays detect even a single BB spirochete. A BB-specific PCR assay was developed for the insert of one genomic clone of chromosomal origin. The assay is sensitive to fewer than 5 copies of the BB genome even in the presence of a 106-fold excess of eukaryotic DNA.

Another PCR assay for BB, when **combined with dot hybridization** detection, recognizes 10 organisms per milliliter of blood or urine. This BB PCR protocol uses amplification of the ***OSP*-A gene** of the spirochete similar to **yet another PCR assay**. With respect to the infection of *Ixodes dammini* tick with BB, the PCR assay offers an advantage over **direct**

fluorescent antibody staining, which is diagnostic only with live ticks.

A **different BB PCR assay** using sets of primers for the *OSP*-A gene is combined with hybridization of the amplified product, using radiolabeled oligonucleotides; it is reported to detect fewer than 50 spirochetes without cross-reactivity to any other organisms tested.

Direct comparison of *in vitro* culture and PCR assays for the detection of BB in spleen, kidney, and bladder from infected gerbils showed that the PCR assay had at least a comparable diagnostic sensitivity, and was much faster.

Chapter 6.7
Parasites

6.7.1 *PLASMODIUM FALCIPARUM* (p. 559)

Specific **DNA probes** may be used to distinguish between past and present *P. falciparum* infections by spotting lysed blood specimens from infected patients onto nitrocellulose filters.

Hybridization of the parasite DNA with a **species-specific repetitive sequences probe** distinguishes *P. falciparum* from other *Plasmodium* species with a sensitivity of 10 pg of purified *P. falciparum* DNA, which is equivalent to 100 parasites.

This **dot blot** test, requiring minimal equipment and sample handling, was shown to detect approximately 40 parasites per microliter of blood. **Spot assays** detecting 5 parasites per microliter in a 10-µl sample of blood have also been reported. However, nonspecific cross-reactivities were observed.

An oligonucleotide probe for the repetitive sequences detects *P. falciparum* in clinical specimens in a markedly simplified assay. It proved to be highly sensitive (70 parasites per microliter) and specific for *P. falciparum*.

The spot hybridization methods, although convenient in field studies, are semiquantitative and cannot distinguish between different clinical isolates of *P. falciparum*.

Primers to sequences of the **polymorphic major merozoite surface antigen (*MSA* 1) gene** allow targeting of the repetitive elements in the genomes of *P. falciparum* isolates.

Using **hybridization of squash blots** of the *Anopheles quadrimaculatus* with **species-specific repetitive DNA sequence probes**, a simple and rapid identification of large numbers of individual mosquitoes can be accomplished under field conditions. Although described for *A. quadrimaculatus*, this technique can be adapted for rapid identification of other species in field collections.

Primers targeted to a 206-bp *P. falciparum* DNA sequence can detect as little as 0.01 pg of DNA, equivalent to one-half of a parasite, in blood without prior extraction.

6.7.2 *SCHISTOSOMA MANSONI* (p. 560)

A short, **0.64-kb DNA probe** containing 121-bp repeats is used in a **hybridization assay** of high **species specificity** for *S. mansoni*. Because this probe occupies at least 12% of the total *S. mansoni* genome, this repeat motif is highly representative, thereby ensuring a highly sensitive level of detection. Cross-reactivity with *S. haematobium* suggests the presence of this repeat in this species as well.

The specificity of the **pSm1-7 insert** for *S. mansoni* is over 1000 times higher than for *S. haematobium*, and even greater with respect to *S. magrebowiei*. The pSm1-7 0.64-kb probe could detect schistosomal DNA in 1 ng infected snail DNA, allowing the identification of *S. mansoni* in individual infected snails without prior purification of DNA.

6.7.3 *BABESIA* SPECIES (p. 560)

Babesiosis is a tick-transmitted infection caused by malaria-like parasites of the genus *Babesia*, which invade and destroy erythrocytes. The infrequent infections in humans are caused mostly by *B. microti*, although *B. bovis* and *B. divergens* also have been reported in humans.

Laboratory diagnosis is usually made by **microscopy** of Giemsa-stained thin blood smears.

Species identification cannot be made on the basis of morphology only and requires **serological tests** and **animal inoculation**.

A **DNA probe** was shown to be capable of detecting 100 pg of the *B. bovis* DNA; however, it displayed cross-reactivity with other species, limiting its utility for species-specific diagnosis.

Better **species-specific probes** for two *B. bovis* repetitive DNA sequences have been constructed, one of which, Bo25, was capable of distinguishing geographic isolates of *B. bovis* at a sensitivity of 100 pg for *B. bovis* Mexican isolate DNA, and of 1 ng for the Australian isolate DNA.

Two **repetitive DNA probes** for the other *Babesia* species, *B. equi* (affecting horses), are capable of identifying the parasite when present at low levels (limit of detection, 0.49 and 0.97 ng *B. equi* DNA for the two probes, respectively).

6.7.4 *LEISHMANIA* SPECIES (p. 560)

Conventional identification of *Leishmania* strains is performed by **isoenzyme analysis.** Even better identification is possible by using **recombinant DNA probes** cloned from *L. infantum* DNA that recognize characteristic patterns on Southern blots of different isolates.

This approach appears to offer a better tool than previously used **restriction analyses** and **probing with kinetoplast DNA**.

Using **circular amplicons**, which hybridize with amplified DNA cloned from a tunicamycin-resistant strain of *Leishmania*, consistent identification of the drug-resistant isolates is possible. This assay can be used not only for the analysis of the mechanism of drug resistance of *Leishmania*, but also for the identification of drug-resistant strains in clinical situations.

6.7.5 *TRYPANOSOMA* SPECIES (p. 560)

The conventional detection of trypanosome infection in **tsetse flies** is laborious, requiring **dissection** and **examination of midgut, salivary glands**, and **proboscids** of the flies.

DNA probes can be used in **slot blot** hybridization formats for the detection of *T. cruzi*, the etiological agent of **Chagas' disease.**

PCR assay is capable of processing a large number of samples of the insect vector or mammalian blood. This assay, designed for the **repetitive motif** in the nuclear DNA of *T. cruzi*, the most abundant sequence in this organism, is sensitive enough to detect 1/200 of the DNA in a single parasite when a radiolabeled probe is hybridized to the amplified product.

A further refinement in the detection and classification of *T. cruzi* can be accomplished using **PCR amplification of the kinetoplast DNA** combined with **restriction analysis of the amplified product.**

6.7.6 FILARIAL SPECIES (p. 561)

A number of **DNA probes** for filarial species have been developed so far. Emphasis has been placed on the **need for simplification of testing procedures** using nonradioactive probes, better transportation, and preservation protocols for the material under study, and on the need to use gene level diagnostics in conjunction with the established morphologically based methods.

6.7.7 *TOXOPLASMA GONDII* (p. 561)

The first **PCR** assay for *Toxoplasma* DNA in clinical specimens was based on the identification of the **repetitive *B1* gene**, and reportedly could detect the DNA of a **single organism** in a crude cell lysate. In purified DNA preparations this assay could detect as few as 10 parasites in the presence of 100,000 WBCs, which is comparable to the **maximal pleocytosis** encountered in CSF of patients with **toxoplasmic encephalitis.**

The detection of **specific IgM** in fetal blood and inoculation of amniotic fluid into **tissue culture** are the two standard methods used to diagnose cases of **maternal toxoplasmosis**. PCR identified *T. gondii* in five of five samples of amniotic fluid from proven cases of congenital toxoplasmosis. The other two methods correctly identified only 3 or 4 of the 10 positive samples. Other **(indirect) methods** — inoculation of amniotic fluid and fetal blood into mice — were able to detect 7 of 10 positive samples.

The speed, sensitivity, and specificity of the PCR assay clearly make it superior to the other modalities available for **prenatal diagnosis of congenital toxoplasmosis**.

A different PCR assay is based on amplification of a part of the ***P30* gene** of *T. gondii*. Following amplification, the PCR product can be **directly visualized** in the gel or by **Southern hybridization** with radioactive or nonradioactive probes. This assay has been used on brain biopsies of AIDS patients to confirm cerebral *Toxoplasma* infection, and on specimens from pregnant women and their fetuses and infants.

The assay is capable of detecting a single *T. gondii* organism in the presence of over 10^6 human cells.

A second round of amplification, using a different set of primers, may be used in cases of low-level parasitemia; this allows the product to be directly visualized by ethidium bromide. Unequivocal results can be obtained within 7 to 8 h.

Another PCR assay for the detection of *T. gondii* is also based on amplification of the *P30* gene and is coupled with the identification of the amplified product by Southern analysis with a **210-bp insert probe**.

Part 7
Selected Methodological Aspects of Gene Level Evaluation in Hematology

Chapter 7.1

Hemoglobinopathies

7.1.1 SICKLE CELL DISEASE (p. 605)

Traditionally, **postnatal diagnosis** of SS anemia has been determined by **Hb electrophoresis.** Molecular diagnostics allow **prenatal** diagnosis to be performed on amniocytes, chorionic villi, or fetal blood cells obtained percutaneously from the umbilical cord early in pregnancy.

Restriction analysis (REA) has demonstrated **polymorphisms in the β^S-gene cluster** related to the observed variability in the manifestation of the disease.

- Certain **haplotype combinations**, such as the **Senegal chromosome**, are associated with milder forms of SS disease.
- In **SC** individuals the presence of the **Central African Republic (CAR) polymorphism** of the β^S chromosome carries the risk of greater morbidity.
- In the presence of the **Benin haplotype**, the condition of SS and SC patients is intermediate between that of patients with the Senegalese and CAR haplotypes.

The **potential drawbacks** of REA for the SS mutation are

- **incomplete digestion** of the sample
- **plasmid contamination**

Initially, **indirect linkage analysis** has been used, whereas later **direct identification** of mutations was developed using REA and oligonucleotide probes, and most recently, the diagnosis became possible by **PCR**.

The **original PCR amplification procedure** was combined with the analysis of the β-globin amplified product by **solution hybridization** with specific oligonucleotide probes and **subsequent digestion** with a restriction enzyme to determine the genotype. This procedure, in fact, became the prototype for the evaluation of other genetic and infectious diseases by PCR.

Subsequently, a **PCR-ASO** protocol with **horseradish peroxidase-**labeled probes was applied to the detection of **sickle** and **β-thalassemia mutations**. An additional advantage of the PCR-ASO method is in the **ease** of identifying the Hb A allele in the presence of high amounts of Hb F.

PCR-ASO combined with Hb electrophoresis can provide an **accurate early diagnosis** of S or C traits and β^o-thalassemia without the need to electrophorese the parental samples.

PCR provides a means to analyze and definitively diagnose abnormal electrophoretic variants. Moreover, PCR allows simultaneous screening for several genetic abnormalities in one blood sample.

An ingenious use of **two different fluorochromes** to label oligonucleotide primers for a PCR assay allows **direct color detection** of a mutant globin sequence.

This procedure, which can be automated, is also applicable to screening for the sickle gene in **dried blood spot** samples.

The **A**, **S**, and **C alleles** of the β-globin gene in dried blood collected on **Guthrie cards** for **phenylketonuria** screening have been detected by PCR. Hybridization of the amplified product with **antisense ASOs** produced a **sevenfold greater signal** than did sense ASO probes. Only one quarter of a Guthrie card blood spot was utilized per assay. Virtually complete agreement of the PCR-ASO dot blot with Hb electrophoresis was observed.

7.1.2 THALASSEMIAS

7.1.2.1 α-THALASSEMIA (p. 606)

α-Thalassemia is caused by defective **α-globin synthesis**, and the **deletions** of either one or both α-globin genes on **chromosome 16** account for over 95% of α-thalassemia cases. One type of deletion is characteristically prevalent in a particular geographic region.

Prenatal diagnosis of α-thalassemia at the molecular genetic level is indicated in pregnancies in which both parents are carriers of a double α-gene deletion chromosome. This is accomplished mostly by **Southern analysis** with a **ζ-globin probe**. Understandably, an α-globin probe fails to hybridize with DNA of an **α^0-thalassemia** sample (**Bart's Hb in hydrops fetalis**) due to the absence of the target α-globin DNA in that gene.

PCR assays for α-thalassemia simultaneously amplify both the β-globin and α-globin sequences:

- the assay is indicative of α-thalassemia if no amplification products of the α-globin gene are produced
- in **α-thalassemia trait**, in which a **reduced number** (rather than complete absence) of α-globin genes is characteristically present, **even a single** α-globin gene per diploid genome will support PCR amplification. To establish the diagnosis of α-thalassemia trait by PCR, **quantitation** of the amplified product is required
- another confusing situation may be encountered in the **large Filipino deletion** (–,–/α^0, thal-1), which eliminates the entire α-globin gene complex so that Southern analysis in such persons may be indistinguishable from the normal pattern

A protocol combining PCR amplification with **dual-restriction enzyme analysis** has been developed to address this diagnostic challenge. The assay targets an **identical sequence** in both the α_1- and α_2-globin genes, and **the unique sequences** in each α-globin gene region. Analysis of the amplified products distinguishes fetuses with no α-globin genes from those in which one of the α-globin genes is present.

The preliminary Hb electrophoresis on parents allows the detection of **Hb H**, that is, the **presence of a single functional α-globin gene** in one of the parents. Using two different restriction enzymes (*Bgl*II and *Asp*718, an isoschizomer of *Kpn*I), and testing with two different ζ-globin probes, the ambiguities in assigning the observed haplotypes can be resolved.

By using only PCR amplification with allele-specific α-globin primers, a distinction between the homozygous α^+ thal-2 (–,α/–,α) from heterozygous α^o thal-1 (–,–/α,α) haplotypes can be established.

Differentiation between the large deletion haplotype and a normal haplotype by **PCR-dual restriction analysis** amounted to the identification of the normally present 12-kb 3′ ζ-globin fragment absent in at-risk Filipino fetuses.

The double-digest REA enables unambiguous differentiation between the most common α-thalassemia haplotypes to be made.

7.1.2.2 β-THALASSEMIA (p. 607)

β-Thalassemia syndromes are caused by defective production of **β-globin chains**. Essentially two conditions — **trait** and **disease** — are recognized in β-thalassemia, compared to the four α-thalassemia states.

The presence of one nonfunctional β-globin gene causes mild hematological abnormalities (elevated Hb A_2, reduced MCH and MCV) compatible with good health and a normal life span. In contrast, **β-thalassemia major**, or **Cooley's anemia**, manifested by a spectrum of symptoms of varying clinical severity, occurs when both β-globin genes are defective.

Abnormalities of **β-globin gene expression** have been due largely to **deletions** or **substitutions** of either single nucleotides or short or long nucleotide fragments within or near the β-globin gene.

The particular **geographic distribution** of β-thalassemia has been linked to the preferential genetic selection for the trait: affected persons experience lower morbidity in malaria infections.

Over **90 point mutations** resulting in β-thalassemia have been recognized so far. Only a few ethnic group-specific alleles are known to account for over 90% of β-thalassemia genes.

Variations in the clinical phenotype of β-thalassemia have been traced to

- mutations in the **regulatory sequences** outside the b-globin gene, which control the **transcriptional efficiency** of the b-globin gene
- mutations affecting RNA processing:
  a. **RNA modification** at the **cap (7-methylguanosine) site**
  b. **RNA cleavage**
  c. **polyadenylation**
  d. **splicing**
- mutations affecting **translation** of the mRNA into globin

Abnormalities in γ-globin gene regulation accounting for cases of **hereditary persistence of fetal hemoglobin (HPFH)** have been defined by PCR-ASO analysis to arise from base changes in the A_γ promoter. **Point mutations** in the γ promoter itself affect both γ- and β-globin gene expression. Mutations in the **regulatory sequences** of upstream promoters of the γ-globin gene have been implicated in the HPFH phenotypes.

Prenatal diagnosis of β-thalassemia can be made by PCR-ASO analysis of the β-globin gene. This approach is informative predominantly in common Mediterranean alleles, but only some of the β-thalassemia alleles of other groups.

When PCR is combined with **dot blot** hybridization and/or **REA** of the amplified product, the result can be obtained within a week of fetal sampling.

Amplifications using **nonradioactive probes** with or without direct sequencing have been applied successfully to the characterization of mutations and prenatal diagnosis of β-thalassemia all over the world.

Other diagnostic procedures for β-thalassemia include

- **restriction primer extension** of oligonucleotide probes
- the **amplification refractory mutation system (ARMS)**

The ARMS, developed for the **identification of multiple mutations**, is based on **specific priming** of PCR as opposed to PCR followed by ASO hybridization.

- To diagnose a specific mutation, the two oligonucleotide primers used are identical in sequence except for the **terminal 3′ nucleotides**.
- In the absence of a perfect match, that particular nucleotide will not support amplification.
- The ARMS method, using seven primers, was capable of detecting mutations in all 73 at-risk cases, proving to be as reliable as haplotype analysis.

When unknown mutations occur, **restriction fragment length polymorphism (RFLP) analysis** or **direct DNA sequencing** must be used.

REA with probes for tandem repeats of different hypervariable regions (HVRs) can, following engraftment in bone marrow transplantation in homozygous β-thalassemia patients, demonstrate donor-specific fragments in the recipients.

Sequential PCR amplification with two pairs of primers has been applied in testing unfertilized human **oocytes** and the first **polar bodies** isolated from them. The genetic defect in the β-globin gene responsible for sickle cell disease and β-thalassemia could be diagnosed.

Analysis of thalassemic phenotypes caused by **Hb E** mutations common in Southeast Asia has been described in great detail.

PCR with direct sequencing can be used to define a mutation.

Chapter 7.2

Coagulopathies

7.2.1 HEMOPHILIA A (HA) (p. 611)

HA is caused by defects in **factor VIII (FVIII)**, accounting for about 85% of inherited coagulation disorders. The *FVIII* gene on **chromosome Xq28** has been cloned and over 500 hemophilic *FVIII* genes have been studied so far.

The molecular basis of FVIII abnormalities have been traced to

- **gross deletions** in the *FVIII* gene
- **insertional mutations**
- **point mutations** resulting in a **stop codon** (frequently due to the mutation of C to T in **CpG islands**), **nonsense mutations**, or in substitutions leading to **missense mutations**
- **small aberrations** leading to **alterations in FVIII function, stability**, and regulation of its **activity** through interaction with other coagulation factors

The major emphasis in the practical application of gene level analysis of FVIII dysfunction lies in assessment of **carrier status** and **prenatal diagnosis.**

RFLP analysis, in evaluating the carrier state, uses

- **probes within the *FVIII* gene** and restriction enzymes (*BclI*, *XbaI*, and *BglI*)
- **extragenic probes** (*BglII*/DX13, and *TaqI*/St14)

The *BclI*, *XbaI*, and *TaqI*/St14 RFLPs allow **100% carrier detection** in some HA families.

Despite similar phenotypic manifestations, the molecular basis of FVIII deficiency is heterogeneous.

- The molecular basis of **individual cases** of HA can be established by **PCR with direct sequencing.**
- Structural alterations in the *FVIII* gene in large HA **pedigrees** are evaluated by the **PCR-ASO** approach.

The **large size** and relative **complexity** of the *FVIII* gene somewhat limit the diagnostic utility of the PCR-direct sequencing approach, denaturing gel electrophoresis, and mismatch analysis by chemical cleavage.

New RFLPs at the **DXS115 locus** have improved characterization of the carrier status of HA.

A new family of **polymorphic markers** — the **$(CA)_n$ repeats** in intron 13 of the *FVIII* gene — has been described.

The intron 13 $(CA)_n$ repeat is considered to be the most informative marker so far available for *FVIII* gene analysis, which can be performed within a day using PCR amplification.

Other **dinucleotide repeats** noticed in the *FVIII* gene clones may prove useful for HA testing.

Limitation of REA following PCR amplification may be misleading since

- **restriction of short fragments** of amplified DNA is **less reproducible** and **predictable** than that of total genomic DNA
- the restriction is **frequently incomplete**

To overcome these problems, the incorporation of an **internal control** of restriction, based on coamplification of a segment of DNA containing a **nonpolymorphic restriction site**, has been proposed. A fragment generated by *Bcl*I restriction of the β-globin gene cluster has been suggested as a **nonpolymorphic internal control**. Fragments resulting from the polymorphic diagnostic system using *Bcl*I migrate in the gel above or below those of the nonpolymorphic control and thus interference by this control is avoided.

RFLP analysis is the method of choice at the present time for prenatal diagnosis and the detection of heterozygous carriers of HA.

7.2.2 HEMOPHILIA B (HB) (p. 612)

HB is caused by a **diminished level** or **dysfunction** of **factor IX (FIX)**, and in at least one third of cases a specific **point mutation** accounts for the defective FIX circulating in the blood. The following situations can be encountered:

- a **missense mutation** that does not impair either synthesis or plasma-clearing kinetics of FIX, but **alters the function** of the FIX antigen, may occur
- **gross deletions** within the *FIX* gene lead to diminished levels or even absence of FIX from the circulation
- even **small deletions** are known to produce severe HB phenotypes.
- **insertion mutations** are recognized in HB
- the **CpG islands** appear to be **mutational hot spots** although no specific regions of instability can be identified in the *FIX* gene

The ***FIX* gene**, located on **chromosome Xq26-27.3**, has been cloned and various probes have been produced

Intragenic *FIX* probes are sometimes uninformative (as in Japanese HB patients).

REA can be used to define **partial and complete gene deletions** that account for the majority of HB mutations.

PCR amplification with **direct sequencing** allows fast and precise characterization of *FIX* gene alterations. [*See Supplements 1 and 2.*]

The direct sequencing of PCR-amplified genomic DNA fragments has allowed the identification of independently occurring mutations in four unrelated Chinese patients with HB; all mutations were found in exon 8 of the *FIX* gene.

7.2.3 VON WILLEBRAND DISEASE (vWD) (p. 612)

vWD is the most common inherited bleeding disorder of humans, with prevalence estimates ranging from 0.5 per million to 1% of the general population.

More than **20 distinct** clinical and laboratory **subtypes** of vWD are now recognized, presenting essentially as **two forms** of abnormality of plasma **von Willebrand factor (vWF)**

- a **quantitative (type II) form**
- a **qualitative (type I) form**

Type III is a recessive form of vWD with variable presentation. Diagnosis by conventional laboratory methods suffers from

- **low sensitivity** and **specificity**
- **high variability**

The ***vWF* gene** was cloned in 1985 and its structure subsequently defined.

Numerous ***vWF* gene RFLPs** have been described; however, the existence of a **pseudogene** may complicate the interpretation in linkage studies. The pseudogene has about 3% divergence in nucleotide sequence from the **authentic gene**.

Southern analysis in the majority of vWD patients is normal, although in some cases **large deletions** of the entire *vWF* gene have been noted. Only a few vWD pedigrees have been studied by RFLP analysis so far.

The differentiation of these loci by PCR is based on the difference between the **lengths of simple repeats** in the gene and pseudogene.

In most vWD patients studied to date, the **molecular defect** appears to reside **within the gene itself**, although extragenic sites may be defined in the future.

Point mutations affect subtle conformational characteristics of vWF, influencing its functional properties.

Defects in the assembly of large multimeric forms of vWF in the IIA subtype may account for 10 to 15% of cases of vWD.

A **comparative DNA** and **RNA PCR-RFLP** method distinguishes between **mRNA expression** from the two *vWF* alleles on the basis of DNA sequence polymorphisms within exons of the *vWF* gene. Using peripheral blood platelet RNA, this approach can be applied for the analysis of type I and type III vWD patients.

So far the molecular defect accounting for the most common variant of vWD (type I), characterized by a quantitative deficiency of vWF, still remains unclear.

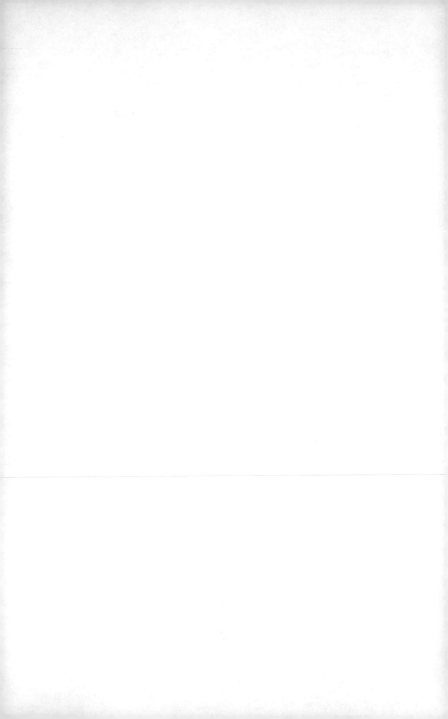

Chapter 7.3

Lymphoid and Myeloid Malignancies

Only **selected methodological issues** in the application of molecular biological tools to the diagnosis and monitoring of lymphoid and myeloid malignancies will be briefly discussed.

7.3.1 DNA CONTENT AND CELL CYCLE ANALYSIS (p. 615)

The evaluation of DNA and RNA content either by **flow cytometry (FCM)** or **image cytometry (ICM)** complements immunophenotyping and traditional histopathological evaluation of **lymphoma**, and monitoring of **minimal residual disease (MRD)** in lymphoma and **acute lymphoblastic leukemia (ALL)**; in some cases FCM and ICM offer a superior alternative.

Although FCM demonstrates the **heterogeneity** of **acute myeloblastic leukemia (AML)**, the measurement of DNA content is subject to significant error due to **variability of sample composition**.

The close correlation between **DNA hyperploidy** and expression of the common ALL antigen (**CALLA**) emphasizes the validity of **immunological phenotyping** of ALL for patient management.

Diagnosis of CML and CLL using traditional morphological and immunophenotyping methods does not present a particular challenge. DNA ploidy analysis may not be indicated.

Refinements in FCM methods for estimating the **DNA**, **RNA**, as well as **double-stranded RNA (dsRNA) content** in **leukemias** have better defined the cell cycle kinetics in AML and ALL, but the **practical utility** of these measurements is not apparent yet. Some of the **limitations of DNA FCM** include

- its inherent **insensitivity to balanced translocations**
- **nonuniform dye binding** by chromatin DNA due to **stereohindrance effects**
- the apparent **lack of prognostic significance** of DNA aneuploidy in AML
- inferiority compared to the utility of **cytogenetic evaluation**
- DNA aneuploidy shows **no relation** to **FAB subtype**, **WBC** count, **S-phase index**, and the **amount of blast forms** in the bone marrow in AML and ALL patients

- **RNA indices**, in contrast, show marked differences between AML and ALL

RFLP analysis and **ISH** with **chromosome-specific probes** have convincingly demonstrated the clonal origin of leukemic cells defined by monoclonal antibodies or morphology.

The presence of **multiple lineages** in acute leukemias necessitates the combined use of DNA and RNA FCM analysis, together with immunophenotyping.

CML is often characterized by multiple lineages and cytogenetically is identified by the **t(9,22) translocation** in about 90 to 95% of adult cases carrying a favorable prognosis. FCM of CML cells shows a characteristic pattern, supporting the practical value of FCM analysis in the early detection of transformation events:

- **myeloblastic** transformation is accompanied by a **high RNA index**
- **lymphoblastic** transformation is accompanied by a **low RNA index**

Immunophenotyping of **non-Hodgkin's lymphomas (NHL)** by FCM based on **stage-specific** and **lineage-specific** antigenic determinants is an established diagnostic modality that utilizes a vast number of **monoclonal markers**.

DNA FCM of lymphoma, similar to the situation in leukemia, detects DNA aneuploidy reflecting numerical cytogenetic abnormalities.

A serious **limitation of FCM** is that it cannot detect translocations characteristic of specific morphological types of lymphoma:

- **t(8;14), t(8;22),** and **t(2;8)** in **diffuse small noncleaved cell lymphoma**
- **t(14;18)** in **follicular lymphomas**
- **deletion 6(q21)** of **diffuse large lymphoma**
- **trisomy 12** and **t(11;14)** in **diffuse small lymphocytic (well-differentiated) lymphomas**

Nevertheless, DNA/RNA FCM in lymphomas, when combined with immunophenotyping, can be used to screen for submicroscopic bone marrow involvement with lymphoma cells.

Cell cycle kinetics determined in lymphoma by DNA/RNA FCM may help in the prognostic evaluation of patients.

The detection of MRD in lymphoma based on **B cell clonal excess** detects 1 to 5% of lymphoma cells in BM specimens, comparable in sensitivity to Southern analysis, but certainly inferior to **PCR**.

Only limited data are available on the use of FCM in lymphoma for the evaluation of oncogene expression which demonstrates cell cycle variation in **c-*myc***, ***p21^{ras}***, ***bcl*-2**, and **c-*myb***.

7.3.2 MYELODYSPLASTIC SYNDROME (MDS) AND MINIMAL RESIDUAL DISEASE (MRD) (p. 616)

The fundamental biological characteristic of MDS lies in the **impairment of differentiation capacity** at the level of the multi- or pluripotent stem cells, which accounts for the diverse hematological manifestations.

The significant cellular heterogeneity of MDS, usually involving two or more hematopoietic cell lineages, has been supported by FCM studies.

Multiple chromosomal abnormalities consistently involving **chromosomes 5, 7,** and **8** have been correlated with a particular clinical course and overall survival. **DNA hypoploidy** is associated with the poorer survival prognosis. It appears, however, that chromosomal aberrations per se do not initiate MDS but should be viewed as markers of a broader **genomic instability**.

MRD evaluation constitutes part of the monitoring of leukemia, traditionally performed by DNA ploidy determinations, or by immunological assessment of the persistence of the abnormal phenotype. Among other techniques for MRD monitoring are:

- the detection of chromosomal abnormalities caused by **premature chromosome condensation**
- the identification of the specific translocations **t(9,22), t(8,14),** and **t(4,11)** by the **PCR**.

7.3.2.1 PCR IN MONITORING MRD (p. 617)

PCR amplification with primers specific for the *bcr/abl* rearranged sequences and their corresponding transcripts is the most sensitive measure of the persistent presence and, therefore, expression of malignant cells.

- In patients in whom expression of the *bcr/abl* rearrangement is not detectable by **reverse transcription** of total RNA followed by **PCR amplification**, complete remission could be observed 5 to 7 years following BMT.
- Multiple PCR analyses should be performed in individual patients following BMT because, despite an initial negative result, some CML patients have been shown to become PCR positive and eventually demonstrate clinical relapse.
- The presence of ***bcr-abl*** **mRNA** identifiable by PCR in Ph[1]-ALL patients predicts hematological relapse, apparently due to the high proliferative potential of the Ph[1] clone.

PCR analysis for the *bcr* gene can also be used in **ALL**. The majority of ALL cases show

- **T cell receptor (TCR)-δ gene rearrangements**
- specific **immunogenotypes** recognized in the area of the ***V-J*** **junction**
- **imprecise** ***VDJ*** **joining**
- extensive **insertion of N-region nucleotides**

Children in complete clinical, hematological, as well as Southern blot-documented remission were tested by PCR using **patient-specific TCR-δ probes**, and 10^{-3} to 10^{-5} neoplastic cells were detected in the bone marrow of those children who were PCR positive.

Possible clonal variations of the immunogenotype that occur in ALL patients may escape detection by PCR using **clonospecific probes**.

Detection of MRD by PCR during complete clinical remission **precedes**

clinical relapse in some cases. Residual leukemic cells can be detected 6 to 18 months following successful induction of remission, emphasizing the need for maintenance therapy.

A single PCR evaluation is of **limited value** because it may fail to detect MRD due to the focal nature of the disease and the absence of leukemic cells in a single bone marrow (BM) sample.

Caution is advocated in extrapolating the so far limited experience in the PCR monitoring of MRD to the development of patient management strategies until large-scale studies firmly establish the clinical significance of the detection of residual leukemic cells at the level of PCR amplification.

It is suggested that this modality be used in conjunction with clinical, morphological, and immunological evaluation of patients. In fact, a comparative study by **double-color immunofluorescence analysis** using **cytoplasmic CD3/TdT markers** and PCR amplification of rearranged TCR-δ genes reaffirmed the **absence of false-positive** results in predicting relapse by this technique. It may fail, however, in 25 to 30% of cases. The importance of using combinations of different techniques for monitoring MRD cannot be overemphasized.

A different approach to PCR detection of residual leukemic cells is based on the identification of the **complementarity-determining region III (CDRIII)** sequences.

In this assay, the primers are constructed homologous to consensus sequences in the **variable (V_H)** and **joining (J_H)** segments that flank the **intervening diversity (*D*)** segment in the rearranged heavy-chain immunoglobulin locus. Amplified CDRIII segments specific for leukemic cells are sequenced and diagnostic oligonucleotide probes are constructed that do not cross-hybridize with the CDRIII sequences of normal B lymphocytes.

So far, **leukemia-specific chromosomal aberrations** have been characterized at the molecular level only in a limited number of patients, and the development of these **novel B lineage probes** complements the PCR assays developed earlier for lymphoblastic leukemias. **The sensitivity of these PCR protocols is comparable to that using T cell-specific rearrangements,** being on the order of 10^{-5} cells.

Quantitation of leukemic cells in BM was done by multiplying the number of PCR-positive cells by the ratio of lymphocytes estimated in the specific BM sample by morphological analysis.

The **CDRIII-PCR** with **quantitation** of the positive sequences detects 1 leukemic cell among 100,000 normal cells, representing 2 to 3 orders of magnitude improvement over **TdT**, **cytofluorimetric** analysis with lineage-specific monoclonal antibodies, and **Southern analysis** of heavy-chain immunoglobulin rearrangements.

The **potential limitation** of the CDRIII-PCR protocol is related to the specific **primer sequence homology** to a V_H region and may lead to **amplification of normal B cell sequences**. The specificity of PCR amplification of the leukemia-specific sequences apparently can be improved by using primers completely homologous to the **subgroup 5 variable-region** segments.

In conclusion, monitoring of MRD by PCR amplification offers a highly sensitive technique capable of detecting and **predicting the clinical behavior** of residual leukemic cells.

Fine aspects of primer sequence selection still require experimental and clinical **verification in larger series** of patients. Used in conjunction with other evaluation methods, PCR quantitation of MRD in leukemia will likely become a routine monitoring tool, significantly improving patient management.

7.3.3 CYTOGENETIC ANALYSIS

7.3.3.1 MYELOID MALIGNANCIES (p. 619)

The well-documented **t(9;22)(q34.1;q11.21) translocation** found in 90 to 95% of adult CML (Ph-positive) cases is associated with rearrangement of c-*abl* and formation of an ***abl-bcr* fusion gene** that gives rise to the aberrant **8.5-kb chimeric *bcr-abl* mRNA** and the **210-kDa Bcr-Abl fusion protein**, rather than the **145-kDa Abl product**.

The presence of t(9;22) and the *bcr* rearrangement is associated with reportedly **more favorable prognosis** for CML patients. Numerous contradictory reports, hoewever, fail to demonstrate a significantly different clinical course, as well as question the diagnosis of CML and of ALL in such patients.

Karyotyping alone appears to be inferior to **Southern analysis** combined with **dilution studies** and **densitometric scanning of autoradiographs** for the diagnosis of CML and in monitoring treated patients. The latter approach allows the **quantitation** in an individual patient of the dynamics in the relative number of cells affected by the *bcr*/*abl* recombination events.

In addition to the t(9;22) translocation, other chromosomal aberrations can be detected in myeloid malignancies by cytogenetic techniques, such as

- **duplication of the Ph chromosome**
- **trisomy 8, 19,** and **21**
- **isochromosome (17q)**
- **t(15,17)**
- **loss of the Y chromosome** and many other chromosomal aberrations

In the course of progression of MDS, sequential cytogenetic studies revealed a **karyotypic evolution paralleling the clinical course** of the patients, supporting the notion that an **unstable karyotype** can be associated with a poor prognosis.

7.3.3.2 LYMPHOID MALIGNANCIES (p. 619)

Chromosomal translocations involving **antigen receptor complexes** of **T cells (14q11)** or **B cells (14q32)** are seen in two thirds of **NHL** and one-third of the **lymphocytic leukemia** cases.

A combination of **Ig heavy-chain** and **TCR gene rearrangements** found in some NHL and lymphocytic leukemia cases reflects their origin at the early (pre-B and pre-T) stage of lymphoid differentiation.

A characteristic cytogenetic aberrations in **Burkitt's lymphoma t(8,14)** involves the juxtaposition of c-*myc* from **8q24** to the **joining** or **switch region** of the **immunoglobin heavy chain**, resulting in **dysregulation of c-*myc* gene expression**.

Conflicting observations have been made on associating an adverse prognosis with **t(1;19)** in **ALL** patients with the **pre-B immunophenotype**. In a recent and extensive study **neither this nor other translocations were associated with a worse outcome**.

However, a larger series of pre-B ALL cases emphasized the worse prognosis for **non-T, non-B cell childhood ALL** with chromosomal translocations, the t(1;19) being largely responsible for the poor prognosis of the pre-B subgroup.

In contrast, **T cell childhood ALL** appears to have a nonrandom occurrence, frequency, and degree of immunophenotype association with **t(1;14) (p34;q11)**.

In patients with **B cell CLL**, chromosomal analysis offers prognostic indicators of poorer survival when **chromosome 14q** rather than **chromosome 13q** is present, as well as when **trisomy 12** and a higher percentage of cells in metaphase with chromosomal abnormalities are detected.

Acquisition of aberrations of the short arm of **chromosome 17** correlates with a rapidly progressive course with short survival. Allelic loss of the ***p53* gene** may be involved and evaluation of *p53* offers a useful prognostic tool.

The most common **follicular-type lymphoma** constitutes up to 40% of NHL and the majority of patients have a **t(14,18) (q32;q21)** that is accompanied by a rearrangement of the ***bcl*-2 oncogene** from **18q21** to the IgH region on **chromosome 14q32**.

A number of other chromosomal aberrations recognized by cytogenetic techniques have been described in lymphomas.

The best laboratory approach is to combine cytogenetic analysis with other modalities such as morphological, immunological, and molecular biological evaluation of specific gene rearrangements.

7.3.3.3 HODGKIN'S DISEASE (HD) (p. 620)

Although cytogenetic studies in HD are few, the emerging impression is that **chromosomes 14q, 11q,** and **6q** are each involved in slightly over 30% of cases, and **8q** is involved in 18%.

The **most frequent breakpoint** is at **14q32** — the site of the Ig heavy-chain genes; others contend, however, that the cytogenetic abnormality that sets HD apart from other lymphomas appears to be that involving **chromosome 11q23**.

The weight of the data does not support the frequent occurrence of monoclonal rearrangements of immunoglobulin or TCR genes in HD.

No correlation could be noted between the presence of rearranged bands and the number of **Reed-Sternberg (RS)** cells.

7.3.4 ISH IN HEMATOLOGICAL DISORDERS (p. 621)

ISH analysis is capable of demonstrating **individual cell involvement** with a pathological process. In fact, ISH allows the detection of Epstein-Barr virus (EBV) in lymphocytes of patients with infectious mononucleosis and Burkitt's lymphoma, in RS cells in HD, and in hairy cell leukemia.

The long-recognized capacity of EBV to transform and/or immortalize T cells and B cells has been ascribed to the activation of **EBV latent membrane protein (LMP)**.

Inherent **antigenic cross-reactivities** between EBV, herpes simplex virus 1 and 2 (HSV1 and 2), and cytomegalovirus (CMV) plague even the better monoclonal antibody methods. The more direct gene level detection of EBV in hematological diseases is, therefore, highly desirable.

ISH for EBV DNA has repeatedly demonstrated the viral nucleic acids in RS cells in HD, lymphoproliferative conditions associated with heart-lung transplants, hairy leukoplakia, B and T cell lymphomas, as well as benign lymphadenitis and HIV-associated persistent generalized lymphadenopathy.

EBV gene expression can be visualized in interphase nuclei of a latently infected human lymphoma cell line.

ISH combined with confocal microscopy is capable of defining a **spatial topography** of specific chromosomal regions and genes in **hematopoietic cells**. FCM analysis of **FISH**-stained hematopoietic cells can be used for evaluation of **β-globin expression**.

Detecting numerical chromosomal abnormalities (**monosomy 9** or **trisomy 9**) in **hematological neoplasias** by interphase cytogenetic analysis with ISH in BM or peripheral blood cells appears to be a clinically useful method.

ISH demonstrated chromosomal aberrations and specific gene rearrangements in HD-derived T cells, using DNA probes for TCR-α, TCR-β, the *met* oncogene, and rRNA.

With the help of specific probes for the *bcr/abl* fusion gene, the ISH technique and Southern analysis help to elucidate the **variant translocations** in CML.

ISH using **antisense RNA/mRNA hybridization** is clinically useful in detecting low levels of expression of the c-*abl* oncogene in NHL. It allows a semiquantitative estimate of the v-*abl*, *bcr/abl*, and c-*abl* gene expression correlated with the predicted relapse. It is suggested that evaluations by anti-oncoprotein antibodies be carried out in parallel with ISH.

EBV can be easily identified by **PCR** under the same conditions in which ISH is informative as well as when ISH does not detect the virus. Some of the disorders already **tested for EBV by PCR** include

- **infectious mononucleosis** (spleen, lung, and lymph nodes)
- **immunoproliferative disorders** of immunosuppressed patients
- **nasopharyngeal carcinoma**
- **lymphomatoid granulomatosis**

- **angioimmunoblastic lymphadenopathy**
- **lymphoepithelial carcinoma of the stomach**

ISH has also been helpful in delineating fine events in **megakaryocyte gene expression**.

7.3.5 GENE REARRANGEMENTS (p. 622)

Molecular biological characterization of antigen receptor gene rearrangements in the ontogeny of T and B cells provided a foundation for the use of defined gene probes in the clinical evaluation of lymphomas and leukemias. [*See also Supplements 1 and 2.*]

The predominant proliferation of an individual cell carrying a **unique rearrangement** of its antigen receptor genes gives rise to a clone that can be identified and distinguished from cells with a nonrearranged configuration.

This recognition is based on **analysis of restriction patterns** of DNA from respective cell populations digested with the **same restriction enzyme**:

- **The presence of fragments (bands) differing from the predominant germline band identifies clones of cells with rearrangements.**

In Southern analysis clonal populations can be identified only if they constitute **no less than 1%** of the total cells.

Southern analysis of gene rearrangements is used essentially in three situations:

- to assess the **clonality** (monoclonality vs. oligo- or polyclonality)
- to determine **cell lineage** (B vs. T vs. nonhematopoietic differentiation)
- to evaluate the existence of a **multiclonal** or **multilineage origin** of tumor cells in some hematopoietic disorders

Identification of rearrangements is independent of the target **gene expression**, and of the presence of **clonal antigenic markers** detectable by immunophenotyping.

In evaluating cases with possible **multilineage origins**, specific immunoglobulin or TCR probes may suggest the tumor cell origin at an early stage **prior to gene rearrangement** when malignant transformation took place.

Multiclonal derivation can be established on the basis of the specific pattern of rearranged bands assignable to different clones of lymphocytes.

Evidence of **monoclonality *per se* does not imply malignancy**, because a number of clinically **benign conditions** may have clonal populations of T and/or B cells, and **precise distinction between benignity and malignancy in these conditions is not straightforward**.

DNA analysis has been found to help in resolving uncertainties arising in the diagnosis of **peripheral T cell lymphomas (PTCLs)**: DNA hybridization helped in differentiating tumors with polymorphous large cell proliferations that included PTCL, SIg-negative large cell lymphoma of the B cell type, and HD, lymphocyte depletion type.

The **interpretive problems** in Southern analysis of gene rearrangements may include

- situations in which the **bands in restriction patterns do not represent rearrangements**. These situations arise when **genetic polymorphisms** of immunoglobulin or TCR genes exist, and can be resolved by **digesting nonlymphoid DNA** and **comparing the patterns**
- other sources of irrelevant bands, such as **cross-hybridization** with pseudogenes, bacterial plasmid contamination, and incomplete digestion; these should be excluded when analyzing Southern blots by incorporation of **appropriate controls**
- the **absence of expected rearranged bands**, which may be due to
  a. **incorrect diagnosis**
  b. the presence of **target cells below 1%** of the total cell population
  c. **comigration** of the rearranged band with the germline band
  d. **failure to reveal clonal gene rearrangements by Southern analysis**; this occurs in some cases, at least of T cell lineage, despite the overtly malignant characteristics of the cells

A large multi-institutional study of gene rearrangements in the diagnosis of lymphomas and leukemias demonstrated a high correlation with conventional immunotyping and morphological diagnosis.

A **set of guidelines** have been proposed, such as:

- bands on Southern blots must **equal the intensity of the 3% positive sensitivity control** to be considered as rearranged
- the use of **three restriction enzymes** (*Eco*RI, *Hin*dIII, and *Bam*HI) is proposed in every case, to **exclude partial digestion** and **cross-hybridization**
- a rearrangement should be established with **at least two** of the three enzymes or **two rearrangements** must be observed with one single enzyme (two rearranged alleles)
- any $\mathbf{J_k}$ **rearrangement** was regarded as **evidence of B cell lineage** (it is rarely seen in T cell neoplasms), whereas $\mathbf{J_H}$ and $\mathbf{C_T\ \beta}$ **rearrangements** can be observed in both common (pre-B) and T cell acute lymphoblastic leukemia and lymphoblastic lymphoma

Other **suggestions for improving diagnosis of gene rearrangements** include the following:

- to achieve **better coordination** between conventional morphological and immunohistochemical studies and DNA hybridization, one of the sections prepared as **fresh-frozen** preparations can be used to **extract DNA** for subsequent **Southern analysis**
- the quality and quantity of DNA recorded from such sections can be affected by **sample size** and **artifacts produced by ice crystals**
- in addition to ensuring **greater sharpness of bands**, the alkaline transfer method saves time compared to the standard Southern high-salt transfer method
- hybridization can be performed on **dried gels** instead of on filters, the main benefit being the elimination of two overnight transfer and overnight prehybridization
- the **yields** of high molecular weight DNA suitable for Southern analysis

of antigen receptor gene rearrangements from frozen and ethanol-fixed, paraffin-embedded tissues are comparable

- the DNA extracted from **formalin-fixed, paraffin-embedded tissue** is more **degraded** and produces variable results in Southern analysis, with less than half of the results being interpretable
- the yields and results on restriction analysis of DNA extracted from **ethanol-fixed tissue** stored for 2 years are similar to those of freshly fixed tissue
- the **length of formalin fixation** is a contributing factor to progressive DNA degradation
- under optimal fixing conditions Southern blots are essentially similar in both fixed and unfixed tissue, except for somewhat reduced electrophoretic mobility of DNA from formalin-fixed tissue
- suboptimally fixed tissue apparently leads to autolysis, resulting in degraded DNA
- **biotinylated DNA probes** are acceptable for gene rearrangement studies in a routine diagnostic laboratory

Although the traditional approach to the analysis of gene rearrangements has been **Southern blotting**, **PCR**-based techniques progressively complement, and in some cases may even replace, Southern analysis.

7.3.6 ANALYSIS OF GENE REARRANGEMENTS AND CHROMOSOME TRANSLOCATIONS BY PCR (p. 624)

At the junction of rearranged gene segments **unique sequences** are produced, which represent **more specific clonal markers** than positions of bands on Southern blots.

Using **primers** targeted for sequences spanning the junction region, and those containing nonspecific DNA sequences flanking the junctional region, a specific amplification of the clonal γT cell receptor gene rearrangement can be obtained in bone marrow cells with a sensitivity of one neoplastic cell out of 10^5 to 10^6 cells.

Thus,

- combining PCR with **denaturing gradient gel electrophoresis** of the amplified product, the clonality of lymphoid cells encountered at low levels in skin biopsies from early **mycosis fungoides** has been demonstrated by a rapid, nonradioactive approach

Although by far exceeding Southern analysis, PCR may also be prone to **sampling artifacts**.

PCR has been used to detect, for example,*

- **residual *bcr/abl* translocation** in CML, CML following BMT, and in Ph-positive ALL
- the **translocation t(14;18)** at both *mbr* and *mcr* in lymphoma tissue and even in the peripheral blood.

* See also Section 7.3.2.1.

- the **precise size of the amplified product** can be used as a clonality marker
- ***bcr-abl*** **fusion sequences**; these are rapidly and reliably detected by a **hybridization protection assay (HPA)** combined with PCR as an alternative to Southern analysis of amplified products. HPA uses **acridinium ester-labeled oligonucleotides** synthesized by Gen-Probe, Inc (San Diego, CA) complementary to the *bcr-abl* junction sequences. [*See also Supplement 1*.] One set of probes detects amplified transcripts encoding **p210**$^{BCR\text{-}ABL}$, whereas the other set detects **p190**$^{BCR\text{-}ABL}$. The resulting hybrids are detected by **chemoluminescence** following appropriate incubation and washing steps
- **rare neoplastic B lymphocytes** in a population of normal cells, using **consensus primers** of the **chain determining region 3 (CDR3)** of the variable and joining regions of the immunoglobulin heavy chain gene
- **leukemia-associated translocations**, such as **t(6;11)**; this novel approach uses PCR amplification of specific target sequences from small numbers of **microdissected chromosome fragments**
- **lymphoma-specific t(14;18) translocation** in occult cases of follicular small cleaved cell lymphoma with a relapse in the central nervous system

7.3.7 EVALUATION OF ONCOGENES IN HEMATOPOIETIC NEOPLASIA

Multiple oncogenes and growth factors cooperate in the course of malignant transformation of hematopoietic cells, and the **relative prominence** of a given set of oncogenes and growth factors can be related to the clinically relevant behavior of a malignancy.

Most of the evidence on oncogene involvement in hematological malignancies has been gathered in experimental systems, and extrapolations to clinical situations in many cases await further confirmation.

7.3.7.1 c-*abl* (p. 625)

Oncogene activation is most frequently associated with chromosomal translocations leading to the **juxtaposition** of a modified protooncogene and one of the antigen receptor genes.

The prototypical c-*abl* oncogene forming a **fused *abl/bcr* gene** in **t(9;22)** translocation is diagnosed and monitored in Philadelphia chromosome-associated malignancies.

Measuring ***abl*** **mRNA** by **northern analysis** and the respective protein by **anti-Abl antibody** is helpful in CLL, because the majority of Abl-positive cases display a more aggressive course of disease.

7.3.7.2 *bcl*-2 (p. 626)

Translocation **t(14;18)** leads to activation of the *bcl*-2 (B cell leukemia-lymphoma) protooncogene, the product of which is present in transformed cells harboring this translocation.

Chromosome 18 DNA probes identify the *bcl*-2 gene in virtually all follicular neoplasms and in about a third of diffuse large cell lymphomas.

The Bcl-2 protein can be easily detected in frozen sections of lymphoid tissues and tumors using monoclonal antibodies in routine immunohistochemical procedures.

In lymphomas other than follicular type (e.g., diffuse lymphomas of small B lymphocytes, B-CLL, and T cell neoplasms), Southern analysis identifies only germline configuration of *bcl*-2.

It can also reveal sequential *bcl*-2 and c-*myc* activation in fine needle aspirate (FNA) specimens during follicular lymphoma progression, particularly when tumors lack significant sclerosis or cell degeneration. Rearrangements of the c-*myc* oncogene accumulate along the transition of initial NHL to a high-grade malignancy.

bcl-2 is not strictly specific for the t(14;18) translocation — it is also expressed in mucosa-associated lymphoid tissue as well as in follicular lymphomas without the t(14;18) translocation. While progressive reduction of *bcl*-2 expression is seen with induction of differentiation, the high expression of *bcl*-2 in myeloma cells leaves the issue of *bcl*-2 expression in the context of differentiation not yet resolved.

7.3.7.3 *bcl*-1 (p. 626)

Although *bcl*-1 is associated with the **t(11;14)** translocation located at 11q13 and is usually found in diffuse small and large cell lymphomas, CLL, and multiple myeloma, the *bcl*-1 locus seems to be only rarely (ca. 4%) rearranged in B cell CLLs and NHLs.

When it is rearranged in B cell neoplasms, it occurs predominantly in low-grade neoplasms, and is not associated with clinical aggressiveness, advanced clinical stage, or large cell transformation (Richter's syndrome). These observations do **not support the clinical usefulness** of evaluating *bcl*-1 rearrangements.

7.3.7.4 c-*myc* (p. 626)

The **t(8;22)** translocation in **Burkitt's lymphoma** results in alteration of c-*myc* and leads to its activation.

Activation of c-*myc* resulting from a chromosomal translocation is another classic example of involvement of oncogenes in tumor development.

The **augmented expression** of c-*myc* following transformation in experimental systems occurs at the transcriptional level. The **transcriptional deregulation** of c-*myc* in Burkitt's lymphoma may result from the position and number of mutations produced in and around the *myc* locus.

The level of expression of the c-Myc protein in lymphomas studied by **immunohistochemical techniques** and **FCM** was found to correlate with the overall DNA synthetic and ploidy characteristics.

More clinical studies are needed to establish whether the c-Myc protein level should be used as a practical tool in managing patients with lymphoid malignancies.

At the gene level, the pattern of **c-*myc* rearrangements** in primary gastrointestinal lymphomas is different from those usually seen in node-based follicle center-cell lymphomas and is characteristic of aggressive lymphomas. In acute myeloid leukemia, dysregulation of c-*myc* leads to enhanced stability of c-*myc* mRNA (as well as c-*myb* mRNA).

c-*myc* amplification reveals a **sequential transition** of these amplified sequences from submicroscopic circular extrachromosomal DNA (**episomal form**), to **double minutes**, and later to a **specific chromosomal site**. This shift of the c-*myc* sequences is associated with a more rapid proliferation of the cells carrying the amplified sequences.

7.3.7.5 *ras* (p. 627)

Point mutations in N-*ras* or Ki-*ras* genes are found in one third of leukemia cells of patients with AML, some cases of ALL, and occasionally in Burkitt cell lines. H-*ras* rearrangements associated with activation of the oncogene are present in T cell tumors, and in some cases of myeloid leukemia (AML and CML).

N-*ras* mutations in **codon 12** have been reported in lymphoid leukemias more frequently than other mutations, whereas the mutations at **codon 13** are rare in acute and chronic leukemias, but occur in MDS.

In general, *ras* mutations are rare events that occur at a late stage in CML during myeloid blast crisis. N- or Ki-*ras* mutations in codons 12 or 13 appear to be frequent in CML, which may be helpful in prognosis if confirmed in large studies.

The detection of point mutations of N- and Ki-*ras* oncogenes, in particular by **oligonucleotide hybridization** or by newer **PCR**-based assays, may be useful in following the progression of the leukemic clone during and after **therapy** of acute nonlymphocytic leukemia and in acute myelomonocytic leukemia.

Conflicting reports do not support the notion that the presence of N-*ras* point mutations clearly identifies a unique clinical or biological subset of AML patients.

ras mutations, particularly those of N-*ras*, are a well-recognized feature of **MDS** and are found to appear at different stages of the disease.

In aggressive diffuse **CALLA-positive lymphomas**, a **coexpression** of $p21^{N\text{-}ras}$ and c-*erb*B-2 (*neu*) oncogene products has been detected.

A **synergistic action** of c-*myc* and Ha-*ras* oncogenes has been demonstrated in EBV-immortalized human B lymphocytes and the transformed phenotype correlated with the level of Ras oncoprotein expression.

7.3.7.6 OTHER ONCOGENES (p. 628)

c-*raf*-1 protooncogene has been studied in **NHL** and **ALL** patients and tumor derived cell lines.

c-*myb* has been implicated in the development of **T lymphocyte proliferation** and expression of the c-*myb* gene is altered in some **leukemias** and **NHL**, and in the erythropoietin-induced **erythroid differentiation**. The enhanced expression of c-*myb* and **B-*myb*** (but not A-*myb*) is correlated with the proliferative activity of T and B lymphocytes, but not monocytes and granulocytes.

c-*fms* is associated with **monocytic lineage differentiation** induced in cultured cells. No c-*fms* expression is found in acute lymphoid leukemias whether of T or B origin, but it appears to be a **specific marker of leukemogenesis** in the myeloid cells. In AML c-*fms* reaches high levels of expression at the M5 stage.

c-*fes* mRNA can be demonstrated **in AML** and **CML**, and at much lower levels **in ALL**. It is believed that an **RNA protection assay** for the c-*fes* mRNA may be **used clinically to diagnose myeloid leukemia**, or to predict eventual **conversion** of ALL cases to AML.

The ***pim*** protooncogene has been reported in human **myeloid leukemias**, whereas **c-*ets*-1** is rearranged in some cases of **ALL**.

c-*fos* is

- highly expressed in acute leukemias with **monocytic phenotype** (FAB M4/M5)
- highly expressed in some **B lymphoid leukemias**
- expressed at low levels in **AML**, the majority of B and all T **lymphoid leukemias**, as well as in **erythroleukemia**

c-*fos* can be used as an additional marker of **myelomonocytic** forms of leukemia.

A number of protooncogenes are expressed in **Hodgkin's disease** (**c-*myc***, ***p53***, **c-*jun***, ***pim*-1**, ***lck***, **c-*syn***, **c-*raf***, **N-*ras***, and **c-*met***), the specific combinations of which are not found to be expressed in untransformed hematopoietic cells. In adult T cell leukemias, ***p53* mutations** are associated with the acute phase and are apparently involved in **transition from chronic to acute phase.**

Another gene, ***blk***, has been described that codes for **B lymphoid kinase** and is specifically expressed in the B cell lineage.

In **T cell lymphomas/leukemias**, the protooncogenes ***TCL*-1**, ***TCL*-2**, ***TCL*-3**, and ***TCL*-5** have been found in association with specific chromosomal aberrations.

The expression of protooncogenes in hematological malignancies not only involves **more than one oncogene**, but the interaction of hematopoietic cells with the **bone marrow stroma** and growth factors released from it apparently affects the **levels of protooncogene expression**.

Evidence of some variation at the **L-*myc*** locus has been obtained and correlated with survival to old age and **susceptibility to NHL** and **ALL**, whereas no such correlation could be found for c-*myb* or c-*mos*. Deletions of the ***RB* gene** with concomitant absence of the **RB110 product** are seen in **primary leukemias.**

7.3.8 MULTIPLE MYELOMA (MM) (p. 629)

Approximately one third of aspirated or biopsied bone marrow cells in MM are aneuploid and a significantly higher **RNA content** has been observed by FCM.

Althogh **DNA aneuploidy** appears to be correlated with adverse prognosis, RNA content cannot reliably predict response to chemotherapy depending on the specific regimen used.

The most common chromosomal abnormalities observed in patients with malignant **monoclonal gammopathies** were in **chromosomes 1**, **14**, and **12**.

Analysis of established human myeloma cell lines **revealed *no* alterations** of the ***met***, ***raf***, ***abl***, ***erb*B**, ***HER*-2 (*neu*)**, ***fos***, ***myb*-7**, ***fms***, **L-*myc***, ***sis***, or ***myb*-1** genes.

The expression of c-*myc* mRNA, however, was notably enhanced, but not due to amplification of the gene.

The **level of expression of some growth-regulated protooncogenes** (c-*myc*, c-*myb*, and ***p53***) **is markedly increased in myelomatous plasma cells.**

Comparison of the expression of $p21^{Ha\text{-}ras}$ in **monoclonal gammopathies of undetermined significance (MGUS)** with that in MM cells revealed a significantly higher level of expression in MM cells as determined by FCM.

Determination of oncogene expression or DNA ploidy offers an additional tool for **predicting the clinical course** of MM patients, complementing the coventional analysis of performance status, infections before diagnosis, renal impairment, serum calcium, anemia, Bence-Jones proteinuria, and other traditional parameters.

The utility of molecular genetic analysis in clinical evaluation of hematolymphoid lesions, the high sensitivity and specificity of genotyping is acknowledged, **although little additional help is thought to be gained from it in unequivocally malignant cases** recognized by conventional analysis. It is recommended that its use be restricted to cases in which the diagnosis is uncertain after conventional analysis.

It seems, however, that the role of genotyping and the identification of malignant cells present at low levels, as in monitoring MRD, and using simplified, nonradioactive formats, will be growing.

Supplement 1
Manufacturers

Commercial Sources of Molecular Biological Reagents and Diagnostics

The following information has been provided by companies on products currently available for or potentially useful for gene level laboratory evaluations. Some of the companies are familiar to primary care physicians and clinical laboratorians, a number of companies are less familiar to this audience and have been asked to summarize their efforts in manufacturing molecular biology products. Some of the "purely" molecular biology companies have also begun to develop products for human diagnostic use.

The inclusion of any company in this listing should not be misconstrued as suggesting the products featured are for human diagnostic use unless specifically stated so by the company. The **listing presented here is only a sampling of companies and products** one should know about when considering gene level evaluation of diseases. No endorsement of the products or companies listed here is implied by the information presented, nor should this information be taken as a recommendation by the author to use the products listed over those of manufacturers not represented herein. In all cases, the interested reader should consult a company's catalog for more detailed information. A number of other companies and further information on related products can be found, for example, in annual editions of *Science*, "The Guide to Biotechnology Products and Instruments."

APPLIED BIOSYSTEMS
850 Lincoln Centre Drive
Foster City, CA 94404
(415) 570-6667

Applied Biosystems supplies equipment and tools for life sciences research and related applications. The company's automated systems can be grouped into two categories: instruments and consumable products; these can be further subdivided based on the specific application for which they are used i.e., DNA or protein studies or separation of biomolecules.

INSTRUMENTS

DNA Chemistry Instruments

DNA Chemistry Instruments are used in the laboratory to automate the isolation, preparation, analysis, and synthesis of DNA and RNA samples.

The instruments listed are employed in chemical synthesis or modification of DNA fragments **for use in research and for products used in diagnosis of disease**.

The main instruments and their applications are listed below:

Model 341A Nucleic Acid Extractor — for automating the isolation of DNA and RNA from biological samples.

Model 800 CATALYST Molecular Biology Labstation — for the preparation and labeling of DNA samples prior to DNA sequence analysis.

Model 675 INHERIT Data Analysis System — for the interpretation of DNA sequence data generated on the model 373 DNA sequencer, and for matching these data with "libraries" of published DNA sequences.

Model 373A DNA Sequencer — for the automated analysis of DNA produced from dideoxy sequencing reactions. This instrument accommodates a wide range of chemistries and includes the model 673.

Model 362 GeneScanner Fluorescent Fragment Analyzer — for the automated analysis of DNA fragments that have been labeled with the company's proprietary fluorescent labels.

Model 673 MacIntosh Upgrade for the Model 370A — for the operation, data collection, and base calling of the model 370A DNA sequencer (replaced by model 373A). This upgrade enhances model 370A by providing 50% more throughput in the number of bases that can be identified, accommodating a wider range of chemistries and offering greater ease of operation.

Model 394 DNA Synthesizer — for the simultaneous production of up to four different DNA or RNA fragments and automatic cleavage and deprotection of these fragments, using any one of several accepted chemistries.

Model 392 DNA Synthesizer — for the simultaneous production of up to two different DNA or RNA fragments and automatic cleavage and deprotection of these fragments, using any one of several accepted chemistries.

Model 391A PCR-Mate DNA Synthesizer — for the automated synthesis of DNA fragments, especially primers for the **polymerase chain reaction (PCR)** technique.

Model 391-EP DNA Synthesizer — for the automated synthesis of DNA fragments used in a variety of applications. This instrument has an enhanced software program with increased versatility compared to the 391A.

Protein Chemistry Instruments

Protein chemistry instruments are designed for isolation, purification, and analysis of proteins, for relating the function of a protein to its amino acid sequence (its primary structure), as well as for the chemical synthesis or modification of protein fragments. The latter is used in "tailoring" protein molecules in research, and for the manufacture of diagnostic and clinically related products.

Analytical Chemistry

Chromatographic techniques separate chemical compounds from each other as they pass through a column packed with adsorbent based on their affinities relative to the adsorbent. Applied Biosystems chromatography products include **high-pressure liquid chromatography (HPLC) systems**, components, and related accessories. Electrophoresis products include **preparative** and **analytical** instrument systems.

The main products are:

Model 477A Protein Sequencer — for the stepwise amino acid sequence determination of proteins and peptides for expert users. This instrument accommodates a wide range of chemical methods, and offers the greatest sensitivity, most comprehensive data analysis, and greatest convenience of the protein sequencing instrument family.

Model 473A Protein Sequencer — for the stepwise amino acid sequence determination of proteins and peptides.

Model 610A Data System — for the analysis of protein sequence information that is generated from models 477A, 473A, and 470A (replaced by model 477A) protein sequencers.

Model 430A Peptide Synthesizer — for the automated synthesis of peptides, using any one of several accepted chemistries.

Model 431A Peptide Synthesizer — a simplified version of model 430A for the automated synthesis of peptides; designed for users with more routine synthesis needs.

Model 420A Amino Acid Analyzer — for the automated direct assessment of the purity and amount of protein in a sample, and for the determination of the amino acid composition of a protein. Results generated are stored in a data analysis module for further optimization of data.

Model 420H Amino Acid Analyzer with Hydrolyzer — for the chemical dissolution of a protein and subsequent assessment of the purity and amount of protein in a sample, and for the determination of the amino acid composition of a protein. Results generated are stored in a data analysis module for further optimization of data.

Bio-Ion 20 — for the molecular weight determination of biopolymers such as proteins, peptides, polysaccharides, and carbohydrates.

Bioseparations Products

Bioseparations products are used in the laboratory to isolate and purify biological compounds. They may be used independently or in conjunction with Applied Biosystems protein and DNA products. The main products are:

Model 270A and Model 270HT Capillary Electrophoresis Systems — for the automated electrophoretic separation and quantitation of chemical components in a mixture.

Model 230A HPEC System — for the automated electrophoretic separation, quantitation, and recovery of chemical components in a mixture.

Model 700 Series Absorbance Detectors and Model 900 Series Fluorescence Detectors — for detecting trace compounds with high resolution, sensitivity, and convenience in liquid chromatography applications.

Model 1000S Diode Array Detector — for detecting trace amounts of chemicals within a mixture, based on absorption of light at multiple wavelengths.

Model 130A Separation System — for the isolation, purification, and recovery of small samples of proteins and peptides by liquid chromatography.

Model 150 Series of Separation Systems — for the isolation, purification, and recovery of samples of protein, peptide, and/or DNA products.

Chromatography Accessories — enhancement and replacement products for Applied Biosystems HPLC systems, which include delivery systems, postcolumn reaction systems, and data systems.

CONSUMABLE PRODUCTS

In addition to standard chemistries for each instrument, a multitude of peptide synthesis, DNA synthesis, and DNA sequencing chemicals are available that allow one to devise chemical strategies for making a desired compound to meet specific research requirements.

Examples of Applied Biosystems optimized chemistries are as follow:

- the patented **phosphoramidite chemistry** for DNA synthesis
- the **fast oligonucleotide deprotection (FOD)** phosphoramidite chemistry
- **cycle sequencing chemistry** for the DNA sequencer

Applied Biosystems also supplies **DNA analogs** in volume to customers exploring the potential applications of these compounds as **antisense drugs**. Research has indicated that DNA analogs may be useful as treatment for a wide variety of diseases, including viral, some forms of cancer, and inflammatory disorders.

Applied Biosystems also manufactures and markets a wide range of **chromatography columns** and **components**, some designed for independent use and many designed for use with Applied Biosystems instruments.

BIO-RAD LABORATORIES
Genetics Systems Division, Life Science Building
2000 Alfred Nobel Drive
Hercules, CA 94547
(510) 741-6700

INTRODUCTION

Over the last several years, molecular biology instruments, reagents, and protocols have been refined to the point where they are now suitable for use in clinical laboratories. As more DNA sequence information is gained about disease-causing genes and infectious agents, an increasing number of well-characterized DNA probes will become available. These probes may be used in DNA-based diagnostic assays to reveal information at the molecular level with exquisite speed, accuracy, and sensitivity. As such, **DNA probe tests soon will be established as a routine component of laboratory diagnostics.**

Although potentially useful DNA probes have been available for several years, their use in the clinical laboratory has been limited, primarily due to the difficulty of implementing simple formats for DNA-based assays. Until the recent development of the PCR procedure (Mullis and Faloona, 1987; Saiki et al., 1985), most DNA assays involved Southern blots and radioactive probes, techniques more suitable for research laboratories. Target amplification methods such as PCR have stimulated renewed interest in DNA probe technology and increased the speed at which these tests will reach routine use in the clinical laboratory. In addition, a variety of **other techniques are gaining increasing acceptance as clinical methods,** including pulsed-field gel electrophoresis (PFGE), nonisotopic *in situ* hybridization, as well as other emerging technologies. A major task in moving these and other research methods to the clinical laboratory is the reduction of techniques to practice in a way that is sufficiently simple and cost effective to permit routine use.

Bio-Rad Laboratories is involved in the research and development of **products for biochemistry and molecular biology research laboratories.** In part, we have focused on instruments and reagent systems for DNA mapping, DNA sequencing, and image analysis. In recent years, these products have seen increased use in clinical and forensic settings, as well as in laboratories that are developing DNA probe assays. This article briefly describes some of these products and their utility in **clinical DNA diagnostics** (see Table 1 for catalog information). In addition, we will mention some of our current research efforts to develop **products with a potential impact on clinical molecular biology**. Bio-Rad products used in the various techniques described in this article are listed with corresponding catalog information in Table 1.

BLOT HYBRIDIZATION METHODS

Several diagnostic tests that utilize radiolabeled probes and Southern blot analysis are currently in routine practice (Grody et al., 1989; see Part 1). The **main clinical application of Southern analysis** is in the identification of chromosomal rearrangements associated with disease states. For example, identification of the reciprocal gene translocations frequently observed in hematopoietic cancers, such as the t(14;18) translocation in follicular lymphoma (Cleary and Sklar, 1985) or the t(9;22) Philadelphia chromosome (Ph′) in chronic myelogenous leukemia (CML) (Blennerhasset et al., 1988), may be accomplished by Southern analysis. In addition to rearrangements, gene amplifications may be detected by Southern analysis. Amplification of the *HER-2/neu* oncogene occurs in approximately 30% of all breast cancers examined (Farkas, 1990) and N-*myc* amplification is found in neuroblastoma (Brodeur et al., 1984). Southern or dot blot analysis (Giulotto et al., 1990) is an appropriate method for detection of these genetic alterations.

A primary problem for clinical laboratories performing Southern blot analysis is the degree of **quality control** required for this multistep process. The interpretability of blot hybridization results depends on the effectiveness of each step, including DNA and probe preparation, nucleic acid transfer, and membrane quality. Table 1 lists products and protocols that we and others (e.g., Church and Gilbert, 1984; Reed and Mann, 1985) have developed, to **increase the reproducibility of Southern and northern (RNA) blots**.

One area of interest for Bio-Rad has been the development of **improved blotting membranes**. Nitrocellulose membranes, initially used for blotting and detection of nucleic acids, have been supplanted by nylon membranes. In particular, Bio-Rad has developed a positively charged nylon membrane, **Zeta-Probe GT**, with properties encompassing the advantages that nylon membranes have over nitrocellulose for blot applications. These include greater strength, increased retention of target DNA to the membrane through the use of ultraviolet light, and the ability to perform multiple probings when using stripping protocols. To address the membrane quality control issue, we have instituted lot-specific functional testing of Zeta-Probe GT using a single-copy detection assay to ensure that each lot of membrane has adequate DNA-binding properties and that nonspecific probe binding is minimal.

Ultraviolet (UV) light is used in blotting applications for two purposes: **cross-linking of DNA** to membranes after transfer (Church and Gilbert, 1984) and **nicking of DNA** in agarose gels after electrophoresis to enhance transfer and subsequent detection (Lee et al., 1991). We have developed a UV chamber, called the **GS Gene Linker UV chamber**, and a series of protocols that optimize the amount of UV energy necessary to cross-link DNA to charged or uncharged nylon membranes. A second application for the UV chamber is to enhance the transfer efficiency of megabase DNA fragments [>50 kilobases (kb)] to membranes following PFGE. The transfer of large DNA molecules without a nicking step is typically inefficient, even by the traditional method of acid depurination. Lee et al. (1991) devised a protocol using UV energy delivered by the GS Gene Linker to optimize and standardize the transfer efficiency of large DNAs. Using their methods, **single-copy genes from mammalian genomes were routinely observed after overnight exposures of blots to X-ray film.**

Blot-based assays are labor intensive and difficult to perform on a routine basis. In addition, the process is difficult to automate to the degree required in clinical laboratories. One commercially available instrument (Oncor) partially automates certain steps of the process, but this product has not gained widespread acceptance. It appears, therefore, that **Southern blot analysis ultimately will find limited use in clinical diagnostic laboratories,** although at present certain assays may be best performed using blots.

TABLE 1
Representative Bio-Rad Products Used in DNA Diagnostics[a]

Product	Comment[b]
Blot hybridization methods	
Sub-cell DNA electrophoresis cell	For high-resolution separations of genomic DNA digests
DNA size standards, low range	88 to 1746 bp
DNA size standards, high range	0.6 to 23.1 kbp
Model 200/2.0 power supply	
Agarose, molecular biology certified	Suitable for genomic DNA blots
Zeta-probe GT membrane	Positively charged nylon membrane for nucleic acid hybridizations
Model 785 vacuum blotter	Rapid DNA transfers to nylon membranes
Klenow random primer labeling kit	For preparation of high specific acitivity DNA probes
Bio-Spin 30 chromatography column	Rapid removal of unincorporated nucleotides in probe preparation
GS gene linker UV chamber	UV cross-linking DNA to membranes
Blot imaging	
Model 620 CCD densitometer	X-ray film densitometry
Phosphorimager system (in development)	For imaging and quantitation of Southern and dot blot assays
Nonisotopic DNA detection	
GS Gene-Lite kit	Chemiluminescent blot detection method
Alpha-probe membrane (in development)	Neutral nylon membrane, suitable for chemiluminescent detection
Gene-Lite marker ladder (in development)	1- to 25-kb ladder, for use with Gene-lite kit
Pulsed-field electrophoresis	
CHEF-DR II system	CHEF system with fixed 120° reorientation angle
CHEF Mapper XA chiller system	CHEF with PACE technology, e.g., variable angle, multivector
DNA size standard, lambda ladder	50 kb to 1 Mb
DNA size standard, yeast chromsomal	245 kb to 2.2 Mb
DNA size standard, *S. pombe*	3 to 7 Mb
DNA size standard, 5-kb ladder	4.9 to 100 kb
Low melt preparative grade agarose	For preparation of high molecular weight DNA for PFGE

TABLE 1 (continued)
Representative Bio-Rad Products Used in DNA Diagnostics[a]

Product	Comment[b]
CHEF DNA plug kits (in development)	For preparation of high molecular weight DNA for PFGE
Pulsed-field certified agarose	High-resolution PFGE, used in CHEF mapper algorithm
Chromosomal grade agarose	For PFGE separations of molecules >2 Mb
GS gene linker UV chamber	UV nicking large DNA, enhanced transfer in Southern blots
Amplification-based methods	
Prep-a-gene DNA purification kit	DNA preparation for PCR amplification
GS gene prep manifold system	Apparatus for multiple DNA preparations
Chelex 100 resin	Rapid preparation of DNA for PCR assays
Bio-dot apparatus	Detection of infectious diseases or single-base variants by ASO
Bio-dot SF apparatus	Detection of infectious diseases or single-base variants by ASO
Wide mini-sub cell	For PCR analysis on agarose gels
DNA size standards, low range	88 to 1746 bp
DNA size standards, 100-bp ladder	100 to 2500 bp, suitable for size standards in PCR analysis
Silver stain plus kit	For staining DNA in acrylamide gels
Ethidium bromide tablets	
PROTEAN II xi cell	For PCR or microsatellite analysis on acrylamide gels
Mini-PROTEAN II cell	For PCR analysis on acrylamide gels
Mini-PROTEAN II ready gels	5 and 10% precast gels for PCR analysis
Model 450 microplate reader	For automating detection of PCR assays, such as OLA
Model 3550 microplate reader	For automating detection of PCR assays, such as OLA
Reverse dot-blot apparatus (in development)	Suitable for screening multiple probes
Nonisotopic *in situ* hybridization	
MRC-600 confocal laser scanning system	Protocols have been developed for cytology and gene mapping
Direct gel analysis of sequence variation	
Sequi-Gen cell	May be used in SSCP or microsatellite gel analysis

TABLE 1 (continued)
Representative Bio-Rad Products Used in DNA Diagnostics[a]

Product	Comment[b]
Temperature gradient gel electrophoresis system (in development)	Exploits melting point differences that result from single-base variations
Direct DNA sequence analysis	
Sequi-gen cell	DNA sequencing cell
GS gene loader automated gel loader	Automates loading radioactive or fluorescent samples on sequencing gels
Model 3000Xi power supply	Interactive with Gene Loader, for pulsing samples onto sequencing gels
Prep-a-gene DNA purification kit	For preparation of single-strand or double-strand DNA templates
GS gene prep manifold system	For preparation of samples for radioactive or fluorescent sequencing
Molecular biology workstation (in development)	Pipette station, automates sequencing or PCR protocols
GS gene reader automated film reader	High-accuracy reading of DNA sequencing autoradiograms
Solid-phase DNA seqencing system (in development)	Solid-phase capture system for preparing single-strand templates

[a] Listings do not include all accessories or formats for products.
[b] See text for full details.

For example, Lee et al. (1987) have reported extremely sensitive detection of chromosomal translocations in both follicular lymphoma and CML using PCR. Moreover, they demonstrated PCR-positive results in a number of patients in long-term remission, suggesting that the clinical significance of the positive PCR results requires further evaluation. In this case, and possibly in some cases of infectious disease probes, **PCR may be overly sensitive**. In time, target amplification methods will most likely replace most blot assays, but further work is needed before amplification assays are fully interpretable (see Part 6).

BLOT IMAGING

An important development in nucleic acid blotting involves the use of **erasable phosphor imaging plates** in place of X-ray film autoradiography for detecting radioactive hybrids (Amemiya and Miyahara, 1988). This technique involves the use of **radiation-sensitive phosphor screens,** which are capable of "storing" the image of a radioactive blot or gel. The latent image is subsequently read from the phosphor screens by laser scanning. The use of these imaging plates has a number of advantages over traditional autoradiographic methods. First, the phosphor plates are **10- to 20-fold more sensitive** to β radiation than is X-ray film, resulting in reduced exposure times with a concomitant increase in throughput of blots. Second, the **linear dynamic range of the system is several orders of magnitude greater** than X-ray film densitometry.

Increased linearity of response to signal greatly enhances the ability to perform **quantitative radioactive assays** and creates the opportunity to develop new applications. For example, an important method of mutation analysis is by **quantitation of gene expression on two-dimensional protein gels**. Although this technique is problematic when using X-ray film to image the resolved proteins, the dynamic range inherent in phosphor imaging plates is well suited for this application. Similarly, carrier screening, which requires the **quantitative analysis of hybridization signals** to determine the copy number of a particular allele, would benefit from the linearity of response available in phosphor imaging plates, as would quantitation of the gene amplifications described earlier in this article.

NONISOTOPIC DNA DETECTION

The use of radioactive probes is problematic in clinical laboratories. In addition to potential health hazards and increasingly stringent disposal requirements, the short half-life of ^{32}P requires repeated probe preparations. Moreover, the rate and extent of the blot hybridization are dependent on the specific activity and length of the probe sequence, both of which are difficult to control rigorously. To alleviate the need for radioactive probes, a variety of nonisotopic DNA detection systems have been developed. The majority of these methods are based on the enzymatic incorporation of **biotin**- or **digoxigenin**-labeled nucleotides into DNA probes (Martin et al., 1990; Muhlegger et al., 1988; see Part 1). The detection of the labeled probe-target duplex is accomplished by incubation of the hybridized blots with streptavidin or antidigoxigenin-alkaline phosphatase conjugates followed by incubation with the **alkaline phosphatase-cleavable chemiluminescent substrate AMPPD** [3-(2′-spiroadamantane)-4-(3″-phosphoryloxy)phenyl-1,2-dioxetane] (Bronstein et al., 1989; see Part 1). Blot images are obtained using X-ray film. Although these approaches work well for colony or plaque screening or for Southern and dot blots, situations in which the target sequence is present in high amounts (>100 pg), a number of problems make these systems unreliable when applied to genomic Southern blots (discussed in Tumolo et al., 1992).

A second approach to nonisotopic detection involves hybridization of covalently cross-linked **oligonucleotide-alkaline phosphatase conjugates**, followed by chemiluminescent detection (Jablonski et al., 1986). Although the use of direct conjugates allows high-sensitivity detection with low backgrounds, a **major drawback** is the requirement to prepare a custom oligonucleotide for each probe-target system. In cases where the same assay is routinely repeated, such as clinical or forensic applications (e.g., DNA typing with repeat sequence probes) direct conjugates may be useful.

We have developed a **two-step hybridization procedure** for genomic Southern applications that eliminates the background problems associated with other chemiluminescent techniques, as well as the need for the custom preparation of oligonucleotide-alkaline phosphatase conjugates (Nguyen et al., 1992). The method employs a **primary probe hybridization** of the Southern blot with single-stranded phagemid or phage vector DNA carrying the probe sequence, followed by a **secondary hybridization** with a covalently cross-linked oligonucleotide-alkaline phosphatase conjugate that is complementary to the universal primer sequence present on the primary hybridization vector. Signal is detected on X-ray film after the blot is incubated with AMPPD. Because **no labeling of the primary probe is required**, large amounts of probe may be prepared and stored for long periods, eliminating the variability associated with enzymatic labeling methods. In addition, the secondary probe, which can be synthesized and conjugated in large quantities, is the same regardless of the target sequence. In practice, we routinely obtain **picogram (single copy) detection from genomic samples, with low backgrounds**. In recent work, we have demonstrated the ability to generate single-strand primary probe by **solid-phase capture of PCR products** where one of the primers is biotinylated, eliminating the need for a single-strand generating vector (Nguyen, in preparation). The two-step blot detection system is marketed as the Bio-Rad **Gene-Lite kit**.

PULSED-FIELD GEL ELECTROPHORESIS

In some cases, the use of **pulsed-field gel electrophoresis (PFGE)** to separate genomic restriction fragments before Southern transfer can give information beyond what may be obtained on conventional agarose gels (see Part 1). Because PFGE is suitable for resolution of DNA fragments greater than 20 kb in length, this technique permits construction of **long-range restriction maps.** PFGE is presently a standard technique and high-resolution separations of DNA molecules as large as 10 Mb (10×10^6 bp) are routine. Although the electrophoretic separations are relatively slow, **diagnostic applications of PFGE stem from the size of the DNA molecules that may be resolved on agarose gels.**

Chronic myelogenous leukemia (CML), which is associated with a t(9;22) exchange of chromosomal material, results in a reciprocal translocation of the c-*abl* gene from the long arm of chromosome 9 to the long arm of chromosome 22. The hybrid chromosome is called the Philadelphia chromosome (Ph′). The **translocation breakpoint** on chromosome 22 typically occurs in a small region (about 6 kb) called the breakpoint cluster region (*bcr*) and leads to **a *bcr-abl* fusion gene** (Blennerhasset et al., 1988). Molecular detection of Ph′ in CML is straightforward by conventional Southern analysis (see Part 7).

In contrast, detection of Ph′ in acute lymphoblastic leukemia (ALL) is complicated because the chromosome 9 breakpoints are dispersed over a large region (Hooberman et al., 1989). PFGE was used by these authors to detect the Ph′ in ALL and suggests the utility of this technique in the detection of other clinically significant chromosomal aberrations. In addition to rearrangements, **large chromosomal deletions** may be revealed by PFGE. For example, a 90-kb deletion was identified by PFGE in a patient with neurofibromatosis type 1 (Upadhyaya et al., 1990).

PFGE is playing an increasingly important role in typing strains of pathogens (see Section 6.1.1.5). For example, simple **restriction digestion** of *E. coli* strains with rare cutting restriction enzymes followed by PFGE reveals differences in band patterns after ethidium bromide staining (Arbeit et al., 1990; P. Schad, unpublished results). Similarly, *Not*I digests have been used to generate fingerprints of *Streptococcus mutans* (the major causative agent of dental caries) chromosomes, which may be useful for epidemiological studies of dental caries (Okahashi et al., 1990). In addition, chromosome typing of pathogenic strains of the yeast *Candida* has been accomplished by CHEF electrophoresis, a form of PFGE described below. These workers determined an electrophoretic karyotype for 130 strains of *C. albicans* (Monod et al., 1990).

A variety of PFGE methods are currently in practice (reviewed in Lai et al., 1989) and Bio-Rad has developed two different pulsed-field gel electrophoresis systems based on the **CHEF (contour clamped homogeneous electric field)** technology. CHEF uses the principles of electrostatics to determine the voltages necessary to generate homogeneous electric fields using multiple electrodes placed in a closed contour. The CHEF-DR II system has 24 electrodes arranged in a hexagonal array, and creates a highly uniform electric field in the gel by clamping (holding at intermediate potentials) certain electrodes. This system results in straight lanes during electrophoresis and was the first commercial product to incorporate individually driven electrodes. Hence, this system allowed easy expansion to a multi-angle system.

A further extension of this work resulted in the CHEF Mapper system, based on **PACE (programmable, autonomously controlled electrodes)** technology (Clark et al., 1988) and using the individually driven electrodes of the CHEF-DR II. While the CHEF-DR II is limited to two electrical vectors and a 120° pulse angle, the CHEF Mapper allows use of up to 15 vectors and variable pulse angles. This latter feature is important, because the ability to alter the reorientation angle between the alternating pulsed fields permits reduced time for separation of large DNAs. For example, changing the reorientation angle from 120° to 90° reduced the time to separate *Schizosaccharomyces pombe* chromosomes from several days to 24 h (Clark et al., 1988). Finally, the CHEF Mapper contains an embedded algorithm that automatically determines optimal conditions for separations of large DNA molecules. Variables that affect migration of DNAs subjected to PFGE were quantitatively examined and used to create the algorithm, which allows the researcher to input only the size range of DNAs to be separated. Optimal pulsed-field conditions then are automatically computed and implemented by the instrument.

AMPLIFICATION-BASED METHODS

Probe assays based on the direct detection of amplified target nucleic acids are particularly well suited for clinical laboratories. A variety of formats, both those that require thermal cycling and as well as isothermal methods, have been devised. At present, **PCR** and a related technology, **ligase chain reaction (LCR)**, are the closest to being used routinely in DNA diagnosis, and only these two methods will be discussed here (see also Part 1).

PCR reactions are simple to perform, relatively robust, and the quality of the starting material is of little importance. Furthermore, the assays are typically nonradioactive, with reaction products simply run on an agarose or acrylamide gel and detected by ethidium bromide or silver stain. Internal controls are easy to design and assays may be multiplexed. Finally, the nature of the PCR reaction is such that automated formats for performance and analysis are relatively easy to design.

The potential applications for PCR assays are unlimited, as virtually any infectious disease or genetic alteration may be diagnosed by this method. For example, several studies have demonstrated the utility of PCR for early detection of HIV infections (e.g., Ou et al., 1988). Other work has proved the feasibility of allele-specific PCR reactions to detect single-base variations in genomic DNA samples (Sarkar et al., 1990; Ugozzoli and Wallace, 1991). Moreover, appropriate selection of primers can reveal chromosomal deletions, insertions, or rearrangements.

The enormous amplification obtained by PCR, and the fact that the expected reaction product will be obtained only if both primers anneal on a contiguous DNA piece, means that the starting material for preparation of template DNA may be of relatively low quality. Furthermore, template purification may be done rapidly, with a minimal number of steps, and a variety of matrices have been developed for this purpose. For example, Singer-Sam et al. (1989) and Walsh et al. (1991) have reported on the use of **Chelex 100 resin** (Bio-Rad) in methods to extract DNA for PCR analysis of forensic samples from virtually any source. They reported that use of Chelex resulted in increased signal relative to other preparation methods, possibly due to the high affinity of this resin for polyvalent metal ions that, if present, can degrade DNA. Purification of DNA for PCR, as well as single- and double-strand DNA templates for restriction or sequence analysis, may also be performed using **Prep-A-Gene** (Bio-Rad), a patented purification matrix that binds DNA in the presence of chaotropes. We have also developed a **vacuum filtration device**, called the **GS Gene Prep manifold**, that allows convenient processing of a large number of samples.

One key feature of PCR is the ease by which reaction products may be analyzed. These products typically are in the range of 100 to 5000 bp and analysis is performed on acrylamide gels, for products <1500 bp, and on agarose gels for the larger molecules. Reproducible electrophoresis and good molecular weight markers are required for accurate analysis. Furthermore, precast acrylamide gels (Ready-Gels; Bio-Rad) and a convenient electrophoresis device (e.g., Mini-PROTEAN II and PROTEAN II xi; Bio-Rad) can simplify analysis. In addition to standard electrophoresis techniques, Nu-Seive agarose, which can be formed at much higher concentrations than standard agarose, can increase the resolution of DNA in the 100- to 2000-bp range. Finally, Allen and co-workers (Budowle et al., 1991) have reported on the use of enhanced buffer systems and silver staining for high-resolution acrylamide gel separations of PCR products. The products marketed by Bio-Rad for electrophoresis and detection of PCR products are listed in Table 1.

Ultimately, fully automated amplification assays will be required in diagnostic laboratories. Among the techniques currently in use for the analysis of sequence variations are **allele-specific PCR** (see above), dot blot hybridizations using **allele-specific oligonucleotides (ASOs)** (Saiki et al., 1986), and gel analysis following **restriction digestion of PCR products** (Chehab et al., 1987; see also Part 1). A variety of approaches have been devised that may allow more automated analysis of PCR products for point mutations or deletions, including fluorescence-based DNA analysis methods, reverse dot blotting in which multiple probes are immobilized and screened with target DNA (Saiki et al.,1989), and direct gel or DNA sequence analysis (discussed below).

One important method for automated analysis of mutations or polymorphisms is the combination of PCR and an **oligonucleotide ligation assay (OLA)** for detection of single base variants (Nickerson et al., 1990; see also Section 1.2.2.6). This strategy involves amplification of a specific target sequence by PCR, followed by analysis of allelic variation with OLA. Two short oligonucleotides complementary to the region of interest are hybridized to the amplified genomic DNA. The oligonucleotides anneal at sites adjacent to each other on the target sequence, one containing a biotin molecule at the 5′ end and the second a reporter moiety at its 3′ end. Examples of reporter molecules include digoxigenin or fluorescent tags. The 3′ end of the biotinylated probe sits at the nucleotide to be assayed. The two oligonucleotides are hybridized to the target sequence and if there is perfect homology, DNA ligase will covalently connect the biotinylated probe and the reporter probe. If the 3′ end is mismatched, the probes will not be ligated. The biotinylated probe is bound by streptavidin and the presence or absence of the reporter molecule is typically determined by an alkaline phosphatase-conjugated antibody to digoxigenin.

OLA assays are performed in a microtiter plate format, with signal generated by alkaline phosphatase-cleavable colorimetric or chemiluminescent substrates, and detected using ELISA or luminometer microplate readers. A major advantage of OLA is that the process may be fully automated: PCR reactions, ligation assays, capture of the ligation products, and readout may be accomplished on workstations. For example, the

Biomek 1000 (Beckman Instruments) automates the pipetting steps, while a **workstation under development at Bio-Rad is aimed at automating the pipetting and amplification steps.**

The **ligase chain reaction (LCR)**, a DNA-level diagnostic method, is similar to OLA because short oligonucleotides are used to probe for sequence variations (Barany, 1991; see also Section 1.2.2.8). The probes are typically in the same format as in OLA, with adjacent short capture (biotinylated) and reporter oligonucleotides. Again, the 3′ end of the biotinylated oligonucleotide is an allele-specific diagnostic reagent. This method, however, employs a heat-stable DNA ligase, which permits multiple rounds of hybridization and ligation of the probes, combining the amplification and detection steps. In contrast to OLA, which probes the products of a PCR reaction, LCR directly amplifies the target species. LCR, similar to OLA, is amenable to complete automation for performance and analysis of reactions.

NONISOTOPIC *IN SITU* HYBRIDIZATION (NISH)

In situ hybridization methods are used to locate nucleic acid probe sequences at specific locations on chromosomes. Initially, *in situ* hybridization was performed using radioactive probes with detection accomplished by X-ray film autoradiography. Recent work has produced a number of approaches to **nonisotopic *in situ* hybridization (NISH)**, as well as a variety of applications including cytogenetics, prenatal diagnostics, disease carrier status, cancer diagnostics, and infectious disease detection. More information on NISH methods, and clinical applications for the technique, may be found in Lawrence (1990) and Lichter et al. (1991), and also Sections 1.2.4.3, 1.2.7.3, 6.1.2.1, 6.4.7.4.1, 6.4.7.8 and 7.3.4 of this Guide.

One major advantage of NISH is the amenability of the technique to multiple simultaneous assays and automated analysis. Furthermore, NISH generates quantitative information, which is particularly significant when one is evaluating the level of an infectious disease or the course of cancer therapies. For example, information that is difficult to obtain with PCR, such as the percentage of cells in a population that contain a particular chromosomal defect, may be determined using NISH. As discussed above, amplification-based methods are so sensitive that quantitative analysis is problematic and the clinical significance of a positive result may be hard to determine. Furthermore, NISH is suitable to following the course of chemotherapies and in remission monitoring.

Bio-Rad has developed the **MRC-600 confocal laser scanning system**, targeted for use in a large number of applications, including gene mapping, through *in situ* hybridization assays based on **fluorescent detection (FISH)** (e.g., Albertson et al., 1991; see also Part 1). Using this instrument, the sensitivity and resolution of FISH is ever improving, and current applications such as ordering of cloned DNAs is becoming routine (Lichter et al., 1990). Interphase mapping methods, for example, have been used to resolve probes as closely spaced as 30 kb, permitting the direct ordering of cosmid or yeast artificial chromosome (YAC) clones on human chromosomes.

DIRECT GEL ANALYSIS OF SEQUENCE VARIATION

To simplify analysis of single-base variations, several laboratories have reported on the use of direct gel-based methods. Orita et al. (1989) described the use of **single-strand conformation polymorphism (SSCP)** to analyze point mutations in PCR-amplified products (see also Section 1.2.3.3.2). The method involves the denaturing of double-strand DNA, followed by electrophoresis of the denatured products on a nondenaturing acrylamide gel. Mobility shifts of mutants in comparison to wild-type sequences may be observed by this method. Although simple to perform, SSCP gels have not been demonstrated to be generally useful in screening for mutations or polymorphisms, because some single-base alterations do not give mobility shifts.

Two additional techniques, **denaturing gradient gel electrophoresis (DGGE)** (Fischer and Lerman, 1985) and **temperature gradient gel electrophoresis (TGGE)** (Wartell et al., 1990) separate DNA fragments based on their melting properties (see also Section 1.2.3.3.1). Two double-strand DNA fragments of almost identical sequence (differing by only a single base change), will have unique melting temperatures. DGGE is based on forming a chemical gradient from the top to the bottom of the gel, while TGGE forms a temperature gradient either from top to bottom or across the gel. As double-strand fragments enter these gradient gels, they migrate according to their molecular weight. When they reach the portion of the gel that contains sufficient denaturant or temperature to unwind the molecule partially, migration is slowed as the single-strand portion of the molecule becomes entangled in the gel matrix. The amount of denaturant or temperature required to partially denature the DNA molecule is sequence dependent and therefore PCR-amplified genomic regions may be rapidly screened for sequence variations. The primary applications for this technique include direct detection of single-base changes that cause disease, as well as screening for polymorphisms. Although it appears that the use of gradient denaturing gels will be generally useful, their acceptance has not been widespread because the methods are not easy to implement. At present, we are in the process of investigating methods to simplify the performance of TGGE.

DIRECT DNA SEQUENCE ANALYSIS

The advent of the PCR has opened up the possibility of performing DNA sequence analysis as a clinical tool. Any genomic region of interest may be directly amplified from chromosomal DNA and the amplification products directly sequenced by a variety of different methods (see also Section 1.2.3.7). Methods developed for direct sequencing of genomic DNA include **asymmetric PCR** to create an abundance of one of the two strands of the PCR products (Gyllensten and Erlich, 1988), **AMV reverse transcriptase sequencing of RNA** made *in vitro* from phage promoters introduced into genomic DNA by PCR (Stoflet et al., 1988), and **solid-phase capture of PCR products** created in the presence of one biotinylated primer and elution of the noncaptured strand (Mitchell and Merril, 1989). This latter technique is particularly useful in that it facilitates sequencing of both strands of an amplified region.

In addition, **linear amplification sequencing** is becoming an increasingly important technique for both genomic or clone bank sequencing (Craxton, 1991). This method is primarily used for fluorescent sequencing and combines linear template amplification and the dideoxy sequencing step to eliminate the need for extensive template preparation. A virtue of this method is the small amount of input template required to achieve a large signal-to-noise ratio, important in fluorescent sequencing. Direct genomic sequencing methods, in combination with fluorescent sequencing strategies, should facilitate the development of clinical DNA sequencing assays.

We currently are developing a series of **instruments for the automation of sequence analysis**. First, as described previously, we have designed a manifold device that semiautomates multiple template preparation, including solid-phase DNA capture and conventional template preparation for radioactive protocols (E. Mardis, unpublished results), as well as DNA for fluorescent-based linear amplification sequencing (B. Roe, unpublished results). Second, we have under development a workstation that automates the pipetting steps involved in setting up radioactive or fluorescent DNA sequencing reactions, including the performance of linear amplification sequencing reactions. Finally, we have begun marketing an instrument that automates the process of loading samples on DNA sequencing gels (called the **GS Gene Loader**). All of these instruments are based on a microtiter plate format and are designed to generate high sample throughput.

SUMMARY

It is clear that a large number of techniques originally developed for research applications are finding use in laboratories focused on performing or developing DNA-based clinical assays. As more genes and mutations that cause disease states are identified, and with further refinement of methodologies, the potential of DNA probes in clinical laboratories will be realized. **It is widely anticipated that the impact of DNA probes on the practice of clinical medicine will be dramatic**. For example, prenatal or infectious disease diagnosis using amplification will take hours instead of days or weeks. The ability to rapidly identify specific mutations in disease-causing genes will give prognostic information with a precision that is presently unavailable. However, **the instruments and protocols described in this article are research tools, and implementation in clinical settings will undoubtedly present its associated problems**. For example, difficulties with blot-based assays were described, and while methods based on direct amplification of target sequences appear relatively simple to implement, interpretation of results and cross-contamination between samples are among the issues that must be addressed. Solutions to these and other potential pitfalls must be effected before DNA diagnostics will move from specialized reference laboratories into routine use.

REFERENCES

Albertson, D. G. et al. (1991). Mapping nonisotopically labeled DNA probes to human chromosome bands by confocal microscopy. Genomics *10*:143–150.

Amemiya, Y. and Miyahara, J. (1988). Imaging plate illuminates many fields. Nature (London) *336*:89–90.

Arbeit, R. D. et al. (1990). Resolution of recent evolutionary divergence among *Escherichia coli* from related lineages: the application of pulsed field electrophoresis to molecular epidemiology. J. Infect. Dis. *161*:230–235.

Barany, F. (1991). Genetic disease detection and DNA amplification using cloned thermostable ligase. Proc. Natl. Acad. Sci. U.S.A. *88*:189–193.

Blennerhasset, G. T. et al. (1988). Clinical evaluation of a DNA probe assay for the Philadelphia (Ph1) translocation in chronic myologenous leukemia. Leukemia *2*:648–657.

Brodeur, G. M. et al. (1984). Amplification of N-*myc* in untreated human neuroblastomas correlates with advanced disease state. Science *224*:1121–1124.

Bronstein, I. et al. (1989). Chemiluminescent 1,2-dioxetane based enzyme substrates and their application in the detection of DNA. Photochem. Photobiol. *49*:9.

Budowle, B. et al. (1991). Analysis of the VNTR locus D1S80 by the PCR followed by high-resolution PAGE. Am. J. Hum. Genet. *48*:137–144.

Chehab, F. F. et al. (1987). Detection of sickle cell anemia and thalassemias. Nature (London) *329*:293–294.

Church, G. M. and Gilbert, W. (1984) Genomic sequencing. Proc. Natl. Acad. Sci. U.S.A. *81*:1991–1995.

Clark, S. M. et al. (1988). A novel instrument for separating large DNA molecules with pulsed homogeneous electric fields. Science *241*:1203–1205.

Cleary, M. L. and Sklar, J. (1985). Nucleotide sequence of a t(14;18) chromosomal breakpoint in follicular lymphoma and demonstration of a breakpoint-cluster region near a transcriptionally active locus on chromosome 18. Proc. Natl. Acad. Sci. U.S.A. *82*:7439–7443.

Craxton, M. (1991). Linear amplification sequencing, a powerful method for sequencing DNA, in *A Companion to Methods in Enzymology*, Vol. 3, Roe, B., Ed., Academic Press, San Diego, pp. 20–26.

Farkas, D. H. (1990). The establishment of a clinical molecular biology laboratory. Clin. Biotechnol. *2*:87–96.

Fischer, S. G. and Lerman, L. S. (1983). DNA fragments differing by single base pair substitutions are separated in denaturing gradient gels: correspondence with melting theory. Proc. Natl. Acad. Sci. U.S.A. *80*:1579–1583.

Giulotto, E. et al. (1990). Amplification of the HER-2/neu proto-oncogene in breast cancer detected by a rapid method. *J. Cell. Biochem.*, Suppl. 14B (abstr.), 334.

Grody, W. W. et al. (1989). Diagnostic molecular pathology. Mod. Pathol. *2*:553–568.

Gyllensten, U. B. and Erlich, H. A. (1988). Generation of single-stranded DNA by the polymerase chain reaction and its application to direct sequencing of the HLA-DQA locus. Proc. Natl. Acad. Sci. U.S.A. *85*:7652–7656.

Hooberman, A. L. and Westbrook, C. A. (1990). Molecular methods to detect the Philadelphia chromosome. Clin. Lab. Med. *10*:839–855.

Jablonski, E. et al. (1986). Preparation of oligodeoxynucleotide-alkaline phosphatase conjugates and their use as hybridization probes. Nucleic Acids Res. *14*:6115–6128.

Lai, E. et al. (1989). Pulsed field gel electrophoresis. BioTechniques *7*:34–42.

Lawrence, J. B. (1990). A fluorescence *in situ* hybridization approach for gene mapping and the study of nuclear organization, in *Genome Analysis*, Vol. 1, Genetic and Physical Mapping, Cold Spring Harbor Laboratory Press, Cold Spring Harbor, NY, pp. 1–38.

Lee, H. et al. (1991). Ultraviolet nicking of large DNA molecules from pulsed-field gels for Southern transfer and hybridization. Anal. Biochem. *199*:29–34.

Lee, M.-S. et al. (1987). Detection of minimal residual cells carrying the t(14;18) by DNA sequence amplification. Science *237*:175–179.

Lichter, P. et al. (1991). Analysis of genes and chromosomes by nonisotopic *in situ* hybridization. GATA *8*:24–35.

Martin, R. et al. (1990). A highly sensitive, nonradioactive DNA labeling and detection system. BioTechniques *9*:762–768.

Mohabeer, A. J. et al. (1991) Non-radioactive single strand conformation polymorphism (SSCP) using the Pharmacia "PhastSystem." Nucleic Acids Res. *19*:3154.

Monod, M. et al. (1990). The identification of pathogenic yeast strains by electrophoretic analysis of their chromosomes. J. Med. Microbiol. *32*:123–129.

Mitchell, L. G. and Merril, C. R. (1989) Affinity generation of single-stranded DNA for dideoxy sequencing following the polymerase chain reaction. Anal. Biochem. *178*:239–242.

Muhlegger, K. et al. (1988). Synthesis and use of new digoxigenin-labeled nucleotides in nonradioactive labeling and detection of nucleic acids. Nucleosides and Nucleotides *8*:1161–1163.

Mullis, K. B. and Faloona, F. A. (1987). Specific synthesis *in vitro* via a polymerase-catalyzed chain reaction, in *Methods in Enzymology*, Vol. 155, Wu, R., Ed., Academic Press, San Diego, pp. 335–350.

Nickerson, D. A. et al. (1990). Automated DNA diagnostics using an ELISA-based oligonucleotide ligation assay. Proc. Natl. Acad. Sci. U.S.A. *87*:8923–8927.

Nguyen, Q. et al. (1992). A two-step hybridization method for chemiluminescent detection of single copy genes. BioTechniques (in press).

Okahashi, N. et al. (1990). Construction of a *Not*I restriction map of the *Streptococcus mutans* genome. J. Gen. Microbiol. *136*:2217–2223.

Orita, M. et al. (1989). Detection of polymorphisms of human DNA by gel electrophoresis as single-strand conformation polymorphisms. Proc. Natl. Acad. Sci. U.S.A. *86*:2766–2770.

Ou, C.-Y. et al. (1988). DNA amplification for direct detection of HIV-1 in DNA of peripheral blood mononuclear cells. Science *239*:295–297.

Reed, K. C. and Mann, D. A. (1985). Rapid transfer of DNA from agarose gels to nylon membranes. Nucleic Acids Res. *13*:7207–7220.

Saiki, R. K. et al. (1985). Enzymatic amplification of beta-globin genomic sequences and restriction site analysis for sickle cell anemia. Science *230*:1350–1354.

Saiki, R. K. et al. (1986). Analysis of enzymatically amplified β-globin and HLA-DQa DNA with allele-specific oligonucleotide probes. Nature *324*:163–166.

Singer-Sam, J. et al. (1989). Use of Chelex to improve the PCR signal from a small number of cells. Amplifications: A Forum for PCR Users, *3*(September):11.

Sarkar, G. et al. (1990). Characterization of PCR amplification of specific alleles (PASA). Anal. Biochem. *186*:64–68.

Stoflet, E. S. et al. (1989). Genomic amplification with transcript sequencing. Science *239*:491–494.

Tumolo, A. et al. (1992). Detection of DNA on membranes with alkaline phosphatase labeled probes and chemiluminescent AMPPD substrate, in *Non-Isotopic DNA Probe Techniques*, Kricka, L. J., Ed., Academic Press, San Diego, pp. 127–145.

Walsh, P. S. et al. (1991). Chelex 100 as a medium for simple extraction of DNA for PCR-based typing from forensic material. BioTechniques *10*:506–513.

Wartell, R. M. et al. (1990). Detecting base pair substitutions in DNA fragments by temperature-gradient gel electrophoresis. Nucleic Acids Res. *9*:2699–2705.

Upadhyaya, M. et al. (1990). A 90 kb DNA deletion associated with neurofibromatosis type 1. J. Med. Genet. *27*:738–741.

Ugozzoli, L. and Wallace, R. B. (1991). Allele-specific polymerase chain reaction, in *A Companion to Methods in Enzymology*, Vol. 2, Arnheim, N., Ed., Academic Press, San Diego, pp. 42–48.

Zhang, Y. et al. (1991). Single-base mutational analysis of cancer and genetic diseases using membrane bound modified oligonucleotides. Nucleic Acids Res. *19*:3929–3933.

CLONTECH LABORATORIES, Inc.
4030 Fabian Way
Palo Alto, CA 94303-4607
(800) 662-2566

Products of potential interest for gene level evaluation in human disease produced by CLONTECH include a range of DNA labeling (nonisotopic systems included), cDNA, and PCR research tools, including those for analysis of oncogenes, *Alu* repeats, cystic fibrosis, etc.

Some of the innovative products regularly featured in the company's publication "CLONTECHniques" are described below.

G3PDH AND TRANSFERRIN RECEPTOR CONTROLS FOR USE IN NORTHERN BLOT AND RT-PCR ANALYSIS

Like **β-actin, glyceraldehyde-3-phosphate dehydrogenase (G3PDH)**, a "housekeeping" gene, and **transferrin receptor**, required for iron uptake, are ubiquitously expressed genes. Traditionally, the abundantly expressed β-actin gene has been used as a transcription reference. However, because levels of β-actin mRNA can vary between tissues and in response to many common inducing agents, including tumor-promoting phorbol esters and various cytokines (Elder et al., 1984, 1988; Siebert and Fukuda, 1985), **β-actin may not be the optimal control** for all experiments.

In contrast to β-actin, G3PDH transcript levels are refractory to a number of inducing agents (Zentella et al., 1991; Bosma and Kooistra, 1991) and, unlike β-actin, transcript levels remain relatively consistent among most tissues. In contrast to G3PDH and β-actin, which are expressed at high levels in most tissues, the transferrin receptor gene is transcribed at moderately low levels. This makes it useful in **multiplex RT-PCR** (reverse transcriptase PCR) experiments, in which two sets of primers are present in the same tube.

Whereas β-actin and G3PDH can overwhelm the target gene because of their abundant expression, the lower level of expression of transferrin receptor makes it more suitable as a reference in many RT-PCR experiments. Clontech has developed **human G3PDH and transferrin receptor control probes** and **primers**, ideal for use as transcription references in northern blot and RT-PCR analysis. cDNA probes used alone or in combination with Clontech **control primers** can

- **document equal loading** of mRNA on northern and RNA dot blots
- determine **relative levels of gene expression** by multiplex northern or PCR analysis
- **verify the efficiency** of first-strand cDNA synthesis in RT-PCR
- **assess the integrity** of an RNA sample

The cDNA probes may also be **used as controls** in nuclear **transcription run-off assays.** The G3PDH and transferrin receptor cDNA probes join our popular β-actin cDNA probe as valuable transcription references.

Were Similar Amounts of RNA Loaded in Each Lane of the Gel?

Whether examining the expression of a gene in different tissues or following the regulation of a gene, one can probe a northern blot containing multiple RNA samples with the human G3PDH cDNA probe. The presence of approximately equal amounts of G3PDH in each lane is assurance that the regulated expression observed for the target gene is genuine.

Was the Efficiency of First-strand cDNA Synthesis Similar in Several Different Reactions?

RT-PCR has been performed on RNA isolated from treated and untreated cells and it has been determined that expression of the target gene may be regulated. G3PDH or transferrin receptor control amplimers in a side-by-side PCR amplification can be used with the target gene primers to verify the observed regulation. **If the amount of amplified control cDNA is uniform in both reactions, then the efficiencies of the cDNA syntheses were similar** and the observed synthetic levels reflect regulated expression. In some cases, control amplimers can serve as controls for both cDNA synthesis efficiency and PCR amplification in multiplex PCR experiments.

How Can Control Probes and Primers Be Used in Quantitative RT-PCR Experiments?

To quantitate changes in the level of expression of a gene in response to different treatments or in different tissue types, Clontech control amplimers and target gene primers can be used to amplify cDNA reverse transcribed from mRNA in a multiplex PCR experiment. Blot the RT-PCR products to a nylon membrane. Probe with a control probe or the target gene probe. Then either expose an autoradiogram and measure the band intensities by densitometry or excise the bands from the membrane and determine the counts. By **comparing the ratios of target gene signal to control probe signal**, the degree of change in target gene expression can be determined.

REFERENCES

Bosma, P. J. and Kooistra, T. (1991). J. Biol. Chem. *266*:17845.
Elder, P. K. et al. (1984). PNAS *81*:74.
Elder, P. K. et al. (1988). Mol. Cell Biol. *8*:480.
Siebert, P. and Fukuda, M. (1985). J. Biol. Chem. *260*:3868.
Zentella, A. et al. (1991). Mol. Cell Biol. *11*:4952.

BIOTIN-ON™ PHOSPHORAMIDITE

Biotinylated oligonucleotides are used in a variety of applications, such as **direct solid-phase sequencing of PCR products** (Hultman et al., 1989, 1991; Mitchell and Merril, 1989; Uhlen et al., 1989), **quantitative analysis of PCR products** (Syvanen et al., 1988; Landgraf et al., 1991; Balaguer et al., 1991), **chemiluminescent sequencing** (Martin et al., 1991; Creasey et al., 1991), and **genomic walking** (Rosenthal and Jones, 1990). Other applications include **dot blots, Southern blots**, and ***in situ* hybridizations** (Misiura et al., 1990; Nelson et al., 1989).

Biotin-ON phosphoramidite is a unique cyanoethyl (CE) phosphoramidite that directly attaches biotin onto oligonucleotides in a single step during automated DNA synthesis, which may be performed on any commercial DNA synthesizer. Biotin-ON allows the incorporation of biotin at the 5′ terminus, any internal position, or both. Single or multiple biotins may be incorporated at any position during the synthesis process with high coupling efficiency: >97%.

Incorporated biotin label is stable to **NH_4OH deprotection**. Ideal for rapid, nonisotopic oligonucleotide labeling, Biotin-ON couples as any other CE phosphoramidite, and the **dimethoxytrityl** (DMT) group enables easy determination of coupling efficiency.

The biotin moiety is attached through a seven-atom linking arm that minimizes steric hindrance around the biotin group, thus giving improved streptavidin binding. Due to the unique structure of Biotin-ON, the three-carbon natural distance between internucleotide phosphate groups is maintained in order to mimic the three-carbon backbone of a normal nucleotide. This feature is designed to give better hybridization and annealing properties.

REFERENCES

Balaguer, P. et al. (1991). Anal. Biochem. *195*:105.
Creasey, A. et al. (1991). BioTechniques *11*:102.
Hultman, T. et al. (1989). Nucleic Acids Res. *18*:4345.
Hultman, T. et al. (1991). BioTechniques *10*(1):84.
Landgraf, A. et al. (1991). Anal. Biochem. *193*:231.
Martin, C. et al. (1991). BioTechniques *11*:110.
Misiura, K. et al. (1990). Nucleic Acids Res. *18*:4345.
Mitchell, L. and Merril, C. (1989). Anal. Biochem. *178*:239.
Nelson, P. et al. (1989). Nucleic Acids Res. *17*:7179.

Rosenthal, A. and Jones, D. S. (1990). Nucleic Acids Res. *18*:3095.
Syvanen, A. C. et al. (1988). Nucleic Acids Res. *16*(11):327.
Uhlen, M. et al. (1989). Nature (London) *340*:733.

DNA SEQUENCING WITH MULTIPLE POLYMERASES

DNA sequencing is an indispensable tool in molecular genetic analysis. The acquisition of accurate sequence information depends largely on the method chosen for sequencing; the ideal sequencing system should be highly accurate and reliable. Currently, the two commonly used methods for DNA sequencing are the **Maxam and Gilbert chemical cleavage technique** (Maxam and Gilbert, 1980) and the enzymatic **chain-termination method of Sanger** et al. (1977).

In the more versatile and popular Sanger method, an oligonucleotide primer is annealed to the single-stranded DNA template and extended by a DNA polymerase. The synthesis of the complementary DNA strand is terminated randomly by the incorporation of a **chain-terminating nucleotide analog**, typically a 2′,3′-dideoxynucleoside 5′-triphosphate (**ddNTP**). Four separate reactions, each containing a different terminating nucleotide (ddATP, ddCTP, ddGTP, or ddTTP), are performed, generating four sets of dideoxy-terminated DNA fragments. The reaction products form a series of fragments of different sizes, which can be separated on adjacent lanes of a polyacrylamide gel. The DNA sequence information is deduced from the pattern of bands in the lanes.

Visualization of the sequencing bands is usually accomplished by autoradiography and, therefore, requires radiolabeling of the sequencing products. There are three common techniques used to label the DNA. Originally, labeling was done by a pulse-chase reaction using Klenow DNA polymerase (Sanger et al., 1977). The polymerase extends, radioactively labels the primer, and terminates some extensions with a ddNTP in four separate reactions. A chase step with high concentrations of all four dNTPs follows.

In the second method, **polynucleotide kinase** is used to label a primer at the 5′ end. DNA polymerase then extends the labeled primer and terminates the strand with a ddNTP.

The **third and most commonly used method involves two reactions** (Tabor and Richardson, 1987). In the **labeling step**, the primer is radioactively labeled by extension with a DNA polymerase under conditions of low dNTP concentration. In **the termination step**, the labeling mix is divided into four separate termination reactions, and the primer is extended further in the presence of a single ddNTP and high concentrations of dNTPs.

Choice of Sequencing Polymerase

The quality of sequence information obtained is heavily dependent on the type of DNA polymerase used. Some enzymes, such as **Klenow**, are sensitive to local secondary structure in the template and tend to terminate extension prematurely when they encounter these regions. This generates artifactual "**false stops**," in which bands appear at the same position across all four lanes of the sequencing gel (Kusukawa et al., 1990).

Enzymes also differ in their abilities to incorporate nucleotide analogs. For example, a modified form of **T7 DNA polymerase** (Engler et al., 1981; Tabor and Richardson, 1987) incorporates nucleotide analogs efficiently, thereby helping to alleviate the gel compressions sometimes caused by GC-rich regions in the template. However, the T7 enzyme, like Klenow, also can generate false stops, particularly when sequencing double-stranded templates (Fawcett and Bartlett, 1990). These artifact bands often can be eliminated by performing the extension reaction at 65 to 70°C to reduce template secondary structures; however, this requires a thermostable polymerase.

Taq, **heat-TUFF**, and ***BstI* heat-stable DNA polymerases**, among others, have all been used for this purpose. These polymerases too are not without problems, because some may not produce clear sequencing ladders near the primer (Kusukawa et al., 1990), nor can all false stops be eliminated (Khambaty and Ely, 1990). The polymerization rate depends both on the sequencing polymerase and the concentration of the dNTPs. Any one enzyme in a defined mix of ddNTP/dNTP will generate sequencing products in a certain size range.

Using current sequencing technology, one set of sequencing reactions must be performed in one concentration of labeling mix in order to read sequence close to the primer, and a second set of reactions must be carried out in a labeling mix containing higher concentrations of dNTPs in order to read many hundreds of bases away from the primer. Under the first set of conditions, the sequencing gel will show dark bands close to the primer, but faint bands approximately 150 nucleotides or more from the primer. In the second case darker bands, hundreds of nucleotides away from the primer, will be seen but bands within 100 nucleotides of the primer will be faint and often illegible.

A Mixed Enzyme Approach to Sequencing

Clontech has developed a unique sequencing system that combines the advantages of several DNA polymerases. The enzyme mix in our new **Multi-Pol™ DNA sequencing system** includes both thermostable and nonthermostable polymerases and acts by minimizing the idiosyncratic problems of each enzyme. The initial labeling step is carried out at low temperature, with all enzymes participating to varying extents, depending on their different temperature optima. The temperature is then raised to 70°C during the extension step to maximize the benefits of the thermostable component.

During the transition from low temperature to high, the activity of the nonthermostable component peaks and then declines, whereas the thermostable component becomes increasingly active. When one enzyme terminates polymerization because of sensitivity to a particular sequence or template secondary structure, another polymerase continues extension of the DNA strand past the premature stop, a key advantage when sequencing double-stranded templates.

Finally, because the polymerases in the Clontech mix incorporate ddNTPs at different frequencies, DNA bands both close to the primer and far away (>500 nucleotides) can be detected with only one set of sequencing reactions. Therefore, unlike most single enzyme systems, band intensity does not vary greatly with distance from the primer, resulting in DNA sequence that is easier to read over a greater size range.

REFERENCES

Engler, M. J. et al. (1981). J. Biol. Chem. *258*:11165–11173.

Fawcett, T. W. and Bartlett, S. G. (1990). BioTechniques *9*:46–48.

Khambaty, F. and Ely, B. (1990). BioTechniques *9*:714–715.

Kusukawa, N. et al. (1990). BioTechniques *9*:66–72.

Maxam, A. M. and Gilbert, W. (1980). in *Methods Enzymology*, Vol. 15, pp. 499–560.

Sanger, F. et al. (1977). Proc. Natl. Acad. Sci. U.S.A. *74*:5463–5467.

Tabor, S. and Richardson, C. C. (1987). Proc. Natl. Acad. Sci. U.S.A. *84*:4767–4771.

MESSAGE AMPLIFICATION PHENOTYPING (MAPPING™): HUMAN CYTOKINE MAPPING™ AMPLIMERS

The MAPPing (**message amplification phenotyping**) system has been used in **cytokine analysis**, and PCR cytokine MAPPing amplimers are designed to analyze the cellular expression of cytokines and related growth factors. MAPPing may be used for the study of cytokine-specific mRNA in lymphocytes, natural killer (NK) cells, granulocytes, monocytes, fibroblasts, and tumor cell lines. Depending on research needs, MAPPing can be used instead of, or in addition to, northern blot or immunoassay methods for cytokine protein analysis.

The MAPPing technique involves an easy-to-follow procedure for (1) RNA isolation, (2) reverse transcription of RNA to cDNA, and (3) amplification of cytokine-specific cDNA fragments using MAPPing amplimers and the PCR.

Cytokine MAPPing amplimers are sensitive, specific primers designed to amplify only the cDNA region specific to the cytokine of interest. Cytokine MAPPing amplimers are able to amplify from single-cell starting materials with no loss in efficiency. Amplification proceeds efficiently even though the cytokine of interest is expressed only at the picogram level. Cytokine MAPPing amplimers for use with PCR allow simultaneous analysis of human cytokines in 3 to 4 h.

The PCR process is covered by U.S. patents owned by Hoffman-La Roche and by pending and issued patents in non-U.S. countries owned by Hoffman-LaRoche AG, licensed by Clontech Laboratories.

Other recently featured CLONTECH products of potential interest for molecular diagnostic evaluations include:

- The **CLONTECH Transformer™ Site-Directed Mutagenesis Kit** helpful in introducing simultaneously several site-specific mutations on the same double-stranded template without the need for successive rounds of mutagenesis. It can also generate large, precise deletions in a dsDNA plasmid. Although the upper limit on the possible size of deletions has not been determined, those ranging in size from a single bp to ca. 500–900 bp have been observed.

- **RT-PCR using QUICK-Clone™ cDNAs** allows to determine which sequences are expressed providing sometimes a better tool than ***northern blot*** or even the more sensitive ***RNase protection assay*** (see Part 1).
- **CLONTECH p53 ONCO-LYZER™ Kit** allows a fast and convenient detection of base pair changes in the human *p53* gene as well as the determination of the sequence of all exons (and most of the intron/exon junctions) within the amplified region of the gene.

REFERENCE

Brenner, C. et al. (1989). BioTechniques *7*(10):1096.

DIGENE DIAGNOSTICS

2301-B Broadbirch Drive
Silver Spring, MD 20904
(301) 680-9680

Digene Diagnostics, located in Silver Spring, Maryland, is a biotechnology firm specializing in the development of nucleic acid hybridization tests for infectious disease agents and molecular biology reagents, serving the medical, healthcare, pharmaceutical, and research industries. Digene is the premier supplier for human papillomavirus (HPV) analysis systems worldwide.

Digene was founded in 1985 to develop proprietary nucleic acid probe technologies. The company is one of the largest American-owned DNA diagnostics companies. The Digene ViraPap and ViraType HPV dot blot assays were the first and remain the only **HPV analysis products that have been cleared by the U.S. Food and Drug Administration (FDA).**

Digene's product offering consists of a number of infectious disease analysis systems including **HPV, hepatitis B virus, cytomegalovirus, Epstein-Barr virus**, and **mycobacteria**, along with hybridization-grade reagents and bacterial growth supplies for molecular biology applications. Digene has also developed a nucleic acid hybridization assay system, **Hybrid Capture**, that utilizes **chemiluminescent amplification** and combines the simplicity of enzyme immunoassay with the sensitivity typical of isotopic systems.

ISOTOPIC DOT BLOT HPV ASSAYS

A recent study of 2627 women published by Lorincz et al. (1992) concluded that:

- HPV types 6/11/42/43/44 **are not associated** with invasive cancer.
- HPV types 31/33/35/51/52 **are strongly associated** with high-grade squamous intraepithelial lesions (SIL 7 CIN2/CIN3).
- HPV type 16 **is strongly associated** with high-grade SIL and invasive cancer.
- HPV types 18/45/56 are **strongly associated** with invasive cancer.

ViraPap and ViraType are capable of detecting and typing the seven **most prevalent** anogenital HPV types 6, 11, 16, 18, 31, 33, and 35 in cervical swabs, scrapes, or fresh cervical biopsies. ViraPap detects all seven HPV types using a one-probe cocktail; ViraType differentiates among low-risk (6/11), intermediate-risk, (31/33/35), and high-risk (16/18) HPV types using three probe cocktails. **Both assays employ ^{32}P-labeled RNA probes.** The use of RNA probes enhances specificity because an enzymatic digestion step reduces nonspecific binding of probes, thus minimizing background signal. Digene offers a specifically designed **specimen collection set**.

Assay protocol is simple, requiring less than 3 h of hands-on time: (1) **disrupt** infected cells and viral particles; (2) **denature** viral DNA; (3) **immobilize** DNA on a solid filter support; (4) **hybridize** ^{32}P-labeled nucleic acid probe to immobilized target; (5) **wash** to remove unhybridized probe; and (6) **detect** hybridized probe by autoradiography. Results can be interpreted as early as 3 days from set-up. Extensive documentation of ViraPap and ViraType performance demonstrates that sensitivity and specificity are greater than 95% relative to Southern blot.

To address the availability of additional, more recently characterized HPV types, Digene has developed a **research use assay**, the **HPV profile**, that extends detection capability to 14 HPV types: 6, 11, 16, 18, 31, 33, 35, 42–45, 51, 52, and 56. In addition, the probe cocktails have been reduced to two for typing specimens into low-risk (6/11/42/43/44) and high/intermediate risk (16/18/31/33/35/45/51/52/56) HPV types. This configuration yields test results that coincide with appropriate patient management options. The HPV profile employs a testing format similar to ViraType and results can be reported with the same turnaround time. In a study conducted on 100 women, HPV profile sensitivity and specificity relative to Southern blot were 100 and 96%, respectively. Further study of the performance characteristics of the HPV profile is underway in preparation for its submission to the FDA. **The detection limit of all Digene HPV dot blot assays is approximately 10^5 DNA copies per specimen aliquot.** This level has been shown to correlate with clinically significant levels of HPV infection.

HYBRID CAPTURE SYSTEM

The Digene Hybrid Capture system is a new molecular hybridization assay that utilizes **chemiluminescent detection** rather than isotopic detection. Specimens containing the target DNA hybridize with a specific RNA probe cocktail. The resultant RNA:DNA hybrid is captured onto the surface of a tube coated with an anti-RNA:DNA hybrid antibody. Immobilized hybrid is then reacted with an anti-hybrid antibody-alkaline phosphatase conjugate and detected with a chemiluminescent substrate. As the substrate is cleaved by the bound alkaline phosphatase, light is emitted, which is measured as **relative light units (RLUs)** on a luminometer. The intensity of the light emitted is proportional to the amount of target DNA in the specimen and, depending on the assay, qualitative or quantitative results can be obtained. The Hybrid Capture system is available for **research use** to detect HPV and HBV DNA.

VIRATYPE PLUS HPV DNA CHEMILUMINESCENT ASSAY

The Digene Hybrid Capture ViraType Plus HPV DNA assay is an *in vitro* chemiluminescent test for the analysis of HPV DNA groups in cervical swabs or scrapes collected with Digene's specimen collection set. As with HPV Profile, 14 HPV types can be detected: 6, 11, 16, 18, 31, 33, 35, 42–45, 51, 52, and 56. A two-probe configuration differentiates low-risk (6/11/42/43/44) from high/intermediate risk (16/18/31/33/35/45/51/52/56) HPV infections. In contrast to the HPV dot blot assays, Hybrid Capture is a simple immunoassay-like procedure that reduces hands-on time by at least 30%. Results are reportable the same day that the assay is set up. **The analytical sensitivity of the assay is 10 pg HPV DNA per milliliter**.

HBV DNA CHEMILUMINESCENT ASSAY

The Digene **Hybrid Capture HBV DNA assay** is an *in vitro* chemiluminescent test for the detection and quantitation of HBV DNA in serum. The detection of HBV DNA is a direct marker of active viral replication and a quantitative marker of infectivity. Measurement of HBV DNA can be used to monitor the response of patients undergoing antiviral therapy. **The analytical sensitivity of the assay is 10 pg HBV DNA per milliliter of serum**. The linear range of detection is 10 to 2000 pg HBV DNA per milliliter of serum.

IN SITU HYBRIDIZATION PRODUCTS

In situ hybridization (ISH) permits the precise localization and identification of a specific nucleic acid target within individual cells in tissue. Most ISH procedures resemble routine immunohistochemical staining protocols. The basic steps of an ISH procedure include (1) sample preparation, in which a formalin-fixed, paraffin-embedded tissue section is placed on a microscope slide, deparaffinized, and treated with a protease to render the target nucleic acid accessible to the probe; (2) hybridization, in which the sequence-specific tagged probe is added, denatured in the presence of the target nucleic acid, and allowed to reanneal in probe-target hybrids; and (3) detection. If the probe is tagged with biotin, as are the Digene HPV probes, then steps to associate a reporter enzyme with the biotin-labeled probe-target DNA hybrid must be taken. The Digene system uses the affinity of biotin/avidin to bind the alkaline phosphatase enzyme to biotin-labeled

DNA via an avidin-enzyme conjugate. The enzyme is then exposed to the precipitating substrate system of BCIP and NBT, resulting in a dark purple-blue precipitate in the nucleus of a positive cell that can be observed with a light microscope.

BIOTINYLATED HPV PROBES

The use of ISH for HPV detection can achieve the following: evaluation of histologically equivocal lesions, typing virus in confirmed lesions, differentiation of latent vs. active infection, and general quality control for histological diagnosis of HPV infection. Digene offers HPV-specific probe sets that can be used interchangeably with the Digene tissue hybridization kit.

HPV OmniProbe can differentiate equivocal from true HPV-related lesions by detecting the 14 most commonly encountered HPV types: 6, 11, 16, 18, 31, 33, 35, 42–45, 51, 52, and 56. This can be accomplished with a single HPV-specific probe cocktail. In a recent study, the sensitivity of the HPV OmniProbe assay for detection of HPV DNA in low-grade CIN was found to be essentially equivalent to Southern blot (96%) and the specificity relative to a histological diagnosis was 100%.

The ViraType *in situ* probes type and differentiate among low-risk (6/11), high-risk (16/18), and intermediate-risk (31/33/35) HPV infections using three HPV cocktails. Both probe sets include positive and negative DNA control probes. Each control probe is recommended to be run on each biopsy to control for adequate tissue processing and technical performance of the assay and for nonspecific background, respectively.

DIGENE TISSUE HYBRIDIZATION KIT

The Digene tissue hybridization kit is a complete reagent system for detecting biotinylated DNA *in situ* probes and contains the reagents necessary for sample preparation through detection of the probe-target hybrids. The kit is intended to be used with Digene biotinylated probes and can be adapted for use with experimental probe cocktails made with the Digene biotin-probe diluent (discussed below).

IN SITU ACCESSORIES

Digene offers a complete hybridization solution, **biotin-probe diluent**, for preparation of biotinylated DNA *in situ* probe cocktails specific for targets other than HPV DNA. Biotin-probe diluent can be used with any biotinylated DNA that meets specific parameters, such as mean nucleotide sequence length and percentage of biotin-labeled nucleotide substitution. DNA meeting those requirements and dissolved in TE or water can be mixed with the Digene biotin-probe diluent to make a cocktail compatible for detection by the Digene tissue hybridization kit. The positive and negative controls probes provided in Digene probe sets are available separately for complete experimental design. The positive DNA control probe is specific for human genomic DNA and should always give a positive result on human tissues. The negative DNA control probe is specific for plasmid sequences and should not give positive hybridization results.

Silanated slides, coated with 3-aminopropyltriethoxysilane (AES), are available in a variety of formats for convenience. Slides coated with AES have been shown to have superior tissue retention in *in situ* assays when compared with more common slide treatments such as poly-L-lysine, poly-D-lysine, and gelatin.

MOLECULAR BIOLOGY PRODUCTS

As the foundation of its business, Digene produces an extensive line of solutions for nucleic acid blot and hybridization analysis that provide increased reliability, assured performance, and convenience to the research molecular biologist. In addition, a complete line of liquid and agar media is offered, as well as microbiological plates specially prepared and tested for the propagation and maintenance of bacterial strains.

HYBRIDIZATION-GRADE REAGENTS

Hybridization-grade reagents include preformulated hybridization buffers optimized for use on nylon and nitrocellulose filters in both isotopic and nonisotopic systems. Buffer components such as Denhardt's solution, formamide, SSPE, dextran sulfate, and SDS are available for in-laboratory preparation of hybridization solutions. Sonicated genomic DNAs (calf thymus, herring sperm, and salmon sperm), ideally suited for prehybridization of filters used in Southern hybridizations, are provided as purified, denatured 10-mg/ml solutions.

BACTERIAL GROWTH SUPPLIES

A variety of liquid and agar growth media is offered, as well as plates formulated specifically for propagating and maintaining *E. coli*. All-purpose LB medium is suitable for growing any *E. coli* strain whereas a variety of other specialized formulations (NZCYM, SOC, Superbroth, TBMM, and YT) are ideal for propagating *E. coli* harboring different bacteriophages and/or plasmids.

REFERENCE

Lorincz et al. (1992). Obstet. Gynecol. 79:328–337.

ENZO DIAGNOSTICS
325 Hudson Street
New York, NY 10013
(800) 221-7705

Enzo Diagnostics markets the research products developed by Enzo Biochem. Since its establishment in 1976, Enzo Biochem has developed expertise in the chemical modification of nucleic acids and proteins. In 1982

Enzo introduced the first of its BioProbe systems. BioProbe labeling introduces a biotin label into nucleic acid probes without interfering with the ability of these probes to hybridize. DETEK signal-generating systems bind to the biotin label, providing a sensitive, nonradioactive hybridization assay.

This unique BioProbe technology (Langer et al., 1981) has been the major emphasis of the company. Enzo Diagnostics currently markets a full range of proprietary BioProbe products, allowing scientists and the medical community throughout the world to detect the presence of small amounts of specific biological molecules nonradioactively. The products offered, including a complete line of biotinylated nucleotides and nucleic acid-labeling and detection kits, provide for a wide variety of applications. These products can be used with both *in situ* procedures and membrane-based assays, such as Southern, Northern, and dot blots.

In addition to the do-it-yourself labeling and detection products, Enzo maintains an expanding line of biotinylated products for the research laboratory, including labeled probes of diverse specificities and complete biotinylated DNA probe based kits for *in situ* use. This technology has changed DNA probe hybridization from a complex, labor-intensive, and time-consuming procedure into a rapid, simple, specific methodology. The Enzo PathoGene and BioPap DNA probe assays provide the advantages of a DNA probe assay without the hazards or instabilities of radioisotopes.

Enzo's newest microplate hybridization assay technology represents a second generation of DNA probe assays in kits for both HIV and *Mycobacterium tuberculosis* complex. With this easy-to-use, nonradioactive system, HIV DNA can be detected independent of the presence or absence of antibodies, and the MTB organisms can be specifically identified under conditions that are unrelated to the state of growth of the culture or organism. The straightforward assay procedure, which is easily automatable, is carried out in a microtiter plate or microwell strips. The generation of a color, which can be easily measured, indicates a positive reaction. This proprietary new method has wide application to research in both AIDS and tuberculosis epidemiology, diagnosis, and treatment. Future applications include the detection and identification of any number of microorganisms.

IN SITU HYBRIDIZATION AND DETECTION ASSAYS

Enzo has simplified the process of nonradioactive nucleic acid hybridization and detection. The Enzo PathoGene identification kits, with a sophisticated technology packaged as a simple technique, make it easy to detect and identify a wide range of infectious agents. Results are read with a light microscope and are based on enzymatic color reactions that identify both the pathogen and its location within the infected cell or tissue. The method is nonradioactive, DNA probes provide high specificity, and the stable reagents provide a long shelf life.

DNA hybridization assays make use of the fundamental chemical and physical properties of DNA molecules to detect and identify specific DNA sequences and, thus, specific organisms. The DNA of most organisms and viruses is double stranded, i.e., composed of two complementary strands. When double-stranded DNA is heated, the complementary strands separate (denature) to form single strands. Under appropriate conditions, separated complementary strands of DNA rejoin to form double-stranded DNA molecules. This process is called hybridization or renaturation. The hybridization of complementary DNA strands is an extremely faithful process. If DNA strands are not complementary, they will not hybridize to each other. Furthermore, the presence of unrelated DNA will not affect the hybridization of complementary DNA strands.

A DNA probe is a segment of DNA that is specific for and complementary to the DNA of the organism or virus that is to be detected and identified. Thus, hybridization with a DNA probe may be used to detect and identify specific DNA sequences. Procedures for hybridization of a DNA probe to a DNA target include three steps: (1) denaturation of the DNA probe and the DNA of the sample, (2) hybridization of the probe with the specimen DNA, and (3) measurement of the amount of hybridization between the DNA probe and the DNA in the specimen. The amount of DNA probe hybridized is directly related to the amount of complementary DNA in the specimen. If there is no DNA complementary to the DNA probe, no probe DNA will hybridize.

The Enzo PathoGene kits have been formatted for an *in situ* hybridization procedure in which the hybridization occurs within cells fixed to a microscope slide. After addition and hybridization of the specific DNA probe, the presence of infectious agent within the cells is detected enzymatically. Detection is based on the binding between the biotin of the probe and avidin, a biotin-binding protein. An avidin-horseradish peroxidase detection complex is used to recognize and bind to the biotin of the DNA probe. The presence of hybridized biotinylated probe is determined by the addition of hydrogen peroxide and aminoethylcarbazole. The DNA-bound horseradish peroxidase enzyme reacts with these reagents to yield a localized red-colored precipitate indicating a positive reaction. At the completion of this reaction, cellular morphology is intact and the cellular location of the target DNA can be visualized with a light microscope using ×100 and ×400 magnification.

Each PathoGene kit contains a detailed protocol and all the necessary reagents for processing 20 slides. PathoGene kits are available for adenovirus, cytomegalovirus, herpes simplex virus, Epstein-Barr virus, hepatitis B virus, *Chlamydia trachomatis*, and HPV. Products for detection and identification of HPV in both biopsy specimens and exfoliated cervical cells will be discussed in more detail in the next section.

GENERAL REFERENCES

Brigati, D. J. et al. (1983). Virology *126*:32–50.

Gall, J. G. and Pardue, M. L. (1971). in *Methods in Enzymology*, Vol. 38, Academic Press, New York, *38*:470–480.

Grody, W. W. et al. (1987). Hum. Pathol. *18*:535–543.
Langer, P. R. et al. (1981). Proc. Natl. Acad. Sci. U.S.A. *78*:6633–6637.
Macario, A. J. L. and Macario, E. C. (Eds.). *Gene Probes for Bacteria*, Academic Press, New York, 1990.
Piper, M. A. and Unger, E. R. (1989). *Nucleic Acid Probes: a Primer for Pathologists*, ASCP Press, Chicago, 1989.
Pollice, M. and Yang, H. L. (1985). Clin. Lab. Med. *5*:463–473.

IDENTIFICATION OF HUMAN PAPILLOMAVIRUS

Infection with specific types of HPV has been associated with an increased risk of developing cervical neoplasia. Although certain types of the virus, e.g., types 6 and 11, have been found in relatively benign diseases such as genital warts, other specific types, e.g., types 16 and 18, are associated with cervical, vulvar, and vaginal malignancies (Durst et al., 1983; Koutsky et al., 1988; Kurman et al. 1983). This strong association of HPV with cervical neoplasia has made the detection of this virus an important tool for the pathologist.

Conventional screening for cervical neoplasia relies on the cytological evaluation of the Pap smear for the presence of atypical cells such as koilocytes and/or dyskeratocytes. An abnormal Pap smear or a history of abnormal Pap smears is frequently followed up by a colposcopically directed cervical biopsy, which is examined for histological abnormalities such as koilocytosis, papillomatosis, acanthosis, and/or hyperkeratosis. Although the morphological criteria used to evaluate both the Pap smears and cervical biopsies can, in certain instances, indicate the presence of an HPV infection, in other cases it is impossible to determine with confidence whether or not the virus is present. Moreover, morphology cannot determine HPV type, i.e., low risk or high risk. The only reliable methods for determining the presence and type of HPV associated with a lesion are molecular hybridization methods.

Histological interpretation of anogenital biopsies can be quite difficult. For example, a panel of histopathologists reviewing 100 cervical biopsies (Robertson et al., 1989) reached good agreement in distinguishing benign lesions from high-grade lesions. They were, however, unable to agree on the diagnosis of low-grade lesions. Direct tests for HPV, such as *in situ* hybridization assays, complement existing morphological methods. In a series in which *in situ* hybridization was used to analyze HPV in lesions clinically suspicious of condyloma, Wright and Felix (1991) found that a large percentage (48%) of lesions that were suspicious but not diagnostic of condyloma were positive for HPV. Their study underscores the usefulness of HPV detection in biopsy specimens.

The false-negative rate for Pap smear screening has been estimated to be anywhere from 8 to 50% (Coppelson and Brown, 1974; Richart and Valiant, 1965). It has been reported (Ritter et al., 1988) that women either positive for HPV DNA or with abnormal Pap smears were more likely to have a confirmed cervical lesion than HPV-negative women with negative cytology results. Other studies (Lorincz et al., 1986; Ward et al., 1991) support the concept that reevaluation of smears from HPV-positive women with normal cytology may result in a revision of the cytological diagnosis.

Until now, HPV detection in clinical specimens has been a complicated, time-consuming process involving the use of radioisotopes. The development of simple and rapid *in situ* hybridization tests for the detection of HPV in both biopsy sections and exfoliated cervical cells has become an important new aid to the pathologist. The Enzo PathoGene and BioPap assays for HPV permit rapid colorimetric detection of HPV in specimens fixed to microscope slides. The assays are based on Enzo proprietary nonradioactive DNA hybridization technology and can be done on both formalin-fixed, paraffin-embedded tissue sections and on exfoliated cervical cells. Each kit contains one or more HPV-specific DNA probe(s) labeled with biotin. The specific probe will recognize and hybridize only to the DNA that is unique to HPV. By assaying for hybridization between the HPV-specific probe and the DNA in the specimen, HPV infection can be visualized within the nucleus of the infected cells.

The Enzo PathoGene and BioPap assays for HPV provide the pathologist with easy-to-use, cost-effective alternatives to the labor-intensive and radioactive Southern and dot-blot tests. The Enzo tests can be used to identify and type HPV, not only in specimens with evidence suggestive of the virus, but also in those specimens with no cytological or morphological indications of infection.

PathoGene DNA probe assay kits are available for both screening and typing of HPV DNA in formalin-fixed, paraffin-embedded tissue sections.

BioPap DNA probe assay kits are available for both screening and typing of HPV in exfoliated cervical cells.

REFERENCES

Coppelson, L. W. and Brown B. (1974). Am. J. Obstet. Gynecol. *119*:953.
Durst, M. et al. (1983). Proc. Natl. Acad. Sci. U.S.A. *80*:3812.
Koutsky, L. A. et al. (1988). Epidemiol. Rev. *10*:122.
Kurman, R. J. et al. (1983). Am. J. Surg. Pathol. *7*:39.
Lorincz, A. et al. (1986). Obstet Gynecol. *68*:508.
Richart, R. M. and Valiant, H. W. (1965). Cancer *18*:1474.
Ritter et al. (1988). Am. J. Obstet. Gynecol. *159*:1517.
Robertson, A. J. et al. (1989). J. Clin. Pathol. *42*:23.
Ward, B. E. et al. (1991). Int. J. Gynecol. Pathol. *9*:297.
Wright, T. and Felix, J. C. (1991). Lab. Invest. *64*:56A.

BIOPROBE LABELED PROBES

For those researchers who wish to design their own hybridization procedures, Enzo offers biotin-labeled probes. These BioProbe labeled probes are specific DNA sequences labeled with a modified nucleotide,

Bio-11-dUTP. These probes can be used for positive identification of pathogens and genetic material in a variety of formats based on hybridization to complementary nucleic acid sequences. Biotinylated DNA probes can be used for hybridization to DNA immobilized on a solid matrix such as Southern blot or dot blot analysis. The probes can be used for ISH to DNA in fixed cells or tissues. Biotinylated probes have been shown to hybridize to homologous DNA at the same rate and to the same extent as nonbiotinylated probes (Langer et al., 1981). The hybridized, biotinylated DNA probe is detected by its interaction with biotin-binding proteins, such as avidin, streptavidin, or antibodies coupled to fluorescent dyes or color-producing enzymes.

BioProbe labeled probes are available for a variety of organisms and oncogenes, including adenovirus, cytomegalovirus, Epstein-Barr virus, herpes simplex virus, hepatitis B virus, hepatitis A virus, JC virus, BK virus, *C. trachomatis, Mycoplasma pneumoniae, Campylobacter jejuni,* c-Ha-*ras*, c-*myc*, N-*myc*, lambda, pBR322, SV40, and Blur 8 (human *Alu* repeat).

REFERENCE

Langer, P. R. et al. (1981). Proc. Natl. Acad. Sci. U.S.A. *68*:6633.

THE ENZO MICROPLATE HYBRIDIZATION ASSAY

The Enzo microplate hybridization assay system is an easy-to-use, rapid, and nonradioactive kit for detecting DNA in a microtiter well. The straightforward assay procedure is based on a 2-probe hybridization method and is carried out entirely in the 96-well microtiter plate or, alternatively, microwell strips. The generation of a color, which can be easily measured, indicates a positive reaction. No specialized equipment, other than a standard microplate reader, which is present in most medical research laboratories, is required. In fact, for plus/minus determinations the results can be read by eye. The assay format uses either a microtiter plate, where up to 96 assays (including controls) can be run, or well strips, where 8 assays per strip (including controls) can be run. The assay requires less than 3 h.

This nonradioactive procedure involves hybridization of target nucleic acid to a well-bound capture probe followed by binding of a biotinylated signaling probe to the captured target. Immobilized target DNA is then visualized by reaction with a biotin-binding signal generating complex of **streptavidin (SA)** and **horseradish peroxidase (HRP)**. A positive reaction is indicated by generation of color that can be measured by a microplate reader commonly used in the laboratory. Using this format, 5×10^7 to 1×10^8 copies of target sequences can be detected.

The assay utilizes several unique features. **The use of probe pairs increases the specificity of the assay**. Two independent hybridization events are required to generate a signal. This type of assay has been found to be insensitive to the presence of cellular components other than the target

DNA sequence. Because the assay is nonradioactive, the components are safe and stable, and the assay can be carried out manually or using devices developed for the **automated processing** of microtiter plate-based assays.

The Enzo microplate hybridization assay is available for HIV-1 and for *M. tuberculosis*.

The Enzo HIV microplate hybridization assay kit provides materials for the colorimetric detection of nucleic acid containing the *gag* region. HIV DNA can be assayed directly if there is sufficient target DNA present, or it can be assayed in procedures employing target amplification.

The microplate assay detects HIV proviral DNA with excellent sensitivity; **fewer than 10 proviral sequences can be detected**. Furthermore, when the assay is performed using a standard curve of HIV DNA copies, quantitative virus measurements can be done. Thus, the microplate assay is readily applicable to studies measuring the effect of drug treatments on virus concentration, virus concentration during the course of infection, virus concentration in animal model studies, and numerous other studies where virus quantitation is a critical parameter.

Direct and quantitative HIV detection offers a **distinct advantage over serological assays**. The presence of provirus DNA can be determined independent of the presence or absense of antibodies, which appear several months after initial infection. Thus, this technique offers researchers the potential for identifying proviral DNA in instances where HIV positivity cannot be identified adequately by any of the current means available as, for example, in samples from HIV-positive individuals who have not seroconverted.

The Enzo MTB hybridization assay kit provides materials for the identification of *Mycobacterium tuberculosis* complex. The assay detects only the members of the MTB complex, that is, *M. africanum, M. bovis, M. tuberculosis, M. microti,* but not 25 other species of mycobacteria, including *M. avium, M. intracellulare, M. kansasii,* and *M. chelonae*. Furthermore, 63 other bacterial species were found to be negative, including a variety of respiratory pathogens. MTB complex DNA can be assayed directly if there is sufficient target DNA present, or it can be assayed in procedures employing target amplification or culture.

While the presence of acid-fast bacteria in a specimen can be highly suggestive of a mycobacterial infection, it is necessary to identify the species present. The identification of mycobacteria is often more difficult than many other pathogens because the clinically relevant species do not grow rapidly in laboratory culture. Using the microplate assay, the researcher can specifically identify members of the MTB complex under conditions that are unrelated to the state of growth of the culture or specimen.

REFERENCES

Cook, A. F. et al. (1988). Nucleic Acid Res. *16*:4077.

Picken, R. N. et al. (1988). Mol. Cell. Probes *2*:111.

DETEK SIGNAL GENERATING SYSTEMS

Enzo offers a wide variety of systems for the detection of biotinylated nucleic acids. Detection is based on recognition of biotin by biotin-binding proteins, either **streptavidin, avidin**, or **anti-biotin antibodies.**

The DETEK signal generating systems currently available are either fluorescent or enzymatic. The streptavidin molecule, which has multiple biotin-binding sites, is complexed with an enzyme such as HRP or alkaline phosphatase, and the entire complex can bind to a biotinylated macromolecule such as DNA, RNA, or protein. With addition of the appropriate enzyme substrate, the biotinylated nucleic acid or protein is visualized by the formation of a colored product.

The fluorescent DETEK signal generating systems are based on a biotin-binding protein and a fluorescent signal. DETEKI-f, for example, a double-antibody system, contains rabbit anti-biotin as a primary antibody and a fluoresceinated goat anti-rabbit antibody as the secondary antibody. The signal is visualized microscopically under ultraviolet light. DETEK-fav, in contrast, contains a fluorescein-conjugated avidin and is designed for direct fluorescent detection.

EPICENTRE TECHNOLOGIES
1202 Ann Street
Madison, WI 53713
(800) 284-8474

Epicentre Technologies is located near the University of Wisconsin in Madison. The company manufactures high-quality reagents, enzymes, and kits for molecular and cellular biology. Epicentre distributes directly in the United States as well as internationally through 17 distributors around the world.

Some of the products of potential interest for clinical research are listed below.

GELASE™ GEL-DIGESTING PREPARATION

GELase is a unique enzyme preparation developed by Epicentre for simple and quantitative **recovery of intact DNA** or **RNA** from low-melting (LMP) agarose. GELase completely digests molten agarose gels in ordinary Tris-acetate, Tris-borate, MOPS, or sodium phosphate electrophoresis buffers to yield a clear liquid that does not become viscous or gel even on cooling to 0°C. Because the oligosaccharide products of agarose gel digestion with GELase are alcohol soluble, **nucleic acids can be precipitated directly from the digested gel solution**. Nucleic acids recovered from gels using GELase are ready for use in restriction mapping, cloning, sequencing, labeling, or other molecular biological manipulations. Rigorous tests are performed by Epicentre to assure that GELase is free of DNA exo- or endonuclease, RNase, and phosphatase activities.

The protocol for the use of GELase calls for simply cutting the band of interest out of the gel, incubating it with the GELase at 70°C and then at 45°C, followed by alcohol precipitation of the released nucleic acids.

AMPLIGASE™ THERMOSTABLE DNA LIGASE

Ampligase thermostable DNA ligase, derived from an extreme thermophile, catalyzes NAD-dependent ligation of DNA at temperatures up to 85°C. Its half-life is about 48 h at 65°C. The introduction of Ampligase has led to improvements in a variety of techniques such as **ligation amplification**, also known as "**ligation chain reaction**" or "**LCR**" (see Part 1). Ligation amplification can be used to detect any defined DNA sequence. The researcher needs two pairs of oligodeoxynucleotides complementary to a contiguous dsDNA sequence that one desires to detect. Because Ampligase does not ligate blunt ends, a pair of oligodeoxynucleotides will be ligated only if after annealing they are adjacent to a target sequence in the sample. **If there is any mismatch at the ligation junction, the oligodeoxynucleotides will not be ligated, so assays can be developed for specific point mutations or analysis of RFLPs.**

Because the product of each ligation becomes a template for subsequent reactions, thermocycling between annealing and hybridization temperatures results in a geometric increase in the amount of product, as with PCR. Under optimal conditions, **attomole quantities of a target DNA sequence can be detected in a mixed population**.

Ampligase also has greatly simplified gene synthesis using overlapping oligodeoxynucleotides. High-temperature ligation with Ampligase has enabled researchers to obtain functional genes after screening only a few recombinant colonies.

Epicentre rigorously assays Ampligase to assure the absence of DNA exo- and endonuclease or phosphatase activities.

READY-LYSE™ LYSOZYME SOLUTION

Ready-Lyse is a stabilized lysozyme preparation for use in lysis of *Escherichia coli*. It is supplied as a ready-to-use solution that is stable at 20°C, which eliminates the need to prepare fresh enzyme solutions prior to each use or to freeze aliquots that are used once and then discarded. **Ready-Lyse can be frozen and thawed repeatedly without loss of activity**.

It has optimal activity at neutral pH, compared to an optimum activity at pH 9.2 for egg white lysozyme. At pH 6.5 to 7.5, Ready-Lyse has a specific activity about 200 times higher than egg white lysozyme. Positively charged proteins, such as lysozyme, may bind to and precipitate DNA or negatively charged proteins and reduce yield. Using similar amounts of protein, Ready-Lyse results in less precipitation of DNA than egg white lysozyme and can be used for efficient lysis at concentrations 200-fold less to further reduce the risk of losses.

RNA SYNTHESIS *IN VITRO:* TRANSCRIPTION PAC-KITS™

Transcription Pac-Kits are for scientists who want the convenience and economy of optimized and pretested transcription kits without being obligated to buy unneeded components.

Pac-Kits are modular kits, meaning that with each module one buys only what is needed, without buying an excess of any one reagent relative to the kit system. The quantities in each kit are sufficient for 500 standard transcription reactions.

"GELase," "Ampligase," "Ready-Lyse," and "Pac-Kits" are trademarks of Epicentre Technologies Corporation.

GENEMED BIOTECHNOLOGIES

458 Carlton Court, Suite B
South San Francisco, CA 94080
(800) 344-5337

Genemed Biotechnologies provides oligonucleotide primers, probes, and custom services useful in the detection and study of infectious organisms, viruses, cytokines, CD markers, and other genes and proteins of importance. The range of applications for these products includes gene amplification by polymerase chain reaction (GeneAmp PCR), Southern blotting, and ISH. Genemed carries a number of products not otherwise commercially available and has developed a high-sensitivity kit for ISH.

OLIGONUCLEOTIDE PRIMERS FOR POLYMERASE CHAIN REACTION

Genemed offers matched sets of oligonucleotide (DNA) primers that are designed and pretested for performance by GeneAmp PCR. A wide range of **primers and probes to different infectious bacteria and viruses** is offered. Furthermore, for many of these infectious diseases, a number of different primer sets are available. For instance, 15 distinct primer sets are offered for the **hepatitis B virus**, allowing the researcher to amplify the genes encoding a variety of proteins, including core and precore antigens, surface antigen, pre-S, pre-S1, pre-S2, S-X, and X. Fifteen primer sets are also available for **human papillomavirus**, which allows detection of more than 10 viral subtypes.

OLIGONUCLEOTIDE PROBES FOR SOUTHERN BLOT AND *IN SITU* HYBRIDIZATION

Unlabeled DNA Probes

For most primer sets, a matching DNA probe is available. This probe is used in Southern blots of the PCR products to confirm the success of the amplification, or to determine whether the amplified sequence is present in larger gene fragments. The probes are also used for ISH to demonstrate the presence of the complementary DNA sequences in intact cells. Probes may be labeled with fluorochromes, enzymes, biotin or ^{32}P for direct detection.

Biotinylated DNA Probes

Biotinylated probes are synthesized with a minimum of two biotin molecules per DNA oligomer. The probes are labeled on both ends with a "long arm" biotin derivative, which does not interfere with hybridization.

Custom-Labeled DNA Probes

Genemed will provide DNA probes conjugated directly to BrdU, Texas Red, fluorescein isothiocyanate (FITC), or alkaline phosphatase for ISH.

KITS FOR *IN SITU* HYBRIDIZATION

Enzymatic Kit

The Probestain-AP kit is designed for detection of DNA sequences *in situ*. The principle is as follows. A biotinylated DNA probe is hybridized to complementary sequences **in intact cells**. A modified high-affinity avidin is added, binding to the biotinylated probe. This is followed with a biotinylated monoclonal antibody to avidin. The complex is completed by addition of avidin conjugated to alkaline phosphatase. Visualization of the complex is accomplished by the addition of BCIP/NBT solution. The presence of less than 1 ng of biotinylated probe has been demonstrated by this system.

Fluorescent Kit

The **Probestain-FITC kit** is designed for detection of DNA sequences *in situ*. The principle is as follows. A biotinylated DNA probe is hybridized to complementary sequences **in intact cells**. An FITC-conjugated monoclonal antibody to biotin is added. The primary antibody is then bound by a biotinylated anti-mouse secondary antibody. Finally, FITC-conjugated streptavidin added. The cells are subsequently observed under the fluorescent microscope.

ONE-STEP NONRADIOACTIVE LABELING REAGENTS

Genemed offers three **phosphoramidite labeling reagents** for use with commercial DNA synthesizers: FITC, rhodamine, and biotin linked via a long arm spacer to CPG polymer. They are provided in either Applied Biosystems, Inc. or Milligen synthesis columns, each containing 0.2 μmol of fluorochrome.

GEN-PROBE
9880 Campus Point Drive
San Diego, CA 92121
(800) 523-5001

Gen-Probe is a San Diego based company whose initial focus has been infectious disease diagnostic assays. The majority of Gen-Probe products are based on **detection of ribosomal RNA**, using the natural amplification of RNA (10,000 copies) matched with the inherent specificity of probes in identifying infectious agents. We have linked the probe-based technology to a chemiluminescence detection system that provides sensitivity even greater than isotopes.

PACE 2 ASSAYS FOR DIRECT TESTING OF SEXUALLY TRANSMITTED DISEASES (STDs)

The PACE 2 system identifies both *Chlamydia* and *Neisseria gonorrhoeae* from a single urogenital swab. These assays were the first to employ nonisotopically labeled DNA probes and a simple protocol has fewer steps and requires less time than all other currently available, immunologically based enzyme-linked immunosorbent assays (ELISAs). The assay takes only 2 h, and 96 specimens can be fully processed in under 3 h.

PACE 2 for *N. gonorrhoeae* is the first definitive direct specimen assay for *N. gonorrhoeae*, with an overall sensitivity of 95.4% and a specificity of 99.8%. The PACE 2 *Chlamydia* assay also shows superior performance with an overall sensitivity of 92.5% and a specificity of 99.6%.

The Gen-Probe PACE specimen collection system eliminates the need to maintain organism viability, and allows for maximum RNA recovery. Specimens are stable for up to 7 days at room temperature to maximize flexibility in sample collection and laboratory testing.

ACCUPROBE CULTURE IDENTIFICATION

Mycobacterial Species

The AccuProbe mycobacteria products provide **rapid identification of clinically significant mycobacteria from primary culture** in less than 1 h. The system utilizes a homogeneous nonisotopic DNA probe assay that is performed in five simple steps. The overall sensitivity and specificity of all AccuProbe mycobacteria products is greater than 99%. AccuProbe will allow significantly earlier diagnosis of mycobacterial infection than routine biochemical methods, and the acridinium ester-labeled DNA probes provide greater shelf life than isotopic labels.

The system includes specimen-lysing vessels, precoated tubes containing acridinium ester-labeled DNA probe, and three reagents necessary to run the assay. Individual probe kits are available for the identification of *M. tuberculosis* complex, *M. avium* complex, *M. intracellulare*, *M. avium*, *M. gordonae*, and *M. kansasii*. A universal reagent kit allows convenient, efficient batching of different tests. Shelf life is 18 months from date of manufacture.

Fungi

AccuProbe products for the four systemic fungi have also been developed. These probe products can identify these fungal species from a culture sample before it becomes morphologically distinct, thus saving days to weeks in diagnostic identification time. Like other AccuProbe products, overall sensitivity and specificity are greater than 99%, and the fungal DNA probe tests can be completed in less than 30 min from the time of sample preparation. The four systemic fungal products available are for *Blastomyces dermitiditis*, *Coccidioides immitis*, *Cryptococcus neoformans*, and *Histoplasma capsulatum*.

Bacterial Species

AccuProbe test assays are also available for identification of a variety of bacterial organisms. The assays once again use a homogeneous nonisotopic assay format that allows results in 20 min. Overall sensitivity and specificity are greater than 99%. The probe tubes are packaged in a foil-lined pouch and have an open pouch stability of 2 months at 28 to 88°C.

Bacterial assays are available for identification of *Campylobacter*, *Enterococcus*, group B *Streptococcus*, *Haemophilus influenzae*, *Neisseria gonorrhoeae*, *Streptococcus pneumoniae*, *Staphylococcus aureus*, and *Listeria monocytogenes*.

RAPID DIAGNOSTIC SYSTEM ASSAYS

Mycoplasma pneumoniae

The Gen-Probe rapid diagnostic system assay for *Mycoplasma pneumoniae* was the first of several DNA probe assays for the diagnosis of infectious diseases. Real-time results allow physicians to begin proper patient therapy the same day. Gen-Probe technology eliminates the problem of subjective interpretation of cross-reactivity seen with other serological methods. It does not require cell viability or a large sample volume. Clinical studies have shown the *M. pneumoniae* DNA probe assay to be 100% specific and 100% sensitive compared with rigorous culture procedure.

Legionella

The Gen-Probe rapid diagnostic system for *Legionella* is a rapid, sensitive, and specific way to diagnose atypical pneumonia caused by this hard-to-detect, hard-to-grow organism. The probe detects all known species and serogroups of *Legionella*. A positive result confirms the presence of *Legionella* in 2 h or less, direct from the patient specimen.

INSTRUMENTATION

Gen-Probe also provides several chemiluminescence detection systems that provide sensitivity, speed, accuracy, and ease of use with Gen-Probe DNA probe technology. **Chemiluminescence is the ideal detection system, sensitive to 10^{-19} mol (as compared to radioisotopes at 10^{-18} mol).** Specific direct labeling of probes maintains the proven specificity of DNA probes and eliminates the specimen interference associated with indirect labeling systems, and the stable nonradioactive format provides reliable performance and extended shelf life.

The different available systems are the LEADER 50 and the LEADER

450, our automated chemiluminescent system, and the AccuLDR, for use only with the culture identification systems.

INVITROGEN
3985 B Sorrento Valley Boulevard
San Diego, CA 92121
(800) 955-6288

Invitrogen manufactures reagents and kits for molecular biology research. Several products may have immediate clinical diagnostic application: the DipStick kit, the Micro-FastTrack kit, and the GenePrep kit.

The **DNA DipStick kit** quantifies DNA and RNA at concentrations between 0.1 and 10 ng/μl within minutes. The assay is designed to process multiple samples quickly and accurately. The DNA DipStick assay measures single- and double-stranded DNA and RNA greater than 6 bp long and each kit is supplied with reagents to quantify up to 150 samples. It provides a permanent record without photography or special equipment.

The **Micro-FastTrack kit** rapidly isolates high-quality poly(A)$^+$ mRNA directly from small samples of tissues or cells (10 to 250 mg of tissue, 100 to 3×10^6 cells) suitable for PCR, northern blots, and cDNA synthesis. The typical yield from 10^8 cells ranges from 0.1 to 3 μg of mRNA. The kit does not employ toxic chemicals and multiple samples can be processed simultaneously. The isolation is complete within 60 min and each kit is supplied with reagents to process 20 samples.

The **GenePrep kit** is designed to isolate genomic DNA rapidly from cells and tissue samples, particularly blood samples. Similar to the Micro-FastTrack kit, it can process multiple samples simultaneously. Each GenePrep kit is supplied with reagents for 20 isolations.

ONCOGENE SCIENCE
106 Charles Lindbergh Boulevard
Uniondale, NY 11553-3649
(800) 662-2616

Oncogene Science is a biopharmaceutical company that utilizes its core technologies in oncogenes, anti-oncogenes, and gene transcription drug development to create innovative products for the treatment and detection of cancer and other cell control-linked diseases. The company is involved in several strategic partnerships with leading pharmaceutical and health care corporations and is itself conducting research on proprietary drugs in the cardiovascular area.

Through its acquisition of Applied BioTechnology and a new, multiyear collaboration with Becton Dickinson, Oncogene Science will become a significant participant in developing a novel class of cancer diagnostics.

These products are derived from the forerunning position of the company in oncogene and anti-oncogene technology.

Oncogene Science manufactures and sells a full line of over 900 products to the oncogene-related and gene transcription research markets. One of the products, **TransProbe-1, is an *in vitro* diagnostic assay** used for detecting the Philadelphia translocation found in chronic myelogenous leukemia (CML). **All other products are to be used for research purposes only.**

IN VITRO DIAGNOSTIC ASSAY: TRANSPROBE-1

Philadelphia Translocation in CML

TransProbe-1, also described as *phl/bcr*-3, is a DNA probe used in Southern blot analysis as an aid in the diagnosis of CML — a pluripotent stem cell disorder characterized by the presence of a cytogenetic abnormality termed the Philadelphia (Ph^1) chromosome. This abnormality, present in approximately 90% of CML patients, results from a reciprocal translocation. The resulting Ph^1 chromosome is characterized by a fusion of c-*abl* protooncogene sequences from chromosome 9 to *bcr* (also known as *phl*) sequences on chromosome 22.

Detection

The primary method for detection of the Ph^1 translocation in CML has routinely been karyotype analysis. Recent studies indicate that classification of the disease is better defined by events at the molecular level, i.e. Southern blot analysis.

Results with TransProbe-1

The Philadelphia (Ph^1) translocation involves only one allele each of chromosomes 9 and 22. All CML patients, therefore, contain at least one normal allele from each chromosome. *Bgl*II digestion yields three fragments from the normal (germline) *bcr*-210 region of chromosome 22, indicated by three bands on the Southern blot. **A positive result is demonstrated by the presence of one or two additional bands**. Results are available within 5 days.

TransProbe-1 vs. Karyotype Analysis

In a multisite study of 340 individuals, including 216 clinically diagnosed CML patients, 99.5% of karyotype Ph^1-positive CML patients were TransProbe-1 positive, and 44% of the 27 karyotype Ph^1-negative CML patients were TransProbe-1 positive.

No false positives were found using TransProbe-1.

TransProbe-1 Kit *phl/bcr*-3 Probe

TransProbe-1, or the *phl/bcr*-3 probe, spans the 5.8-kb *bcr*-210 region of the *bcr* gene. It is provided as a gel-isolated DNA fragment ready for labeling using the random primer or nick-translation labeling methods. A negative control HL-60 DNA and a positive control K562 DNA (15 g each in TE buffer) are included in each TransProbe-1 kit.

High-volume testing: Up to 20 samples (total), includes positive- and negative-control DNAs

Low-volume testing: Up to five samples (total), includes positive- and negative-control DNAs

Control DNAs: Positive- and negative-control DNAs are also offered separately from the TransProbe-1 kit

A **sensitivity control** sample can be prepared, representing 5% leukemic cells, by mixing the control DNAs

Reagent systems are available for optimal use of the TransProbe-1, and include a DNA extraction system, random primer labeling system, probe purification system, DNA molecular weight markers, and hybridization system.

PRODUCTS FOR RESEARCH USE ONLY

With the exception of TransProbe-1, Oncogene Science products are for research use only. These products are listed by category below.

Molecular Biology Products

Leukemia and Lymphoma Phenotyping by Polymerase Chain Reaction

TransPrimer leukemia and lymphoma phenotyping reagents use GeneAmp polymerase chain reaction (PCR) technology for the detection of chromosomal translocations in human leukemic cells. TransPrimer synthetic oligonucleotides are designed for use with GeneAmp PCR to amplify specific chromosomal translocations found in CML, acute lymphocytic leukemia (ALL), pre-B leukemia (pre-B), and follicular lymphoma (*bcl*-2/mbr and *bcl*-2/mcr). The unique design of our TransPrimers, along with the high sensitivity of the GeneAmp PCR techniques, allows the detection of minimal residual disease.

Leukemia Phenotyping

Each leukemia (CML, ALL, and pre-B) TransPrimer set contains "nested" primers to amplify the junction region of the translocation at the mRNA level. Nonspecific side reactions are eliminated by the use of nested primers. Specifically amplified DNA is rapidly detected by simple gel visualization.

TransPrimer	Rearrangement	Genes involved	Phenotype
CML primers[a]	t(9;22) q34;q11.2)	*bcr*/*abl*	CML
ALL primers[a]	t(9:22)(q34;q11.2)	*bcr*/*abl*	ALL
Pre-B primers[a]	t(14;18)(q32;p13)	E2A/PRL	Pre-B

Lymphoma Phenotyping

Each lymphoma (MBR and MCR) TransPrimer set contains one set of genomic DNA amplification primers and one additional oligonucleotide to confirm the amplification product by Southern hybridization. Translocations can be detected by simple gel visualization of the PCR amplification products. Additional sensitivity and confirmation can be achieved by Southern blotting.

Lymphoma TransPrimer sets include 5′ and 3′ primers for both normal and rearranged alleles, and an oligonucleotide probe for Southern blot confirmation.

TransPrimer	Rearrangement	Genes involved	Phenotype
MBR primers[a] lymphoma	t(14;18)(q32;q21)	*bcl*-2(mbr)/IgH	Follicular
MCR primers[a] lymphoma	t(14;18)(q32;q21)	*bcl*-2(mcr)/IgH	Follicular

[a] U.S. Patent No. 4,681,840, U.S. Serial Nos. 749,178 and 863,250, and related foreign patent applications.

GeneAmp is a trademark of the Cetus Corporation.

DNA and RNA Purification Systems

These include: **QuickClean DNA** extraction system, QuickClean probe purification system, QuickClean RNA isolation system, QuickClean poly(A)$^+$ RNA isolation system, QuickClean PCR purification system, RadStop radiation shield

Low Molecular Weight DNA Standards

Lambda *Hind*III markers, X174/*Hae* markers, Lambda *Hind*III-X174/*Hae*III markers, pUC19/*Sau*3AI-pUC, 19/*Taq*I markers, X174/*Hind*I markers. **Restriction endonucleases, DNA/RNA modifying enzymes, primers, linkers, adapters**

Probe Labeling and Hybridization Systems

Random primer labeling kit, SP6 RNA polymerase, T7 RNA polymerase, T4 polynucleotide kinase, terminal transferase, DNase I (RNase free), RNase cocktail, RNase inhibitor, hybridization system

cDNA and Genomic DNA Probes in SP6/T7 Vectors

abl, *bcr*, *fes*, *fms*, G-CSF, GM-CSF, *myc* (c,L,N), *neu*/*erb*B-2, *p53*, *raf*, *ras* (H,K,N), *RB*, *sis*

Mobility Shift Analysis

AP-1, NFB, Oct-1, NF/CTF, CRE, SP-1

40-mer DNA Probes

Over 400 probes for human, mouse, and rat genomic and mRNA analysis, Southern or northern blots, or *in situ* analysis

ras Mutation Detection by PCR Amplification and Differential Hybridization using MutaProbe

Ha-*ras* set for codon 12, Ha-*ras* set for codon 61, Ki-*ras* set for codon 12, Ki-*ras* set for codon 61, N-*ras* set for codon 12, N-*ras* set for codon 13, N-*ras* set for codon 61

ELISA ASSAY PRODUCTS

ELISA assays for fast, nonisotopic, quantitative analysis of *neu*, *p53*, EGF, G-CSF, and GM-CSF are available.

Immunohistochemistry Systems

Systems for high quality staining of formalin-fixed and paraffin-embedded tissue sections: anti-*neu* System, anti-*ras* systems, anti-*p53* system, anti-*fos/jun* AP-1 systems, anti-phosphotyrosine system, anti-TGF, -EGF, and -EGF receptor systems, anti-PTHLP, -vimentin, and -desmosomal glycoprotein systems, LN4/macrophage, G-CSF, activated B lymphocytes and eosinophil peroxidase systems, enhanced second step systems for mouse, rat, and rabbit Ig

Antibodies and Reagents

anti-*p53*, RB, HSP 70, and DNA tumor viruses, anti-$p21^{ras}$, transcription factors and tyrosine kinases, anti-phosphotyrosine anti-cell cycle proteins and serine/threonine kinases, anti-growth factors and receptors, anti-bacterial proteins and mdr, anti-hemopoietic factors and interleukins-human and mouse blood cell panning, anti-CD antigens and blood cell antigens, anti-nuclear antigens, calpactins, integrins, cytoskeletal proteins and intermediate filaments, cardiovascular and neuroscience

Immunoreporter Reagents

Proteins and peptide antigens, molecular weight markers, western blotting standards, silver staining and western blotting systems, streptavidin, second-antibody conjugates, chemiluminescent substrates and normal sera second-antibody fluorescent conjugates, direct antibody $F(ab)_2$ fragments, and immunoprecipitation reagents

---

ONCOR
209 Perry Parkway
Gaithersburg, MD 20877
(301) 963-3500

Oncor develops, manufactures, and markets DNA probe test systems and related products for use in the diagnosis, monitoring, and management of cancer and other genetic diseases. The company has developed a broad, proprietary portfolio of DNA probes to specific human genes and was the first to market **DNA probes specific to each human chromosome**.

Oncor believes that the use of DNA probes will become one of the principal means of diagnosing, monitoring, and managing the treatment of cancer and inherited genetic disorders. Furthermore, DNA probe test systems may be useful throughout the cycle of cancer management, from initial screening for predisposition to cancer, through detection and typing of cancer and selection of therapy, to monitoring of treatment and detection of residual disease or relapse.

ONCOR DNA PROBES

DNA probes offer a direct means of detecting the presence or absence of these specific genes or gene abnormalities. Traditional immunochemistry, on the other hand, only measures the presence or absence of specific proteins produced by the cell after the genetic mutation has occurred.

The Oncor DNA probes are capable of identifying a wide variety of genetic abnormalities in a cell. These genetic abnormalities include (1) **aneuploidy**, or an abnormal number of a particular chromosome; (2) **translocations**, in which genetic material migrates from one chromosome to another; (3) **rearrangements**, in which genetic material moves from one position to another on the same chromosome; (4) **amplifications**, in which increased number of copies of a specific gene are being produced; (5) **deletions**, in which a specific genetic segment is not present; (6) the presence or absence of a particular chromosome; (7) the presence of **oncogenes**, mutated forms of normal genes that can initiate uncontrolled cellular growth and malignancy; and (8) the absence of **tumor suppressor genes**, genes that are thought to block abnormal cell growth and malignant transformation. DNA probes are also capable of detecting certain viruses and bacteria by their characteristic nucleic acid sequences.

ONCOR DNA PROBE TEST SYSTEMS

Oncor DNA probe test systems incorporate both "Southern" (membrane-based) hybridization techniques and *in situ* hybridization technology. Oncor has obtained three U.S. patents relating to the integration of electrophoresis and transfer that is the basis for Southern analysis of DNA. This technology is incorporated into the Oncor Probe Tech 2™ system for **automated** Southern analysis, which integrates all necessary reagents into a single system.

The new DNA Oncor probe test systems generally incorporate technologies and reagent chemistries developed by the company that enable *in situ* (microscope slide-based) DNA analysis. This process (1) permits a more definitive diagnosis than current diagnostic methods, (2) allows the analysis of single cells from virtually any tissue source, (3) does not require any instrumentation other than a microscope, (4) yields test results within 24 h rather than the week or more required by current standard methods of chromosome analysis, and (5) allows the Oncor DNA probe test systems to be integrated easily into the current routine of microscopic, morphology-based tissue analysis.

In an *in situ* analysis, a cell sample is immobilized on a glass slide and the chromosomes are denatured, after which the DNA probe is added to the slide and allowed to hybridize to the target DNA.

After hybridization, the labeled probes can be located through examination with a microscope. Classic cytogenetic analysis performed under a microscope by a pathologist is subjective, is limited to genetic abnormalities that can be seen with a microscope, and can be performed only on actively dividing, "metaphase" cells.

However, through about 99% of their life cycle, cells are in a resting, "interphase" state. For this reason, difficult and time-consuming culturing techniques must be used to obtain additional metaphase cells. By contrast, ISH techniques allow the analysis of interphase cells, thus bypassing the need for cell culturing and providing the capability of direct tissue analysis. When used in samples in which not all the cells are abnormal, e.g., when cancer cells are mixed with normal cells, *in situ* analysis can indicate the number or concentration of the cells that harbor the genetic abnormality in question.

ONCOR CANCER PRODUCTS

The recognition of the role of genetic abnormalities in causing cancer, and the continual discovery of specific genetic abnormalities associated with particular cancers along with the unique sensitivity and specificity of DNA probes to detect those genetic abnormalities, give rise to the attractiveness of DNA probe technology as a potential cancer diagnostic tool.

LEUKEMIA AND LYMPHOMA: ONCOR B AND T CELL GENE REARRANGEMENT TEST SYSTEM

Oncor manufactures and markets a DNA probe test system, the **B/T gene rearrangement nonisotopic test system** for ***in vitro* diagnostic use**, for the use in the diagnosis and monitoring of leukemia and lymphoma. This test system combines biotin-labeled DNA probes and the Southern hybridization technique to detect gene rearrangements specific to B cells and T cells, which allows for the early identification and characterization of leukemia and lymphoma and subsequent monitoring for recurrence.

The test allows the identification of malignant cells at any stage of their maturation. Moreover, most leukemias or lymphomas that could not previously be typed as B cell or T cell, an important distinction in determining the optimal therapy, can now be accurately differentiated. This test also permits the detection of changes in the character of a tumor during the course of treatment or the development of a different type of tumor.

LEUKEMIA AND LYMPHOMA: CHROMOSOMAL TRANSLOCATIONS (FOR RESEARCH USE ONLY)

Chromosomal translocations, representing breakage and recombination of specific chromosomes, serve as useful genetic markers for the identification of subtypes of leukemia and lymphoma. In addition, individual cells that contain these translocations serve as markers of recurrence of the disease following therapy. The present method for detecting these chromosomal translocations, conventional cytogenetics, is expensive, time consuming, and often impossible.

The **products under development by Oncor for 1992** include DNA probe test systems specific to these chromosomal translocations, which will enable the analysis to be performed directly on blood cells with results obtained within 24 h. The Oncor translocation test system will facilitate the determination of the proper treatment protocol, which can vary dramatically depending on the subtype of leukemia or lymphoma.

For example, the translocation of a particular gene from chromosome 9 to chromosome 22 indicates a particularly lethal subtype of leukemia known as chronic myelogenous leukemia (CML), for which one possible treatment is bone marrow transplants.

CERVICAL CANCER: HUMAN PAPILLOMAVIRUSES (FOR RESEARCH USE ONLY)

Oncor has developed a nonisotopic DNA probe test system that provides simultaneous detection and typing of the clinically significant genital HPVs, which show promise as markers for assessing the risk of cervical cancer.

ONCOR GENETIC PRODUCTS

Chromosome Analysis: Chromosome *in Situ* Systems

Nonisotopic **chromosome *in situ* hybridization** has many applications in molecular cytogenetics that enable both basic and clinical investigators to examine the cytogenetic status of pre- and postnatal samples, tumor cell populations, and other benign or neoplastic cell and tissue types. Some of these include **chromosome identification**, **aneuploidy detection**, **interphase cytogenetics**, and **heteromorphism analysis** for the determination of parental origin of individual chromosomes. Also of interest is the identification of marker chromosomes that are extremely difficult to classify using standard karyotypic analysis.

Precise localization of specific unique DNA sequences to individual chromosomes is now easily performed, providing a tool for human gene mapping studies and molecular cytogenetic analyses of chromosomal deletion syndromes.

Oncor has developed **chromosome *in situ* systems** that utilize reagents and protocols for use with metaphase chromosomes, interphase nuclei, whole cells, and formalin-fixed, paraffin-embedded tissue sections. The procedures can be used with either fluorescence or light microscopy.

These cytogenetic analysis systems are complemented by a full selection of the following probes — **α, β**, and **classical satellite probes, total genomic** and **chromosome-specific painting probes, telomere probes, unique sequence cosmid DNA probes** for microdeletion syndrome research (**Miller-Dieker, cri-du-chat**, and **Wolf-Hirschhorn**), and the **fragile X probe** utilizing the Southern hybridization technique to detect a particular genetic amplification associated with the fragile X gene, the gene that causes fragile X syndrome.

Other Oncor DNA Probe Products

In addition to the DNA probe test systems for specific cancers and genetic diseases, Oncor manufactures and sells **over 100 other DNA probes to specific genes** for research purposes. Oncor's DNA probes to specific genes are currently used for (1) **paternity testing**, in which distinctive banding patterns in the chromosomes of a child are compared with such patterns in the father to confirm paternity, and (2) **tissue typing**, which involves the detection of human leukocyte antigens, for the purpose of matching organ donors with potential recipients.

Other Oncor Products for Use with DNA Probes

Oncor manufactures and markets **SUREBLOT** reagent kits for Southern analysis, which contain highly qualified nylon membrane and all the reagents necessary for DNA extraction, electrophoresis, and transfer to perform Southern analysis with the Oncor DNA probes. Other products include DNA probe **labeling kits, Hybrisol I** and **II**, and other reagents and solutions for use in hybridization.

PROMEGA
2800 Woods Hollow Road
Madison, WI 53711-5399
(608) 274-4330

Promega Corporation is a biotechnology company that applies biochemistry and molecular biology for the development of novel, high-value products. Founded in 1978 with a staff complement of 1 person, Promega corporation now employs over 230 people. Since 1980, the biological research products developed and manufactured at Promega include both the "tools" needed for basic research and sophisticated, easy-to-use products that utilize biotechnology for many diverse applications. The broad international scope of the company includes affiliates in the Netherlands, United Kingdom, Australia, and Switzerland, complemented by an outstanding distributor network in 26 foreign countries and a joint venture in the People's Republic of China. Most recently, additional collaborations include a strategic alliance with Boehringer-Ingelheim, a multibillion dollar German pharmaceutical concern, which has made a substantial investment in Promega and opened additional marketing channels in Europe.

Today, Promega markets hundreds of products for microbiology and fermentation technology, cell culture, protein purification, gene cloning and engineering, nucleic acid purification and sequencing, oligonucleotide synthesis, monoclonal and polyclonal antibody production and purification, and rapid, nonisotopic systems. The expertise and capabilities of Promega in these areas can be applied to

1. isolation of high-value compounds using immunopurification and other proprietary separation technologies
2. production of "custom" proteins and enzymes for pharmaceutical development and food processing
3. DNA typing for forensic and parentage testing, bone marrow transplantation monitoring, cell line authentication, and military identification
4. developing novel, rapid tests to detect food pathogens and environmental contaminants
5. bioremediation: isolation and genetic modification of microorganisms for large-scale cleanup of toxic wastes
6. identifying natural sources of flavors, fragrances, and selected food additives

PRODUCT SUPPORT

Promega publishes and provides free copies of a protocols and applications guide for biological research products. This guide is one of the most comprehensive reference books for the laboratory. Additionally, Promega provides toll-free telephone service to customers requiring technical assistance when using, but not limited to, Promega products during conduct of research experiments.

PRODUCTS

Human Identification Products

The human identity division has primarily focused on the development of products for **individual identification**. Promega has acquired the exclusive rights to probes/loci for many applications, including parentage testing, forensic analysis, bone marrow transplantation monitoring, familial disease studies, and human genome analysis.

Most recently, we have focused a large portion of our research and development efforts toward developing tests for human identity and disease diagnosis based on the polymerase chain reaction technology.

Food Diagnostics

Rapid, simple test systems for detecting microbial pathogens in food, such as the "total viable organisms (TVO)" assay for raw milk, which allows the measurement of fluid milk quality in 30 min or less. Additional products that will be introduced in 1992 include an environmental assay for detecting bacteria, fungi, and organic "dirt" on food processing equipment and surfaces, and rapid assays for the detection of *Listeria*, *Escherichia coli* 0157:H7, and *Salmonella*.

Research Products

The research and development efforts at Promega are organized into sections defined by technical expertise or business unit focus. The following sections can be defined.

Bioluminescence

The main focus of the efforts of this group is the development of products from **beetle luciferase technology**. The project has two main objectives: (1) to advance the ability to use different colors of luminescence as multiplexed molecular reporters and (2) to enhance the stability and other enzymological properties to promote their use in existent and new applications.

Products include the **luciferase assay system**, **pGEM-*luc* DNA** as a source of the gene coding firefly luciferase, and **GeneLight™ reporter vectors** designed to support rapid and convenient analysis of promoter and enhancer sequences.

Detection and Purification

The detection and purification group has been involved in two areas of product development: (1) the development of **nonisotopic detection methods** for nucleic acids and (2) the development of **isolation methods for RNA using magnetic particle technology**.

Products include **Magic™ DNA purification products** for fast and easy purification of DNA, a **PolyATRact™ mRNA system** utilizing MagnaSphere magnetic particle technology, **Lightsmith™ luminescence engineering** for oligonucleotides, a **reverse transcription system** to reverse transcribe RNA in 15 min, and **RNasin ribonuclease** and **angiogenesis inhibitor**.

Complex Genome Analysis

This group is less than a year old and is focused on developing and supporting systems used in DNA sequencing and mapping, especially targeting the human genome project and related gene mapping efforts.

Products include the **fmol™ DNA sequencing system** for sequencing PCR products of 2 ng or less, **ProMega markers™** for pulsed-field gel electrophoresis, **LambdaGEM superior cloning vectors, genome qualified restriction enzymes**, and **TaqTrack™ DNA sequencing systems**.

Protein Analysis

This group was formed in 1989 and began developing systems for peptide preparation for **Edman protein sequence analysis**, including a probe design peptide separation system, sequencing grade proteases, and a protein fingerprinting system. The focus of the group has expanded to include *in vitro* protein synthesis and Western blotting-electrophoresis technology.

Products include the **Chromaphor™ protein staining system, TnT™ lysate systems** for *in vitro* coupled transcription/translation, and **PepTag™ nonradioactive protein kinase systems**.

Molecular Biology: Eukaryotic Gene Expression

The efforts of this group concentrate on developing products that apply to the study of *in vivo* and *in vitro* systems. A key focus includes the development of reagents for the study of transcriptional regulation.

Products include **HeLaScribe™ nuclear extract** for *in vitro* transcription, purified transcription factors, and **TRANSFECTAM** for highly efficient transfection of eukaryotic cells.

Molecular Biology: Cellular Regulation

This product area is focusing the majority of its efforts on developing products for investigators studying signal transduction pathways and cellular adhesion mechanisms.

Products include human recombinant lymphokines, natural and recombinant growth factors, antisera to growth factors, high-activity kinases, and **CellTiter 96™**, which provides a rapid and convenient nonradioactive alternative to [^{3}H]-thymidine incorporation for determining viable cell number in proliferation or cytotoxicity assays.

FUTURE PROSPECTS

Recently, Promega has begun to develop **products for human health care**.

ROCHE MOLECULAR SYSTEMS
Hoffmann-La Roche and Perkin-Elmer
1080 U.S. Highway 202
Branchburg, NJ 08876-1760

POLYMERASE CHAIN REACTION PRODUCTS

Roche Molecular Systems (RMS), is a unit of Hoffmann-La Roche's worldwide diagnostics group, dedicated to the development and commercialization of the GeneAmp polymerase chain reaction (PCR) process. RMS will sell a broad range of PCR-based products for human diagnostics worldwide through Roche Diagnostic Systems, the sister company of RMS. The rights to the PCR process are owned by Hoffmann-La Roche (Nutley, NJ) and F. Hoffmann-La Roche (Basel, Switzerland), the parent companies of RMS.

The Perkin-Elmer Corporation (Norwalk, CT) has previously been involved in the PCR process with Cetus Corporation and has entered into a strategic alliance with Roche under which Perkin-Elmer has rights to the PCR process in nondiagnostic fields such as research, human identity, and environmental testing, and is the exclusive distributor of RMS-manufactured PCR reagent products in those fields. Perkin-Elmer also develops, manufactures, and sells instrument systems for the PCR process. The established line of Perkin-Elmer PCR reagent products and instrument systems is distributed worldwide and is rapidly expanding the technologies and products offered to the life sciences research and industrial communities.

HUMAN DIAGNOSTICS

Roche Molecular Systems is developing complete PCR-based kits for the detection of infectious disease agents and for genetic analysis to detect disease and disease susceptibility. These kits will be introduced following clinical trials and FDA approval. Tests currently being developed include those detecting **HIV-1**, **mycobacteria**, **HTLV-1** and **-2**, **Lyme disease**, **chlamydia** and **gonorrhea**, and **cystic fibrosis**.

Kits will include sample preparation, nucleic acid amplification, and detection modules. Products for establishing tissue matches for transplantation purposes are also in development.

VETERINARY DIAGNOSTICS

In this field, Roche has licensed several entities to perform services, and has granted licenses to manufacture products. PCR-based products are sold by the following companies:

IDEXX Corporation (Portland, Maine): A kit for the diagnosis of Johne's disease (*Mycobacterium paratuberculosis*)

AB Technologies (Australia): A single-tube reaction mix for the determination of sex in bovine embryos.

Both of these licensees plan to add additional diagnostic tests.

RESEARCH REAGENTS

Perkin-Elmer sells enzymes, reagent kits, nucleic acids (standards, primers, and probes), and buffers for the GeneAmp PCR process. These products are developed and manufactured by Roche Molecular Systems. Representative products in each category are described below.

PCR Enzymes

AmpliTaq DNA polymerase, a heat-stable recombinant DNA polymerase from the bacterium *Thermus aquaticus*, is the enzyme of choice for PCR amplification. For those few applications of the GeneAmp PCR process that may benefit from the use of a DNA polymerase with somewhat different characteristics, there are several currently available and others that are being developed.

Recombinant *Tth* DNA polymerase, a heat-stable DNA polymerase from the bacterium *Thermus thermophilus*, will accomplish reverse transcription and PCR amplification at high temperature.

PCR Kits

GeneAmp PCR reagent kits provide all of the components necessary for the PCR process. A core kit contains AmpliTaq DNA polymerase, buffer, and deoxynucleoside triphosphates. A complete kit includes those reagents plus control template bacteriophage DNA, and primers to amplify a segment of the control.

RNA-PCR kits provide the components for reverse transcription of RNA to cDNA and subsequent amplification of the cDNA. One kit uses a separate enzyme, **Moloney MLV RT**, for reverse transcription, and AmpliTaq DNA polymerase for PCR amplification. A second kit uses recombinant *Tth* DNA polymerase to perform both steps at elevated temperature, allowing the efficient amplification of sequences that are difficult to reverse transcribe at normal temperatures due to secondary structure.

Kits for DNA sequencing employ AmpliTaq DNA polymerase at elevated temperatures, allowing efficient sequencing of DNA that has a high melting temperature. One kit provides components for **Sanger dideoxy sequencing**, a second kit provides the reagents for "**cycle sequencing**," an enhancement of the Sanger method that allows direct sequencing of double-stranded templates, specifically PCR products.

Many of the individual components of these Perkin-Elmer PCR reagent kits are also sold separately.

PCR Primers and Probes

Synthetic oligonucleotides for use in the GeneAmp PCR process are directed at a variety of genetic loci of special interest to researchers. Products include primers for quantitation of mRNAs of common human cytokines and lymphokines, primers and probes for much-studied viruses such as HIV-1, HPV, HTLV-I, HTLV-II, and primers and probes for the analysis of certain polymorphic regions of the human histocompatibility loci.

HUMAN IDENTITY

DNA analysis is rapidly gaining worldwide recognition as a tool in the investigation of human identity. DNA analysis products for PCR-based testing of biological evidence samples can also be used to establish parentage, and develop large-scale databases for both armed forces personnel and convicted sex offender programs. Available from Perkin-Elmer are two different types of PCR-based typing products for these purposes. One is based on sequence-specific oligonucleotide probe hybridization to amplified polymorphic PCR product (AmpliType), and the second is based on the amplification of polymorphic genetic loci with **variable number of tandem repeats** also considered length polymorphisms (**AMP-FLPs**).

The **AmpliType HLA-DQa forensic DNA amplification and typing kit** is the first complete kit developed for human identification using GeneAmp PCR instrument systems. The kit analyzes the human leukocyte antigen class II (HLA-D) sequence variable genes located on chromosome 6 and is validated for use on forensic casework.

The **D1S80 forensic DNA amplification reagent set** is the first complete reagent set available for use in AMP-FLP DNA typing of the D1S80 locus.

The collective discrimination power of DNA typing tests is dependent both on the number of loci analyzed and the extent of polymorphic variation identified at each locus. Highly polymorphic variable number of tandem repeat (VNTR) loci are commonly analyzed by restriction fragment length polymorphism (RFLP)/Southern blotting methodology. Analysis of VNTR loci after PCR amplification, referred to as **amplified fragment length polymorphism (AMP-FLP) analysis**, provides an additional method for exploiting length dependent polymorphisms (see also Part 1). GeneAmp PCR technology offers advantages over RFLP analysis because it is faster, has smaller sample size requirements, is easier to interpret, and offers nonisotopic detection of alleles. To date, the AMP-FLP loci have not yet been fully validated for forensic casework.

Additional products are being developed that will increase the power of discrimination using PCR-based DNA typing methodologies.

ENVIRONMENTAL TESTING

Testing of water and soil samples for microorganisms is a technical challenge that is facilitated by use of the GeneAmp PCR process, as often a low level of organism must be detected in a large quantity of sample.

EnviroAmp *Legionella* kits (a set of three) are now available to detect both the genus *Legionella* and the species *L. pneumophila* in environmental water samples. These kits employ the reverse dot-blot method of detection, which entails specific DNA probes bound to membranes and hybridized with PCR products. Additional targets for environmental testing will be developed by RMS and sold by Perkin-Elmer.

INSTRUMENT SYSTEMS

The GeneAmp PCR instrument systems are the most sophisticated, reliable, and widely used instruments for PCR amplification. Perkin-Elmer offers a range of instruments to meet the variety of requirements faced by the broad spectrum of bioresearchers around the world who use the

GeneAmp PCR process daily. From basic research needs (e.g., molecular genetic analysis) to routine analytical applications (e.g., environmental water quality testing), all GeneAmp PCR instrument systems are designed to provide advanced superior performance with convenient sample handling, accurate state-of-the-art temperature cycling, and integrated software systems.

Among the products offered are the **GeneAmp PCR system 9600**, the **DNA thermal cycler 480**, and the **DNA thermal cycler**. Each is described below.

GeneAmp PCR System 9600

The GeneAmp PCR system 9600 combines integrated instrument design, thin-walled MicroAmp reaction tubes, and a 96-well format to offer greater speed, oil-free operation, lower reaction volumes, unsurpassed cycle time reproducibility, and high-performance PCR protocols. Based on the demands of second-generation PCR technology, the innovative engineering and advanced system design of the GeneAmp PCR system 9600 introduces new standards of accuracy, precision, reproducibility, efficiency, and convenience for PCR amplification.

DNA Thermal Cycler 480

The DNA thermal cycler 480, based on our proven DNA thermal cycler standards for quality and excellence, combines guaranteed PCR performance with reduced reagent requirements and improved temperature uniformity for quantitative PCR.

DNA Thermal Cycler

First introduced in 1987, the original DNA thermal cycler is the industry standard for performing the GeneAmp PCR process. Well-characterized and documented PCR protocols have proved that automation of PCR amplification is flexible, reliable, and easy with the original DNA thermal cycler.

UNITED STATES BIOCHEMICAL CORPORATION
P.O. Box 22400
Cleveland, OH 44122
(800) 321-9322

United States Biochemical Corporation (USB), founded in 1973, is a manufacturer and supplier of a full range of Life Science Biochemicals and Molecular Biology Products. The global client base of USB includes customers in the pharmaceutical, clinical diagnostic, agricultural, food sciences, and genetic engineering industries as well as government institutions and more than 100 of the leading institutions of higher learning in the U.S. USB reaches over 200,000 researchers throughout the world.

USB is instrumental in the supply of components to develop new diagnostic assays and technologies as well as in fulfilling manufacturing requirements for existing diagnostic kits, and easy-to-use research kits such as for DNA and RNA sequencing, cDNA synthesis, DNA cloning and mutagenesis, nucleic acid purification, DNA and RNA sequence and protein detection, and RNA cleavage.

DNA SEQUENCING

DNA sequencing is a critical step in positive identification of defined segments of DNA produced from DNA manipulations or diagnostic PCR.* In 1987, USB introduced **Sequenase** T7 DNA polymerase** for **dideoxy-DNA sequencing** utilizing a novel two-step method that quickly revolutionized the previous method utilizing **Klenow polymerase**. The original Sequenase version 1.0 enzyme, along with the more recent genetically engineered Sequenase version 2.0, are offered in complete sequencing kits with control tested reagents, assuring optimal results. The Sequenase version 2.0 DNA sequencing kit contains the same tested nucleotide mixtures as those found in the version 1.0 kit, however, the version 2.0 kit also offers (1) Mn buffer*** for more reliable sequencing close to the primer, (2) Sequence extending mixes, which can be used to extend sequence-specific terminations to more than 3000 b from the primer, and (3) inorganic pyrophosphatase,† which will prevent occasional weak band intensity problems, especially seen with dITP.

These Sequenase kits are backed with a large variety of complementary sequencing kits offered by USB, a few of which are the new presequencing kit for linear, dsDNA, the reagent kit for DNA sequencing with Sequenase T7 DNA polymerases, **the dGTP, 7-deaza-dGTP**, and **dITP nucleotide kits**, and **the sequencing primer kit**.

In addition, USB offers DNA sequencing kits based on ***Taq* DNA polymerase**, **TAQuence,‡** and TAQuence version 2.0 kits, and a new

---

* PCR is covered by patents issued to Cetus Corporations. This article does not provide any required license to use PCR. If interested in performing PCR, contact Cetus for information on obtaining an appropriate license.

** Sequenase is a registered trademark of the United States Biochemical Corporation. This kit (reagent) is covered by or suitable for use under one or more U.S. Patents: 4,795,699, 4,946,786, 4,942,130, 4,962,020, and 4,994,372. Patents pending in the U.S. and other countries.

*** Mn buffer: This reagent is suitable for use under one or more U.S. Patents: 4,795,699, 4,962,020, and 4,994,372. Patents pending in the U.S. and other countries.

† Pyrophosphatase: Patents pending in the U.S. and other countries.

‡ TAQuence, T7-Gen, USBioclean, and RNAzyme are registered trademarks of United States Biochemical Corporation. *FTaq*, ClonStruct, Base F, Codon F, Multi-Base F, REX, SurePure, and SureCheck are trademarks of United States Biochemical Corporation.

modified form of *Taq*, ***FTaq***[‡‡] version 2.0 DNA polymerase. **The F*Taq* cycle sequencing kit** is the newest kit from USB, featuring F*Taq* polymerase, which offers the ability to sequence small quantities of template DNA. However, the results obtained using Sequenase T7 DNA polymerase are noted for low background and uniform band intensity.

An even newer segment within the USB infrastructure is dedicated to **custom sequencing** for company clients. This service affords research laboratories the opportunity to utilize the extensive facilities and resources of USB to alleviate a complex, delicate, critical, and costly step in genetic research.

RNA SEQUENCING

For the analysis of RNA sequences, USB now offers two kits. The newest kit, **RT RNA sequencing kit**, is designed to allow the sequencing of RNA by the dideoxy chain-termination method. Up to 300 b of sequences can be analyzed from a single sequencing reaction. Unlike our **nuclease RNA sequencing kit**, some sequence information is required to design sequencing primers for use with **the RT (reverse transcriptase) kit**. However, the new RT kit has the advantage that the RNA to be sequenced need not be pure because the sequencing primer will be specific for the desired template.

cDNA CONSTRUCTION

The new **cDNA ClonStruct**[‡‡] **Kit** is designed to promote efficient cloning of full-length cDNAs resulting in a large, highly representative cDNA library. This new kit can be used for transcription or expression. The USB cDNA synthesis kit is ideal for producing quality first-strand cDNA synthesis, essential for generating PCR products from mRNA. This kit employs **M-MLV reverse transcriptase**, ensuring the highest efficiency for making full-length transcripts. Second-strand synthesis may be carried out in such a manner as to retain sequence information at the 5′ end of the mRNA.

CLONING AND MUTAGENESIS

For cloning into M13 vectors, USB offers the **M13 cloning kit**, which contains all the necessary reagents, DNAs, host strains, and ligase to construct clones in M13mp18 and M13mp19. These vectors have multiple cloning sites oriented in opposite directions so that the orientation of the cloned fragment can be determined by proper choice of vector. The creation of precise nucleotide changes in cloned genes can easily be accomplished with the **Base F**,[‡] **Codon F**,[‡] and **Multi-Base F**[‡] excision linker kits. The **T7-Gen**[‡] ***in vitro*** **mutagenesis kit** employs a more powerful technique, oligonucleotide-directed *in vitro* mutagenesis, for generating site-specific mutations in any cloned DNA segment of interest. This kit utilizes native T7 DNA polymerase, a superior enzyme for oligonucleotide-directed site-specific mutagenesis.

‡‡ F*Taq*: Patents pending in the U.S. and other countries. Cycle Sequencing: Patent pending.

NUCLEIC ACID PURIFICATION

When working with DNA, steps often require extraction from agarose gels, removal of unincorporated nucleotides, removal of organic solvents such as phenol, isolation of plasmid DNA from minipreparations, and/or concentration without ethanol precipitation. The **USBioclean‡ kits** from USB provide a rapid and convenient method for all the above. While the original USBioclean kit contains a **TAE buffer** concentrate, the USBioclean MP kit can be used for both **TAE** and **TBE gels** due to the inclusion of the buffer convertor. Both USBioclean kits result in extracted and/or purified DNA, suitable for restriction digestion and many other enzymatic manipulations, including DNA sequencing with Sequenase T7 DNA polymerase.

For the extraction of RNA, USB offers **the REX‡ total RNA extraction kit**. The RNA obtained with REX is suitable for use in Northern blots, *in vitro* translation, or as a source of mRNA that can be isolated using the USB **poly(A) RNA isolation kit**. The mRNA obtained with this isolation kit is suitable for use as a template for cDNA synthesis and the preparation of cDNA libraries, such as with USB cDNA ClonStruct‡ and cDNA synthesis kits.

Oligonucleotides can be purified rapidly from failed sequences and contaminating salts using the USB new **SurePure‡ oligonucleotide purification kit**. The oligonucleotides purified using this kit are ready for use in DNA sequencing or PCR reactions. To determine the purity of oligonucleotide samples and the necessity of using the SurePure kit, USB offers **the SureCheck‡ oligonucleotide analysis kit**.

NUCLEIC ACID AND PROTEIN DETECTION

USB offers kits for detection of specific DNA/RNA sequences and proteins, using either radioactivity or chemiluminescence. Purified DNA can be readily and conveniently labeled with ^{32}P at the 3′ end of **Maxam and Gilbert sequencing** or for other applications using the 3′ end labeling kit from USB. The **USB nick translation kit** offers a simple yet inexpensive method for generating uniformly radioactively labeled DNA of high specific activity for use as a hybridization probe in techniques such as Southern blots, dot blots, and plaque screening. Probes radioactively labeled using the new Sequenase** random primed DNA labeling kit or the original random primed DNA labeling kit from USB are suitable for all types of hybridization techniques, including Southern and northern blots, and for *in situ* hybridization.

For human genome mapping, USB offers low-frequency probes. The **USB human genome kits** allow the investigator to find continuous stretches of DNA in any human library — **rodent/human hybrids**, chromosomes specific, or whole genomic.

RNA CLEAVAGE

USB offers the first commercially available sequence-specific ribonucleases — **RNAzyme‡ Tet 1.0, RNAzyme RCH 1.0,** and **RNAzyme RCH 1.1**.

Three kinds of analysis are facilitated by the sequence-specific cleavage activity of these ribozymes: direct physical mapping, nucleotide sequencing using base-specific ribonucleases or chemical agents, and empirical determination of the higher-order structure of RNA. These ribozymes are offered in kits, allowing convenience and reliable results.

Other **representative companies** offering reagents and instruments for molecular biological evaluation are listed below. Inclusion in this listing does not represent any form of endorsement or advertisement.

AMBION, INC.: 2130 Woodward St., #200, Austin TX 78744-1832; (800) 888-8804
AMERSHAM CORPORATION: 2636 South Clearbrook Dr., Arlington Heights, IL 60005; (800) 323-9750
AMICON (W. R. Grace & Co.): 72 Cherry Hill Dr., Beverly, MA 01915; (508) 777-3622
AT BIOCHEM, INC.: 30 Spring Mill Dr., Malvern, PA 19355; (215) 889-9300
BACHEM BIOSCIENCE, INC.: 3700 Market St., Philadelphia, PA 19104; (800) 634-3183
BECKMAN INSTRUMENTS, INC.: 2500 Harbor Blvd., Fullerton, CA 92634; (800) 742-2345
BETAGEN CORPORATION: 100 Neaver St., Waltham, MA 02154; (617) 899-3400
BIO 101, INC.: P.O. Box 2284, La Jolla, CA 92038-2284; (800) 424-6101
BIOPROBE SYSTEMS: 26 Bis Rue Kleber, 93100 Montereuil-Sous-Bois, France; TEL 33/1 48 51 66 22
BIOVENTURES, INC.: 848 Scott St., Murfreesboro, TN 37129; (800) 235-8938
BOEHRINGER MANNHEIM BIOCHEMICALS: 7941 Castelway Dr., Indianapolis, IN 46250; (800) 428-5433
CALBIOCHEM: P.O. Box 12087, San Diego, CA 92112-4180; (800) 854-3417
CAMBRIDGE RESEARCH BIOCHEMICALS, INC.: Wilmington, DE 19897; (800) 327-0125
COY CORPORATION: 14500 Coy Dr., Grass Lake, MI 49240; (313) 475-2200
CRUACHEM, INC.: 45150 Business Court, Suite 500, Sterling, VA 22170; (703) 689-3390
DYNAL, INC.: 475 Northern Blvd., Great Neck, NY 11021; (516) 829-0039
FMC BIOPRODUCTS: 191 Thomaston St., Rockland, ME 04841; (800) 341-1574
GENOSYS BIOTECHNOLOGIES, INC.: 8701A New Trials Dr., The Woodlands, TX 77381-9979; (800) 2345-DNA

GIBCO BRL/LIFE TECHNOLOGIES, INC.: P.O. Box 8418, Gaithersburg, MD 20898; (800) 828-6686

E-C APPARATUS CORPORATION: 3831 Tyrone Blvd, North, St. Petersburg, FL 33709; (813) 344-1644

EG & G BERTHOLD ANALYTICAL INSTRUMENTS, INC.: 472 Amherst St., Nashua, NH 03063; (603) 889-3309

ELCHROM AG: Spatzstrasse 18, CH-8810 Horgen, Switzerland; Fax (411) 725 61 21

INTEGRATED SEPARATION SYSTEMS: One Westinghouse Plaza, Hyde Park, MA 02136; (800) 433-6433

INTERNATIONAL BIOTECHNOLOGIES, INC.: Subsidiary of Eastman Kodak Company, P.O. Box 9558, New Haven, CT 06535; (800) 243-2555

ISCO, INC.: P.O. Box 5347, Lincoln, NE 68505; (800) 228-4250

JANSSEN BIOCHIMICA: RDA, Pleasant Hill Rd., Flanders, NJ 07836; (201) 584-7093

KRONEM SYSTEMS, INC.: 6850 Goreway Dr., Mississauga, Ontario, Canada L4V 1P1; (416) 612-0411

MACHENERY-NAGEL GMBH & CO.: P.O. Box 10 1352, D-5160 Duren, Germany; TEL (02421) 698-0

MEDAC GMBH: Fehlandstrasse 3, D-2000 Hamburg 36; TEL 49/40/350 90 2-0

MICRON SEPARATIONS, INC.: 135 Flanders Rd., P.O. Box 1046, Westborough, MA 01581-6046; (800) 444-8212

MICROPROBE CORPORATION: 7390 Lincoln Way, Garden Grove, CA 92641; (800) 800-4714

MILLIPORE CORPORATION: 397 Williams St., P.O. Box 9162, Marlborough, MA 01752; (800) 225-1380

MOLECULAR BIOPROBES, INC.: 4849 Pitchford Ave., P.O. Box 22010, Eugene, OR 97402-0414; (503) 344-3007

MOLECULAR DYNAMICS: 880 East Arques Ave., Sunnyvale, CA 94086; (408) 773-1222

NEW ENGLAND BIOLABS, INC.: 32 Tozer Rd., Beverly, MA 01915-5599; (800) 632-5227

ORGENICS INTERNATIONAL CORPORATION: 9113 Guilford Rd., Suite 180, Columbia, MD 21046; (301) 776-3151

PERKIN ELMER CETUS: 761 Main Ave., Norwalk, CT 06859-0012; (203) 762-1000

PHARMACIA LKB BIOTECHNOLOGY: 800 Centennial Ave, Piscataway, NJ 08855-1327; (800) 526-3593

QIAGEN, INC.: 9259 Eton Ave., Chatsworth, CA 91311; (800) 426-8157

RESEARCH GENETICS: 2130 South Memorial Pkwy, Huntsville, AL 35801; (800) 533-4363

SAVANT INSTRUMENTS, INC.: 110-103 Bi-County Blvd., Farmingdale, NY 11735; (800) 634-8886

SCHLEICHER & SCHUELL, INC.: 10 Optical Ave., Keene, NH 03431; (603) 352-3810

SERVA: 50 A&S Dr., Paramus, NJ 07652; (201) 967-5900

SIGMA CHEMICAL COMPANY: P.O. Box 14508, St. Louis, MO 63178-9916; (800) 325-8070

SYNTHETIC GENETICS: 10455 Roselle St., San Diego, CA 92121; (800) 562-5544

WASHINGTON BIOTECHNOLOGY, INC.: 7509 Holiday Terrace, Bethesda, MD 20817; (301) 229-8993

WORTHINGTON BIOMEDICAL CORPORATION: Halls Mill Rd., Freehold, NJ 07728; (800) 445-9603

5 PRIME → 3 PRIME, INC.: 5603 Arapahoe Ave., Boulder, CO 80303; (800) 533-5703

Supplement 2
Testing Laboratories

Selected Laboratories Offering Testing Services in Molecular Genetics and Diagnosis of Infectious Diseases at the Gene Level

Only some of the numerous institutions in the U.S. that perform molecular biological procedures related to patient care are represented in this Supplement. At the time of this writing no single directory of such facilities is available although efforts to establish one are underway (see below). Some of the institutions had been approached with a request to share their experience in providing this type of service to the community and the author is grateful for their contribution, which hopefully reflects at least in part the current state of affairs.

Inclusion in this publication of the institutions listed below should in no way be construed either as endorsement or recommendation of these institutions for any services they may offer, or as suggestive of implied preference of the quality of their services and/or qualifications over those of other such institutions.

REGISTRY OF DNA DIAGNOSTIC LABORATORIES

Roberta A. Pagon, M.D.
Associate Professor of Pediatrics
Department of Pediatrics/RD-20
University of Washington School of Medicine
Seattle, WA 98195

By the end of 1991, 170 human genes had been sequences and over 2300 other human genes had been mapped using molecular genetic techniques. Much of this information is or will be useful in the diagnosis of human genetic diseases and in the genetic counseling of individuals at risk. Currently there is no information service that allows the transfer of this clinically relevant or potentially clinically relevant molecular genetic information from the research sphere into the clinical sphere. The approximately 1000 medical geneticists and genetic counselors in the United States must access this information either through a search of the literature or through word of mouth.

Discussions have been held with Dr. David Lipman, the Director of the National Center for Biotechnology Information at the National Library of Medicine, concerning the possibility of establishing an **on-line registry of laboratories performing DNA diagnostic studies** for the purpose of providing an accessible, up-to-date, and comprehensive listing for clinicians. Such a registry would include all laboratories in the United States and Canada that are performing DNA diagnostic studies both for service and research. There would be no cost to the users. The registry will be accessed either by network, modem, or by telephone contact with the registry manager. The registry information would be continuously updated and the entry for each laboratory would be verified routinely.

Funding for this registry has not yet been approved. When approved it is anticipated that there will be a four- to six-month organizational period. The earliest that the registry is hoped to be functional would be early 1993.

COLLABORATIVE DIAGNOSTICS
204 Second Avenue
Waltham, MA 02154
(617) 487-7979

Collaborative Diagnostics is a DNA diagnostic reference laboratory licensed in 1986 to offer a wide range of diagnostic and monitoring services based on molecular techniques. Collaborative Diagnostics provides genetic and cancer testing services from its laboratories in Waltham, Massachusetts to hospitals, clinical laboratories, health maintenance organizations, and physicians located throughout the United States.

GENETIC SERVICES

Molecular Genetics

Cystic fibrosis:
- 10 Mutation analysis by PCR
- Linkage analysis
- Haplotyping

Adult polycystic kidney disease:
- Presymptomatic and prenatal

Y chromosome detection and analysis
Parentage testing
DNA banking (preparation/storage)
DNA preparation

Cytogenetics

Amniotic fluid
Peripheral blood

Skin fibroblasts
Chorionic villi sampling (CVS)
Fragile X analysis (peripheral blood)
Products of conception (POC)
High resolution banding (700 to 800 bands)
Other:
Additional karyotype
Additional cells counted
Additional special banding
Tisssue culture only:
Amniocytes, fibroblasts, POC
Peripheral blood
CVS
No growth

Biochemical Genetics
MS-AFP or AF-AFP
MS-AFP Plus (AFP, HCG, UE3)

ONCOLOGY SERVICES

Gene Rearrangements
T cell/B cell clonality analysis
Individual probes:
Heavy chain (J-heavy)
Immunoglobulin light chain (C-κ)
Immunoglobulin light chain (C-λ)
T-cell receptor (β chain)
T-cell receptor (γ chain)
bcr/abl gene rearrangement:
Southern blot and PCR analysis
Fine mapping of *bcr* breakpoint
bcl-2 gene rearrangement (Southern blot)
Densitometric analysis of autoradiographs

Flow Cytometry
Leukemia/lymphoma immunophenotyping
Comprehensive leukemia/lymphoma analysis
Includes immunophenotyping and gene rearrangements

Bone Marrow Transplant Monitoring
Allogeneic transplant genotyping/engraftment monitoring
Minimal residual disease monitoring (CML and ALL)
PCR analysis of the *bcr/abl* rearrangement

GENESCREEN
2600 Stemmons Freeway, Suite 133
Dallas, TX 75207
(214) 631-8152
Robert C. Giles, Ph.D., Scientific Director

GeneScreen is a DNA reference laboratory located in Dallas, Texas. GeneScreen has been performing DNA analysis for genetic diseases since 1988, making it one of the first commercial laboratories performing such a service. In addition to its testing capabilities for genetic diseases, GeneScreen has become one of the leaders in the area of identity testing, both for establishment of paternity and for investigative purposes in criminal cases.

ISSUES

Some of the problems experienced at GeneScreen in performing medical genetic testing both in the area of ethics and operations such as account collections are briefly outlined below.

In the area of **ethics**, the company has tried to put decision making regarding ethical issues as close to the patient as possible. It would be difficult for a centralized laboratory to know all the relevant facts in a case and provide the appropriate level of individual support (such as counseling) to the patient. Therefore, the company first determines, before offering the test, if there are circumstances that have legitimate reasons for wanting a certain test performed. If not, the test would not be offered. (An example would be prenatal sex determination. Only under unusual circumstances would this be a medically significant result.) If, however, there are regular circumstances where a test is warranted, then that test might be offered by the company. The personnel do not attempt to screen the cases to determine the merits of the case, but we do require that all such tests be referred by a physician, on whom we rely to make the appropriate case-specific recommendation for the patient. Furthermore, results are given only to the physician for proper presentation to the patient.

In the **area of business operations**, the largest problems are common to most medical areas and involve **billing** and **collections**. Some of the specific concerns have been as follows:

- **Insurance companies are often not familiar with the tests or technology**, requiring extra correspondence, documentation, and so on.
- **Insurance forms** provided by the patient are used by the referring hospital, physician, or genetic counselor for their claims and **none are provided to the testing laboratory, complicating reimbursement of its services**.
- **Insurance pays only 80% or less if the deductible has not been met**, and when this is resolved months later, collection of the balance from the patient is difficult.

- The high cost of some tests, $500 or more, is significant and, if denied by insurance, **the patient often cannot afford to pay**.

To prevent some of these problems, GeneScreen has implemented several measures. Each test must have one of **the following billing arrangement set in advance**:

- account **billing to the hospital or physician**
- **complete insurance forms** assigning benefits and **20% copay**
- **payment in full**

or testing will not begin. Wherever possible, **we encourage the hospital to handle insurance reimbursement with account billing to them**.

Other operational areas such as logistics have been relatively straightforward, using overnight carriers and special specimen handling kits that contain necessary informed consent paperwork.

GeneScreen offers the following tests:

α-1-Antitrypsin
  Prenatal diagnosis
Cystic fibrosis
  ΔF508 and other mutations
  Haplotype, ΔF508, and other mutations
  Linkage analysis, family
  Prenatal diagnosis
Duchenne/Becker muscular dystrophy
  Complete deletion analysis
  Family analysis, known deletion
  Family analysis, linkage
HLA-B27
Leukemia/Lymphoma: T and B cell rearrangements, Philadelphia chromosome t(9;22); *BCR* gene rearrangements, t(14;18) rearrangements, *myc* rearrangements
Sickle cell
Cardiac risk: Lp(a), APO B3500, APO E isoforms
Paternity analysis
  Complete DNA-RFLP analysis, trio of samples
    Prenatal diagnosis
    Additional samples
  Zygosity of twins (including both parents)
  Special paternity analysis
  Complete HLA trio of samples, peripheral blood (95%)
    Additional samples
  FasPat, surcharge, trio (accelerated processing of samples)
    DNA — 3 weeks
    HLA — 1 week
Other services
  DNA repository/bank

Includes extraction of DNA and storage for 5 years
This cost may be applied in full toward testing
Cell culturing
From amniotic fluid
From CVS tissue
Cytogenetic analysis (including culturing for DNA analysis)
From amniotic fluid
From CVS tissue
From blood, suspected leukemia

MAYO CLINIC
Rochester MN 55905
(507) 284-2511
Steven S. Sommer, M.D., Ph.D., Laboratory Director,
Department of Biochemistry and Molecular Biology

DIRECT DNA TESTING FOR HEMOPHILIA B

Much of the direct DNA-based testing for hemophilia B in the United States is performed in that Laboratory. At present, mutations have been defined in 290 families with hemophilia B.

Approximately 25% of families have one of three common mutations (Ketterling et al., 1991). **PCR amplification of specific alleles** (also known as **allele-specific amplification**) is used to screen rapidly for the presence of these common mutations. In the remainder of the samples, the promoter sequence, the coding sequences, and the splice junctions are directly sequenced by **GAWTS** (Genomic Amplification With Transcript Sequencing) (Stoflet et al., 1988). Once the mutation is delineated, carrier detection and prenatal diagnosis can be performed by determining if the mutation is present in the at-risk individuals (Bottema et al., 1989).

GAWTS is a technically robust and rapid method of genome sequencing (Bottema et al., 1989). In the method, the regions of interest are amplified with at least one primer containing a phage promoter sequence. The amplified product is then transcribed and sequenced with reverse transcriptase. **The major advantage of GAWTS is that the transcription step generates a great abundance of sequencing template.** It is sequencing the template that is often limiting in other methods of genomic sequencing such as asymmetric PCR and double-stranded DNA sequencing. In GAWTS, typically 2 μl of the PCR is placed in the transcription reaction and 3 μl of the transcription products is placed in the sequencing reaction. No purification is needed and a strong sequencing signal is generally seen in a 6- to 16-h exposure.

REFERENCES

Bottema, C. D. K. et al. (1989). Direct carrier testing in 14 families with hemophilia B. Lancet *ii*:526–529.

Ketterling, R. P. et al. (1991). Evidence that descendants of three founders constitute about 25% of hemophilia B in the United States. Genomics *10*:1093–1096.
Stoflet, E. S. et al. (1988). Genomic amplification with transcript sequencing. Science *239*:491–494.

Stephen N. Thibodeau, Ph.D.
Molecular Genetics Laboratory
970 Hilton Building
Mayo Clinic
Rochester, MN 55905
(507) 284-0043

GENETIC SERVICES

Cystic fibrosis: Direct, ΔF508 (additional mutations in development); indirect, linkage and haplotype testing
Duchenne muscular dystrophy: Direct, detection of gene deletions; indirect, linkage
Fragile X syndrome: Direct, detection of *FMR*-1 gene mutations
α-Thalassemia: Direct, detection of gene deletions
Familial adenomatous polyposis: Indirect, linkage
Familial amyloidosis: Direct, detection of mutations in prealbumin gene
Hemophilia A: Indirect, linkage
Y-DNA analysis: Direct, detection of Y DNA material

HEMATOLOGY/ONCOLOGY SERVICES (SOUTHERN BLOT ANALYSIS)

Immunoglobin gene rearrangement, T-cell receptor gene rearrangement, *BCR* gene rearrangement, N-*myc* gene rearrangement and bone marrow transplantation (DNA Fingerprinting).

IN DEVELOPMENT

Neurofibromatosis Type 1, Myotonic Dystrophy, Multiple Endocrine Neoplasia, Minimal Residual Disease, Detection of Hematologic Neoplasms

---

NICHOLS INSTITUTE
33608 Ortega Highway
San Juan Capistrano, CA 92690
(800) 642-4657

For over 20 years Nichols Institute has led the clinical testing field by transferring advanced medical technology to the clinical laboratory. This process begins with our academic associates, leading research scientists from clinical and university settings who engineer the transfer of technological advancements to our laboratories. It is implemented by our team of medical directors, scientific directors, and certified technologists.

The institute now includes 17 reference laboratories, providing the full spectrum of esoteric clinical testing. The newest laboratory, molecular biology, is using DNA probes for new clinical applications. DNA probes provide an important adjunct method for clinical diagnosis. They also allow analysis for certain conditions that cannot be performed by other methods, thus the need for a molecular biology laboratory.

Specific tests using DNA probes at Nichols Institute include the following:

***bcl*-2**: Analysis involves amplification of the major breakpoint region of the *bcl*-2 gene using PCR primers directed to the *bcl*-2 gene and the immunoglobulin heavy chain gene. The *bcl*-2 gene translocations can also be detected with lower sensitivity using Southern blotting.

BCR: The **breakpoint cluster region (BCR) region** of **chromosome 22** is translocated to the c-*abl* protooncogene locus of **chromosome 9** in most patients with chronic myelogenous leukemia (**CML**) and a subset of individuals with acute lymphocytic leukemia (**ALL**). This translocation usually results in a truncated chromosome 22 designated the Philadelphia chromosome. The BCR translocation can be detected by Southern blotting. Alternatively, the translocation can be detected by analysis of RNA which exist as unique fusion messages of the c-*abl* and *BCR* genes in cells harboring a t(9;22) translocation. RNA analysis by reverse transcription followed by PCR has several advantages over the Southern blot method including greater sensitivity and the ability to distinguish the nature of the fusion junction. RNA analysis of the fusion junction can distinguish between ALL and CML translocations and the different types of CML translocations.

B & T cell gene rearrangement: This test analyzes for the presence of specific rearrangements of the immunoglobulin genes or the T cell receptor genes in lymphoid cells by Southern blotting. DNA probe analysis can provide a useful clinical adjunct to other methods for diagnosis of leukemias and lymphomas, particularly in cases where other diagnostic indicators are inconclusive.

N-*myc* gene amplification: Amplification of the N-*myc* gene occurs in advanced stages of neuroblastomas and can provide a useful prognostic indicator. Amplification of the gene is detected by Southern blotting using a specific probe for the N-*myc* gene.

Cystic fibrosis (CF) gene analysis: With the current panel of mutations being analyzed, approximately 85 to 90% of the CF mutations can be detected in Caucasians of Northern European origin.

Duchenne muscular dystrophy (DMD): Mutation analysis can be used for diagnosis or confirmation of diagnosis in a male suspected of DMD. With the identification of a mutation in an affected individual, carrier detection can be performed for appropriate female relatives of the affected individual. DNA probe analysis can also be performed for prenatal diagnosis and for linkage studies in families with a history of DMD.

Fragile X: The DNA probe test looks for the expansion in size of a particular region of the fragile X locus. The size and methylation status of this expandable region can be used as a diagnostic indicator of fragile X syndrome.

Sickle cell anemia: The use of DNA probes for analysis of sickle cell anemia is mostly limited to prenatal diagnosis. The test is offered to couples who are at risk of having an affected offspring. Analysis is performed on either a chorionic villus sample or on amniotic fluid. As with all prenatal DNA probe diagnoses performed at the Nichols Institute, analyses are performed to verify the absence of maternal or other contamination in the fetal sample.

α-1 antitrypsin: α-1 antitrypsin analysis by DNA probes, like for sickle cell anemia, is largely limited to prenatal diagnosis.

Human papillomavirus: The Nichols Institute offers both a screen and a typing assay. The screen, called ViraPap, utilizes DNA probe technology and detects simultaneously HPV types 6, 11, 16, 18, 31, 33, and 35 from cervical cells. Use of the DNA hybridization technology allows analysis of HPV-infected patients who currently do not present clinical symptoms and who may appear normal by Pap smear screening alone.

***Mycobacterium* identification**: DNA molecular probe technology is used to identify the most common species found in humans, including *M. tuberculosis* complex, *M. avium-M. intracellulare* complex, and *M. gordonae*. Although not normally an etiological agent, *M. gordonae* is a common contaminant found in mycobacteria testing. Proper identification ensures correct diagnosis.

ROCHE BIOMEDICAL LABORATORIES

P.O. Box 2230
Burlington, NC 27216-2230
(800) 334-5161
Joseph A. Chimera, Ph.D., Assistant Vice President

In September of 1989 Roche Biomedical Laboratories, opened the Center for Molecular Biology (CMB) in Research Triangle Park, North Carolina. The mission of the CMB is to develop and validate clinical laboratory tests based on molecular biology techniques such as polymerase chain reaction (PCR), and provide testing services to the medical community.

The CMB is staffed with over 30 scientists who specialize in PCR-based clinical science for the infectious diseases, identity testing, genetics, and oncology disciplines.

The following is a list of PCR-based tests that are either available or in development.

INFECTIOUS DISEASES

1. HIV DNA. The direct detection of HIV proviral DNA in peripheral blood or tissues. One HIV molecule can be detected per 10,000 white blood cells. Clinical applications include:

- resolution of indeterminate western blot test results in patients at risk for HIV infection or blood donors
- detection of HIV infection in infants born to HIV-positive mothers. Serological tests are not useful because maternal anti-HIV antibodies are passed to the newborn
- earlier detection of HIV infection in individuals at "high-risk" for exposure to HIV. It has been demonstarted that PCR-based tests can detect HIV DNA prior to the detection of anti-HIV antibodies
- screening of cadaveric source materials (i.e., muscle, bone, etc.) used in tissue transplantation

2. HTLV 1, 2. The direct detection of HTLV DNA in peripheral blood or tissues. Clinical applications include:

- resolution of indeterminate serology test results in patients suspected of HTLV infection or blood donors
- distinguishing between HTLV type 1 and HTLV type 2 infection. HTLV 1 is associated with adult T cell leukemia
- screening of cadaveric source materials used in tissues transplantation

3. Quantitative analysis of HIV DNA or RNA. A primary application of this test is to assess the effects of antiviral therapy used in the treatment of AIDS. Quantitative detection may also be used with other markers to monitor disease progression.

4. *Borrelia burgdorferi* (Lyme disease). Serological testing for Lyme disease is prone to both false-negative and false-positive results. The direct detection of *B. burgdorferi* DNA may be useful in patients suspected of Lyme disease.

5. Mycobacteria. Conventional laboratory methods such as culture can take up to 8 weeks to confirm a negative result. PCR-based tests may be used to screen for mycobacteria infection in afew days. In addition, DNA probes can be engineered to detect specific pathogenic species such as *M. tuberculosis*, *M. avium*, *M. intracellulare*, and *M. kansasii* from common avirulent species such as *M. gordonae*.

6. Hepatitis B and C.

- determination of the infection status of carriers
- confirmation of serology results in symptomatic patients and blood donors

IDENTITY TESTING

A more recent approach to producing DNA profiles is based on PCR amplification. Millions of copies of a specific gene are produced and the method can be used to generate a DNA profile from the minute quantities of DNA present in a single hair follicle. Applications of RFLP and PCR amplification in the filed of identity testing include:

- comparison of DNA recovered from crime scene evidence (blood, semen stain, hair) with the DNA profile of a suspected felon
- investigating cases of disputed parentage
- establishing databanks of DNA profiles from convicted felons for use in crime investigation (repeat offenders, serial offenders, etc.)
- DNA profile databanks of military and government officials for use in body identification
- immigration
- genetic testing

GENETICS

1. Sickle cell anemia. A developing fetus can be examined for sickle cell disease or sickle cell carrier status using cells collected by amniocentesis. PCR-based testing can detect this disease or identify carriers before aberrant proteins can be detected.

2. Cystic fibrosis. DNA detection methods are the only assays specific for identifying carriers or affected individuals. PCR-based testing may be used for rapid prenatal diagnosis or carrier screening in adults.

3. Fragile X syndrome. While 20% of males carrying the fragile X chromosome do not clinically or cytogenetically express the fragile X mutation, the PCR-based identification of the fragile X site at the molecular level has a sensitivity as high as 98%, regardless of clinical or cytogenetic expression.

ONCOLOGY

1. Follicular lymphoma. Approximately 90% of all cases of follicular lymphoma are associated with a chromosomal translocation t(14;18) that results in the deregulation of the *bcl*-2 oncogene. PCR-based analysis is a more rapid and sensitive technique compared with conventional procedures such as karyotyping. The high specificity of this assay can be used to resolve equivocal diagnoses in patients suspected of follicular lymphoma. Moreover, because only minute quantities of cancer cells are required for early diagnosis and the detection of metastatic disease, the test can be performed from fine needle aspirates, bone marrow, or peripheral blood.

2. Chronic myelogenous leukemia (CML) and acute lymphocytic leukemia (ALL). Traditional methods of diagnosing CML and ALL, such as cytogenetic analysis, are time consuming and suffer from a general lack of sensitivity. PCR-based testing may detect fewer than 10 cancer cells, and, therefore, is the most sensitive method for the earliest detection of these diseases. Also, due to sensitivity of PCR peripheral blood can be used as a specimen source eliminating the need for bone marrow sampling.

Supplement 3
PCR Licensees

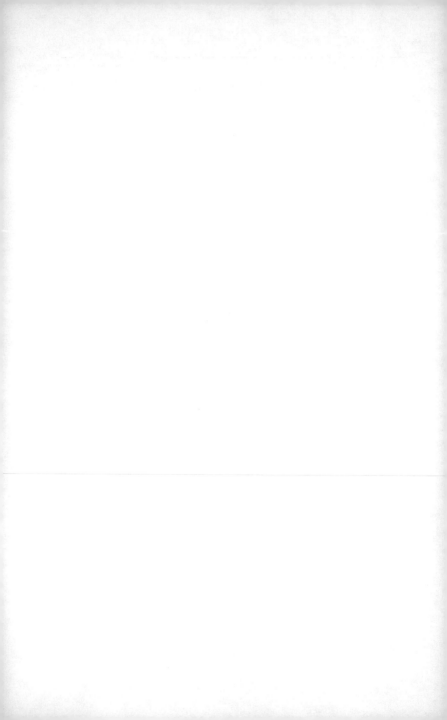

Institutions in the U.S. Licensed by Roche Molecular Systems to Perform Human Diagnostic Services by Polymerase Chain Reaction Technology

All Children's Hospital: 801 - 6th Street South, St. Petersburg, FL
Baylor College of Medicine: One Baylor Plaza, Houston, TX 77030
The Blood Center of Southeastern Wisconsin: 1701 W. Wisconsin Ave., Milwaukee, WI 53233
The Cleveland Clinic Foundation: One Clinic Center, 9500 Euclid Ave., Cleveland, OH 44195-5108
Clinical Reference Laboratory: 11850 W. 8th St., Lenexa, KS 66214
Collaborative Research, Inc.: Two Oak Park, Bedford, MA 01730
Diagnostic Services, Inc.: 340 Goodlette Road South, Naples, FL 33940
Duke University: Durham, NC 27710
Gen-Care Biomedical Research Corporation: 271 Sheffield St., Mountainside, NJ 07092
Genica Pharmaceuticals Corporation: Two Biotech Park, 373 Plantation Street, Worcester, MA 01605
GeneScreen, Inc.: 2600 Stemmons Fwy #133, Dallas, TX 75207
George Washington University Medical Center, The University Hospital: 901 Twenty-Third St., NW, Washington, D.C. 20037
IG Laboratories: One Mountain Rd., Framingham, MA 01701
Immungenex, Inc.: 797 San Antonio Rd., Palo Alto, CA 94303
Immunological Associates of Denver, Inc.: 101 University Blvd. Suite 330, Denver, CO 80206
Irwin Memorial Blood Bank: 270 Masonic Ave., San Francisco, CA 94118
Johns Hopkins Hospital: 500 N. Wolfe St., Baltimore, MD 21205
Johns Hopkins University: 720 Rutland Ave., Baltimore, MD 21205
Mayo Foundation: 200 First St., S.W., Siebens 6, Rochester, MN 55095
Mayo Medical Laboratories: Rochester, MN
MediGene, Inc.: 100 Corporate Dr., Yonkers, NY 10701-6807
MetPath, Inc.: 1 Malcolm Ave., Teterboro, NJ 07608-1070

Michigan State University, Department of Medicine: East Lansing, MI 48824-1317

Midwest Organ Bank: 1900 West 47th Place, Suite 400, Westwood, KS 66205

Nichols Institute: P.O. Box 92797, Los Angeles, CA 90009-2797

Oncore Analytics: 4900 Fannin, Houston, TX 77004

The Scripps Research Institute: 11107 Roselle St., Ste. A, San Diego, CA 92121

Simmons GeneType Diagnostics, Inc.: 851 Burlway Rd., Ste. 704, Burlingame, CA 94010

SmithKline Beecham Clinical Laboratories, Inc.: 620 Freedom Business Center, Ste. 400, King of Prussia, PA 19406

Specialty Laboratories, Inc.: 2211 Michigan Ave., Santa Monica, CA 90404-3900

University of California at Davis, School of Medicine, Pathology Department: Davis, CA 95616

University of California, San Diego, School of Medicine, Center for Molecular Genetics: La Jolla, CA 92093

University of California, San Francisco, School of Medicine, Reproductive Genetics Unit 6: San Francisco, CA 94143-0720

University of Florida, College of Medicine, Department of Pathology: P.O. Box J-275, JHMHC, Gainesville, FL 32610

University of Medicine and Dentistry of New Jersey: 110 ADMC Bergen St., Newark, NJ 07107

University of New Mexico: Dept. of Microbiology, Albuquerque, NM 87131-5276

University of Virginia, Health Science Center: Medical Center Box 168, Charlottsville, VA 22908

University Pathology Associates, University of Southern California School of Medicine: 204 Hoffman Memorial Research Bldg., 2100 Zonal Ave., Los Angeles, CA 90033

Vivigen, Inc.: 2000 Vivigen Way, Santa Fe, NM 87505

ViroMed Laboratories, Inc.: 5500 Feiti Rd., Minnetonka, MN 55343-9806

Wuesthoff Hospital: 110 Longwood Ave., Rockledge, FL 32955

Roche Molecular Systems is offering human diagnostic licenses broadly and is responding to over 180 requests worldwide for additional licenses. For information on obtaining a license, please contact:

Within the U.S.: Roche Molecular Systems, Inc., 1145 Atlantic Ave., Alameda, CA 94501, Attn: Licensing Manager

Outside the U.S.: F. Hoffman-La Roche AG, CH-4002 Basle, Switzerland, Attn: Licensing Manager

Supplement 4
Efforts of Professional Organizations

Medical Genetics: the Current State of the Field

Michael S. WATSON, Ph.D.
Chairman, Laboratory Practices Committee
American College of Medical Genetics
Assistant Professor of Pediatrics and Genetics
Director, Clinical Cytogenetics
Washington University School of Medicine
St. Louis, MO 63110

The past decade has been one of rapid development and expansion of the specialty of medical genetics and its subspecialty of clinical genetics and the laboratory genetics subspecialties of clinical cytogenetics, clinical biochemical genetics, and clinical molecular genetics. While the interaction of the various traditional medical specialties with medical genetics will vary, it is likely that these interactions will increase over the next decade.

Presently, the different practice specialties have differing perceptions of how genetic information will impact their field. From the point of view of the obstetrical community, genetic testing involves prenatal diagnosis of chromosomal and single gene disorders. The pediatric perpective is one of newborn screening as well as the diagnosis of chromosomal and single gene disorders. The clinical pathologists tend to recognize the powerful contributions that molecular genetics will make to the area of infectious disease testing and cancer diagnosis.

Although molecular genetics has tremendous technical potential for the detection of disease that is often treatable, there is still little information available to address associated issues such as quality control, public and provider education, and cost-benefit. As these types of testing become more utilized, they will have an increasing impact on many different specialty areas.

As of 1991, there were already more than 5000 mendelizing phenotypes identified in McKusick's *Mendelian Inheritance in Man*. While many of these are rare, single-gene disorders that traditionally fall into the area of expertise of the clinical geneticist, other diseases such as cystic fibrosis and sickle cell disease are relatively common and may be dealt with by referral for laboratory services only. As newer areas of testing become available for disease predisposition genes such as those predisposing to cancer, atherosclerosis, and heart disease, or for presymptomatic testing for diseases such as Huntington disease and multiple endocrine neoplasia, the primary physician will become the focal point for medical management. Hence, what once may have been a specialty focused on an individual's disease will become one more oriented to a family-based medical perspective.

Of immediate concern is the relative paucity of medical geneticists to provide the services related to this technology. At present there are only 600–700 clinical geneticists and only 400–500 clinical laboratory geneticists in the U.S. This manpower deficiency will force the primary care providers to be the focal point of utilization and, to some extent, interpretation of such tests. Compounding the manpower problem is an existing deficiency in education and continuing education in human and medical genetics in medical schools. As of 1985, nearly 50% of medical schools either lacked teaching or had deficiencies in their teaching of medical genetics. When combined with the speed at which the Human Genome Project is providing information of potential diagnostic importance it is clear that the number of trained individuals will have to be increased and that there will have to be a drastic change in the understanding of genetics by health care workers.

The practitioners of the subspecialties of medical genetics historically have been primarily recognized professionally through other existing specialties such as pediatrics and obstetrics. However, during the past year there have been several significant changes in the organization of the services sector of medical genetics in the U.S. In recognition of the growing impact of this field on the practice of medicine, the American Board of Medical Genetics (ABMG) has become the 24th member of the American Board of Medical Specialties, the first new primary board to be admitted since 1979. The ABMG now is the only board that certifies clinical and laboratory geneticists in the specific subspecialty areas of clinical genetics, clinical cytogenetics, clinical biochemical genetics and clinical molecular genetics, and includes both M.D.s and Ph.D.s.

Early in 1991, laboratory geneticists further organized themselves to address issues that particularly concerned them. Simultaneously, a committee of the American Society of Human Genetics (ASHG) strongly recommended and supported the creation of the American College of Medical Genetics (ACMG), charged with the settting and maintenance of standards of clinical and laboratory practice. With the creation of the ACMG, the discipline of medical genetics is now organized with a format similar to that of other clinical specialties. The ASHG now serves as the scientific/research arm; the ABMG being the accreditation arm and the new ACMG responsible for standards and practices. Fellows of the ACMG will include doctoral level individuals (M.D. and Ph.D.) who devote a significant portion of their time to the practice of medical genetics. It includes both the clinical geneticists and the clinical laboratory geneticists. However, varying levels of membership exist for individuals whose interests overlap those of the college.

The major purpose of the ACMG is to stimulate and support patient care, education, and research in the field of medical genetics. More specifically, the purpose of the college will be to

- advance the art and science of medical genetics
- foster the development and implementation of methods of diagnosis, treatment, and prevention of genetic diseases
- promote uniform standards of laboratory quality assurence and proficiency testing
- increase access to medical genetics services and, therefore, improve public health

The ACMG already has initiated efforts to draft national standards and guidelines for each of the laboratory practice subspecialties. Similarly, the ACMG will develop standards for clinical diagnosis including dysmorphology, inborn errors of metabolism, prenatal diagnosis, and common genetic disorders, as well as guidelines for the diagnosis, treatment, and prevention of genetic disorders.

Of paramount importance will be the development of continuing education programs for clinicians already in practice and the expansion of teaching of medical genetics in medical schools. The rapid acquisition of knowledge related to genetic disorders and the evolving technology for diagnosing those disorders will also require that continuing education programs for medical genetics laboratory personnel be developed. There will also need to be a concerted effort to better educate the general public about genetics and its impact on human health. An awareness of the indications for testing and the availability of testing will improve public access to the system.

Quality Assurance in Clinical Laboratory Molecular Pathology Testing

Robert M. Nakamura, M.D.
Chairman, Department of Pathology
Scripps Clinic & Research Foundation
San Diego, CA 92037
Chairman, Molecular Pathology Committee
College of American Pathologists
325 Waukegan Road
Northfield, IL 60093-2750

In 1961, the College of American Pathologists (CAP) initiated the development of a laboratory accreditation program (Duckworth, 1987). The CAP inspection and accreditation program began operations in 1962.

The Commission on Laboratory Accreditation is composed of the Chairman, Vice Chairman, and Regional Commissioners, who operate the program under the auspices of the CAP Board of Governors.

The CAP Accreditation Program was later recognized by the Joint Commission on Accreditation of Hospitals (JCAH). In the CAP program, after initial accreditation of the laboratory, the laboratory is reaccredited with a program that requires on-site inspection every 2 years, successful participation in a self-evaluation in the interim year, and a laboratory proficiency testing comparison system (Duckworth, 1987).

Accrediation by the CAP is voluntary. However, laboratories found noncompliant with the CAP standards are reported to the JCAH and to the Health Care Financing Administration (HCFA) (if it is an interstate laboratory).

Technical expertise to the Commission on Laboratory Accreditation is provided by the CAP Technical Resource Committees. The primary goal is continuous laboratory improvement through peer review and education. The CAP program encompasses all aspects of quality control in the laboratory, including methodology, reagents, control media, equipment, specimen handling, procedure manuals, reports, proficiency testing, personnel safety, and the overall management principles that distinguish a quality laboratory.

QUALITY ASSURANCE, QUALITY CONTROL, AND PROFICIENCY TESTING

In regards to the clinical laboratory, the following terms are defined:

1. **Quality assurance** is a term used for the total system designed to ensure the quality of the result (Howanitz and Howantiz, 1987).
2. **Quality control** is a more restrictive term and refers only to the procedures that must be completed before a particular batch of results is reported.
3. **Proficiency testing** and **surveys** are terms synonymous with interlaboratory comparisons and external quality control (Rippley, 1987). Proficiency testing is an external quality control mechanism that reflects the performance of the laboratory regarding accuracy whereas internal quality control represents the performance of the laboratory regarding precision (Rippley, 1987).

MOLECULAR PATHOLOGY

Molecular pathology, which has been defined as the use of nucleic acid probes to diagnose and study disease, is a new emerging discipline of importance (Matthews and Krika, 1988; Mifflin, 1989; Lowell, 1989; Antonarakis, 1989; Grody et al., 1989). DNA and RNA probes are being used in the following areas:

- infectious diseases (use of DNA probes for bacteria, viruses, and parasites)
- neoplastic diseases (detection of gene rearrangement, tissue-specific gene transcription, and oncogene activation)
- hereditary diseases (detection of specific mutated genes or linked DNA polymorphisms)
- DNA identity and fingerprinting (for purposes of donor/recipient identification in transplant and paternity testing, or forensic investigations)

ACTIVITIES OF THE COLLEGE OF AMERICAN PATHOLOGISTS MOLECULAR PATHOLOGY RESOURCE COMMITTEE

The CAP Molecular Pathology Resource Committee was formed in 1989. It is the successor to an ad hoc presidential advisory committee headed by G. Glenn, M.D. The CAP ad hoc Committee on Molecular Pathology was created by W. Zeiler, M.D., to advise the CAP leader on new developments in molecular biologic techniques, their emerging roles in diagnostic medicine, and how they can be best incorporated into pathology practice.

The Molecular Pathology Resource Committee was developed when it became apparent that there was an immediate need to produce a laboratory accreditation checklist, as well as plans for proficiency surveys.

The goals and mission of the CAP Molecular Pathology Resource Committee have been defined as follows:

1. to coordinate the scientific resources and expertise in molecular pathology both within the CAP and with other interested professional organizations and to facilitate communications among them
2. to develop appropriate guidelines for utilization of molecular biology techniques in the clinical laboratory and in the practice of medicine
3. to monitor performance of diagnostic molecular pathology testing
4. to monitor developments in the discipline of molecular biology, particularly its practical diagnostic and therapeutic applications

Committee members have identified seven areas of interest:

- molecular oncology testing
- molecular genetic disease testing
- molecular HLA/histocompatibility testing
- parentage testing by DNA polymorphisms
- forensic identity testing by DNA polymorphisms
- *in situ* hybridization testing
- molecular infectious disease testing

The CAP Molecular Pathology Resource Committee has developed a checklist that, in general, covers these areas of interest as follows:

1. The checklist covers extent of service, proficiency testing, quality control, procedure manuals, records, test requisitions, specimen handling, reagents, controls and standards, procedures and tests, personnel, physical facilities, laboratory safety, and quality assurance.
2. The laboratory accreditation checklists have generic questions that monitor the various categories and will cover the following nucleotide probe assays: Southern blot analysis, dot blot analysis, sandwich hybridization, *in situ* hybridization, and amplification methods (polymerase chain reaction, Q-beta replicase, and ligase chain reaction).

A CAP proficiency survey for molecular oncology that includes T cell and B cell gene rearrangement studies, oncogene detection, and amplification was developed and made available to subscribers in 1992.

The CAP in cooperation with the American Society of Human Genetics has developed a survey for human genetic donors, which will be available to subscribers.

The CAP Molecular Pathology Resource Committee is in the process of developing an *in situ* hybridization survey for viral antigens that should become available in 1993.

A CAP survey for parentage testing by DNA polymorphism is jointly planned with the American Association of Blood Banks and should become available in 1993.

A DNA forensic identity testing survey is in the planning stage in cooperation with the American Society of Crime Laboratory Directors and several other law enforcement organizations.

REFERENCES

Antonarakis, S. E. (1989). Recombinant DNA technology in the diagnosis of human genetic disorders. Clin. Chem. *35*:B4–B6.

Duckworth, J. K. (1987). in *Laboratory Licensure and Accreditation*, Chapter 15, Howanitz, P. J. and Howanitz, J. H., Eds., New York, McGraw-Hill, 334–353.

Grody, W. A. et al. (1989). Diagnostic molecular pathology. Mod. Pathol. *2*:553–568.

Howanitz, J. H. and Howanitz, P. J. (1987). in *Laboratory Quality Assurance*, Howanitz, P. J. and Howanitz, J. H., Eds., McGraw-Hill, New York, 1–19.

Lowell, M. A. (1989). Molecular genetics of leukemia and lymphoma. Clin. Chem. *35*:B43–B47.

Matthews, J. A. and Krika, L. J. (1988). Analytical strategies for use of DNA probes. Anal. Biochem. *169*:1–25.

Mifflin, T. E. (1989). Use and applications of nucleic acid probes in the clinical laboratory. Clin. Chem. *35*:1819–1825.

Rippley, J. H. (1987). in *Proficiency Testing*, Howanitz, P. J. and Howanitz, J. H., Eds., McGraw-Hill, New York, 317–333.

Supplement 5
DNA in Forensic Practice

DNA Technology in Forensic Science: Executive Summary

Committee on DNA Technology in Forensic Science
Board of Biology
Commission on Life Sciences
National Reasearch Council
Washington, D.C. 1992
(Reprinted by permission)

Committee on DNA Technology in Forensic Science:

Victor A. McKusick, Chairman, The Johns Hopkins Hospital, Baltimore, MD
Paul B. Ferrara, Division of Forensic Sciences, Department of General Services, Richmond, VA
Haig H. Kazazian, The Johns Hopkins Hospital, Baltimore, MD
Mary-Claire King, University of California, Berkeley, CA
Eric S. Lander, Whitehead Institute for Biomedical Research, Cambridge, MA
Henry C. Lee, Connecticut State Police, Meriden, CT
Richard O. Lempert, University of Michigan Law School, Ann Arbor, MI
Ruth Macklin, Albert Einstein College of Medicine, Bronx, NY
Thomas G. Marr, Cold Spring Harbor Laboratory, Cold Spring Harbor, NY
Philip R. Reilly, Shriver Center for Mental Retardation, Waltham, MA
George F. Sensabaugh, Jr., University of California, Berkeley, CA
Jack B. Weinstein, U.S. District Court, New York, Brooklyn, NY

Former Members

C. Thomas Caskey (resigned December 21, 1991), Baylor College of Medicine, Houston, TX
Michael W. Hunkapiller (resigned August 17, 1990), Applied Biosystems, Forest City, CA

National Research Council Staff

Oskar R. Zaborsky, Study Director; Director, Board on Biology
Norman Grossblatt, Editor
Marietta E. Toal, Administrative Secretary
Mary Kay Porter, Senior Secretary

PREFACE

In recent years, advances in the techniques for mapping and sequencing the human genome have contributed to progress in both basic biology and medicine. The applications of these techniques have not been restricted to biology and medicine, however, but have also entered forensic science. Today, methods developed in basic molecular biology laboratories can potentially be used in forensic science laboratories in a matter of months.

On the basis of its study of the mapping and sequencing of the human genome (reported in 1988), the Board on Biology and several federal agencies recognized the potential of DNA typing technology for forensic science. In particular, the Federal Bureau of Investigation, the preeminent organization in the U.S. for the development and application of forensic techniques, initiated an effort to develop and evaluate DNA typing in forensic applications in the mid-1980s. The first case work was performed in December of 1988. Several private-sector laboratories entered the field early, and state government crime laboratories also began to offer services on DNA typing. However, as DNA typing entered the courtrooms of this country, questions appeared about its reliability and methodological standards and about the interpretation of population statistics.

By the summer of 1989, a crescendo of questions concerning DNA typing had been raised in connection with some well-publicized criminal cases, and calls for an examination of the issues by the National Research Council of the National Academy of Sciences came from the scientific and legal communities. As a response, this study was initiated in January 1990.

Because of the broad ramifications of forensic DNA typing, a number of federal agencies and one private foundation provided financial support for this study: the Federal Bureau of Investigation, the National Institutes of Health National Center for Human Genome Research, the National Institute of Justice, the National Science Foundation, the State Justice Institute, and the Alfred P. Sloan Foundation.

Many persons offered assistance to the committee and staff during this complex study. In particular the following deserve recognition and praise for their efforts: John Hicks, Federal Bureau of Investigation; Elke Jordan and Eric Juengst, National Institutes of Health National Center for Human Genome Research; James K. Stewart and Bernard V. Auchter, National Institute of Justice; John C. Wooley, National Science Foundation; David I. Tevelin, State Justice Institute; and Michael S. Teitelbaum, Alfred P. Sloan Foundation.

I also thank the many who offered their advice to the committee during its briefings and open meetings. The names of those who offered testimony are given in the appendix. Additionally, I want to thank the many who wrote to me or to the National Research Council and provided valuable data and suggestions to the Committee; much was gained from their input. We also acknowledge the efforts of Robert Kushen, Columbia Law School, in assisting Judge Weinstein. I also thank Della Malone, my secretary, for her help throughout. The committee thanks the reviewers of our report for many valuable comments and suggestions. Although the reviewers are anonymous to us, I personally want to thank them for their constructive comments and suggestions.

The staff of the Board on Biology deserve special praise for their efforts during the many months of intense activity. Oskar R. Zaborsky, Study Director and Director of the Board on Biology, deserves recognition for his administrative and technical contributions and for handling many complex matters. Marietta Toal, Administrative Secretary, served in the committee well in logistics and the preparation of the report. The committee also thanks Mary Kay Porter for her assistance. Norman Grossblatt edited the report.

Last but not least, I thank my colleagues on the committee who served so well and unselfishly to address key issues from the perspective of their special expertise and to prepare this report in a timely fashion.

DNA typing for personal identification is a powerful tool for criminal investigation and justice. At the same time, the technical aspects of DNA typing are vulnerable to error, and the interpretation of results requires appreciation of the principles of population genetics. These considerations and concerns arising out of the felon DNA databanks and the privacy of DNA information made it imperative to develop guidelines and safeguards for the most effective and socially beneficial use of this powerful tool. We hope that our efforts will enhance understanding of the issues and serve to bring together people of good will from science, technology, law, and ethics. We hope that our report will serve well the sponsors and the general public.

Victor A. McKusick
Chairman
Committee on DNA Technology
in Forensic Science

SUMMARY

Characterization, or "typing," of deoxyribonucleic acid (DNA) for purposes of criminal investigation can be thought of as an extension of the forensic typing of blood that has been common for more than 50 years; it is actually an extension from the typing of proteins that are coded for by DNA to the typing of DNA itself. Genetically determined variation in proteins is the basis of blood groups, tissue types, and serum protein types. Developments in molecular genetics have made it possible to study the person-to-person differences in parts of DNA that are not involved in coding for proteins, and it is primarily these differences that are used in forensic applications of DNA typing to personal identification. DNA typing can be a powerful adjunct to forensic science. The method was first used in casework in 1985 in the United Kingdom and first used in the United States by commercial laboratories in late 1986 and by the Federal Bureau of Investigation (FBI) in 1988.

DNA typing has great potential benefits for criminal and civil justice; however, because of the possibilities for its misuse or abuse, important questions have been raised about reliability, validity, and confidentiality. By the summer of 1989, the scientific, legal, and forensic communities were calling for an examination of the issues by the National Research Council of the National Academy of Sciences. As a response, the Committee on DNA Technology in Forensic Science was formed; its first meeting was held in January of 1990. The committee was to address the general applicability and appropriateness of the use of DNA technology in forensic science, the need to develop standards for data collection and analysis, aspects of the technology, management of DNA typing data, and legal, societal, and ethical issues surrounding DNA typing. The techniques of DNA typing are fruits of the revolution in molecular biology that is yielding an explosion of information about human genetics. The highly personal and sensitive information that can be generated by DNA typing requires strict confidentiality and careful attention to the security of data.

DNA, an active substance of the genes, carries the coded messages of heredity in every living thing: animals, plants, bacteria, and other microorganisms. In humans, the code-carrying DNA occurs in all cells that have a nucleus, including white blood cells, sperm, cells surrounding hair roots, and cells in saliva. These would be the cells of greatest interest in forensic studies.

Human genes are carried in 23 pairs of chromosomes, long thread-like or rod-like structures that are a person's archive of heredity. Those 23 pairs, the total genetic makeup of a person, are referred to as the human "diploid genome." The chemistry of DNA embodies the universal code in which the messages of heredity are transmitted. The genetic code itself is spelled out in strings of nucleotides of four types, commonly represented by the letters A, C, G, and T (standing for the bases adenine, cytosine, guanine, and thymine), which in various combinations of three nucleotides spell out the codes for the amino acids that constitute the building blocks of proteins. A gene, the basic unit of heredity, is a sequence of about 1000 to over 2 million nucleotides. The human genome, the total genetic makeup of a person, is estimated to contain 50,000 to 100,000 genes.

The total number of nucleotides in a set of 23 chromosomes — 1 from each pair, the "haploid genome" — is about 3 billion. Much of the DNA, the part that separates genes from one another, is noncoding. Variation in the genes, the coding parts, are usually reflected in variations in the proteins that they encode, which can be recognized as "normal variation" in blood type or in the presence of such diseases as cystic fibrosis and phenylketonuria; but variations in the noncoding parts of DNA have been most useful for DNA typing.

Except for identical twins, the DNA of a person is for practical purposes unique. That is because one chromosome of each pair comes from the father and one from the mother; which chromosome of a given pair of a parent's chromosomes that parent contributes to the child is independent of which chromosome of another pair that parent gives to that child. Thus, the different combinations of chromosomes that one parent can give to one child is 2^{23}, and the number of different combinations of paired chromosomes a child can receive from both parents is 2^{46}.

The substitution of even one nucleotide in the sequence of DNA is a variation that can be detected. For example, a variation in DNA consisting of the substitution of one nucleotide for another (such as the substitution of a C for a T) can often be recognized by a change in the points at which certain biological catalysts called "restriction enzymes" cut the DNA. Such an enzyme cuts DNA whenever it encounters a specific sequence of nucleotides that is peculiar to the enzyme. For example, the enzyme *Hae*I cuts DNA whenever it encounters the sequence AGGCCA. A restriction enzyme will cut a sample of DNA into fragments whose lengths depend on the location of the cutting sites recognized by the enzyme. Assemblies of fragments of different lengths are called "restriction fragment length polymorphisms" (RFLPs), and RFLPs constitute one of the most important tools for analyzing and identifying samples of DNA.

An important technique used in such analyses is the "Southern blot," developed by Edwin Southern in 1975. A sample of DNA is cut with a restriction enzyme, and the fragments are separated from one another by electrophoresis (i.e., they are separated by an electrical field). The fragments of particular interest are then identified with a labeled probe, a short segment of single-stranded DNA containing a radioactive atom, which hybridizes (fuses) to the fragments of interest because its DNA sequence is complementary to those of the fragments (A pairing with T, C pairing with G). Each electrophoretic band represents a separate fragment of DNA, and a given person will have no more than two fragments derived from a particular place in his or her DNA — one representing each of the genes that are present on the two chromosomes of a given pair. The forms of a given gene are referred to as alleles. A person who received the same allele from the mother and the father is said to be "homozygous" for that allele; a person who received different alleles from the mother and the father is said to be "heterozygous." Many RFLP systems are based on a change in a single nucleotide. They are said to be "diallelic," because there are only two common alternative forms. And there are only three genotypes: two kinds of homozygous genotypes and a heterozygous genotype. Another form of RFLP is generated by the presence of variable number tandem repeats (VNTRs). VNTRs are sequences, sometimes as small as two different nucleotides (such as C and A), that are repeated in the DNA. When such a structure is subjected to cutting with restriction enzymes, fragments of varied lengths are obtained.

It was variation of the VNTR type to which Alex Jeffreys in the United Kingdom first applied the designation "DNA fingerprinting." He used probes that recognized not one locus, but multiple loci, and "DNA fingerprinting" has come to refer particularly to multilocus, multiallele systems. A locus is a specific site of a gene on a chromosome. In the United States, in particular, single-locus probes are preferred, because their results are easier to interpret. "DNA typing" is the preferred term, because "DNA fingerprinting" is associated with multilocus systems. Discriminating power for personal identification is achieved by using several — usually at least four — single-locus, multiallelic systems.

After the bands (alleles) are visualized, those in the evidence sample and the suspect sample are compared. If the bands match in the two samples, for all three or four enzyme-probe combinations, the question is: What is the probability that such a match would have occurred between the suspect and a person drawn at random from the same population as the suspect?

Answering that question requires calculation of the frequency in the population of each of the gene variants (alleles) that have been found, and the calculation requires a databank where one can find the frequency of each allele in the population. On the basis of some assumptions, so-called **Hardy-Weinberg ratios** can be calculated. For a two-allele system, the ratios are indicated by the expressions p^2 and q^2 for the frequency of the two homozygotes and $2pq$ for heterozygotes, p and q being the frequencies of the two alleles and $p + q$ being equal to 1. Suppose that a person is heterozygous at a locus where the frequencies of the two alleles in the population are 0.3 and 0.7. The frequency of that heterozygous genotype in the population would be $2 \times 0.3 \times 0.7 = 0.42$. Suppose, further, that at three other loci the person being tested has genotypes with population frequencies of 0.01, 0.32, and 0.02. The frequency of the combined genotype in the population is $0.42 \times 0.01 \times 0.32 \times 0.02$, or 0.000027, or approximately 1 in 37,000.

That example illustrates what is called the **product rule**, or **multiplication rule**. Its use assumes that the alleles at a given locus are inherited independently of each other. It also assumes that there are no subpopulations in which a particular allele at one locus would have a preferential probability of being associated with a particular allele at a second locus.

Techniques for analyzing DNA are changing rapidly. One key technique introduced in the last few years is the **polymerase chain reaction (PCR)**, which allows a million or more copies of a short region of DNA to be made easily. For DNA typing, one amplifies (copies) a genetically informative sequence, usually 100 to 2,000 nucleotides long, and detects the genotype in the amplified product. Because many copies are made, DNA typing can rely on methods of detection that do not use radioactive substances. Furthermore, the technique of PCR amplification permits the use of very small samples of tissue or body fluids — theoretically even a single nucleated cell.

The PCR process is simple; indeed, it is analogous to the process by which cells replicate their DNA. It can be used in conjunction with various methods for detecting person-to-person differences in DNA.

It must be emphasized that new methods and technology for demonstrating individuality in each person's DNA are being developed. The present methods explained here will probably be superseded by others that are more efficient, error-free, automatable, and cost-effective. Care should be taken to ensure that DNA typing techniques used for forensic purposes do not become "locked in" prematurely, lest society and the criminal justice system be unable to benefit fully from advances in science and technology.

TECHNICAL CONSIDERATIONS

The forensic use of DNA typing is an outgrowth of its medical diagnostic use — analysis of disease-causing genes based on comparison of a patient's DNA with that of family members to study inheritance patterns of genes or comparison with reference standards to detect mutations. To understand the challenges involved in such technology transfer, it is instructive to compare forensic DNA typing with DNA diagnostics.

DNA diagnostics usually involves clean tissue samples from known sources. Its procedures can usually be repeated to resolve ambiguities. It involves comparison of discrete alternatives (e.g., which of two alleles did a child inherit from a parent?) and thus includes built-in consistency checks against artifacts. It requires no knowledge of the distribution of patterns in the general population.

Forensic DNA typing often involves samples that are degraded, contaminated, or from multiple unknown sources. Its procedures sometimes cannot be repeated, because there is too little sample. It often involves matching of samples from a wide range of alternatives in the population and thus lacks built-in consistency checks. Except in cases where the DNA evidence excludes a suspect, assessing the significance of a result requires statistical analysis of population frequencies.

Each method of DNA typing has its own advantages and limitations, and each is at a different state of technical development. However, the use of each method involves three steps:

1. Laboratory analysis of samples to determine their genetic marker types at multiple sites of potential variation

2. Comparison of the genetic marker types of the samples to determine whether the types match and thus whether the samples could have come from the same source
3. If the types match, statistical analysis of the population frequencies of the types to determine the probability that a match would have been observed by chance in a comparison of samples from different persons

Before any particular DNA typing method is used for forensic purposes, precise and scientifically reliable procedures for performing all three steps must be established. It is meaningless to speak of the reliability of DNA typing in general — that is, without specifying a particular method.

Despite the challenges of forensic DNA typing, it is possible to develop reliable forensic DNA typing systems, provided that adequate scientific care is taken to define and characterize the methods.

RECOMMENDATIONS

1. Any new DNA typing method (or a substantial variation of an existing method) must be rigorously characterized in both research and forensic settings, to determine the circumstances under which it will yield reliable results.
2. DNA analysis in forensic science should be governed by the highest standards of scientific rigor, including the following requirements:
   - Each DNA typing procedure must be completely described in a detailed, written laboratory protocol.
   - Each DNA typing procedure requires objective and quantitative rules for identifying the pattern of a sample.
   - Each DNA typing procedure requires a precise and objective matching rule for declaring whether two samples match.
   - Potential artifacts should be identified by empirical testing, and scientific controls should be designed to serve as internal checks to test for the occurrence of artifacts.
   - The limits of each DNA typing procedure should be understood, especially when the DNA sample is small, is a mixture of DNA from multiple sources, or is contaminated with interfering substances.
   - Empirical characterization of a DNA typing procedure must be published in appropriate scientific journals.
   - Before a new DNA typing procedure can be used, it must have not only a solid scientific foundation, but also a solid base of experience.
3. The committee strongly recommmends the establishment of a National Committee on Forensic DNA Typing (NCFDT) under the auspices of an appropriate government agency or agencies to provide expert advice primarily on scientific and technical issues concerning forensic DNA typing.

4. Novel forms of variation in the genome that have the potential for increased power of discrimination between persons are being discovered. Furthermore, new ways to demonstrate variations in the genome are being developed. The current techniques are likely to be superseded by others that provide unambiguous individual identification and have such advantages as automatability and economy. Each new method should be evaluated by the NCFDT for use in the forensic setting, applying appropriate criteria to ensure that society derives maximal benefit from DNA typing technology.

STATISTICAL BASIS FOR INTERPRETATION

Because any two human genomes differ at about 3 million sites, no two persons (barring identical twins) have the same DNA sequence. Unique identification with DNA typing is, therefore, possible, in principle, provided that enough sites of variation are examined. However, the DNA typing systems used today examine only a few sites of variation and have only limited resolution for measuring variability at each site. There is a chance that two persons have DNA patterns (i.e., genetic types) that match at the small number of sites examined. Nevertheless, even with today's technology, which uses three to five loci, a match between two DNA patterns can be considered strong evidence that the two samples came from the same source. Interpreting a DNA typing analysis requires a valid scientific method for estimating the probability that a random person by chance matches the forensic sample at the sites of DNA variation examined. To say that two patterns match, without providing any scientifically valid estimate (or, at least, an upper bound) of the frequency with which such matches might occur by chance, is meaningless. The committee recommends approaches for making sound estimates that are independent of the race or ethnic group of the subject.

A standard way to estimate frequency is to count occurrences in a random sample of the appropriate population and then use classical statistical formulas to place upper and lower confidence limits on the estimate. (Because forensic science should avoid placing undue weight on incriminating evidence, an upper confidence limit of the frequency should be used in court). If a particular DNA pattern occured in 1 of 100 samples, the estimated frequency would be 1%, with an upper confidence limit of 4.7%. If the pattern occurred in 0 of 100 samples, the estimated frequency would be 0%, with an upper confidence limit of 3%. (The upper bound cited is the traditional 95% confidence limit, the use of which implies that the true value has only a 5% chance of exceeding the upper bound). Such estimates produced by straightforward counting have the virtue that they do not depend on theoretical assumptions, but simply on the samples having been randomly drawn from the appropriate population. However, such estimates do not take advantage of the full potential of the genetic approach.

In contrast, population frequencies often quoted for DNA typing analyses are based not on actual counting, but on theoretical models based on the principles of population genetics. Each matching allele is assumed to provide statistically independent evidence, and the frequencies of the individual alleles are multiplied together to calculate a frequency of the complete DNA pattern. Although a databank contains only 500 people, multiplying the frequencies of enough separate events might result in a estimated frequency of their all occurring in a given person of 1 in a billion. Of course, the scientific validity of the multiplication rule depends on whether the events (i.e., the matches at each allele) are actually statistically independent.

Because it is impossible or impractical to draw a large enough population to test directly calculated frequencies of any particular DNA profile much below 1 in 1000, there is not a sufficient body of empirical data on which to base a claim that such frequency calculations are reliable or valid. The assumption of independence must be strictly scrutinized and estimation procedures appropriately adjusted if possible. (The rarity of all the genotypes represented in the databank can be demonstrated by pairwise comparisons, however. Thus, in a recently reported analysis of the FBI databank, no exactly matching pairs were found in 5-locus DNA profiles, and the closest match was a single 3-locus match among 7.6 million pairwise comparisons.)

The multiplication rule has been routinely applied to blood group frequencies in the forensic setting. However, that situation is substantially different. Because conventional genetic markers are only modestly polymorphic (with the exception of human leukocyte antigen, HLA, which usually cannot be typed in forensic specimens), the multilocus genotype frequencies are often about 1 in 100. Such estimates have been tested by simple empirical counting. Pairwise comparisons of allele frequencies have not revealed any correlation across loci. Hence, the multiplication rule does not appear to lead to the risk of extrapolating beyond the available data for conventional markers. But highly polymorphic markers exceed the informative power of protein markers and so multiplication of their estimated frequencies leads to estimates that are far less than the reciprocal of the size of the databanks, the $1/N$, N being the number of entries in the databank.

The multiplication rule is based on the assumption that the population does not contain subpopulations with distinct allele frequencies — that each person's alleles constitute statistically independent random selections from a common gene pool. Under that assumption, the procedure for calculating the population frequency of a genotype is straightforward:

- Count the frequency of alleles. For each allele in the genotype, examine a random sample of the population and count the proportion of matching alleles — that is, alleles that would be declared to match according to the rule that is used for declaring matches in a forensic context.
- Calculate the frequency of the genotype at each locus. The frequency of a homozygous genotype a1/a1 is calculated to be p_{a1}^2, where p_{a1} denotes the frequency of allele a1. The frequency of a heterozygous genotype a1/a2 is calculated to be $2p_{a1}p_{a2}$, where p_{a1} and p_{a2} denote the frequencies

of alleles a1 and a2. In both cases, the genotype frequency is calculated by multiplying the two allele frequencies, on the assumption that there is no statistical correlation between the allele inherited from one's father and the allele inherited from one's mother. When there is no correlation between the two parental alleles, the locus is said to be in **Hardy-Weinberg equilibrium**.

- Calculate the frequency of the complete multilocus genotype by multiplying the genotype frequencies at all the loci. As in the previous step, this calculation assumes that there is no correlation between the genotypes at the individual loci; the absence of such correlation is called **linkage disequilibrium**. Suppose, for example, that a person has genotype a1/a2, b1/b2, c1/c1. If a random sample of the appropriate population shows that the frequencies of allels a1, a2, b1, b2, and c1 are approximately 0.1, 0.2, 0.3, 0.1, and 0.2, respectively, then the population frequency of the genotype would be estimated to be [2(0.1)(0.2)][2(0.3)(0.1)][(0.2)(0.2)] = 0.000096, or about 1 in 10,417.

The validity of the multiplication rule depends on the assumption of the absence of population substructure. Population substructure violates the assumption of statistical independence of alleles. In a population that contains groups each with different allele frequencies, the presence of one allele in a person's genotype can alter the statistical expectation of the other alleles in the genotype. For example, a person has one allele that is common among Italians is more likely to be of Italian descent and thus more likely to carry additional alleles that are common among Italians. The true genotype frequency is thus higher than would be predicted by applying the multiplication rule using the average frequency in the entire population.

To illustrate the problem with a hypothetical example, suppose that a particular allele at a VNTR locus has a 1% frequency in the general population, but a 20% frequency in a specific subgroup. The frequency of homozygotes for the allele would be calculated to be 1 in 10,000 according to the allele frequency determined by sampling the general population, but would actually be 1 in 25 for the subgroup. That is a hypothetical and extreme example, but illustrates the potential effect of demography on gene frequency estimation.

The key question underlying the use of the multiplication rule — that is, whether actual populations have significant substructure for the loci used for forensic typing — has provoked considerable debate among population geneticists. Some have expressed serious concern about the possibility of significant substructure. They maintain that census categories — such as North American Caucasians, blacks, Hispanics, Asians, and Native Americans — are not homogeneous groups, but rather that each group is an admixture of subgroups with somewhat different allele frequencies. Allele frequencies have not yet been homogenized, because people tend to mate within their subgroup.

Those population geneticists also point out that, for any particular genetic marker, the actual degree of population differentiation cannot be predicted in advance, but must be determined empirically. Furthermore, they doubt that the presence of substructure can be detected by the application of statistical tests to data from large mixed populations. Population differentiation must be assessed through direct studies of allele frequencies in ethnic groups.

Other population geneticists, while recognizing the possibility or likelihood of population substructure, conclude that the evidence to date suggests only a minimal effect on estimates of genotype frequencies. Empirical studies concerning VNTR loci detected no deviation from independence within or across loci. Moreover, as pointed out earlier, pairwise comparisons of all five-locus DNA profiles in the FBI database showed no exact matches; the closest match was a single 3-locus match among 7.6 million pairwise comparisons. These studies are interpreted as indicating that multiplication of gene frequencies across loci does not lead to major inaccuracies in the calculation of genotype frequency — at least not for the specific polymorphic loci examined.

Although mindful of those opposing views, the committee has chosen to assume for the sake of discussion that population substructure may exist and to provide a method for estimating population frequencies in a manner that would adequately account for it. Our decision is based on four considerations:

1. It is possible to provide conservative estimates of population frequency, without giving up the inherent power of DNA typing.
2. It is appropriate to prefer somewhat conservative numbers for forensic DNA typing, especially because the statistical power lost in this way can often be recovered through typing of additional loci, where required.
3. It is important to have a general approach that is applicable to any loci used for forensic typing. Recent empirical studies pertain only to the population genetics of the VNTR loci in current use. However, we expect forensic DNA typing to undergo much change over the next decade — including the introduction of different types of DNA polymorphisms, some of which might have different properties from the standpoint of population genetics.
4. It is desirable to provide a method for calculating population frequencies that is independent of the ethnic group of the subject.

The committee is aware of the need to account for possible population substructure, and it recommends the use of the **ceiling principle**. The multiplication rule will yield conservative estimates even for a substructured population, provided that the allele frequencies used in the calculation exceed the allele frequencies in any of the population subgroups. The ceiling principle involves two steps:

1. For each allele at each locus, determine a ceiling frequency that is the upper bound of the allele frequency that is independent of the ethnic background of a subject.

2. To calculate a genotype frequency, apply the multiplication rule according to the ceiling allele frequencies.

To determine ceiling frequencies, the committee strongly recommends the following approach:

1. Draw random samples of 100 persons from each of 15 to 20 populations that represent groups that are relatively homogeneous genetically.
2. Take as ceiling frequency the largest frequency in any of those populations or 5%, whichever is larger.

Use of the ceiling principle yields the same frequency of a given genotype, regardless of the suspect's ethnic background, because the reported frequency represents a maximum for any possible ethnic heritage. Accordingly, the ethnic background of the individual suspect should be ignored in estimating the likelihood of a random match. The calculation is fair to suspects, because the estimated probabilities are likely to be conservative in their incriminating power.

Some legal commentators have pointed out that frequencies should be based on the population of possible perpetrators, rather than on the population to which a particular suspect belongs. Although that argument is formally correct, practicalities often preclude use of that approach. Furthermore, the ceiling principle eliminates the need for investigating the perpetrator population, because it yields an upper bound to the frequency that would be obtained by that approach.

Although the ceiling principle is a conservative approach, we feel that it is appropriate. DNA typing is unique, in that the forensic analyst has an essentially unlimited ability to adduce additional evidence: whatever power is sacrificed by requiring conservative estimates can be regained by examining additional loci. (Although there might be some cases in which the DNA sample is insufficient to permit typing additional loci with RFLPs, this limitation is likely to disappear with the eventual use of PCR).

That no evidence of population substructure is demonstrable with the markers tested so far cannot be taken to mean that such does not exist for other markers. Preservation of population DNA samples in the form of immortalized cell lines will ensure that DNA is available for determining population frequencies of any DNA pattern as new and better techniques become available, without the necessity of collecting fresh samples. It will also provide samples for standardization of methods across laboratories.

Because of the similarity in DNA patterns between relatives, databanks of DNA of convicted criminals have the ability to point not just to the individuals but to entire families—including relatives who have committed no crime. Clearly, this raises a serious issue of privacy and fairness. It is inappropriate, for reasons of privacy, to search databanks of DNA from convicted criminals in such a fashion. Such uses should be prevented both by limitations of the software for searching and by statutory guarantees of privacy.

The genetic correlation among relatives means that the probability that a forensic sample will match a relative of the person who left it is considerably greater than the probability that it will match a random person.

Especially for technology with high discriminatory power, such as DNA typing, laboratory error rates must be continually estimated in blind proficiency testing and must be disclosed to juries.

Recommendations

- As a basis for the interpretation of the statistical significance of DNA typing results, the committee recommends that blood samples be obtained from 100 randomly selected persons in each of 15 to 20 relatively homogeneous populations; that the DNA in lymphocytes from these blood samples be used to determine the frequencies of alleles currently tested in forensic applications; and that the lymphocytes be "immortalized" and preserved as reference standard for determination of allele frequencies in tests applied in different laboratories or developed in the future. The collection of samples and their study should be overseen by a National Committee on Forensic DNA Typing.
- The ceiling principle should be used in applying the multiplication rule for estimating the frequency of particular DNA profiles. For each allele in a person's DNA pattern, the highest allele frequency found in any of the 15 to 20 populations or 5% (whichever is larger) should be used.
- In the interval (which should be short) while the reference blood samples are being collected, the significance of the findings of multilocus DNA typing should be presented in two ways:
    1. If no match is found with any sample in a total databank of *N* persons (as will usually be the case), that should be stated, thus indicating the rarity of a random match.
    2. In applying the multiplication rule, the 95% upper confidence limit of the frequency of each allele should be calculated for separate United States "racial" groups and the highest of these values or 10% (whichever is the larger) should be used. Data on at least three major "races" (e.g., Caucasians, blacks, Hispanics, Asians, and Native Americans) should be analyzed.
- Any population databank used to support DNA typing should be openly available for scientific inspection by parties to a legal case and by the scientific community.
- Laboratory error rates should be measured with appropriate proficiency tests and should play a role in the interpretation of results of forensic DNA typing.

STANDARDS

Critics and supporters of the forensic uses of DNA typing agree that there is a lack of standardization of practices and a lack of uniformly accepted methods for quality assurance. The deficiencies are due largely to the rapid emergence of DNA typing and its introduction in the U.S. through the private sector.

As the technology developed in the U.S., private laboratories using widely differing methods (single-locus RFLP, multilocus RFLP, and PCR) began to offer their services to law-enforcement agencies. During the same period, the FBI was developing its own RFLP method, with a different restriction enzyme and different single-locus probes. The FBI method has become the one most widely used in public forensic science laboratories. Each method has its own advantages and disadvantages, databanks, molecular weight markers, match criteria, and reporting methods.

Regardless of the causes, practices in DNA typing vary, and so do the educational backgrounds, training, and experience of the scientists and technicians who perform the tests, the internal and external proficiency testing conducted, the interpretation of results, and approaches to quality assurance.

It is not uncommon for an emerging technology to go without regulation until its importance and applicability are established. Indeed, the development of DNA typing technology has occurred without regulation of laboratories and their practice, public or private. The committee recognizes that standardization of practices in forensic laboratories in general is more problematic than in other laboratory settings; stated succinctly, forensic scientists have little or no control over the nature, condition, form, or amount of sample with which they must work. But it is now clear that DNA typing methods are a most powerful adjunct to forensic science for personal identification and have immense benefit to the public — so powerful, so complex, and so important that some degree of standardization of laboratory procedures is necessary to assure the courts of high-quality results. DNA typing is capable, in principle, of an extremely low inherent rate of false results, so the risk of error will come from poor laboratory practice or poor sample handling and labeling; and, because DNA typing is technical, a jury requires the assurance of laboratory competence in test results.

At issue, then, is how to achieve standardization of DNA typing laboratories in such a manner as to assure the courts and the public that results of DNA typing by a given laboratory are reliable, reproducible, and accurate.

Quality assurance can best be described as a documented system of activities or processes for the effective monitoring and verification of the quality of a work product (in this case, laboratory results). A comprehensive quality-assurance program must include elements that address education, training, and certification of personnel; specification and calibration of equipment and reagents; documentation and validation of analytical methods; use of appropriate standards and controls; sample handling procedures; proficiency testing; data interpretation and reporting; internal and external audits of all the above; and corrective actions to address deficiencies and weigh their importance for laboratory competence.

RECOMMENDATIONS

Although standardization of forensic practice is difficult because of the nature of the samples, DNA typing is such a powerful and complex technology that some degree of standardization is necessary to ensure high standards.

- Each forensic science laboratory engaged in DNA typing must have a formal, detailed quality-assurance and quality control program to monitor work, on both an individual and a laboratory-wide basis.
- The Technical Working Group on DNA Analysis and Methods (TWGDAM) guidelines for a quality-assurance program for DNA RFLP analysis are an excellent starting point for a quality-assurance program, which should be supplemented by the additional technical recommendations of this committee.
- The TWGDAM group should continue to function, playing a role complementary to that of the National Committee on Forensic DNA Typing (NCFDT). To increase its effectiveness, TWGDAM should include additional technical experts from outside the forensic community who are not closely tied to any forensic laboratory.
- Quality-assurance programs in individual laboratories alone are insufficient to ensure high standards. External mechanisms are needed, to ensure adherence to the practices of quality assurance. Potential mechanisms include individual certification, laboratory accreditation, and state or federal regulation.
- One of the best guarantees of high quality is the presence of an active professional organization committee that is able to enforce standards. Although professional societies in forensic science have historically not played an active role, the American Society of Crime Laboratory Directors (ASCLD) and the American Society of Crime Laboratory Directors-Laboratory Accreditation Board (ASCLD-LAB) have shown substantial interest in enforcing quality by expanding the ASCLD-LAB accreditation program to include mandatory proficiency testing. ASCLD-LAB must demonstrate that it will actively discharge this role.
- Because private professional organizations lack the regulatory authority to require accreditation, further means are needed to ensure compliance with appropriate standards.
- Courts should require that laboratories providing DNA typing evidence have proper accreditation for each DNA typing method used. Any laboratory that is not formally accredited and that provides evidence to the courts — for example, a nonforensic laboratory repeating the analysis of a forensic laboratory — should be expected to demonstrate

that it is operating at the same level of standards as accredited laboratories.

- Establishing mandatory accreditation should be a responsibility of the Department of Health and Human Services (DHHS), in consultation with the Department of Justice (DOJ). DHHS is the appropriate agency, because it has extensive experience in the regulation of clinical laboratories through programs under the Clinical Laboratory Improvement Act and has extensive expertise in molecular genetics through the National Institutes of Health. DOJ must be involved, because the task is important for law enforcement.
- The National Institute of Justice (NIJ) does not appear to receive adequate funds to support proper education, training, and research in the field of forensic DNA typing. The level of funding should be reevaluated and increased appropriately.

DATABANKS AND PRIVACY OF INFORMATION

DNA typing in the criminal justice system has so far been used primarily for direct comparison of DNA profiles of evidence samples with profiles of samples from suspects. However, that application constitutes only the tip of the iceberg of potential law-enforcement applications. If DNA profiles of samples from a population were stored in computer databanks (databases), DNA typing could be applied in crimes without suspects. Investigators could compare DNA profiles of biological evidence samples with profiles in a databank to search for suspects.

In many respects, the situation is analogous to that of latent fingerprints. Originally, latent fingerprints were used for comparing crime scene evidence with suspects. With the development of the Automated Fingerprint Identification Systems (AFIS) in the last decade, the investigative use of fingerprints has dramatically expanded. Forensic scientists can enter an unidentified latent fingerprint pattern into an automated system and within minutes compare it with millions of patterns contained in a computer file. In its short history, automated fingerprint analysis has been credited with solving tens of thousands of crimes.

The computer technology required for an automated fingerprint identification system is sophisticated and complex. Fingerprints are complicated geometric patterns, and the computer must store, recognize, and search for complex and variable patterns of ridges and minutiae in the millions of prints on file. Several commercially available but expensive computer systems are in use around the world. In contrast, the computer technology required for DNA databanks is relatively simple. Because DNA profiles can be reduced to a list of genetic types (hence, a list of numbers), DNA profile repositories can use relatively simple and inexpensive software and hardware. Consequently, computer requirements should not pose a serious problem in the development of DNA profile databanks.

Confidentiality and security of DNA-related information are especially important and difficult issues, because we are in the midst of two extraordinary technological revolutions that show no signs of abating: in molecular biology, which is yielding an explosion of information about human genetics, and in computer technology, which is moving toward national and international networks connecting growing information resources.

Even simple information about identity requires confidentiality. Just as fingerprint files can be misused, DNA profile information could be misused to search and correlate criminal record databanks or medical record databanks. Computer storage of information increases the possibilities for misuse. For example, addresses, telephone numbers, social security numbers, credit ratings, range of incomes, demographic categories, and information on hobbies are currently available for many of our citizens in various distributed computerized data sources. Such data can be obtained directly through access to specific sources, such as credit-rating services, or through statistical disclosure, which refers to the ability of a user to derive an estimate of a desired statistic or feature from a databank or a collection of databanks. Disclosure can be achieved through query or a series of queries to one or more databanks. With DNA information, queries must be directed at obtaining numerical estimates of values or at deducing the state of an attribute of an individual through a series of Boolean (yes-no) queries to multiple distributed databanks.

Several private laboratories already offer a DNA banking service (sample storage in freezers) to physicians, genetic counselors, and, in some cases, anyone who pays for the service. Typically, such information as name, address, birth date, diagnosis, family history, physician's name and address, and genetic counselor's name and address is stored with samples. That information is useful for local, independent bookkeeping and record management. But it is also ripe for statistical or correlative disclosure. Just the existence in a databank of a sample from a person, independent of any DNA-related information, may be prejudicial to the person. In some laboratories, the donor cannot legally prevent access by outsiders to the samples, but can request its withdrawal. A request for withdrawal might take a month or more to process. In most cases, only physicians with signed permission of the donor have access to samples, but typically no safeguards are taken to verify individual requests independently. That is not to say that the laboratories intend to violate the rights of donors; they are simply offering a service for which there is a recognized market and attempting to provide services as well as they can.

Recommendations

- In the future, if pilot studies confirm its value, a national DNA profile databank should be created that contains information on felons convicted of particular violent crimes. Among crimes with high rates of recidivism, the case is strongest for rape, because perpetrators typically leave biological evidence (semen) that could allow them to be identified. Rape is the crime for which the databank will be of primary use. The case is somewhat weaker for violent offenders who are most likely to commit homicide as a recidivist offense, because killers leave biological evidence only in a minority of cases.

- The databank should also contain DNA profiles of unidentified persons made from biological samples found at crime scenes. These would be samples known to be of human origin, but not matched with any known persons.
- Databanks containing DNA profiles of members of the general population (as exist for ordinary fingerprints for identification purposes) are not appropriate, for reasons of both privacy and economics.
- DNA profile databanks should be accessible only to legally authorized persons and should be stored in a secure information resource.
- Legal policy concerning access and use of both DNA samples and DNA databank information should be established before widespread proliferation of samples and information repositories. Interim protection and sanctions against misuse and abuse of information derived from DNA typing should be established immediately. Policies should explicitly define authorized uses and should provide for criminal penalties for abuses.
- Although the committee endorses the concept of a limited national DNA profile databank, it doubts that existing RFLP-based technology provides an appropriate wise long-term foundation for such a databank. We expect current methods to be replaced soon with techniques that are simpler, easier to automate, and less expensive — but incompatible with existing DNA profiles. Accordingly, the committee does not recommend establishing a comprehensive DNA profile databank yet.
- For the short term, we recommend the establishment of pilot projects that involve prototype databanks based on RFLP technology and consisting primarily of profiles of vilent sex offenders. Such pilot projects could be worthwhile for identifying problems and issues in the creation of databanks. However, in the intermediate term, more efficient methods will replace the current one, and the forensic community should not allow itself to become locked into an outdated method.
- State and federal laboratories, which have a long tradition and much experience in the management of other types of basic evidence, should be given primary responsibility, authority, and additional resources to handle forensic DNA testing and all the associated sample-handling and data-handling requirements.
- Private-sector firms should not be discouraged from continuing to prepare and analyze DNA samples for specific cases or for databank samples, but they must be held accountable for misuse and abuse to the same extent as government-funded laboratories and government authorities.

DNA INFORMATION IN THE LEGAL SYSTEM

To produce biological evidence that is admissible in court in criminal cases, forensic investigators must be well trained in the collection and handling of biological samples for DNA analysis. They should take care to minimize the risk of contamination and ensure that possible sources of DNA are well preseved and properly identified. As in any forensic work, they must attend to the essentials of preserving specimens, labeling, and the chain of custody and must observe constitutional and statutory requirements that regulate the collection and handling of samples. The Fourth Amendment provides much of the legal framework for the gathering of DNA samples from suspects or private places, and court orders are sometimes needed in this connection.

In civil (noncriminal) cases — such as paternity, custody, and proof-of-death cases — the standards for admissibility must also be high, because DNA evidence might be dispositive. The relevant federal rules (Rules 403 and 702–706) and most state rules of evidence do not distinguish between civil and criminal cases in determining the admissibility of scientific data. In a civil case, however, if the results of a DNA analysis are not conclusive, it will usually be possible to obtain new samples for study.

The advent of DNA typing technology raises two key issues for judges: determining *admissibility* and explaining to jurors the appropriate standards for *weighing* evidence. A host of subsidiary questions with respect to how expert evidence should be handled before and during a trial to ensure prompt and effective adjudication apply to all evidence and all experts.

In the United States, there are two main tests for admissibility of scientific information through experts. One is the **Frye test**, enunciated in *Frye* v. *United States*. The other is a **"helpfulness" standard** found in the Federal Rules of Evidence and many of its state counterparts. In addition, several states have enacted laws that essentially mandate the admission of DNA typing evidence.

The test for admissibility of novel scientific evidence enunciated in *Frye* v. *United States* is still probably the most frequently invoked test in American case law. A majority of states profess adherence to the *Frye* rule, although a growing minority have adopted variations on the helpfulness standard suggested by the Federal Rules of Evidence.

Frye predicates the admissibility of novel scientific evidence on its general acceptance in a particular scientific field: "While courts will go a long way in admitting expert testimony deduced from a well-recognized scientific principle or discovery, the thing from which the deduction is made must be sufficiently established to have gained general acceptance in the particular field in which it belongs." Thus, admissibility depends on the quality of the science underlying the evidence, as determined by scientists themselves. Theoretically, the court's role in this preliminary determination is narrow: it should conduct a hearing to determine whether the scientific theory underlying the evidence is generally accepted in the relevant scientific community and whether the scientific techniques used are reliable for their intended purpose.

In practice, the court is much more involved. The court must determine the scientific fields from which experts should be drawn. Complexities arise with DNA typing, because the full typing process rests on theories and findings that pertain to various scientific fields. For example, the underlying theory of detecting polymorphisms is accepted by human geneticists and molecular biologists, but population geneticists and other statisticians might differ as to the appropriate method for determining the population frequency of a genotype in the general population or in a particular geographic, ethnic, or other group. The courts often let experts on a process, such as DNA typing, testify to the various scientific theories and assumptions on which the process rests, even though the experts' knowledge of some of the underlying theories is likely to be at best that of a generalist, rather than a specialist.

The *Frye* test sometimes prevents scientific evidence from being presented to a jury unless it has sufficient history to be accepted by some subspecialty of science. Under *Frye*, potentially helpful evidence may be excluded until consensus has developed. By 1991, DNA evidence had been considered in hundreds of *Frye* hearings involving felony prosecutions in more than 40 states. The overwhelming majority of trial courts ruled that such evidence was admissible, but there have been some exceptions.

In determining admissibility according to the helpfulness standard under the Federal Rules of Evidence, without specifically repudiating the *Frye* rule, a court can adopt a more flexible approach. Rule 702 states that, "if scientific, technical or other specialized knowledge will assist the trier of fact to understand the evidence or to determine a fact in issue, a witness qualified as an expert by knowledge, skill, experience, training, or education, may testify thereto in the form of an opinion or otherwise."

Rule 702 should be read with Rule 403, which requires the court to determine the admissibility of evidence by balancing its probative force against its potential for misapplication by the jury. In determining admissibility, the court should consider the soundness and reliability of the process or technique used in generating evidence; the possibility that admitting the evidence would overwhelm, confuse, or mislead the jury; and the proffered connection between the scientific research or test result to be presented and particular disputed factual issues in the case.

The federal rule, as interpreted by some courts, encompasses *Frye* by making general acceptance of scientific principles by experts a factor, and in some cases a decisive factor, in determining probative force. A court can also consider the qualifications of experts testifying about the new scientific principle, the use to which the technique based on the principle has been put, the potential of the technique for error, the existence of specialized literature discussing the technique, and its novelty.

With the helpfulness approach, the court should also consider factors that might prejudice the jury. One of the most serious concerns about scientific evidence, novel or not, is that it possesses an aura of infallibility that could overwhelm the critical faculties of a jury. The likelihood that the jury would abdicate its role as critical fact finder is believed by some to be greater if the science underlying an expert's conclusion is beyond its intellectual grasp. The jury might feel compelled to accept or reject a conclusion absolutely or to ignore the evidence altogether. However, some experience indicates that jurors tend not to be overwhelmed by scientific proof and that they prefer experiential data based on traditional forms of evidence. Moreover, the presence of opposing experts might prevent a jury from being unduly impressed with one expert or the other. Conversely, the absence of an opposing expert might cause a jury to give too much weight to expert testimony, on the grounds that, if the science were truly controversial, it would have heard the opposing view. Nevertheless, if the scientific evidence is valid, the solution to those possible problems is not to exclude the evidence, but to ensure through instructions and testimony that the jury is equipped to consider rationally whatever evidence is presented.

In determining admissibility with the helpfulness approach, the court should consider a number of factors in addition to reliability. First is the significance of the issue to which the evidence is directed. If the issue if tangential to the case, the court should be more reluctant to allow a time-consuming presentation of scientific evidence that might itself confuse the jury. Second, the availability and sufficiency of other evidence might make expert testimony about DNA superfluous. And third, the court should be mindful of the need to instruct and advise the jury so as to eliminate the risk of prejudice.

RECOMMENDATIONS

- Courts should take judicial notice of three scientific underpinnings of DNA typing:
  a. The study of DNA polymorphisms can, in principle, provide a reliable method for comparing samples.
  b. Each person's DNA is unique (except that of identical twins), although the actual discriminatory power of any particular DNA test will depend on the sites of DNA variation examined.
  c. The current laboratory procedure for detecting DNA variation (specifically, single-locus probes analyzed on Southern blots without evidence of band shifting) is fundamentally sound, although the validity of any particular implementation of the basic procedure will depend on proper characterization of the reproducibility of the system (e.g., measurement variation) and inclusion of all necessary scientific controls.
- The adequacy of the method used to acquire and analyze samples in a given case bears on the admissibility of the evidence and should, unless stipulated by opposing parties, be adjudicated case by case. In this adjudication, the accreditation and certification status of the laboratory performing the analysis should be taken into account.
- Because of the potential power of DNA evidence, authorities should make funds available to pay for expert witness, and the appropriate parties must be informed of the use of DNA evidence as soon as possible.

- DNA samples (and evidence likely to contain DNA) should be preserved whenever that is possible.
- All data and laboratory records generated by analysis of DNA samples should be made freely available to all parties. Such access is essential for evaluating the analysis.
- Protective orders should be used only to protect the privacy of individuals.

DNA TYPING AND SOCIETY

The introduction of any new technology is likely to raise concerns about its impact on society. Financial costs, potential harm to the interests of individuals, and threats to liberty or privacy are only a few of the worries typically voiced when a new technology is on the horizon. DNA typing technology has the potential for uncovering and revealing a great deal of information that most people consider to be intensely private. Examples might be the presence of genes involved in known genetic disorders or genes that have been linked to a heightened risk of particular major diseases in some populations.

Although DNA technology involves new scientific techniques for identifying or excluding people, the techniques are extensions and analogs of techniques long used in forensic science, such as serological and fingerprint examinations. Ethical questions can be raised about other aspects of this new technology, but the committee does not see it as violating a fundamental ethical principle.

A new practice or technology can be subject to further ethical analysis by using two leading ethical perspectives. The first examines the action or practice in terms of the rights of people who are affected; the second explores the potential positive and negative consequences (nonmonetary costs and benefits) of the action or practice, in an attempt to determine whether the potential good consequences outweigh the bad.

Two main questions can be asked about moral rights: Does the use of DNA technology give rise to any new rights not already recognized? Does the use of DNA technology enhance, endanger, or diminish the rights of anyone who becomes involved in legal proceedings? In answer to the first question, it is hard to think of any new rights not already recognized that come into play with the introduction of DNA technology into forensic science. The answer to the second requires a specification of the classes of people whose rights might be affected and what those rights might be.

Concerns about intrusions into privacy and breaches of confidentiality regarding the use of DNA technology in such enterprises as gene mapping are frequently voiced, and they are legitimate ethical worries. The concerns are pertinent to the role of DNA technology in forensic science, as well as to its widespread use for other purposes and in other social contexts. A potential problem related to the confidentiality of any information obtained is the safeguarding of the information and the prevention of its unauthorized release or dissemination; that can also be classified under the heading of abuse and misuse, as well as be seen as a violation of individual rights in the forensic context.

Another factor to be weighed in a consequentialist ethical analysis is whose interests are to count and whether the interest of some people should be given greater weight than those of others. For example, there are the interests of the accused, the interests of victims of crime or their families in apprehending and convicting perpetrators, and the interests of society. Whether the interests of society in seeing that justice is done should count as much as the interests of the accused or the victim is open to question.

A major issue is the preservation of confidentiality of information obtained with DNA technology in the forensic context. When databanks are established in such a way that state and federal law-enforcement authorities can gain access to DNA profiles, not only of persons convicted of violent crimes but of others as well, there is a serious potential for abuse of confidential information. The victims of many crimes in urban areas are relatives or neighbors of the perpetrators, and these victims might themselves be former or future perpetrators. There is greater likelihood that DNA information on minority group members, such as blacks and Hispanics, will be stored or accessed. However, it is important to note that use of the ceiling principle removes the necessity to categorize criminals (or defendants in general) by ethnic group for the purposes of DNA testing and storage of information in databanks.

The introduction of a powerful new technology is likely to set up expectations that might be unwarranted or unrealistic in practice. Various expectation regarding DNA typing technology are likely to be raised in the minds of jurors and others in the forensic setting. For example, public perception of the accuracy and efficacy of DNA typing might well put pressure on prosecutors to obtain DNA evidence whenever appropriate samples are available. As the use of the technology becomes widely publicized, juries will come to expect it, just as they now expect fingerprint evidence.

Two aspects of DNA typing technology contribute to the likelihood of its raising inappropriate expectations in the minds of jurors. The first is a jury's perception of an extraordinarily high probability of enabling a definitive identification of a criminal suspect; the second is the scientific complexity of the technology, which results in laypersons' inadequately understanding its capabilities and failings. Taken together, those two aspects can lead to a jury's ignoring other forensic evidence that it should be considering.

As large felon databanks are created, the forensic community could well place more reliance on DNA evidence, and a possible consequence is the underplaying of other forensic evidence. Unwarranted expectations about the power of DNA technology might result in the neglect of relevant evidence.

The need for international cooperation in law enforcement calls for appropriate scientific and technical exchange among nations. As in other areas of science and technology, dissemination of information about DNA typing and training programs for personnel likely to use the technology should be encouraged. It is desirable that all nations that will collaborate in law-enforcement activities have similar standards and practices, so efforts should be furthered to exchange scientific knowledge and expertise regarding DNA technology in forensic science.

RECOMMENDATIONS

- In the forensic context as in the medical setting, DNA information is personal, and a person's privacy and need for confidentiality should be respected. The release of DNA information on a criminal population without the permission of the subjects for purposes other than law enforcement should be considered a misuse of the information, and legal sanctions should be established to deter the unauthorized dissemination or procurement of DNA information that was obtained for forensic purposes.
- Prosecutors and defense counsel should not oversell DNA evidence. Presentations that suggest to a judge or jury that DNA typing is infallible are rarely justified and should be avoided.
- Mechanisms should be established to ensure accountability of laboratories and personnel involved in DNA typing and to make appropriate public scrutiny possible.
- Organizations that conduct accreditation or regulation of DNA technology for forensic purposes should not be subject to the influence of private companies, public laboratories, or other organizations actually engaged in laboratory work.
- Private laboratories used for testing should not be permitted to withhold information from defendants on the grounds that trade secrets are involved.
- The same standards and peer review processes used to evaluate advances in biomedical science and technology should be used to evaluate forensic DNA methods and techniques.
- Efforts at international cooperation should be furthered to ensure uniform international standards and the fullest possible exchange of scientific and technical expertise.

Legislative Guildlines for DNA Databases

Federal Bureau of Investigation
U.S. Department of Justicc
Washington, D.C. 20535

November 1991

The FBI Laboratory is working with state and local forensic laboratories to establish a national DNA identification index system to enable law enforcement agencies to share DNA information when investigating sex offenses and violent crime. Recognizing the need for more uniform state laws to facilitate cooperation among states wishing to participate in the national system, the FBI has developed legislative guidelines for use in drafting or reviewing state DNA database laws.

Fifteen states have already enacted laws authorizing establishment of DNA databases to store DNA identification records for law enforcement purposes. The states include Arizona, California, Colorado, Florida, Kansas, Illinois, Iowa, Louisiana, Michigan, Minnesota, Nevada, Oregon, South Dakota, Virginia, and Washington. Unfortunately, the statutes differ significantly from each other, particularly in the categories of offenders to be included in the database and DNA collection procedures.

The attached guidelines were developed for use by state legislatures, state attorneys general, state police agencies, and other organizations engaged in drafting legislation to establish and operate a state DNA identification database. Most of the suggested provisions appear in one or more current state laws. Most existing state laws, however, were drafted without anticipating national standards or federal requirements for participating in the national DNA index. Other provisions are suggested to address privacy and civil liberties concerns.

The guidelines are heavily influenced by provisions of the proposed "DNA Identification Act of 1991," authored by Congressman Don Edwards. The Edwards Bill would authorize the FBI to set standards for forensic DNA testing and establish specific requirements for states to participate in the national DNA index system. Several provisions are suggested to make state laws consistent with the Edwards Bill. Also attached are a description of the FBI plans to develop the national DNA index system, called CODIS.

For more information, write: Assistant Director in Charge, Laboratory Division, Federal Bureau of Investigation, Washington, DC 20535; call (202) 324-4410; or fax (202) 324-1093.

TABLE OF CONTENTS

TERMS AND DEFINITIONS

State legislation should include standard definitions for key terms, including the following:

CODIS: Originally an acronym for the Combined DNA Index System, now understood to mean the FBI national DNA identification index system which allows the storage and exchange of DNA records submitted by state and local forensic DNA laboratories.

Designated State Agency: The agency or organization within state government responsible for the policy management and administration of the state-level DNA identification record system to support law enforcement, and for liaison with the FBI regarding the state's participation in CODIS.

DNA: Deoxyribonucleic acid. DNA is located in the nucleus of cells and encodes genetic information that is the basis of human heredity and forensic identification.

DNA Record: DNA identification information stored in the state DNA Database or CODIS for the purposes of generating investigative leads or supporting statistical interpretation of DNA test results. The DNA record is the objective form of the DNA analysis test (e.g., numerical representation of DNA fragment lengths, digital image of autoradiographs, discrete allele assignment numbers, etc.) of a DNA sample. The DNA record is composed of the characteristics of a DNA sample that are of value in establishing the identity of individuals.

DNA Sample: DNA is found in any nucleated cell of the body. Blood is a rich source of DNA, although only white blood cells contain DNA. Blood is preferred because of the ease of collection, storage, and processing for DNA typing. In addition, blood is relatively simple to collect and involves a generally acceptable degree of infringed privacy for affected individuals.

FBI: Federal Bureau of Investigation.

State DNA Database: The state-level DNA identification record system to support law enforcement that is administered by the designated state agency and that provides DNA records to the FBI for storage and maintenance in CODIS. The state DNA Database system is the collective capability provided by computer software and procedures administered by a designated state agency to store and maintain DNA records related to forensic casework, convicted offenders required to provide DNA sample under state law, and anonymous DNA records used for research, quality control, and so on.

TWGDAM: Technical Working Group on DNA Analysis Methods.

AUTHORITY TO ESTABLISH A STATE DNA DATABASE

Suggested Provision. The director of the designated state agency is authorized to establish a database of DNA identification records for convicted criminals, crime scene specimens, and close biological relatives of missing persons.

The principal purpose of the state DNA database is to assist federal, state, and local criminal justice and law enforcement agencies in the putative identification, detection, or exclusion of individuals who are subjects of the investigation or prosecution of sex-related crimes, violent crimes, or other crimes in which biological evidence is recovered from the crime scene(s).

Secondary purposes of the state DNA database are (1) to support development of a population statistics database, when personal identifying information is removed, (2) to support identification research and protocol development of forensic DNA analysis methods, (3) for quality control purposes, or (4) to assist in the recovery or identification of human remains from mass disasters, or for other humanitarian purposes including identification of living missing persons.

Commentary. Legislative authority to collect DNA samples and store DNA test results in a state DNA database should be clearly linked to recognized state interests. A major strength of existing state laws is that they consistently specify that the primary reason for establishing a state DNA database is for law enforcement identification purposes. To withstand constitutional challenges on Fourth Amendment grounds, it is important for the legislative record to establish a clear link between the categories of offenses for which DNA samples must be provided and the types of criminal investigations that could be aided in the future if the DNA records of those same individuals are on file. That is why this suggested provision states that the state DNA database is for identification purposes related to the "detection or exclusion of individuals who are subjects of the investigation or prosecution of sex-related crimes, violent crimes, or other crimes in which biological evidence is recovered from the crime scene(s)."

The proposed "DNA Identification Act of 1991" (Edwards bill) provides that DNA samples and test results can be maintained in the national DNA index system (i.e., CODIS) to create a population statistics database, perform identification research and develop DNA testing protocols, and for quality control purposes (where personal identifying information is removed). Each state should consider allowing the same flexibility to forensic laboratories located within the state.

The close biological relatives of missing persons should be included in the scope of the state DNA database to assist police investigators in identifying missing persons or those persons whose whereabouts or identity are the subject of an ongoing law enforcement investigation, not necessarily of a criminal nature.

COMPATIBILITY WITH THE FBI

Suggested Provision. To ensure that DNA records are fully exchangeable between DNA laboratories within a state, or with laboratories in other states, forensic DNA testing should be conducted in a manner that is compatible with procedures specified by the FBI Laboratory, including use of comparable test procedures, laboratory equipment, supplies, and computer software.

Commentary. A national system for exchanging DNA identification records depends on all member DNA laboratories using compatible test procedures, equipment, supplies, and computer software, so that DNA records of known quality can be shared by crime laboratories. To date, however, Washington has the only statute that specifically addresses this issue.

The Washington law specifically requires that "the DNA identification system as established [by the state patrol] shall be compatible with that utilized by the Federal Bureau of Investigation" [Title 43.43.752(2)].

Furthermore, local law enforcement agencies in Washington that establish or operate a DNA identification system are required to ensure that equipment used by the local agency is compatible with the state system, and that the local system is equipped to receive and answer inquiries from the state system, and transmit DNA records to the state system [Title 43.43.758(1)(b)].

Finally, the local system must use "procedure[s] and rules for the collection analysis, storage, expungement, and use of DNA identification data [that] do not conflict with procedures and rules applicable to the state patrol DNA identification system" [Title 43.43.758(1)(c)].

CATEGORIES OF OFFENDERS INCLUDED IN THE STATE DNA DATABASE

Suggested Provision: To be determined by the states.

Commentary. State law should clearly establish the relationship between the need to collect DNA samples and the potential for law enforcement agencies to identify or detect know offenders in future criminal investigations. Certainly, DNA samples should be received from those persons convicted of offenses where biological evidence (e.g., blood, semen, saliva) is typically left by the offender at the crime scene. Violent crimes (e.g., rape and murder) are obvious categories to be covered.

Some states, however, have gone further. Virginia law provides that every person convicted of a felony shall have a sample of blood taken for DNA analysis [Va. Code Ann. § 19.2-310.2]. The first Virginia DNA database statute covered only sex offenses, but the law was expanded in 1990 to cover all convicted felons after it was demonstrated that individuals committing rape or murder often have previous convictions for nonviolent crimes (e.g., burglary).

California statue provides that persons required to register as sex offenders, convicted murderers, or persons convicted of felony assault or battery must provide a blood sample prior to parole or probation [Title 9 § 290.2(a)]. The Sex Registration Unit within the California Department of Justice was established several years before the DNA database law was enacted, and offenses covered by the DNA database statute were tied to the preexisting law.

Washington law requires that persons convicted of a sex offense or violent offense shall have a blood sample drawn for DNA identification analysis before release from detention [Title 43.43.754]. Minnesota provides that convicted sex offenders shall be ordered by the sentencing court to provide a biological specimen for DNA analysis [Section 609.3461].

On the other hand, South Dakota requires a DNA sample to be taken on arrest for certain offenses and the genetic marker information to be stored in a database [Section 23-5-14]. Oregon requires a DNA sample to be provided by persons convicted of "promoting or compelling prostitution" (i.e., pimping), among other sex-related offenses [Enrolled House Bill 3444, Section 2(1)(c), 1991].

It is incumbent on state legislatures to weigh all pertinent considerations to assure that state DNA database statutes bear a close and substantial relation to legitimate state interests. Furthermore, the Edwards Bill would allow storage in the national DNA identification index (i.e., CODIS) of DNA records only from individuals who are convicted of a crime, either felony or misdemeanor. Under the Edwards Bill, DNA identification records from persons not convicted of a crime cannot be stored in CODIS.

COLLECTION OF DNA SAMPLES FOR THE STATE DNA DATABASE

Suggested Provision. State law should specify that persons included under "Categories of Offenders" must, as a minimum, provide a blood sample for DNA typing, rather than some other type of DNA sample.

Commentary. Several states currently require collection of a saliva sample, in addition to a blood sample, to determine secretor status using conventional serology. No known state is planning to include secretor status in its DNA database. It is not clear, therefore, if any purpose is served by including saliva samples in the collection requirement.

Suggested Provision. State law should specify that tests to be performed on the blood samples are (1) to analyze and type the genetic markers contained in or derived from DNA, (2) for law enforcement identification purposes, (3) for research or administrative purposes, including (a) development of a population statistics database, when personal identifying information is removed, (b) to support identification research and protocol development of forensic DNA analysis methods, (c) for quality control purposes, and (4) to assist in the recovery or identification of human remains from mass disasters or for other humanitarian purposes, including identification of living missing persons.

Commentary. It is prudent to state explicitly that blood samples are collected for DNA identification testing. Three existing state laws (i.e., Louisiana, Nevada, and South Dakota) refer to the analysis of blood for "genetic marker grouping" or "genetic markers," without specifically mentioning DNA.

Suggested Provision. The designated state agency should be authorized to issue specific regulations relating to procedures for DNA sample collection and shipment for DNA identification testing. State law should require uniform implementation of DNA sample collection procedures by all criminal justice agencies within the state so DNA samples are reliably collected from all persons convicted of covered offenses.

Commentary. Some existing state laws are detailed in specifying that the designated state agency shall provide blood specimen vials, mailing labels and packaging, and instructions for collection, storage and shipment of DNA samples. It may be sufficient, however, to provide authority to the designated state agency to issue instructions and provide the necessary materials for collection without specifying exactly how collection must be performed.

Suggested Provision. DNA samples for the DNA database should be collected in a medically approved manner by a physician, registered nurse, licensed vocational nurse, licensed clinical laboratory technologist, or other persons trained to properly collect blood samples.

Commentary. Listing job titles in legislation should be for illustration purposes only and should not be worded as to unduly restrict the job categories of individuals authorized to take blood, provided that they are properly trained.

Suggested Provision. State law should provide for expungement of a DNA record in the state DNA database on reversal of conviction for a covered offense, or under other circumstances specified by state law.

Commentary. As a matter of policy, the FBI will honor a request from a designated state agency to delete the DNA record from CODIS.

EFFECTIVE DATE FOR TAKING DNA SAMPLES

Suggested Provision. State law should specify the effective date on which persons convicted of covered felonies and misdemeanors are required to provide DNA samples. The provision should address persons held in jails or prisons on the effective date and whether they are required to provide a DNA sample prior to release. State law should also specify the event(s) that triggers the taking of a DNA sample, and should address the practical issues of whether a DNA sample is to be provided: (1) on conviction, (2) at sentencing as part of a standard sentencing order, (3) on intake to a jail or prison, or (4) prior to release on probation or parole. State law should also provide for a judicial procedure to enforce this provision against any convicted offender who refuses to provide a DNA sample.

Commentary. This suggested provision has both practical and legal issues associated with it. As a practical matter, the designated state agency should receive DNA samples with sufficient lead time to allow DNA testing to be performed and DNA records to be stored in the state DNA database prior to release of each covered offender.

The legal issues associated with this suggested provision relate to the *ex post facto* clause of the Constitution and the liberty interest of prisoners regarding parole and mandatory release from custody. These were central issues in *Jones* v. *Murray*, 763 F Supp 842 (W.D. Va 1991), which challenged the constitutionality of the Virginia DNA database statute. (See the Court's complete opinion, Appendix 2, for a full treatment of these and other issues raised by the plaintiffs.)

AUTHORITY TO CONTRACT FOR DNA TYPING OF OFFENDERS

Suggested Provision: States should consider including authority, to be exercised at the option of the designated state agency, to contract for services to perform DNA typing of convicted offenders. The designated state agency should be required to ensure that (1) the contractor performs DNA typing in accordance with quality assurance standards issued by the FBI Director, and (2) typing results meet acceptance criteria established by the FBI for inclusion of DNA records in CODIS.

Commentary: Limited staff and laboratory capacity may prevent or inhibit in-house typing of all offender samples collected for the state DNA database. The cost to contract for DNA typing services that meet the FBI acceptance criteria is estimated at $100 to $200 per sample when performed in volume. Such tests do not require forensic interpretation (beyond establishing an individual's DNA record), generation of a report of analysis, or testimony.

CONFORMANCE WITH "DNA IDENTIFICATION ACT OF 1991"

Suggested Provisions. State law should contain provisions that ensure conformance with the proposed "DNA Identification Act of 1991" governing participation by states in CODIS.

Proficiency testing standards — Forensic DNA laboratories should implement and follow DNA quality assurance and proficiency testing standards issued by the FBI Director. Until such DNA testing standards are issued by the FBI, quality assurance guidelines issued by the Technical Working Group on DNA Analysis Methods should serve as the standards.

Authorized disclosure of DNA records — Forensic DNA laboratories providing DNA identification records to the state database should have authority to disclose or allow access to DNA samples and DNA analyses collected only in the following circumstances: (1) to criminal justice agencies for law enforcement identification purposes, (2) to a defendant for criminal defense purposes, or (3) for a population statistics database, identification research and protocol development, or quality control purposes, and then only if personal identifying information is removed.

Cancellation of authority to exchange DNA records — The designated state agency should have the authority to revoke the right of a forensic DNA laboratory within the state to exchange DNA identification records with federal, state, or local criminal justice agencies if required quality control and privacy standards specified by the state for the state DNA database are not met.

Criminal penalties — State law should provide criminal penalties for anyone who, by virtue of employment or official position, has possession of, or access to, individually identifiable DNA information contained in the state DNA database system and willfully discloses it in any manner to any person or agency not entitled to receive it. In addition, anyone who, without authorization, willfully obtains individually identifiable DNA information in the state DNA database should also be subject to criminal penalty.

Maintain eligibility for federal grants — Any forensic laboratory in the state that conducts or plans to conduct forensic DNA analysis should maintain eligibility to receive federal grants under the "DNA Identification Act of 1991." To be eligible for such grants, the DNA laboratory is required to certify to the Director of the Bureau of Justice Assistance, U.S. Department of Justice, that:

1. DNA analysis is performed according to quality assurance standards issued by the Director of the FBI.
2. DNA samples and test results are accessible only (a) to criminal justice agencies for law enforcement identification purposes, (b) to the defendant for criminal defence purposes, and (c) for a population statistics database, identification research and protocol development purposes, or for quality control purposes, if personal identifying information is removed.
3. Each DNA analyst undergoes external proficiency testing at least every six months by a program meeting standards issued by the Director of the FBI.

PRIVACY AND CIVIL LIBERTIES ISSUES

Suggested Provisions. Statutes authorizing the establishment of state DNA identification record systems should address privacy concerns regarding collection and storage of genetic information. Law enforcement-related DNA identification record systems should protect the rights of citizens to privacy and protection of civil liberties in the following ways:

1. Only DNA records that directly relate to the identification of individuals should be collected and stored. The information contained in the DNA records should *not* be collected or stored for the purpose of obtaining information about physical characteristics, traits or predisposition for disease, and should not serve any purpose other than to facilitate personal identification of an offender.
2. Personal information stored in the state DNA database system should be limited to the data necessary to generate investigative leads and support statistical interpretation of test results. Only an objective form of the DNA analysis test (e.g., numerical representation of DNA fragment lengths, digital image of autoradiographs, discrete allele assignment numbers, etc.) of a DNA sample should be stored as a DNA record in the state DNA database, together with the identity of the submitting agency so the submitting agency can maintain control of DNA records developed by it.

3. Matching a target DNA record against records contained in the state DNA database should serve only to generate investigative lead information by the forensic DNA laboratory acting on behalf of a client law enforcement agency conducting an official investigation. When a match is found, the index should point the inquiring laboratory to the laboratory that originally entered the matching DNA record. The two laboratories should then exchange detailed technical information to verify the match and establish appropriate coordination between the respective investigating agencies. A putative match should be confirmed through independent examination of a new DNA sample from the suspect, but the same match should not be used as evidence to support a subsequent prosecution.
4. Access to the state DNA database system should be limited to duly constituted federal, state, and local law enforcement agencies through their servicing forensic DNA laboratories. The system should protect against unauthorized access to the system or files containing DNA-related information.
5. Security for the state DNA database should comply with pertinent state-specified policies, if any. Transmissions of DNA identification records within the state should be encrypted and the integrity of DNA identification records should be protected and monitored.
6. The quality of the DNA identification records generated by forensic laboratories within a state should be assured through continuous monitoring of submissions from participating crime laboratories by the designated state agency. No DNA records should be accepted in the state DNA database unless they meet quality assurance standards specified by the state and the FBI. The designated state agency should have authority to audit the procedure and operations of any forensic DNA laboratory within the state which submits DNA records to the state DNA database.
7. The state DNA database should comply with any pertinent state laws and regulations governing confidentiality and privacy of data on individuals.
8. State law should protect DNA samples collected and stored for the state DNA database program to ensure they are not used or disseminated for purposes other than those specified by state statute. DNA samples and nonforensic identification information that can be derived from DNA samples must be protected from examination by employers, insurance companies, and others who may seek to obtain genetic information contained in the DNA samples.

FUNDING AUTHORIZATION

Suggested Provision. State law should provide adequate funding for staff, laboratory space, and equipment to allow for the success of the state DNA database program and to carry out the purposes of the law.

THE FBI'S PLANS FOR CODIS

The FBI Laboratory is establishing CODIS, the Combined DNA Identification System. When completed, CODIS will serve as a national index of DNA identification records so police departments in distant states and cities can tell whether a serial or recidivist rapist or murderer is at work. DNA profiles from blood or semen found at scenes of sex-related or violent crimes can be searched against files of convicted offenders and unknown subjects to provide investigative leads to law enforcement agencies. Matches in CODIS will alert crime laboratories that DNA profiles originally developed in their laboratories are the subject of investigations underway in other jurisdictions — and will allow affected DNA laboratories to exchange detailed information on the suspects involved. CODIS will also include anonymous DNA records from population studies to assist DNA examiners in estimating the statistical significance of a DNA match.

Benefits to Society. Cases involving DNA analysis are typically rapes or murders that cannot be conclusively solved using conventional forensic tests for blood and semen. DNA typing results can be used to associate or exclude biological evidence found at crime scenes with specific individuals. CODIS will provide two major benefits to society. First, it will be possible to detect and identify serial offenders more quickly, thereby allowing law enforcement the opportunity to intervene in a suspect's crime spree and reduce the number of potential victims. Second, by preventing continued violent behavior of serial offenders law enforcement will save time and effort, and courts will have fewer cases to process because investigations can be better focused and coordinated.

CODIS Supports for Criminal Investigations. Traditionally, forensic examinations are conducted by crime laboratories only after evidence is gathered and a suspect is identified. DNA technology, combined with CODIS, will allow crime laboratories to become directly involved in identifying investigative leads and suspects. Crime laboratories will be able to store DNA profiles for convicted sex offenders and violent criminals, as specified by state laws, in state DNA indexes. CODIS will then link all state DNA laboratories using DNA testing procedures compatible with the FBI Laboratory into a national DNA index system.

DNA Research and Training. The Forensic Science Research and Training Center (FSRTC), located at the FBI Academy in Quantico, Virginia, conducts applied research to develop practical methods for forensic DNA analysis and provides DNA training to state and local crime laboratories. Since 1989, a four-week DNA course (combining classroom lectures with hands-on laboratory practice) conducted by FSRTC has trained more than 300 forensic scientists from state and local laboratories. Students from 37 states and 81 crime laboratories in the United States have received DNA training, plus 15 students from 12 foreign countries. By the end of 1991, approximately 25 state and local crime laboratories will be conducting forensic DNA testing in casework using the FBI protocol.

FBI Laboratory's Role in CODIS. Developing CODIS is a natural extension of the role of the FBI Laboratory in providing technical assistance to crime laboratories and the law enforcement community. For a national DNA system to work, however, all participating crime laboratories must use the same methods for DNA testing to ensure that resulting DNA profiles are comparable and can be stored without regard to which crime laboratory submitted a particular DNA profile. Much of the FBI Laboratory effort in developing CODIS is spent fostering standardized DNA testing methods across laboratories. Toward that end, the FBI Laboratory sponsors the Technical Working Group on DNA Analysis Methods (TWGDAM), composed of forensic scientists from state and local crime laboratories that have adopted the FBI method for forensic DNA analysis. TWGDAM has issued guidelines on DNA quality assurance and proficiency testing for use by forensic laboratories in the United States. As provided under the proposed "DNA Identification Act of 1991" (Edwards Bill), the TWGDAM guidelines will serve as the interim standards for the quality of DNA identification records stored in CODIS until the DNA Advisory Board, also authorized by the Edwards Bill, issues updated or revised DNA testing standards.

Pilot DNA Laboratories. Fifteen states* have passed legislation requiring the establishment of a DNA database for law enforcement identification purposes. CODIS software will be provided free of charge to participating crime laboratories. For now, however, implementation of CODIS is limited to ten pilot forensic laboratories in seven states, plus the Laboratory's DNA Analysis Unit. The purpose of the pilot project is to demonstrate the feasibility and operational requirements of CODIS in a limited setting before offering the software nationwide. The pilot laboratories are:

1. Arizona Department of Public Safety, Phoenix, Arizona
2. Broward County Sheriff's Office, Fort Lauderdale, Florida
3. California Department of Justice, Berkeley, California
4. Florida Department of Law Enforcement, Tallahassee, Florida
5. Metro-Dade Police Department, Miami, Florida
6. Bureau of Criminal Apprehension, St. Paul, Minnesota
7. Orange County Sheriff's Office, Santa Ana, California
8. Virginia Department of Forensic Sciences, Richmond, Virginia
9. Washington State Patrol, Seattle, Washington
10. Washoe County Sheriff's Office, Reno, Nevada
11. DNA Analysis Unit, Laboratory Division, FBI Headquarters, Washington, D.C.

Pilot laboratories are directly involved in the design and development of CODIS. Representatives from pilot laboratories meet frequently with the FBI to review and comment on the system design while software is being written. Every effort is being made to ensure that the right system gets

* States that have enacted laws establishing DNA databases are: California, Colorado, Florida, Illinois, Iowa, Kansas, Louisiana, Michigan, Minnesota, Nevada, Oregon, South Dakota, Virginia, and Washington.

developed in the least amount of time possible, while making sure all of CODIS works properly before putting it into nationwide operation.

Phased Development of CODIS Software. The pilot CODIS system is being developed in five phases. During the summer of 1991, pilot laboratories began testing phase 1 software, which provides a basic means for exchanging DNA profiles between pilot DNA laboratories and the national site in Washington, D.C. During phase 1, only anonymous population data will be used to test system operations. Phase 2, scheduled for delivery during the spring and summer of 1992, will provide local laboratories with full database capabilities, improve communications and security, and integrate existing image analysis software, also developed by the FBI, into CODIS. Phase 2 will also use only anonymous population data. During 1992 and 1993, phases 3 to 5 will provide additional capabilities to allow pilot DNA laboratories to compare investigative DNA profiles from unknown suspect cases using casework and convicted offender files in CODIS. After the complete system is installed and tested in all pilot DNA laboratories, participation in CODIS will be expanded to other interested laboratories.

Low-Cost Design. To reduce costs for participating laboratories, CODIS will operate on IBM-compatible personal computers and is being designed using commercially available software packages. CODIS software operates on the same PC that hosts the image analysis software that is used to read autoradiographs. To participate in CODIS, states are required to procure computer equipment and commercial software packages required to operate CODIS application software provided by the FBI.

Privacy and Civil Liberties Issues. The FBI Laboratory is aware of privacy issues regarding collection and storage of genetic information, but also recognizes the need of society to balance privacy concerns with the increasing problem of violent crime. CODIS is being designed to protect the rights of citizens to privacy and protection of civil liberties in the following ways:

Identification purposes only — Only DNA profiles relating directly to the identification of individuals will be collected and stored. Information contained in the DNA profiles will be only for the purpose of facilitating identification of offenders or missing persons.

Minimal data stored — Personal information stored in CODIS will be limited to the minimum data necessary to generate investigative leads and support statistical interpretation of test results. CODIS will not store criminal history information. Only the DNA profile will be stored in CODIS together with the identity of the submitting laboratory so that the submitting laboratory can maintain control of the DNA profile information developed by it.

Match confirmation — Matching a target DNA profile against profiles contained in CODIS will serve only to generate investigative lead information for the inquiring DNA laboratory. When a match is found, the index will point the inquiring crime laboratory to the laboratory that originally entered the matching DNA profile. The two crime laboratories will then exchange detailed technical information to verify the match and establish appropriate coordination between the respective investigating agencies. A putative match must be confirmed through independent examination of a new blood sample to support prosecution.

Law enforcement access only — Access to CODIS will be limited to duly constituted law enforcement agencies through their servicing forensic DNA laboratories. CODIS will be designed and tested to protect against unauthorized access to the system or files containing DNA-related information. Member laboratories will be bound by a CODIS User Agreement specifying responsibilities for ensuring authorized access to and use of DNA records stored in CODIS.

System security — Security for CODIS will comply with requirements specified by the Federal Computer Security Act and policies of the Office of Management and Budget, Department of Justice, and the FBI. DNA transmissions will be encrypted and data integrity, at local, state, and national levels, will be protected and monitored. Furthermore, CODIS is being designed to incorporate proven concepts and procedures from the National Crime Information Center system.

Quality assurance of stored data — The quality of DNA data generated by forensic laboratories will be assured through continuous monitoring of submissions from participating crime laboratories. No data will be accepted unless it meets minimum quality assurance standards established by the FBI. The CODIS User Agreement will require adherence to DNA testing standards and regular proficiency testing for member laboratories.

Privacy act — CODIS will comply with the Federal Privacy Act.

Index

A

B

C

D

H

M

N

S

Printed and bound by CPI Group (UK) Ltd, Croydon, CR0 4YY
22/10/2024
01777624-0001